Ecological Research Monographs

Series Editor: Yoh Iwasa

For further volumes:
http://www.springer.com/series/8852

Shin-ichi Nakano • Tetsukazu Yahara
Tohru Nakashizuka
Editors

Integrative Observations and Assessments

Springer

Editors
Shin-ichi Nakano
Center for Ecological Research
Kyoto University
Otsu, Shiga, Japan

Tohru Nakashizuka
Graduate School of Life Sciences
Tohoku University
Sendai, Miyagi, Japan

Tetsukazu Yahara
Department of Biology Center for Asian Conservation Ecology
Kyushu University
Fukuoka, Japan

ISSN 2191-0707 ISSN 2191-0715 (electronic)
ISBN 978-4-431-54782-2 ISBN 978-4-431-54783-9 (eBook)
DOI 10.1007/978-4-431-54783-9
Springer Tokyo Heidelberg New York Dordrecht London

Library of Congress Control Number: 2014933782

Cover illustration: Front cover: Arachnothera longirostra, little spiderhunter, pollinating torch ginger in Borneo (photo by Shoko Sakai). *Back cover: Left: Oecophylla smaragdina*, weaver ants, in a parking lot in Palembang, Indonesia (photo by Shin-ichi Nakano). *Center: Hydrangea macrophylla* (Japanese: *ajisai*), along a path in Otsu, Shiga, Japan (photo by Shin-ichi Nakano). *Right*: A fish tank in a restaurant near Lake Taihu, China (photo by Shin-ichi Nakano).

Printed on acid-free paper

Springer is part of Springer Science+Business Media (www.springer.com)

Preface

It has been commonly accepted that biological diversity is important as a natural resource and for the functioning of the ecosystem that provides us with benefits essential for human well-being. However, biodiversity loss is continuing on both regional and global scales. To change this trend, the Convention on Biological Diversity at COP10 has adopted the new Strategic Plan for Biodiversity, including the Aichi Biodiversity Targets for the 2011–2020 period. Among the Aichi Biodiversity Targets, no. 19 states, "By 2020, knowledge, the science base and technologies relating to biodiversity, its values, functioning, status and trends, and the consequences of its loss, are improved, widely shared and transferred, and applied." To achieve this target, promotion of biodiversity observations including development of new technologies is needed.

In March 2012, to provide a quantum step forward in advancing science that optimizes the synergy between development and biodiversity conservation in Asia, we published the book *The Biodiversity Observation Network in the Asia-Pacific Region: Toward Further Development of Monitoring*. In that book, we provided an overview of biodiversity observation studies now organized under the Asia-Pacific Biodiversity Observation Network (AP-BON).

Here, we are publishing the second volume from AP-BON and have decided to launch a new book series titled *Asia-Pacific Biodiversity Observation Network* as a subseries in Ecological Research Monographs. Using this series, we will advance knowledge, the science base, and technology relating to biodiversity of the Asia-Pacific Region and thereby contribute to the achievement of the Aichi Biodiversity Targets by 2020.

In the present book, we have collected papers that review the challenges of studying the spatial variability of biodiversity and ecosystems in the Asia-Pacific region. A special focus is on reviews of advances in concepts and methods of biodiversity observation including ubiquitous genotyping, systematic conservation, monitoring of the function and services of ecosystems, and biodiversity informatics. Together with newly developed concepts and methodologies, the contributions of the present book will lead to the establishment of integrative observations and assessments of Asian biodiversity, which are unquestionably required for reporting the status of biodiversity in this region and for contributing to its effective conservation and sustainable use.

We are very grateful to the Ministry of the Environment of Japan for providing administrative and financial support. We would also like to thank the authors for their contributions, the editors of individual parts of the book—Tadashi Miyashita, Hidenori Tachida, Hiroya Yamano, and Hiroyuki Yamamoto—for their constructive criticism of the manuscripts, the Secretariat of DIVERSITAS of the Western Pacific and Asia (DIWPA) for formatting the submitted manuscripts, and Springer Japan for their patience in publishing our delayed edition of this book.

Otsu, Shiga, Japan — Shin-ichi Nakano
Fukuoka, Japan — Tetsukazu Yahara
Sendai, Miyagi, Japan — Tohru Nakashizuka

Contents

Contributors

Mustafa Kamal Abu Hena Department of Animal Sciences and Fishery, Universiti Putra Malaysia Bintulu Sarawak Campus, Sarawak, Malaysia

Institute of Marine Sciences and Fisheries, University of Chittagong, Chittagong, Bangladesh

Shau-Hwai Aileen TAN School of Biological Sciences, Universiti Sains Malaysia, Penang, Malaysia

Munemitsu Akasaka Faculty of Agriculture, Tokyo University of Agriculture and Technology, Tokyo, Japan

Takumi Akasaka Department of Forest Science, Graduate School of Agriculture, Hokkaido University, Sapporo, Japan

Tutu Almonte City Agricultural Office of Pt. Princesa, Puerto Princesa city, The Philippines

Tatsuya Amano Department of Zoology, University of Cambridge, Cambridge, UK

Rohani Ambo-Rappe Hasanuddin University, Makassar, Indonesia

Mukunda Dev Behera Centre for Oceans, Rivers, Atmosphere and Land Sciences (CORAL), Indian Institute of Technology, Kharagpur, India

Japar Sidik Bujang Universiti Putra Malaysia Bintulu Sarawak Campus, Sarawak, Malaysia

Chien-Chung Cheng Chinese Wild Bird Federation, Taipei, Taiwan

Kwang-Sik Choi School of Marine Biomedical Science, Jeju National University, Jeju, Republic of Korea

Dedy Darnaedi Research Center for Biology, Indonesian Institute of Sciences, Bogor, Indonesia

Meng-Wen Fan Endemic Species Research Institute, Nantou, Taiwan

Miguel D. Fortes University of the Philippines, Quezon City, The Philippines

Masahiko Fujii Faculty of Environmental Earth Science, Hokkaido University, Sapporo, Japan

Takehiko Fukushima Faculty of Life and Environmental Sciences, University of Tsukuba, Ibaraki, Japan

Hitoshi Hasegawa Kokushikan University, Tokyo, Japan

Hiroshi Hata Hazama Technical Research Institute, Ibaraki, Japan

Seiji Hayashi National Institute for Environmental Studies, Ibaraki, Japan

Tsutom Hiura Tomakomai Research Station, Field Science Center for Northern Biosphere, Hokkaido University, Tomakomai, Japan

Chuki Hongo Department of Physics and Earth Sciences, University of the Ryukyus, Okinawa, Japan

Masakazu Hori National Research Institute of Fisheries and Environment of Inland Sea, Fisheries Research Agency, Hiroshima, Japan

Xiaoping Huang South China Sea Institute of Oceanology, Guangzhou, China

Kiseong Hyeong Deep-Sea and Seabed Resources Research Department, Korea Institute of Ocean Science and Technology, Seoul, Republic of Korea

Yuji Isagi Graduate School of Agriculture, Kyoto University, Kyoto, Japan

Shinya Ishida Center for Environmental Biology and Ecosystem Studies, National Institute for Environmental Studies, Ibaraki, Japan

Fumiko Ishihama Center for Environmental Biology and Ecosystem Studies, National Institute for Environmental Studies, Ibaraki, Japan

Masae I. Ishihara Field Science Center for Northern Biosphere, Hokkaido University, Sapporo, Japan

Mitsunori Ishihara National Institute for Agro-Environmental Sciences, Ibaraki, Japan

Reiichiro Ishii Research Institute for Global Change, Japan Agency for Marine-Earth Science and Technology (JAMSTEC), Kanagawa, Japan

Takaya Iwasaki Graduate School of Arts and Sciences, The University of Tokyo, Tokyo, Japan

Mari Iwata Mitsubishi Research Institute, Tokyo, Japan

Taku Kadoya Center for Environmental Biology and Ecosystem Studies, National Institute for Environmental Studies, Ibaraki, Japan

Shingo Kaneko Faculty of Symbiotic Systems Science, Fukushima University, Fukushima, Japan

Zin'ichi Karube National Institute for Environmental Studies, Ibaraki, Japan

Ken Kato Energy Electronics Institute, National Institute of Advanced Industrial Science and Technology, Ibaraki, Japan

Hajime Kayanne Department of Earth and Planetary Science, The University of Tokyo, Tokyo, Japan

Eun-Shik Kim Department of Forestry, Environment, and Systems, Kookmin University, Seoul, South Korea

Wawan Kiswara Research Center for Oceanography, Indonesian Institute for Sciences, Bogor, Indonesia

Chie-Jen Ko Institute of Ecology and Evolutionary Biology, National Taiwan University, Taipei, Taiwan

Ryo Kohsaka Graduate School of Human and Socio-Environmental Studies, Kanazawa University, Ishikawa, Japan

Kun-Seop Lee Pusan National University, Busan, South Korea

Pei-Fen Lee Institute of Ecology and Evolutionary Biology, National Taiwan University, Taipei, Taiwan

Ruey-Shing Lin Endemic Species Research Institute, Nantou, Taiwan

Keping Ma Institute of Botany, Chinese Academy of Science, Beijing, China

Evangeline T. Magdaong Graduate School of Environmental Science, Hokkaido University, Sapporo, Japan

Shafique Matin Centre for Oceans, Rivers, Atmosphere and Land Sciences (CORAL), Indian Institute of Technology, Kharagpur, India

Bunkei Matsushita Faculty of Life and Environmental Sciences, University of Tsukuba, Ibaraki, Japan

Toshihiro Miyajima Atmosphere and Ocean Research Institute, The University of Tokyo, Chiba, Japan

Tadashi Miyashita Department of Ecosystem Studies, The University of Tokyo, Tokyo, Japan

Shin-Ichi Morinaga Graduate School of Arts and Sciences, The University of Tokyo, Tokyo, Japan

CREST, Japan Science and Technology Agency, Kyoto, Japan

M.P. Prabhakaran School of Marine Sciences, Cochin University of Science and Technology (CUSAT), Cochin, India

Shin Nagai Research Institute for Global Change, Japan Agency for Marine-Earth Science and Technology (JAMSTEC), Kanagawa, Japan

Shin-ichi Nakano Center for Ecological Research, Kyoto University, Shiga, Japan

Yushin Nakao Overseas Environmental Cooperation Center, Tokyo, Japan

Graduate School of Environmental Science, Hokkaido University, Hokkaido, Japan

Masahiro Nakaoka Akkeshi Marine Station, Hokkaido University, Akkeshi, Japan

Tohru Nakashizuka Graduate School of Life Sciences, Tohoku University, Sendai, Japan

Akira Negishi Energy Electronics Institute, National Institute of Advanced Industrial Science and Technology, Ibaraki, Japan

Ken Nozaki Energy Electronics Institute, National Institute of Advanced Industrial Science and Technology, Ibaraki, Japan

Mifuyu Ogawa National Institute for Environmental Studies, Ibaraki, Japan

Hiroyuki Oguma National Institute for Environmental Studies, Ibaraki, Japan

Kimiko Okabe Forestry and Forest Products Research Institute, Ibaraki, Japan

Noboru Okuda Center for Ecological Research, Kyoto University, Shiga, Japan

Shigefumi Okumura Mitsubishi Research Institute, Tokyo, Japan

Takeshi Osawa National Institute for Agro-Environmental Sciences, Ibaraki, Japan

Anchana Prathep Faculty of Science, Prince of Songkla University, Songkhla, Thailand

Partha Sarathi Roy University of Hyderabad, Hyderabad, India

Yoichirou Sakai Center for Ecological Research, Kyoto University, Shiga, Japan

Jun-ya Shibata Center for Ecological Research, Kyoto University, Shiga, Japan

Yoshihisa Shirayama Ocean Biodiversity Research Program, Japan Agency for Marine-Earth Science and Technology (JAMSTEC), Kanagawa, Japan

Kaoru Sugihara Center for Environmental Biology and Ecosystem Studies, National Institute for Environmental Studies, Ibaraki, Japan

Yoshihisa Suyama Field Science Center, Graduate School of Agricultural Science, Tohoku University, Sendai, Japan

Toru Suzuki Rakuno Gakuen University, Hokkaido, Japan

Satoshi Tachibana Faculty of Life and Environmental Sciences, University of Tsukuba, Ibaraki, Japan

Hidenori Tachida Department of Biology, Center for Asian Conservation Ecology, Kyushu University, Fukuoka, Japan

Shuichiro Tagane Department of Biology, Kyushu University, Fukuoka, Japan

Noriko Takamura Center for Environmental Biology and Ecosystem Studies, National Institute for Environmental Studies, Ibaraki, Japan

Akio Takenaka National Institute for Environmental Studies, Ibaraki, Japan

Tomohiro Takeyama Laboratory of Animal Sociology, Department of Biology and Geosciences, Graduate School of Science, Osaka City University, Osaka, Japan

Hisatomo Taki Forestry and Forest Products Research Institute, Ibaraki, Japan

Masayuki Tamura Department of Urban and Environmental Engineering, Kyoto University, Kyoto, Japan

Hiroyuki Tanouchi Shikoku Research Center, Forestry and Forest Products Research Institute, Kochi, Japan

Ichiro Tayasu Center for Ecological Research, Kyoto University, Shiga, Japan

Hajime Utsugi Hokkaido Research Center, Forestry and Forest Products Research Institute, Sapporo, Japan

Sheila G. Vergara ASEAN Centre for Biodiversity, Laguna, Philippines

Siti Maryam Yaakub Department of Biological Sciences, National University of Singapore, Singapore

Shigeo Yachi Center for Ecological Research, Kyoto University, Shiga, Japan

Tetsukazu Yahara Department of Biology, Center for Asian Conservation Ecology, Kyushu University, Fukuoka, Japan

Takehisa Yamakita Marine Biodiversity Research Program, Institute of Biogeosciences, Japan Agency for Marine-Earth Science and Technology, Kanagawa, Japan

Department of Ecosystem Studies, The University of Tokyo, Tokyo, Japan

Hiroyuki Yamamoto Ocean Biodiversity Research Program, Japan Agency for Marine-Earth Science and Technology (JAMSTEC), Kanazawa, Japan

Hiroya Yamano National Institute for Environmental Studies, Ibaraki, Japan

Fan Yang Graduate School of Life and Environmental Sciences, University of Tsukuba, Ibaraki, Japan

Yumiko Yara Center for Environmental Biology and Ecosystem Studies, National Institute for Environmental Studies, Ibaraki, Japan

Zulfigar Yasin School of Biological Sciences, Universiti Sains Malaysia, Penang, Malaysia

Peidong Zhang Ocean University of China, Qingdao, China

Part I
General Introduction: Perspectives on Spatial Variability Within an Ecosystem and Interdependence Between Various Ecosystems

Chapter 1
Developing a Regional Network of Biodiversity Observation in the Asia-Pacific Region: Achievements and Challenges of AP BON

Tetsukazu Yahara, Keping Ma, Dedy Darnaedi, Tadashi Miyashita, Akio Takenaka, Hidenori Tachida, Tohru Nakashizuka, Eun-Shik Kim, Noriko Takamura, Shin-ichi Nakano, Yoshihisa Shirayama, Hiroyuki Yamamoto, and Sheila G. Vergara

T. Yahara (✉) • H. Tachida
Department of Biology, Center for Asian Conservation Ecology, Kyushu University, 6-10-1 Hakozaki, Fukuoka 812-8581, Japan
e-mail: tet.yahara@gmail.com

K. Ma
Institute of Botany, Chinese Academy of Science, Nanxincun 20#, Xiangshan, Beijing 100093, China

D. Darnaedi
Research Center for Biology, Indonesian Institute of Sciences, Jl Juanda 18, Bogor 16122, Indonesia

T. Miyashita
Graduate School of Agricultural and Life Sciences, University of Tokyo, 1-1-1 Yayoi, Bunkyo-ku, Tokyo 113-8657, Japan

A. Takenaka • N. Takamura
National Institute for Environmental Studies, Onogawa 16-2, Tsukuba, Ibaraki 305-8506, Japan

T. Nakashizuka
Graduate School of Life Sciences, Tohoku University, 6-3 Aoba, Aramaki, Aoba-ku, Sendai 980-8578, Japan

E.-S. Kim
Department of Forestry, Environment, and Systems, Kookmin University, Seoul 136-702, South Korea

S.-i. Nakano
Center for Ecological Research, Kyoto University, 509-3, 2-chome, Hirano, Otsu, Shiga 520-2113, Japan

Y. Shirayama • H. Yamamoto
Ocean Biodiversity Research Program, Japan Agency for Marine-Earth Science and Technology, Natsushima-cho 2-15, Yokosuka, Kanazawa 237-0061, Japan

S.-i. Nakano et al. (eds.), *Integrative Observations and Assessments*, Ecological Research Monographs, DOI 10.1007/978-4-431-54783-9_1, © Springer Japan 2014

Abstract Organizing regional networks of biodiversity observation is crucial to establishing the knowledge base for biodiversity conservation. Here we review the history and activities of AP-BON (Asia-Pacific Biodiversity Observation Network), an active network of biodiversity observation in the Asia-Pacific region. AP-BON was organized in 2009, 1 year earlier than CBD COP 10 in 2010, and developed through four workshops and additional related meetings. The most recent, the fourth, AP-BON workshop was held in 2011, wherein implementation for 2012–2015 was confirmed. Major activities planned for 2012–2015 are (1) editing and networking the national biodiversity outlook, (2) publication of additional AP-BON books, (3) promotion of collaborative projects, (4) development of a shared database, and (5) capacity building. As for the third activity, implementation for 2012–2015 has been developed in five working groups (WGs). WG 1 (genetics/phylogenetic diversity) is promoting genetic diversity assessment to contribute to a GEO BON genetic diversity report. WG 2 is carrying out plant diversity assessments using a standardized transect method in Southeast (SE) Asian countries. WG 3 (terrestrial ecosystem) is developing (1) networks of in situ monitoring and a database on terrestrial ecosystems, (2) a database on function and services of terrestrial ecosystems, and (3) maps of biodiversity, function, and ecosystem services. WG 4 (freshwater) is monitoring Asian freshwater biodiversity at the genus, species, and ecosystem levels, focusing on fish and aquatic plants. WG 5 (marine) is developing a database to share biodiversity information and collecting data on kelp forests, seagrass meadows, coral reefs, pelagic water, and the deep sea. An important role of AP BON is to network these projects by providing opportunities for communication and interaction, developing tools for data collection, sharing, analysis, and synthesis, and carrying out assessments helpful for improving ecosystem management and sustainable use of biodiversity.

Keywords Biodiversity • Database • Ecosystem • Freshwater • Marine • Terrestrial

Introduction

Under the increasing human impacts on various habitats and ecosystems, biodiversity is being rapidly lost, driving the deterioration of associated ecosystem services (Díaz et al. 2006; Bradshaw et al. 2007; Leadley et al 2010). This biodiversity loss has been documented in many taxa and areas by observing and assessing states and trends of biodiversity at various scales (Pereira et al 2010; Yahara et al. 2012). As a result, some integrated pictures of biodiversity loss have emerged and been documented (Butchart et al. 2010; Secretariat of the Convention on Biological

---

S.G. Vergara
ASEAN Centre for Biodiversity, UP College of Forestry Campus, College, Third Floor, ERDB Bldg., Los Baños, Laguna, Philippines

Diversity 2010). However, compared with climate, for which thousands of observation sites are distributed on the global scale, and thus an integrated picture of contemporary changes can be documented every year, our ability to document and report biodiversity loss remains much less developed, both at global and regional scales (Scholes et al. 2008). It is now urgently needed to develop an observation system by which we can document ongoing changes adequately and develop better options to halt biodiversity loss.

A difficulty associated with biodiversity observation is that observed data are often stored in many isolated, often personally maintained data files, and it is not always easy to access and use them to draw an integrated picture of biodiversity states and trends. To overcome this difficulty, the GEO BON (Group on Earth Observations Biodiversity Observation Network) was organized in 2008 (Scholes et al. 2008), and its detailed implementation plan was subsequently released (GEO BON 2010). The goal of GEO BON is to establish a coordinated global network that gathers and shares information on biodiversity and ecosystem services. In the Asia-Pacific region, AP-BON, a corresponding regional network, was organized in 2009. Since then, AP-BON has been successfully developed as a network of researchers, institutions, and organizations that enables collaborations to collect, exchange, analyze, and integrate information on biodiversity and ecosystems in the Asia-Pacific region. Here, we summarize how AP-BON activities have been developed and identify major challenges we need to meet in future.

Brief History of AP-BON

The Asia-Pacific Biodiversity Observation Network (AP-BON) is a regional network of biodiversity observation organizations with specific interest in collaborating with a global network, Group on Erath Observation Biodiversity Observation Network (GEO BON; http://www.earthobservations.org/geobon.shtml). After GEO BON was organized in 2008 and its concept document was released, the 3rd GEOSS-AP Symposium (http://www.aprsaf.org/feature/feature_82.html) was held February 4–6, 2009, in Kyoto, and the participants of WG3 (Monitoring changes in ecosystems, biodiversity and ecosystem services) agreed to organize the Japanese Biodiversity Observation Network (J-BON) as a voluntary network of Japanese scientists who are working in biodiversity observations, not only of Japan but also of the Asia-Pacific region.

The first J-BON workshop was held May 8–10, 2009, in Tokyo, and the participants of this workshop agreed to organize AP-BON by inviting scientists in the Asia-Pacific region. Through consultation with the Ministry of the Environment, Japan, it was decided that AP-BON is officially supported by the Ministry of the Environment, Japan. Subsequently, the first AP-BON workshop was held July 21–22, 2009, at Nagoya University under the support of the Ministry of the Environment, Japan. This workshop provided the first opportunity for biodiversity scientists of the Asia-Pacific region to interact with each other, to mutually learn about various biodiversity observations that have been made in the region (Table 1.1), and to discuss various possibilities of collaboration for collecting, exchanging, analyzing, and integrating biodiversity information.

The second AP-BON workshop, held December 10–11, 2009, in Tokyo, resulted in further progress and summarized the following points.

1. The presentations gave participants an improved understanding of the observations and research in this region.
2. Off-line discussions were useful for team building and additional exchange of ideas.
3. An interim Steering Committee was formed, consisting of Tetsukazu Yahara and Rodorigo Fuentes as co-chairs, Keping Ma, Tohru Nakashizuka, Dedy Darnaedi, Eun-Shik Kim, and Yoshihisa Shirayama.
4. Candidate products were discussed and listed.
5. For the purposes of AP-BON, the initial AP boundaries were defined: on the east, bounded by the scope of PBIF, and excluding Russia, New Zealand, and Australia, but allowing these countries to join as observers.
6. A plan to publish a series of books compiling findings from biodiversity observations in the Asia-Pacific region was discussed and updated

In 2010, the year of CBD COP10, some important meetings relevant to AP-BON activities were held. A GEO BON meeting to discuss detailed implementation was held February 23–25 in Asilomar, California, USA, and the detailed GEO BON implementation plan was released in May. The fourth GEOSS-AP symposium was held March 9–11 in Denpasar, Bali, Indonesia. In this symposium, we had an AP-BON session as one of four parallel sessions and discussed the needs of AP-BON draft implementation as a response to the Asilomar meeting of GEO BON. The CBD COP10 preconference was held March 21–22 in Nagoya to discuss post-2010 biodiversity targets, and subsequently the third AP-BON workshop was held on March 23. In this workshop, AP-BON draft implementation was distributed and preliminary discussion on this draft was held; then, participants agreed to revise AP-BON draft implementation until the fourth workshop. Participants agreed to organize the following five working groups in AP-BON.

Working Group 1: Genetics/phylogenetic diversity
Working Group 2: Terrestrial species monitoring
Working Group 3: Terrestrial ecosystem change
Working Group 4: Freshwater ecosystem change
Working Group 5: Marine ecosystem change

During the CBD COP10, a side event of AP-BON and the first meeting of the AP-BON interim Steering Committee were organized. In this SC meeting, an institutional issue (how each person and institute can be a member of AP-BON) was addressed, and the need for memoranda with key institutes and individuals for data sharing, etc., were discussed to formalize the relationship of institutes and individuals to AP-BON.

Table 1.1 Presentation titles of 1st AP-BON workshop

Nation/ Organization	Name	Category	Organization	1st WS, July 21–22, 2009	2nd WS, Dec 10–11, 2009	4th GEOSS-AP, March 10–12, 2010	Precop and 3rd WS, Mar 21–23, 2010	AP-BON side event at COP10, Oct 2010	4th WS, Dec 2–4, 2011	5th GEOSS-AP, April 2–4, 2012
Australia	Barry Brook	Climate change	University of Adelaide	○						
	Daniel P. Faith	Phylogenetic diversity	Australian Museum		○		○			
Brunei Darussalam	Noralinda Ibrahim	Government	Senior Forester Officer, Forestry Department, Ministry of Industry and Primary Resources, Brunei Darussalam					○		
Cambodia	Ung Sam Oeun	Government	Chief Officer, Department of International Convention and Biodiversity, General Department of Administration for Nature Conservation and Protection of the Ministry of Environment					○		
China	Haigen Xu	Terrestrial	Professor and Director, Division of Nature Conservation and Biodiversity, Nanjing Institute of Environmental Sciences, Ministry of Environmental Protection of China					○		
	Keping Ma	AP BON SC (2009-), Terestrial Plant	Director and Secretary General, Institute of Botany, Biodiversity Committee, Chinese Academy of Science		○	○		○	○	
	Liu Huanzhang	Freshwater	Professor, Research Center for aquatic biodiversity and resource conservation Institute of Hydrobiology, Chinese Academy of Sciences				○			
	Gray A. Williams	Marine	Hon Director, The Swire Institute of Marine Science and Division of Ecology and Biodiversity, School of Biological Sciences, The University of Hong Kong		○					
	Hui Ding	Forest	Nanjing Institute of Environmental Sciences, Ministry of Environmental Protection of China						○	
	Xin Gao	Freshwater	Institute of Hydrobiology, Chinese Academy of Sciences						○	
	Xiubo Yu		Chinese Ecosystem Research Network							○

(continued)

Table 1.1 (continued)

Nation/ Organization	Name	Category	Organization	1st WS, July 21–22, 2009	2nd WS, Dec 10–11, 2009	4th GEOSS-AP, March 10–12, 2010	Precop and 3rd WS, Mar 21–23, 2010	AP-BON side event at COP10, Oct 2010	4th WS, Dec 2–4, 2011	5th GEOSS-AP, April 2–4, 2012
Chinese Taipei	Chau Chin Lin	Database	Taiwan Forestry Research Institute, Chinese Taipei		○				○	
	Yuan-Tay Shyu	Plant	Dean of College of Bioresources and Agriculture, National Taiwan University					○		
Fiji	Posa A. Skelton	Marine	Coordinator, PACINET USP/SPC/SPRE, The University of the South Pacific, Fiji		○					
	Akuila Cakacaka	Marine	World Conservation Society	○						
India	Chandramohan Doraira Jasingam	Marine	Environmental Impact Assessment	○						
	Partha Sarathi Roy	Remote sensing	Director, Indian Institute of Remote Sensing						○	
Indonesia	Utami Andayani	Government	Assistant Deputy			○				
	Siti Nuramaliati Prijono	Terrestrial Animal	Director, Research Center of Biology, LIPI			○				
	Dedy Darnaedi	Terrestrial Plant	GTI Focal Point-Indonesia, Research Center for Biology Indonesia, Institute of Science	○	○	○	○	○	○	
	Herwint Simbolon	Terrestrial Forest	Herbarium Bogoriense, Research Center for Biology, Indonesia Institute of Science (LIPI)			○				
	Ahmed Jauhar Arief	Terrestrial Fauna	Head of Zoology Division, Research Center for Biology, LIPI		○	○		○		
	Teguh Triono	Terrestrial Plant	Research Center for Biology Indonesia, Institute of Science			○				
	Dede Irving Hartoto	Freshwater	Senior Scientist, Research Center for Limnology-LIPI, Cibinong Science Center			○	○		○	
	Suharsono Soemorumekso	Marine	Research Center for Oceanography LIPI, Coral Reef Rehabilitation and Management Program (COREMAP)			○			○	
	Susetiono	Marine	Senior Scientist, Research Center for Oceanography, LIPI		○	○	○			
	Lilik Budi Prasetyo	Remote sensing	Bogor Agricultural University							○

Japan	Tetsukazu Yahara	AP BON co-chair (2009-), terrestrial plant	Kyushu University	○	○	○	○	○	○	○
	Yoshihisa Shirayama	AP-BON SC (2009-), marine	Japan Agency for Marine-Earth Science and Technology	○			○	○	○	
	Eiji Suzuki	Forest	Kagoshima University	○	○				○	
	Motomi Ito	Database	University of Tokyo	○	○		○	○	○	○
	Tohru Nakashizuka	AP-BON SC (2009-), forest	Tohoku University	○	○	○	○	○	○	
	Shin-ichi Nakano	AP-BON SC (2009-), freshwater	Center for Ecological Research, Kyoto University	○	○	○	○	○	○	
	Akio Takenaka	Terrestrial	National Institute of Environmental Studies	○	○		○	○	○	
	Hiroyuki Matsuda	Theoretical biology	Yokohama National University	○	○		○	○		
	Yoshiki Yamagata	Carbon	National Institute of Environmental Studies	○	○		○	○		
	Reiichiro Ishii	Remote sensing	Japan Agency for Marine-Earth Science and Technology		○	○	○		○	○
	Hiroyuki Yamamoto	Marine (microbe)	Japan Agency for Marine-Earth Science and Technology	○					○	
	Hiroya Yamano	Marine (coral)	National Institute of Environmental Studies	○					○	
	Hiroyuki Muraoka	Remote sensing	Gifu University				○		○	
	Rikie Suzuki	Remote sensing	Japan Agency for Marine-Earth Science and Technology				○			○
	Ryo Kosaka	Social science	Kanazawa University				○		○	
Republic of Korea	Eun-Shik Kim	AP BON SC (2009–2012), AP BON co-chair (2012-), ILTER, forest	Professor, Department of Forest Resources, Kookmin University, Korea	○	○	○	○	○	○	
	Jeong-Mi Park	Terrestrial Plant	Senior Researcher, National Institute of Biological Resources	○				○		
	Jun-Mi Hur	Insect	Researcher, National Institute of Biological Resources, Korea		○					
	Byoung-Yoon Lee	Terrestrial Plant	Senior Researcher, Vascular Plants Research Division, National Institute of Biological Resources							
	Byeong-Gug Yang	LTER	National Institute of Environmental Research, Korea		○					
	Chang-Seok Lee	Forest	Professor, Seoul Woman's University, Korea		○					
	Chan-Ho Park	Terrestrial Plant	National Institute of Biological Resources						○	

(continued)

Table 1.1 (continued)

Nation/ Organization	Name	Category	Organization	1st WS, July 21–22, 2009	2nd WS, Dec 10–11, 2009	4th GEOSS-AP, March 10–12, 2010	Precop and 3rd WS, Mar 21–23, 2010	AP-BON side event at COP10, Oct 2010	4th WS, Dec 2–4, 2011	5th GEOSS-AP, April 2–4, 2012
Lao PDR	Vongvilay Vongkhamsao	Terrestrial	Deputy Head, Planning and Cooperation Division, NAFRI, Ministry of Agriculture and Forestry, Lao PDR		○			○		
	Somchanh Bounphanmy	Forest, terrestrial plant	Alga Specialist and Vice Dean, Science Faculty, National University of Laos					○		
Malaysia	Abdul Hamid Ahmad	Terrestrial Zoology	Professor and Director, Institute for Tropical Biodiversity and Conservation (ITBC), Universiti Malaysia Sabah		○			○		
	Saw Leng Guan	Terrestrial Plant	Director, Forest Biodiversity Division, Forest Research Institue Malaysia		○	○		○	○	
	Tan Shau Hwai	Marine	USM					○		
	Nik Muhamad Nik Ab. Majid	Terrestrial, Forest	Professor, Faculty of Forestry, Universiti Putra Malaysia	○						
	Jaya Radha Veersamy	Terrestrial, Forest	Forest Research Institute Malaysia (FRIM)	○						
	Zulfigar Bin Yasin	Marine	Marine Science, School of Biological Sciences, Universiti Sains Malaysia						○	
Mongolia	Bayarkhuu Sandagdorj	Government	Sains Malaysia		○			○		
	Gombobaatar Sundev	Terrestrial Birds	Head, Ornithological Laboratory, Faculty of Biology, National University of Mongolia		○					
	Tsolmon Renchin	Database	National University of Mongolia							○
Palau	Noelle Wenty Oldiais	Marine	Palau International Coral Reef Center	○						
Papua New Guinea	Yati A. Bun	Terrstrial Forest	Executive Director, Foundation for People and Community Development Inc., Papua New Guinea		○				○	
Philippines	Edwino S. Fernando	Terrestrial Plant	Director and GTI Specialist Forest Tree, Makiling Center for Mountain Ecosystems, College of Forestry and Natural Resources, University of the Philippines, Los Baños		○			○		

	Nievales, Marie Frances	Marine	Associate Professor, Project Leader, College of Arts and Sciences, University of the Philippines	○						
	Miguel D. Fortes	Marine	Marine Science Institute CS, University of the Philippines						○	
Samoa	Paul Anderson	Marine	Secretariat of the Pacific Regional Environment Programme (SPREP)						○	
Singapore	Benito Tan	Terrestrial Plant	Keeper of the Herbarium, National Parks Board Singapore		○			○		
Thailand	Sirikul Bunpapong	Government	Director, Biological Diversity Division, Office of Natural Resources and Environmental Policy and Planning (ONEP), Ministry of Natural Resources and Environment, Thailand		○			○		
	Benchamaporn Wattanatongchai	Government	Environmental Official Professional Level, Biological Diversity Division, Office of Natural Resources and Environmental Policy and Planning (ONEP), Ministry of Natural Resources and Environment, Thailand					○		
	Somchai Bussarawit	Marine	Senior Marine Biologist, Chief of Museum and Aquarium, Phuket Marine Biological Center, Department of Marine and Coastal Resources, Thailand	○	○		○			
	Dokrak Marod	Terrestrial Forest	Associate Professor, Department of Forest Biology, Faculty of Forestry, Kasetsart University	○	○			○	○	
	Surachai Ratanasermpong	GIS	Geoinformatics and Space Technology Development Agency							○
	Yongyut Trisurat	GIS	Kasesart University							○
Vietnam	Thuy Phung Thu	Government	Team Leader, International Cooperation Team, Biodiversity Conservation Agency, Vietnam Environment Administration, Ministry of Natural Resources and Environment					○		
	Tran Duc Thanh	Marine	Director, Institute of Marine Environment and Resources (IMER)	○						
ACB	Rodorigo U. Fuentes	AP BON co-chair (2009–2012)	Executive Director, ASEAN Center for Biodiversity		○		○	○		
	Sheila Vergara	AP BON co-chair (2012)	Director, Biodiversity Information Management unit, ASEAN Center for Biodiversity			○		○	○	
	Filiberto Pollisco		Program Development Specialist, ASEAN Center for Biodiversity		○			○		

(continued)

Table 1.1 (continued)

Nation/ Organization	Name	Category	Organization	1st WS, July 21–22, 2009	2nd WS, Dec 10–11, 2009	4th GEOSS-AP, March 10–12, 2010	Precop and 3rd WS, Mar 21–23, 2010	AP-BON side event at COP10, Oct 2010	4th WS, Dec 2–4, 2011	5th GEOSS-AP, April 2–4, 2012
	Roland A. Inciong		Head, Communication and Public Affairs Unit, ASEAN Center for Biodiversity		○					
BioNET	Richard Smith		Director, BioNET-INTERNATIONAL Secretariat		○			○		
	Soetikno S. Sastroutomo		BioNET-International ASEANET, Senior Scientist, CABI Southeast and East Asia					○		
CBD	Junko Shimura		GTI Programme Officer, Secretariat of the Convention on Biological Diversity		○					
	David Cooper		Senior Programme Officer, Secretariat of the Convention on Biological Diversity				○			
	Kieran Noonan Mooney		Programme Assistant, Secretariat of the Convention on Biological Diversity						○	
CIFOR	Terry Sunderland	Forest	CIFOR			○				
DIVERSITAS	Anne Larigauderie		Executive Director, DIVERSITAS		○		○			
	Harold Mooney		Chair, DIVERSITAS				○			
	Eva Spehn	Terrstrial plant	Global Mountain Biodiversity Assessment, DIVERSITAS/Institute of Botany, University of Basel						○	
GBIF	Keiichi Matsuura	Fish	National Museum of Nature and Science, Tokyo	○			○	○	○	
	Éamonn O Tuama		Senior Programme Officer for Inventory, Discovery, Access (IDA), GBIF						○	
GEO-BON	Gary N. Geller	GEO BON SC	Deputy Manager, NASA Ecological Forecasting Program, Conservation Liaison, ASTER Science Project Jet Propulsion Laboratory	○	○					
	Bob Scholes	Chair, GEO BON	Council for Scientific and Industrial Research, South Africa				○			
	Vânia Proença	Terrestrial species	Centro de Biologia Ambiental, University of Lisbon						○	
World Fish Center	Eric Baran	Freshwater	Research Scientist, World Fish Center	○						
	Yumi Kura	Freshwater	Researcher, WorldFish Center				○	○		
	Nicolas Bailly	Database	Project Leader, Fish Base Team, Natural Resources Management, World Fish Center		○					

Through discussion in the meetings just described, we agreed on the following AP-BON visions, which are slightly modified from GEO BON visions.

– To establish a coordinated Asia-Pacific network that gathers and shares information on biodiversity and ecosystem services,
– To provide tools for data collection, sharing/exchange, analysis, and synthesis/integration, and
– To contribute to improving ecosystem management, sustainable use of biodiversity, and human well-being.

We also agreed about the following AP-BON missions.

– Observing and analyzing changes in biodiversity over time.
– Improving delivery of biodiversity information and services to users, particularly decision makers.
– Facilitating linkages among many countries, organizations, and individuals contributing to biodiversity observations.
– Identifying gaps between existing biodiversity observation systems and promoting mechanisms/projects to fill them.

The fourth AP-BON workshop was planned to be held March 14–15, 2011, in Tokyo, back-to-back with the fifth GEOSS-AP symposium, March 16–17. However, both meetings were canceled because of the crash of the Fukushima nuclear power station.

In July 2011, a Japanese project, "Integrative Observations and Assessments of Asian Biodiversity" (Leader: Tetsukazu Yahara) was launched. This project is the ninth strategic project (S9) of the Ministry of the Environment, Japan and is supported by the Environment Research and Technology Development Fund for the term from 2011 to 2015. It is expected that this project contributes to the development of AP-BON activities. This project has five themes (modeling and integration, species/genetic diversity, terrestrial ecosystem, freshwater biodiversity, and marine biodiversity) under which approximately 100 Japanese researchers are organized. Since this S9 project started, Japanese scientists have developed a series of observation plans in the Asia-Pacific scale under collaboration with non-Japanese scientists who have been networked through past AP-BON workshops. However, the S9 project is one of many projects ongoing in the Asia-Pacific region, and a key vision of AP-BON is to establish a network by coordinating many projects that have been mostly isolated from each other. This task remains a great challenge.

On December 2–3, 2011, the fourth AP-BON workshop was held in Tokyo. In this workshop, the designs of the S9 project for five themes were introduced to the participants, and AP-BON implementation 2011–2015 was discussed in each working group. The participants agreed that the S9 project can provide various opportunities of collaboration among AP-BON members on observations and assessments of biodiversity in terrestrial, freshwater, and marine ecosystems in the Asia-Pacific region. As a next step, plans to have three workshops in 2012, on genetic diversity, freshwater biodiversity, and marine biodiversity, were discussed and agreed. On December 4, the first AP-BON symposium was held by inviting representatives of CBD secretariat, DIVERSITAS, GBIF, and GEO BON. The symposium was successful to tighten linkage of AP-BON activities with these global mechanisms.

A workshop on genetic diversity observation was held March 14–16, 2012, at Kyushu University, Japan, inviting not only Asian scientists but also American, Australian, and European scientists to link AP-BON activities of genetic observation with global efforts. A plan to edit the GEO BON Genetic Diversity Report was discussed. "An international workshop on freshwater biodiversity conservation in Asia" was held November 26–27, 2012, at Kyushu University, Japan, inviting specialists of freshwater biodiversity from Cambodia, China, Indonesia, Malaysia, Thailand, and Vietnam. First, availability of data on aquatic plants and fish were reviewed for Chinese lakes, Mekong River, and Tonle Sap Lake, rivers of Peninsular Malaysia, and Indonesian lakes. Second, early products of the S9 project for freshwater biodiversity in Japan were introduced. Based on these presentations, plans of further collaboration were discussed. Another international workshop, on "Biodiversity in changing coastal waters of tropical and subtropical Asia," was held at the Amakusa Marine Biology Laboratory (AMBL), Kyushu University, December 1–4, 2012. Here, various changes of marine biodiversity in the coastal ecosystems of the Asia-Pacific region were reported, including northward migration of coral, degradation of the temperate seaweed community, and spread of invasive marine organisms in the West Pacific region. Also, early products of the S9 project for marine biodiversity in the Asia-Pacific region were introduced, and plans of further collaboration were discussed.

In addition to these workshops, AP BON activities have been strengthened through three other meetings held in 2012. First, the 5th GEOSS AP symposium was held April 2–4 in Tokyo, and in its AP BON parallel session, various efforts to integrate satellite and ground observations on status and changes of biodiversity and ecosystems were reported by the participants from various countries including Indonesia, Japan, Korea, Mongolia, and Thailand. Second, "An international workshop to develop key strategies for the establishment and promotion of the Asia-Pacific Biodiversity Observation Network (AP-BON) at local, national, and regional levels" was held on September 8 in Jeju, Korea, as an event of the World Conservation Congress. In this workshop, not only representatives of AP-BON but also those of the IUCN Asian regional office, Wildlife Watch Group (WWG, Nepal), and Pacific Rim Application and Grid Middleware Assembly (PRAGMA) were invited and the network was broadened. Third, the AP BON side event "Asia Pacific Biodiversity Observation Network (AP BON): its aims and implementation" was held on October 15 in CBD COP 11, Hyderabad, India. In this meeting, some key issues including data sharing and integration, documenting traditional knowledge on plant resources, roles of national biodiversity outlook, and scientific assessment to be conducted by IPBES were discussed.

From December 3 to 6, 2012, a GEO BON all-hands meeting was held in Asilomar, California. In this meeting, AP BON was provided an opportunity to introduce its activities in the plenary session, and five participants from AP BON actively contributed to the discussion toward establishing an initial operational system of GEO BON by 2015. While nine working groups of GEO BON are strengthening efforts to deliver some products by 2015, AP BON has already published its first product (Nakano et al. 2012) that compiled reviews on observation networks in China (Xu et al. 2012; Liu and Gao 2012), Indonesia (Parikesit et al. 2012; Simbolon et al. 2012), Korea (Yoon et al. 2012), Thailand (Marod and Kutintara 2012), Mekon Basin (Baran et al. 2012), and LTER sites (Kim 2012), databases of BISMaL (Yamamoto et al. 2012), ESABII (Jinbo and Ito 2012), FDDB (Niiyama et al. 2012), and JaLTER (Ohte et al. 2012), and techniques of DNA barcoding (Ebihara and Kuo 2012; Kato et al. 2012), satellite ecology (Muraoka et al. 2012), and spatial modeling (Fukushima 2012; Trisurat 2012). AP BON has now successfully established an initial operational network of biodiversity observations in the Asia-Pacific region. The next step is to strengthen it by carrying out a series of collaborative observations there through the opportunity provided by the S9 project.

Major Activities Planned for 2012–2015

The 4th AP-BON workshop, held from December 2 to 3, 2011, was particularly successful in agreeing on implementation of AP-BON for 2012–2015, including the following plans of five major activities.

1. Editing and networking national biodiversity outlook
   It is important for each country to develop a national network of biodiversity scientists and edit a national biodiversity outlook. Toward CBD COP10 held in 2010, a national biodiversity outlook was edited in Japan and a regional biodiversity outlook was edited by ACB. Similar efforts in other countries will greatly improve our understanding on states and trends of biodiversity in the Asia-Pacific region. Thus, AP-BON will facilitate editing national biodiversity outlooks in the Asia-Pacific countries by providing standard protocols, databases, and analytical tools.
2. Publication of additional AP-BON books
   The AP-BON book entitled *The Biodiversity Observation Network in the Asia-Pacific Region: Toward Further Development of Monitoring* (Nakano et al. 2012) provided a helpful overview for various biodiversity observations that had been mostly isolated from each other. However, this book could cover only a limited proportion of the activities ongoing in the Asia-Pacific region. First, it includes few chapters on coastal and marine biodiversity. Second, a large amount of work going on in forest plots in the AP region was not reviewed. We will publish successive volumes of the AP-BON book filling these gaps.
3. Promotion of collaborative projects using S9 as a leverage
   The Japanese project "Integrative Observations and Assessments of Asian Biodiversity" (S9) is providing various opportunities for biodiversity scientists

to carry out collaborative observations and thus can be used as leverage to develop collaborative projects in five working groups.

4. Development of shared database
   Data sharing/exchange, analysis, and synthesis/integration are key missions of AP-BON. The GBIF Darwin core and its extension provide a basic tool for this mission for distribution records and other data associated with a taxon name and a GPS record. On the other hand, EML provides a basic tool to database metadata of various ecological observations. The DNA database has been well developed for DNA sequence data. Satellite images have been accumulated in some key institutes responsible for satellite observations in each country. AP BON will work to develop tighter linkages of these databases.
5. Capacity building
   Many tools are available for biodiversity observations, but it is not always easy to learn them personally or to develop the capacity in individual institutes. Thus, AP-BON will provide opportunities of capacity building for such tools as DNA barcoding, distribution modeling, and plot data analyses.

In addition to these general activities, AP BON implementation that was agreed on in the fourth workshop includes the following plans of the five working groups.

Genetic and Phylogenetic Diversity

For monitoring genetic and phylogenetic diversity, the GEO BON Concept Document proposed to assess states and trends in selected target species or taxonomic groups. Responding to this proposal, Global Legume Diversity Assessments (GLDA) are called (Yahara et al. 2013), and its first project is being advanced in the Asia-Pacific region. Some early findings were reported in the 6th International Legume Conference held in January 2013 in Johannesburg, South Africa. In addition to GLDA, assessments of genetic diversity in invasive species (Colautti et al. 2008; Goka et al. 2009), vertebrates under captive breeding and in the wild (Araki et al. 2008), tropical trees under rapid forest loss (Obayashi et al. 2002; Tnah et al. 2010), critically endangered species (Isagi 2013), and crop and crop relatives such as rice (Yamanaka et al. 2003; Londo et al. 2006) and soybean (Lam et al. 2010), were the focus in "Workshop for initiating a global genetic diversity assessment" held March 14–15, 2012, at Kyushu University, Japan. Subsequent to this workshop, a one-day workshop on DNA barcoding in Asia was held on March 16, and recent progress was introduced by participants from China, Indonesia, Japan, Korea, and Thailand. Progress in Korea is notable in that researchers succeeded in barcoding all the native species of flowering plants in Korea. The next step is to integrate findings from these observations and contribute to editing the GEO BON genetic diversity report that will be published by 2015.

Terrestrial Species Monitoring

The GEO BON Concept Document defined two concepts that should be implemented for monitoring of biodiversity at the species level: (1) changes in species distribution and abundance, and (2) distribution range maps for a large number of species. Surveys of national flora and fauna are required as a basis of this monitoring. However, national flora and fauna remain to be compiled in Asia-Pacific countries except for China, Japan, Korea, and Singapore. Projects such as Flora Malesiana, Flora of Thailand, and Flore du Cambodge, du Laos et du Vietnam have never been completed. Thus, it is a big challenge to carry out terrestrial species monitoring in the Asia-Pacific region.

Among various organisms, data of distribution and abundance are relatively richer in vertebrates but much poorer for plants and invertebrates. Because it is difficult to work on various taxa at once, the AP BON terrestrial species working group is concentrating efforts on assessments of vascular plants. Under the rapid loss of tropical forest in the Asia-Pacific region, it is expected that many species of vascular plants are declining and often threatened with extinction. The primary goal of plant diversity assessment is thus to document the threat status for as many species as possible. To achieve this goal, three approaches were proposed by Yahara et al. (2012): specimen-based approach, plot-based approach, and transect-based approach for some candidate hotspots.

In the specimen-based approach, distribution records obtained from herbarium specimens are used for modeling distribution of species and assessing threats under land use change, climate change, and other major threats including overharvesting. The GBIF database is useful for this purpose, but specimen records of many herbaria in the Asia-Pacific region are only partly archived in GBIF. We need more effort to digitize specimen records of herbaria in the Asia-Pacific region: this is one of the major challenges of the terrestrial species working group of AP BON.

Many forest plots have been placed in various countries in the Asia-Pacific region (e.g., Condit et al. 1999, 2000; Blanc et al. 2000; Bunyavejchewina et al. 2003; Slik et al. 2003, 2010, 2011; Sri-Ngernyuang et al. 2003; Kessler et al. 2005; Kitamura et al. 2005; Simbolon et al. 2012; Toyama et al. 2013). Data of species composition, abundance, DBH, height, and other observations in those plots are extremely valuable for documenting states and trends of species in forest ecosystems. However, one of the difficulties of using those data for species monitoring is the inaccuracy of species identification. In their floristic analysis using plot data from 111 locations across Sunderland, Slik et al. (2011) could work only on genera because most of the available plot data contained many unidentified species, making comparisons based on species-level identifications impossible. To improve the accuracy of species identification, the species and genetic diversity team of the S9 project is visiting some plots where good records of forest composition in the past are available, and collecting both herbarium specimens and silica gel-dried samples for DNA sequencing. In this project, not only are plot trees being monitored for growth, but also species of all other life forms (herbs, shrubs, vines, and epiphytes) are being recorded using the transect-based approach (see following). Using both herbarium specimens and DNA sequences, efforts to identify species are being made for plots in Cambodia (Toyama et al. 2013), Indonesia (Simbolon et al. 2012; Slik et al. 2010, 2011), and Thailand (Bunyavejchewina et al. 2003; Sri-Ngernyuang et al. 2003; Kitamura et al. 2005), and these efforts will be extended to Peninsular Malaysia and Vietnam in 2013 and to the Philippines and some other countries in 2014. Available as the first product of this project is a paper on a taxonomic inventory for forest trees in Cambodian plots (Toyama et al. 2013).

In the transect-based approach, a 100 × 5 m belt transect is used as a standard unit of observation (Yahara et al. 2012). One transect is divided into ten sections of 10 × 5 m, and all the vascular plant species in each section are distinguished and recorded as field names because it is difficult to accurately identify them in the field to taxonomic species. Herbarium specimens and silica gel-dried samples are collected for all the species, and DNA sequence is determined and used for identifying species precisely (sterile specimens in particular). The first transect survey was conducted in Gede Pangrango National Park in September 2011 as a collaborative project of Indonesian and Japanese scientists. There, a total of 14 transects were placed from the reserve within the Chibodas Botanical Garden (1,400 m) to the top area (3,030 m). This field survey revealed the utility of transects describing flora and species distribution; by 2 weeks of fieldwork, approximately 75 % of species recorded in the published flora of Gede Pangrango National Park were collected and their occurrences were georeferenced. Furthermore, at least ten species not listed in the published flora were discovered, even though Gede Pangrango National Park is one of the best botanized locations in tropical Asia. Similarly, a series of 100 × 5 m belt transects were placed in different altitudes of Bokor National Park of Cambodia and Doi Inthanon National Park of Thailand. Further, 100 × 5 m belt transects were placed in forest plots of Kalimantan, Sumatra, and Thailand where good past records are available. Identification of specimens collected in those transects is now in progress. Extending this project in collaboration with scientists in various countries in the Asia-Pacific region will result in the first Asia-Pacific-wide database of vascular species distribution that is inevitable for editing a Red List of vascular plants in the Asia-Pacific scale.

Terrestrial Ecosystem Change

The activities of the AP BON Terrestrial Ecosystem Change working group mostly follow the detailed implementation plan of GEO-BON, although it rather has a cross-cutting approach on terrestrial ecosystem change. This working group includes the activities of GEO-BON WG3, together with the perspectives of WG6 (ecosystem services) and the methodology of WG7 (in situ/remote sensing integration): this is a more practical approach considering the limitations of human resources and funding for scientific projects in this region.

The final goal is to draw maps on biodiversity and ecosystem services covering this region and to establish the persistent monitoring system. Particular focus is put on forests and related human land use because the potential vegetation in this region is mostly forest. To enable this, we are working on the following three tasks.

Establishing Networks of In Situ Monitoring and a Database on Terrestrial Ecosystems

There are several international networks that can potentially provide monitoring data on biodiversity, function, and services of terrestrial ecosystems, such as DIWPA, ILTER-EAP (Kim 2012), as well as many domestic networks of biodiversity observation in each country and area. We will establish the network among these existing networks and to increase the interoperability of their database. The first mission is to establish the meta-database of these monitoring activities/sites, and then, develop devices and systems to increase interoperability of these datasets.

Creating a Database on Function and Services of Terrestrial Ecosystems

This task is to create a new dataset or database as a basis for mapping function and services of forest ecosystems that are associated with biodiversity. Among various forest functions, we are focusing on biomass, plant traits, and biological interaction because those provide measures related to regulating services such as carbon sequestration and pollination.

Biomass

We are working to evaluate the relationships between biomass, primary productivity, and plant diversity in the Asia-Pacific region. In many observation sites, tree species composition, biomass, and growth have been monitored and recorded (Slik et al. 2010, 2011). By creating a database of those data, analyses necessary for mapping of primary productivity, carbon sequestration, and other ecosystem services related to forest biomass and productivity can be made.

Plant Traits

A considerable number of data sets about various plant traits have been accumulated in the Asia-Pacific region. Examples are the ecophysiological traits of leaves (Wright et al. 2004), such as photosynthetic ability (e.g., Hikosaka et al. 2002; Kenzo et al. 2006) and decay rate, as well as nitrogen content and phenol content. Also, some ecological traits such as growth rate, maximum plant size, and seed production have been studied in this region, although the data are not really sufficient or convenient to integrate. These traits are related to functional characteristics of species and also largely dependent on species. Thus, by creating a database on plant traits together with geographic distribution and phylogeny, we can link biodiversity with the function and services of the forest ecosystem.

Biological Interactions

Diversity of biological interaction is the basis of regulating services of ecosystem, and it is necessary to elucidate the mechanisms of biological interactions to obtain clearer figures correlating biodiversity and the functions and services of ecosystems. Among ecosystem services that have strong links with biodiversity, we focus on pollination and biological control as pilot examples to show the geographic mapping of services. Particular efforts are made to collect the data on diversity of some potential pollinators for crops and mechanisms controlling pests and disease for crops in dominant forest types. The relationships between pollinators, pests, and disease, and biological diversity, climate, land use, and other information that can be detected at the landscape scale or larger scales are also analyzed. These analyses will help to develop the algorithms to map ecosystem services in larger spatial scales.

Mapping of Biodiversity, Function, and Ecosystem Services

Algorithms for mapping biodiversity and ecosystem services will be developed. Subsequently, maps on biodiversity and ecosystem services will be provided by linking in situ observation data with spatial information at various scales obtained by satellite and other sources. This task is also aiming at establishing biodiversity monitoring in a large spatial scale.

Freshwater Ecosystem Changes

The biodiversity crisis in freshwater systems is the most serious among all natural systems (Millennium Ecosystem Assessment 2005a, b), although international efforts were made to achieve, by 2010, significant reduction of the current rate of biodiversity loss under supervision by the Convention on Biological Diversity (CBD). To identify the reasons for the crisis in freshwater environments, and to conduct better conservation and management, we must urgently collect information about the status quo of freshwater environments through appropriate environmental monitoring, together with compiling the data already collected by and independently stored at individual institutions.

The AP BON Freshwater Monitoring working group focuses on monitoring freshwater biodiversity at the genus, species, and ecosystem levels. Thus, the activities of this working group correspond to the activities of GEO BON WG4. Environmental, and biodiversity monitoring in Asian freshwater systems has been actively conducted by local governments, NGOs, NPOs, and amateurs. We also already have strong worldwide networks that may serve as links among various monitoring activities. However, the data derived from monitoring activities in the Asia Pacific region are not always included in a worldwide network or open resource database and are often individually or independently stored. Thus, integrating these monitoring activities and data is a key challenge to establish freshwater observation systems in the Asia-Pacific region.

As a first step for this integration, "An international workshop on freshwater biodiversity conservation in Asia" was held November 26–27, 2012, at Kyushu University, Japan. In this workshop, data on aquatic plants in major lakes of China, Japan, and Indonesia as well as fish data from the Mekong basin (Laos, Thailand, Cambodia, and Vietnam) and rivers of Peninsular Malaysia were introduced. Efforts to create databases for those data are now in progress. For aquatic plants, some historical records are available, at least in China and Japan, and those should be useful to document changes of species composition over time. For fish in the Mekong basin, a census on fish diversity and catchment has been made under the collaboration of scientists in Laos, Thailand, Cambodia, and Vietnam. A notable achievement is the accumulation of fish specimens in stock centers of four countries that have been developed under the support of Nagao Natural Environment Foundation. These specimens are accompanied by precise distribution records, color photographs taken in the field, and alcohol samples for DNA analyses. Database construction on those data as well as DNA barcoding of fish species are now in progress.

Recently, the Ramsar Convention on Wetlands has proposed the development of a Global Wetland Observing System (GWOS), a network and portal for organizing information on wetlands and their wise use. This goal requires collaboration of relevant Earth observation programs, academic institutions, and NGOs. The concept of the GWOS fits closely with objectives of GEO BON aiming at building global observations systems for biodiversity. Thus, GEO BON WG 4 is going to organize collaboration for the GWOS. Freshwater monitoring in the Asia Pacific region is expected to play a significant role in developing the GWOS.

Marine Ecosystem Changes

The AP BON Marine Monitoring working group focuses on marine biodiversity, including species diversity as well as marine ecosystem dynamics in the tropical, subtropical, temperate, and boreal climate regions in the Western Pacific region. The area includes the area beyond national jurisdiction (often referred as open ocean) as well as the area within the exclusive economic zone (EEZ) defined by the Law of the Sea. The activities of this group thus correspond to the activities of GEO BON WG5.

Hopeful opportunities for AP BON Marine Monitoring are already provided by existing programs and resources. First, some programs such as Reef Check and Seagrass Watch have been carried out by NGOs, NPOs, and volunteers. At the global scale level, international research programs such as Census of Marine Life (CoML) and SeagrassNet have been carried out in the intensive support from funding agencies. Second, a global database of the Ocean Biogeography Information System (OBIS, http://www.iobis.org) has been developed by CoML in which many data have been accumulated since 2000. This database is now transferred to the International Oceanographic Commission (IOC), which is an international organization under UNESCO. Currently, OBIS is thus a very trustworthy database for data sharing on the global scale, because the metadata structure of OBIS is compatible with GBIF. Third, changes and drivers of marine biodiversity have been identified and observed, including disruption of coastal habitats, coral breaching, ocean acidification, decline of primary production, and development of deep-sea resources.

Our major challenge is to develop a regional network of marine biodiversity observation based on these opportunities. Although the global synthesis analyses using the data set in OBIS have revealed the extent of many ecologically or biologically significant areas such as the Asian and Pacific Ocean region (Tittensor et al. 2010), inventories of regional biodiversity based on long-term observations remain incomplete. The data coverage of OBIS is insufficient to analyze regional biodiversity changes in fine scale. Data in OBIS also have a large bias to shallow-water biota (Webb et al. 2010). For example, nekton and plankton data are very rich, but benthos data are not. Data from the open ocean, midwater, and deep sea are very poor. The AP BON Marine Monitoring working group will fill these deficiencies by networking various available opportunities in the Asia-Pacific region and promoting new observations under the collaboration of marine scientists.

To carry out AP BON Marine Monitoring, six research teams have been established under the support of the S9 project. Those teams are focusing on database development to share biodiversity information and are collecting data on kelp forests, seagrass meadows, coral reefs, pelagic waters, and the deep sea. The data sets are collected by specimen-based approach, habitat-targeted approach, and remote sensing approach. For database development, the data sets described in Darwin Core format are accumulated into the Biological Information System for Marine Life (BISMaL, http://www.godac.jamstec.go.jp/bismal/e/) managed by Global Oceanographic Data Centre (GODAC) of JAMSTEC (Yamamoto et al. 2012). BISMaL is a regional node of the OBIS network and has a broader capacity of data content.

Statistical analysis and modeling for marine biodiversity trends are performed to identify the risks for biodiversity decline. The data accumulated in BISMaL are an intelligent resource to analyze the present state of biodiversity decline, and suggest improvements on protocol of marine biodiversity decline, ecosystem-based environmental management methods for sustainable use of marine products, and biodiversity conservation methods for mining activities. The data mining approach from such data sets is also useful to document historical changes of biodiversity and improve inventories of regional biodiversity based on long-term observations.

We already had the first regional workshop on "Biodiversity in changing coastal waters of tropical and subtropical Asia" from December 1 to 4, 2012. There were many good presentations on the status and changes of costal biodiversity in tropical and subtropical Asia, including reports on coral changes, deterioration of seaweed beds, marine nonindigenous species, phylogeography and genetic diversity of marine organisms, and identification of candidates for ecologically or biologically significant areas (EBSA, http://ebsa.cbd.int) using a large dataset of occurrence records. It is notable that the states and trends of coral biodiversity were reported by researchers of Indonesia, Malaysia, Thailand, Philippines, Taiwan, and Japan, and thus we could develop a better integrated picture of coral changes along the environmental gradient from Indonesia to Japan. This workshop illustrated the importance of networking various observations to obtain an integrated picture of marine biodiversity changes and how fruitful further collaboration under AP BON will be.

Next Steps

As is summarized here, AP BON has been successfully organized as a network of biodiversity scientists, academic institutions, and relevant programs. Now, it has entered its second stage, for promoting collaborative observations of biodiversity and ecosystem services in the Asia-Pacific region. Five working groups are now promoting integrated observations supported by the S9 project of the Ministry of the Environment, Japan. In addition, many other projects of biodiversity observations including nationwide monitoring on particular measures and in-depth field studies focusing on particular questions are going on in various countries of the Asia-Pacific region. An important role of AP BON is to network these projects by providing opportunities for communication and interaction, developing tools for data collection, sharing, analysis, and synthesis, and carrying out assessments helpful for improving ecosystem management and sustainable use of biodiversity.

Because participants in AP BON have exceeded 100 and are still increasing, it is now required to establish a better governance of AP BON including membership management, publication of newsletters, and continued updates of the website. Until today, the Biodiversity Center of the Ministry of the Environment Japan (BiodiC-J) has played a significant role as the Secretariat of AP BON. Based on the successful networking to date, in the next step, we can expect a stronger, collaborative secretariat mechanism to support the growing activities of AP BON. Involvement of more countries and institutes to share the increased work of the secretariat would contribute to strengthening collaboration among countries in the Asia-Pacific area. For example, Athe SEAN Center for Biodiversity (ACB) is one of the major hub institutes in this area and provides an opportunity to link various observation activities with its Clearing House Mechanism for ASEAN countries.

Another notable capacity is provided by the Korean scientists who organized Korean BON (K-BON). Because CBD COP 12 will be held in Korea in 2014, it is expected that Korean scientists could play a more significant role in promoting AP BON under the possible support of the Korean government.

Many countries in the Asia-Pacific region are rapidly growing, not only in economic power but also in their scientific capacity. Thus, we can expect that AP BON could be developed with an increasing body of scientific activities in each country. Coordinating these activities to better collaboration and more synergy is a challenging but pivotal and fruitful task that should contribute to the achievement of Aichi Targets, the success of CBD COP12, and the first assessment of IPBES. With this vision in mind, we call for participation and contribution of more biodiversity scientists to AP BON through collaboration in observing states and trends of biodiversity and ecosystem services, and sharing, analyzing, and integrating obtained data to improve our capacity for ecosystem management and sustainable use of biodiversity.

Acknowledgments The development of AP BON has been achieved through discussion in four AP BON workshops and additional meetings including GEOSS AP symposia, as described here. We thank all the participants for their enthusiastic contribution to discussions in these meetings. It is greatly regretted that Rodorigo Fuentes (co-chair), Herwint Simbolon (terrestrial ecosystem change working group), and Dede Irving Hartoto (freshwater ecosystem change working group) passed away since the 4th AP BON workshop in 2011. We are respectfully grateful for their enthusiasm and contribution to the development of AP-BON. We also thank the Biodiversity Center of the Ministry of the Environment of Japan (BiodiC-J) for its continued support as the Secretariat of AP BON. The present study was supported by the Environment Research and Technology Development Fund (S9) of the Ministry of the Environment, Japan.

References

Araki H, Berejikian BA, Ford MJ, Blouin MS (2008) Fitness of hatchery-reared salmonids in the wild. Evol Appl 1:342–355

Baran E, Chum N, Fukushima M, Hand T, Hortle KG, Jutagate T, Kang B (2012) Fish biodiversity research in the Mekong Basin. In: Nakano S, Yahara T, Nakashizuka T (eds) The biodiversity observation network in the Asia-Pacific region: toward further development of monitoring. Springer, Tokyo

Blanc L, Maury-Lechon G, Pascal J-P (2000) Structure, floristic composition and natural regeneration in the forests of Cat Tien National Park, Vietnam: an analysis of the successional trends. J Biogeogr 27:141–157

Bradshaw CJA, Sodhi NS, Peh KSH, Brook BW (2007) Global evidence that deforestation amplifies flood risk and severity in the developing world. Glob Chang Biol 13:2379–2395

Bunyavejchewina S, LaFrankie JV, Baker PJ, Kanzaki K, Ashton PS, Yamakura T (2003) Spatial distribution patterns of the dominant canopy dipterocarp species in a seasonal dry evergreen forest in western Thailand. For Ecol Manag 175:87–101

Butchart SHM, Walpole M, Collen B, van Strien A, Scharlemann JPW, Almond REA, Baillie JEM, Bomhard B, Brown C, Bruno J, Carpenter KE, Carr GM, Chanson J, Chenery AM, Csirke J, Davidson NC, Dentener F, Foster M, Galli A, Galloway JN, Genovesi P, Gregory RD, Hockings M, Kapos V, Lamarque J-F, Leverington F, Loh J, McGeoch MA, McRae L,

Minasyan A, Morcillo MH, Oldfield TEE, Pauly D, Quader S, Revenga C, Sauer JR, Skolnik B, Spear D, Stanwell-Smith D, Stuart SN, Symes A, Tierney M, Tyrrell TD, Vié J-C, Watson R (2010) Global biodiversity: indicators of recent declines. Science 328:1164–1168

Colautti RI, Maron JL, Barrett SCH (2008) Common garden comparisons of native and introduced plant populations: latitudinal clines can obscure evolutionary inferences. Evol Appl 2:187–199

Condit R, Ashton PS, Manokaran N, LaFrankie JV, Hubbell SP, Foster RB (1999) Dynamics of the forest communities at Pasoh and Barro Colorado: comparing two 50 ha plots. Philos Trans R Soc Lond B 354:1739–1748

Condit R, Ashton PS, Baker P, Bunyavejchewin S, Gunatilleke S, Gunatilleke N, Hubbell SP, Foster RB, Itoh A, LaFrankie JV, Lee HS, Losos E, Manokaran N, Sukumar R, Yamakura T (2000) Spatial patterns in the distribution of tropical tree species. Science 288:1414–1418

Díaz S, Fargione J, Chapin FS, Tilman D (2006) Biodiversity loss threatens human well-being. PLoS Biol 4(8):e277

Ebihara A, Kuo L-Y (2012) East and Southeast Asian pteridophyte flora and DNA barcoding. In: Nakano S, Yahara T, Nakashizuka T (eds) The biodiversity observation network in the Asia-Pacific region: toward further development of monitoring. Springer, Tokyo

Fukushima M (2012) Spatially explicit models for freshwater fish for conservation planning. In: Nakano S, Yahara T, Nakashizuka T (eds) The biodiversity observation network in the Asia-Pacific region: toward further development of monitoring, Springer, Tokyo

GEO BON (2010) Group on Earth Observations Biodiversity Observation Network (GEO BON) Detailed Implementation Plan Version 1.0. http://www.earthobservations.org/documents/cop/bi_geobon/geobon_detailed_imp_plan.pdfJinbo and Ito 2012

Goka K, Yokoyama J, Une Y, Kuroki T, Suzuki K, Nakahara M, Kobayashi A, Inaba S, Mizutani T, Hyatt AD (2009) Amphibian chytridiomycosis in Japan: distribution, haplotypes and possible route of entry into Japan. Mol Ecol 18:4757–4774

Hikosaka H, Nagamatsu D, Ishii HS, Hirose T (2002) Photosynthesis–nitrogen relationships in species at different altitudes on Mount Kinabalu, Malaysia. Ecol Res 17:305–313

Isagi Y (2013) Ubiquitous genotyping for conservation of endangered plant species. In: Nakano S, Yahara T, Nakashizuka T (eds) Asia-Pacific biodiversity observation network: integrative observations and assessments. Springer, Tokyo

Kato T, Jinbo U, Ito M (2012) DNA barcoding: a novel tool for observation of biodiversity. In: Nakano S, Yahara T, Nakashizuka T (eds) The biodiversity observation network in the Asia-Pacific region: toward further development of monitoring. Springer, Tokyo

Kenzo T, Ichie T, Watanabe Y, Yoneda R, Ninomiya I, Koike T (2006) Changes in photosynthesis and leaf characteristics with tree height in five dipterocarp species in a tropical rain forest. Tree Physiol 26:865–873

Kessler M, Keßler PJA, Gradstein SR, Bach K, Schmull M, Pitopang R (2005) Tree diversity in primary forest and different land use systems in Central Sulawesi, Indonesia. Biodivers Conserv 14:547–560

Kim E-S (2012) International long-term ecological research network activities in the East Asia-Pacific Region and biodiversity monitoring. In: Nakano S, Yahara T, Nakashizuka T (eds) The biodiversity observation network in the Asia-Pacific region: toward further development of monitoring. Springer, Tokyo

Kitamura S, Suzuki S, Yumoto T, Chuailua P, Plongmai K, Poonswad P, Noma N, Maruhashi T, Suckasam C (2005) A botanical inventory of a tropical seasonal forest in Khao Yai National Park, Thailand: implications for fruit–frugivore interactions. Biodivers Conserv 14:1241–1261

Lam H-M, Xu X, Liu X, Chen W, Yang G, Wong F-L, Li M-W, He W, Qin N, Wang B, Li J, Jian M, Wang J, Shao G, Wang J, Sun SS, Zhang G (2010) Resequencing of 31 wild and cultivated soybean genomes identifies patterns of genetic diversity and selection. Nat Genet 42:1053–1059

Leadley P, Pereira HM, Alkemade R, Fernandez-Manjarrés JF, Proença V, Scharlemann JPW, Walpole MJ (2010) Biodiversity scenarios: projections of 21st century change in biodiversity and associated ecosystem services. Technical Series no. 50. Secretariat of the Convention on Biological Diversity, Montreal

Liu H, Gao X (2012) Monitoring fish biodiversity in the Yangtze River, China. In: Nakano S, Yahara T, Nakashizuka T (eds) The biodiversity observation network in the Asia-Pacific region: toward further development of monitoring. Springer, Tokyo

Londo JP, Chiang Y-C, Hung K-H, Chiang T-Y, Schaal BA (2006) Phylogeography of Asian wild rice, *Oryza rufipogon*, reveals multiple independent domestications of cultivated rice, *Oryza sativa*. Proc Natl Acad Sci USA 103:9578–9583

Marod D, Kutintara U (2012) Biodiversity observation and monitoring in Thailand. In: Nakano S, Yahara T, Nakashizuka T (eds) The biodiversity observation network in the Asia-Pacific region: toward further development of monitoring, Springer, Tokyo

Millennium Ecosystem Assessment (2005a) Ecosystems and human well-being: synthesis. Island Press, Washington, DC

Millennium Ecosystem Assessment (2005b) Ecosystems and human well-being: current state and trends. Island Press, Washington, DC

Muraoka H, Ishii R, Nagai S, Suzuki R, Motohka T, Noda HM, Hirota M, Nasahara KN, Oguma H, Muramatsu K (2012) Linking remote sensing and in situ ecosystem/biodiversity observations by "satellite ecology." In: Nakano S, Yahara T, Nakashizuka T (eds) The biodiversity observation network in the Asia-Pacific region: toward further development of monitoring. Springer, Tokyo

Nakano S, Yahara T, Nakashizuka T (eds) (2012) The biodiversity observation network in the Asia-Pacific region: toward further development of monitoring. Springer, Tokyo

Niiyama K, Takyu M, Kawarasaki S, Ogawa M, Sato T, Masaki T (2012) Long-term monitoring of natural forests and their database (Forest Dynamics Database: FDDB) constructed by the Forestry and Forest Products Research Institute, Japan. In: Nakano S, Yahara T, Nakashizuka T (eds) The biodiversity observation network in the Asia-Pacific region: toward further development of monitoring. Springer, Tokyo

Obayashi K, Tsumura Y, Ihara-Ujino T, Niiyama K, Tanouchi H, Suyama Y, Washitani I, Lee C-T, Lee SL, Muhammad N (2002) Genetic diversity and outcrossing rate between undisturbed and selectively logged forests of *Shorea curtisii* (Dipterocarpaceae) using microsatellite DNA analysis. Int J Plant Sci 163:151–158

Ohte N, Nakaoka M, Shibata H (2012) ILTER and JaLTER: their missions and linkage to database development in the Asia-Pacific region. In: Nakano S, Yahara T, Nakashizuka T (eds) The biodiversity observation network in the Asia-Pacific region: toward further development of monitoring. Springer, Tokyo, pp 205–216

Parikesit, Okubo S, Husodo T, Takeuchi K, Muhamad D (2012) Biodiversity issues in Indonesia, with special reference to biodiversity in human-dominated landscapes. In: Nakano S, Yahara T, Nakashizuka T (eds) The biodiversity observation network in the Asia-Pacific region: toward further development of monitoring. Springer, Tokyo

Pereira HM, Leadley PW, Proença V, Alkemade R, Scharlemann JPW, Fernandez-Manjarrés JF, Araújo MB, Balvanera P, Biggs R, Cheung WWL, Chini L, Cooper HD, Gilman EL, Guénette S, Hurtt GC, Huntington HP, Mace GM, Oberdorff T, Revenga C, Rodrigues P, Scholes RJ, Sumaila UR, Walpole M (2010) Scenarios for global biodiversity in the 21st century. Science 330:1496–1501

Scholes RN, Mace GM, Turner W, Geller GN, Jürgens N, Larigauderie A, Muchoney D, Walther BA, Mooney HA (2008) Toward a global biodiversity observing system. Science 321:1044–1045

Secretariat of the Convention on Biological Diversity (2010) Global Biodiversity Outlook 3. Secretariat of the Convention on Biological Diversity, Montréal

Simbolon H, Suzuki E, Susanti R (2012) Some vegetation types studied with the same methodology in Indonesia. In: Nakano S, Yahara T, Nakashizuka T (eds) The biodiversity observation network in the Asia-Pacific region: toward further development of monitoring. Springer, Tokyo, pp 71–92

Slik JWF, Poulsen AD, Ashton PS, Cannon CH, Eichhorn KAO, Kartawinata K, Lanniari I, Nagamasu H, Nakagawa M, van Nieuwstadt MGL, Payne J, Purwaningsih, Saridan A,

Sidiyasa K, Verburg RW, Webb CO, Wilkie P (2003) A floristic analysis of the lowland dipterocarp forests of Borneo. J Biogeogr 30:1517–1531

Slik JWF, Aiba S-I, Brearley FQ, Cannon CH, Forshed O, Kitayama K, Nagamasu H, Nilus R, Payne J, Paoli G, Poulsen AD, Raes N, Sheil D, Sidiyasa K, Suzuki E, van Valkenburg JLCH (2010) Environmental correlates of tree biomass, basal area, wood specific gravity and stem density gradients in Borneo's tropical forests. Glob Ecol Biogeogr 19:50–60

Slik JWF, Aiba S-I, Bastian M, Brearleye FQ, Cannon CH, Eichhorng KAO, Fredrikssonh G, Kartawinatai K, Laumonierj Y, Mansork A, Marjokorpil A, Meijaardm E, Morley RJ, Nagamasu H, Nilus R, Nurtjahya E, Payne J, Permana A, Poulsen AD, Raes N, Riswan S, van Schaik CP, Sheil D, Sidiyasa K, Suzuki E, van Valkenburga JLCH, Webb CO, Wichu S, Yoneda T, Zakaria R, Zweifel N (2011) Soils on exposed Sunda Shelf shaped biogeographic patterns in the equatorial forests of Southeast Asia. Proc Natl Acad Sci USA 108:12343–12347

Sri-Ngernyuang K, Kanzaki M, Mizuno T, Noguchi H, Teejuntuk S, Sungpalee C, Hara M, Yamakura T, Sahunalu P, Dhanmanonda P, Bunyavejchewina S (2003) Habitat differentiation of Lauraceae species in a tropical lower montane forest in northern Thailand. Ecol Res 18:1–14

Tittensor DP, Mora C, Jetz W, Lotze HK, Ricard D, Berghe EV, Worm B (2010) Global patterns and predictors of marine biodiversity across taxa. Nature (Lond) 466:1098–1101

Tnah LH, Lee SL, Ng KKS, Faridah Q-Z, Faridah-Hanumb I (2010) Forensic DNA profiling of tropical timber species in Peninsular Malaysia. For Ecol Manag 259:1436–1446

Toyama H, Tagane S, Chhang P, Kajisa T, Ichihashi R, Samreth V, Ma V, Sokh H, Katayama A, Itadani H, Tateishi M, Tachiki Y, Mase K, Onoda Y, Mizoue N, Tachida T, Yahara T (2013) Inventory of woody flora in permanent plots of Kampong Thom and Kampong Chhnang Provinces, Cambodia. Acta Phytotax Geobot 64:45–105

Trisurat Y (2012) Long-term monitoring and prediction of ecosystem using remote sensing and the CLUE-S Model: Sakaerat Environmental Research Station. In: Nakano S, Yahara T, Nakashizuka T (eds) The biodiversity observation network in the Asia-Pacific region: toward further development of monitoring. Springer, Tokyo, pp 309–320

Webb TJ, Berghe EV, O'Dor R (2010) Biodiversity's big wet secret: the global distribution of marine biological records reveals chronic under-exploration of the deep pelagic ocean. PLoS One 5:e10223

Wright IJ, Reich PB, Westoby M, Ackerly DD, Baruch Z, Bongers F, Cavender-Bares J, Chapin T, Cornelissen JHC, Diemer M, Flexas J, Garnier E, Groom PK, Gulias J, Hikosaka K, Lamont BB, Lee T, Lee W, Lusk C, Midgley JJ, Navas M, Niinemets U, Oleksyn J, Osada N, Poorter H, Poot P, Prior L, Pyankov VI, Roumet C, Thomas SC, Tjoelker MG, Veneklaas EJ, Villar R (2004) The worldwide leaf economics spectrum. Nature (Lond) 428:821–827

Xu H, DingH, Wu J (2012) Introduction to ecological and biodiversity monitoring in China. In: Nakano S, Yahara T, Nakashizuka T (eds) The biodiversity observation network in the Asia-Pacific region: toward further development of monitoring. Springer, Tokyo, pp 65–70

Yahara T, Akasaka M, Hirayama H, Ichihashi R, Tagane S, Toyama H, Tsujino R (2012) Strategies to observe and assess changes of terrestrial biodiversity in the Asia-Pacific regions. In: Nakano S, Yahara T, Nakashizuka T (eds) The biodiversity observation network in the Asia-Pacific region: toward further development of monitoring. Springer, Tokyo, pp 3–20

Yahara T, Javadi F, Onoda Y, de Queiroz LP, Faith DP, Prado DE, Akasaka M, Kadoya T, Ishihama F, Davies S, Slik JWF, Yi T. Ma K, Bin C, Darnaedi D, Pennington RT, Tuda M, Shimada M, Egan AN, Buerki S, Raes N, Kajita T, Mimura M, Tachida H, Iwasa Y, Smith GF, Victor JE, Nkonki T (2013) Global legume diversity assessments: concepts, key indicators, and strategies. Taxon 62:249–266

Yamamoto H, Tanaka K, Fujikura K, Maruyama T (2012) BISMaL: Biological information system for marine life and role for biodiversity research. In: Nakano S, Yahara T, Nakashizuka T (eds) The biodiversity observation network in the Asia-Pacific region: toward further development of monitoring. Springer, Tokyo, pp 247–256

Yamanaka S, Nakamura I, Nakai H, Sato YI (2003) Dual origin of the cultivated rice based on molecular markers of newly collected annual and perennial strains of wild rice species, *Oryza nivara* and *O. rufipogon*. Theor Appl Genet 50:529–538

Yoon J-D, Jang M-H, Kim H-W, Joo G-J (2012) Fish Biodiversity Monitoring in Rivers of South Korea. In: Nakano S, Yahara T, Nakashizuka T (eds) The biodiversity observation network in the Asia-Pacific region: toward further development of monitoring. Springer, Tokyo, pp 175–191

Chapter 2
Effects of Ecosystem Diversity on Species Richness and Ecosystem Functioning and Services: A General Conceptualization

Tadashi Miyashita, Tatsuya Amano, and Takehisa Yamakita

Abstract Biodiversity has a hierarchical structure, consisting of genetic, species, and ecosystem levels. Although the importance of genetic and species diversities is well recognized from both basic and applied perspectives, there has been virtually no conceptualization of how and why ecosystem diversity matters from an ecological or socioecological perspective. Here, we describe the role of ecosystem diversity with respect to three different properties, namely, species diversity, ecosystem functioning, and ecosystem services. Ecosystem diversity increases species diversity by additive and nonadditive, or emergent, processes. The additive process is self-evident, but the emergent process is not well recognized. Habitat complementation and supplementation are examples of the emergent effect; both these paths facilitate population persistence and enhance species richness under composite ecosystems or in heterogeneous landscapes. Here not only compositional heterogeneity but also configuration heterogeneity matters. Linkage and interactions of different ecosystems are also important for ecosystem resilience and services. Empirical studies show that movement of organisms or nutrients between ecosystems sustains and/or stabilizes ecosystem dynamics, and theoretical food web studies show that the presence of a weak trophic linkage between ecosystems stabilizes the dynamics of the system as a whole. Also, heterogeneous landscapes consisting of multiple landscape elements are buffered against disturbances caused by cyclones, fire, and eutrophication. Maintaining spatial resilience across landscapes is the key to increasing the robustness and sustainability of socioecological systems in a changing world.

T. Miyashita (✉)
Department of Ecosystem Studies, University of Tokyo, Tokyo 113-8656, Japan
e-mail: tmiya@es.a.u-tokyo.ac.jp

T. Amano
Department of Zoology, University of Cambridge, Cambridge CB2 3EJ, UK

T. Yamakita
Institute of Biogeosciences, Japan Agency for Marine-Earth Science and Technology, 2-15, Natsushima-cho, Yokosuka-city, Kanagawa 237-0061, Japan

S.-i. Nakano et al. (eds.), *Integrative Observations and Assessments*, Ecological Research Monographs, DOI 10.1007/978-4-431-54783-9_2, © Springer Japan 2014

Keywords Ecosystem services • Emergent effect • Insurance hypothesis • Intermediate heterogeneity hypothesis • Landscape complementation • Landscape supplementation • Resilience • Satoyama landscape • Spatial resilience

Introduction

Biodiversity has a hierarchical nature, consisting of genetic, species, and ecosystem levels (e.g., Johnson 1993; Gaston 1996; Millennium Ecosystem Assessment 2005). Before the term "biodiversity" was created in the late 1980s, ecologists and evolutionary biologists had long been interested in the causes and consequences of genetic and species diversity (Darwin 1872; Fisher 1930; Elton 1958; May 1973). However, the importance of ecosystem diversity has not been well understood in ecology, partly because ecosystems have often been considered as closed systems within which species interactions and nutrient cycling occur (Hairston et al. 1960; Oksanen and Oksanen 2000). Recent theoretical and empirical studies have indicated that most ecosystems are open, to some degree, and that the linkage of systems is an important determinant of species diversity and ecosystem stability (Polis et al. 2004; Rooney et al. 2008). In the face of growing concern about biodiversity in general, it is time to reconceptualize the role of ecosystem diversity to catch up with the more advanced concepts of genetic and species diversity. Here, we provide a review of why and how ecosystem diversity is important with respect to (1) species diversity, (2) ecosystem functioning, and (3) ecosystem services, with emphasis on the interactions among ecosystems or landscape components.

Species Diversity

Additive and Nonadditive Effects

Species diversity has a hierarchical nature that consists of α, β, and γ. α- and β-diversities measure within- and between-habitat diversity, respectively, and γ-diversity is a total diversity including different habitats. In this context, the importance of ecosystem diversity is evident, because different ecosystems have different community members; this enhances β-diversity and hence γ-diversity. However, the effect of ecosystem diversity on species richness is not limited to additive effects but also includes nonadditive or emergent effects. Many organisms require different ecosystems for completion of their life cycles or persistence of their populations. For instance, anurans and aquatic insects use both freshwater and terrestrial habitats in different life stages, and anadromous fishes use both freshwater and marine habitats. The phenomenon whereby a species requires nonsubstitutable resources found in different habitats is called landscape complementation (Dunning et al. 1992). In contrast, in landscape supplementation there is a lesser degree of restriction: a species uses substitutable resources in different habitats, thus enhancing its population densities (Dunning et al. 1992). The gray-faced buzzard (*Butastur indicus*), a raptor species observed in an agriculture–forest mosaic in East Asia, is a good example: the buzzard usually preys on frogs in rice paddies but also on lepidopterous larvae in deciduous forests in summer, when capture rates in rice paddies decrease (Sakai et al. 2011).

Notably, the nonadditive effect of ecosystem diversity emerges only when different ecosystems are accessible to focal organisms (or when the prey of the focal organisms can migrate between different habitats). In particular, the human-induced disconnection between habitats used by different life history stages of a species is known as habitat split and greatly affects population persistence of the species (Becker et al. 2007). For instance, the number of egg masses of the montane brown frog (*Rana ornativentris*) laid in a rice paddy decreased when there was a large ditch between the paddy and adjacent forests (Kato et al. 2010). Also, the abundance in rice paddies of loaches (*Misgurnus anguillicaudatus*) that reproduce in the paddies and then move to creeks or rivers decreased when there was a large vertical gap between a creek and the rice paddies (Katayama et al. 2011). These results imply that the presence of an ecosystem network that can be used by the focal organism matters; the mere presence of different ecosystems is not sufficient.

The nonadditive effect can be regarded as creating an emergent niche derived from combinations of different ecosystems. This emergent effect is not limited to one level but produces further emergent niches, as is the case with the crested ibis (*Nipponia nippon*), a top predator that inhabits a mosaic of agricultural landscapes in Japan. The crested ibis requires various prey organisms—such as loaches, anurans, insects, and earthworms—that inhabit rice paddies (C. Endo and H. Nagata, personal communication). The loaches reproduce in shallow, still waters such as rice paddies, which harbor abundant zooplankton as prey for the juveniles, but they migrate to adjacent creeks and rivers as they grow. The loaches therefore reach high population densities if there is a linkage between paddies and creeks (Katayama et al. 2011). The populations of amphibians, including the montane brown frog, tree frog (*Rhacophorus arboreus*), and black salamander (*Hynobius nigrescens*), reach high densities in paddies that have forests nearby (Kato et al. 2010; Uruma et al. 2012), as they spend their post-metamorphosis periods on the forest floor. Accordingly, the loaches and amphibians require different combinations of ecosystems and the crested ibis appears to depend on three ecosystems through the ecosystem network and food web network (Fig. 2.1). To our knowledge, such multiple emergent processes as exemplified by top predators have not been explicitly recognized.

The Issue of Spatial Scale

The traditional concept of the ecological niche does not consider spatial scale, but when ecosystem diversity is considered it is necessary to identify the spatial scales

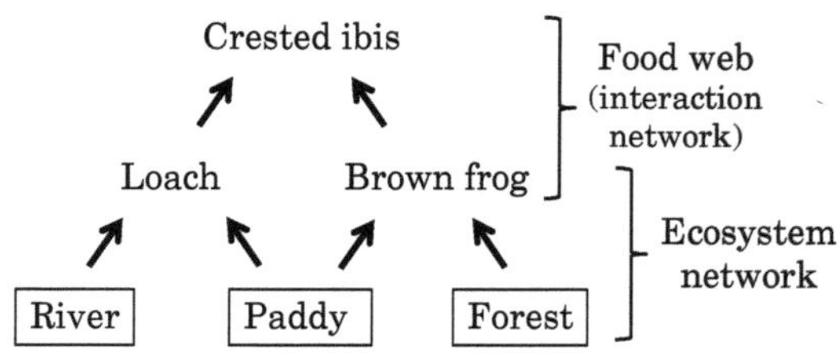

Fig. 2.1 Example of multiple emergent processes of ecosystem diversity. Loaches and montane brown frogs require different combinations of ecosystems to complete their life cycles; through its prey, the crested ibis indirectly requires three ecosystems

effective for persistence or enhancement of the population densities of focal organisms. This is also true for species assemblage or species richness. The need to consider the spatial scale of niches has recently been stressed in conceptual papers (Szabo and Meszena 2006; Holt 2009), but with the development of statistical techniques, such as generalized linear models, Bayesian analysis, and machine learning methods (e.g., Franlkin 2010), it has become empirically feasible to explore the spatial scales determining population densities and species richness by using data on species distribution and land use.

For example, statistical analysis has revealed that the montane brown frog, which is distributed widely in Japanese agricultural landscapes, consisting mostly of rice paddies and forests, produces the highest egg mass density in rice paddies in locations where there is about 60 % forest cover within a radius of 300 m from the paddy (Kato et al. 2010). Interestingly, this spatial scale is roughly consistent with the distance moved by adults of a closely related species Rana japonica from breeding sites (rice paddies) to nonbreeding sites (forests), as measured by individual marking (Osawa and Katsuno 2001).

Similarly, spiders inhabiting grasslands, which are sparsely distributed in this landscape, exhibit their highest species richness in locations where there is about 60 % forest cover within a radius of 400 m from the focal grassland (Fig. 2.2a) (Miyashita et al. 2012). Analysis of abundance responses to landscapes revealed that the abundance of six of eight major spider species exhibited a hump-shaped pattern with respect to the degree of forest cover surrounding the grasslands (Fig. 2.2b). It is not clear why these spider populations responded to the nearby presence of forest cover, but the forest edge might have functioned as a refuge from mowing by humans, or prey subsidies from nearby forests might have enhanced local spider population densities. The spatial scales that best explained the distribution of spiders differed among species, even though they had similar ecosystem diversity requirements (about 40 % surrounding forest cover for maximum abundance; see Fig. 2.2): for example, 200-m radius for *Argiope* and 700-m radius for *Tetragnatha* (Miyashita et al. 2012). This difference may reflect differences in dispersal abilities and hence the effective spatial scales determining the abundance of spiders.

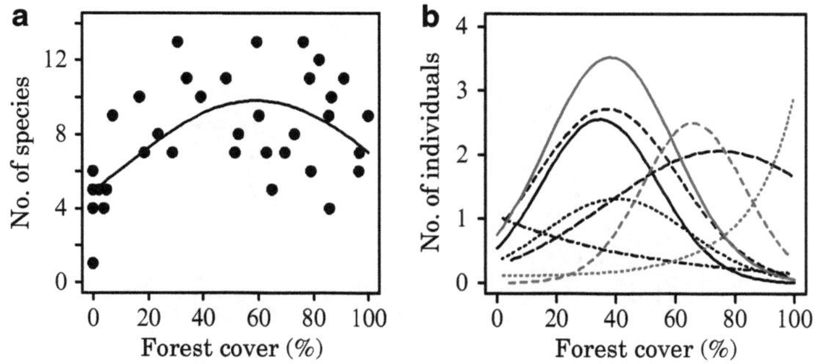

Fig. 2.2 Relationship between species richness of grassland-dwelling spiders (**a**) or abundance of each spider species (**b**) in relationship to forest cover in surrounding landscapes. Each *solid circle* represents a study plot (**a**), and each *line* shows a different species (**b**)

Issues of Compositional and Configuration Heterogeneity

Ecosystem diversity can be envisioned within the framework of landscape heterogeneity, although landscape heterogeneity occasionally implies vegetational or topographic heterogeneity that is found within an ecosystem. Hereafter, we use the term "landscape heterogeneity" synonymously with "ecosystem diversity."

There are two types of measurement of landscape heterogeneity: compositional and configurational. Compositional heterogeneity is measured by using diversity indices such as Simpson's or Shannon's index, or more simply by the number of different landscape elements or the area proportion of a particular landscape element embedded in a landscape (e.g., percent of forest cover, as shown earlier) (Fahrig et al. 2011). Compositional heterogeneity does not include any information on the shapes or arrangements of landscape elements, whereas configuration heterogeneity considers this information. One of the simplest measures of configuration heterogeneity is the perimeter-to-area ratio of a given land cover type or patch, which increases as the patch configuration becomes more complex. There are many other measures of configuration heterogeneity (McGarigal et al. 2002). Earlier studies often included several different configuration measures as independent variables in a particular statistical analysis, probably for exploratory purposes (Pickett and Siriwardena 2011), but we recommend choosing only a couple of appropriate measures to test an a priori hypothesis driven by a particular mechanism.

Compositional and configuration heterogeneities are usually not mutually independent. For instance, in a landscape consisting of two types of land cover (e.g., rice paddies and forests), the ratio of forest perimeter length to forest area (configuration heterogeneity) exhibits a hump-shaped curve with respect to the proportion of forest area (compositional heterogeneity), because too little or too much forested area inevitably produces a shorter forest perimeter. However, it is not well recognized that increasing configuration heterogeneity increases local compositional heterogeneity, and it is even less well known that landscape configuration

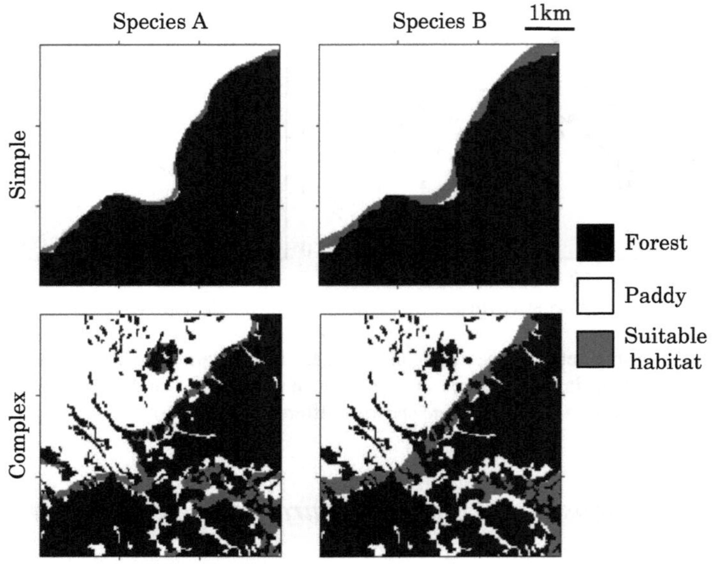

Fig. 2.3 Distributions of suitable habitats for two species of frog that have identical compositional niches (landscape heterogeneities) but different effective spatial scales under simple and complex landscapes. *Species A* and *Species B* have effective scales of 300- and 600-m radius from a focal paddy, respectively, but both have compositional niches of 40–60 % forest cover

heterogeneity and the scale of a species niche interact to yield emergent properties of species distributions.

Consider two species of animal, say frogs A and B, which require both rice paddies and forests to complete their life cycles. The landscape compositions that allow persistence of the two species populations are the same for both species (40–60 % forest cover surrounding a focal rice paddy), but the effective spatial scales for the landscape differ between the species (A, 300-m radius from a paddy; B, 600-m radius from a paddy). What would be the distributions of suitable habitats for the two species under simple and complex landscapes that differ only in the degree of configuration heterogeneity? Comparison of suitable habitats under the two landscapes reveals that the amount of suitable habitat in a complex landscape increases for both species (Fig. 2.3), because increasing configuration heterogeneity increases local-scale compositional heterogeneity, which in turn increases the area of rice paddies that have 40–60 % surrounding forest cover. Another important point to note is that the areas suitable for the two species overlap greatly in a simple landscape, but they are somewhat segregated in a complex landscape (Fig. 2.3). In the complex landscape, species A appears to occupy fine-scale mosaic structures to a greater degree than does species B, as a result of the smaller spatial scale necessary for species A. This difference suggests that, in a landscape with high configuration heterogeneity, coexistence is more readily

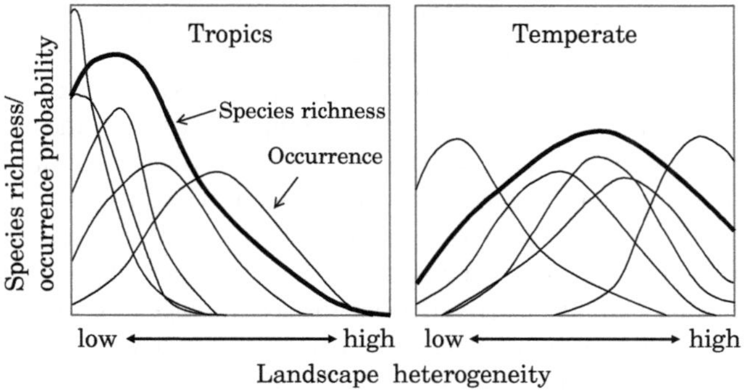

Fig. 2.4 Hypothetical patterns of species richness in relationship to landscape heterogeneity in different climatic zones. *Thick solid lines* indicate species richness; *thin lines* indicate the occurrence probability of each species

realized between species that have different niche scales. In other words, even if the compositional niches of organisms are identical, the combination of configuration heterogeneity and differing niche scales yields an emergent property that results in separate distribution patterns. This argument is principally the same as that for the discontinuous body size distributions of coexisting organisms found in fractal-like landscapes (Holling 1992; Szabo and Meszena 2006).

Landscape Heterogeneity Does Not Always Enhance Species Diversity

The concept that ecosystem diversity or landscape heterogeneity increases the potential number of niches and hence enhances species richness should be viewed as only one side of the coin. It is true that there are many organisms that are sustained by landscape heterogeneity, but for habitat specialists such as forest interior bird species, increased landscape heterogeneity may result in habitat fragmentation, reducing the likelihood of population persistence (Fahrig 2003; Zipkin et al. 2009). For example, Pickett and Siriwardena (2011) found that although the majority of farmland bird species in the UK were more common in heterogeneous landscapes, some species, particularly those under a high level of conservation concern, tended to be less abundant in more heterogeneous landscapes. Whether landscape heterogeneity acts positively or negatively on species richness depends on the balance of habitat complementation versus habitat fragmentation, and this may be related to the nature of the regional species pool that reflects the evolutionary history of the region. For instance, tropical rainforests spreading over large areas appear to have persisted for a long time through glacial periods (Ray and Adams 2001), so that species adapted to homogeneous forest landscapes are expected to be much more common than those adapted to heterogeneous landscapes (Fig. 2.4).

In contrast, as most temperate regions have experienced historical environmental changes since the last glacial period and are still undergoing periodic disturbances such as typhoons and wildfires, there may be many species adapted to heterogeneous landscapes (Romme 2005). However, even in temperate regions, the negative effect of habitat fragmentation will prevail over the positive effect of habitat complementation when the landscape becomes too heterogeneous. Here, a hump-shaped pattern of species richness with respect to landscape heterogeneity is expected (Fig. 2.4); this has been coined the "intermediate heterogeneity hypothesis" (Fahrig et al. 2011). This pattern is derived from the assemblages of species exhibiting varying occurrence responses to landscape heterogeneity. Examining how the relationship between landscape heterogeneity and species richness changes along the gradient of climatic regions or biomes differing in species pools is a challenge for future macroecological studies.

Ecosystem Functioning

During the past two decades, many ecologists have tried to examine the relationship between species diversity and ecosystem functioning (Tilman 1999; Hooper et al. 2005; Naeem et al. 2009). Two properties are used to measure ecosystem function, namely, the "average level" and "temporal stability" of the functions (Tilman 1999). We propose that the effect of ecosystem diversity on ecosystem functioning can be assessed by using the same measures. Here, we describe how the linkage and interactions of different ecosystems or landscape elements enhance ecosystem functioning.

Enhancement of Ecosystem Functioning Through the Movement of Materials and Organisms Between Ecosystems

Riverine and coastal ecosystems are intimately linked with terrestrial ecosystems, such as forests. In mountain streams, more than 90 % of organic matter is derived from forest ecosystems (Begon et al. 2006). The contribution of land ecosystems decreases moving downstream, but it is still about 20 % in estuaries and on continental shelves (Begon et al. 2006). This allochthonous input substantially enhances the production of coastal organisms.

Recent studies have also revealed the importance of reverse direction, from riverine or coastal ecosystems to land ecosystems, via tidal drift and animal movements. In some arid islands, the amount of organic materials coming from the ocean exceeds the in situ productivity (Polis and Hurd 1996). The importance of this allochthonous input can be seen after the invasion of devastating alien predators. In the Aleutian Islands, the introduction of alien arctic foxes (*Alopex lagopus*) decreased the abundance of seabirds, which had acted as agents for the transport of nutrients. This alteration led to the reduction of soil phosphorus concentrations by one third and vegetation biomass to less than one half (Croll et al. 2005). Transport of nutrients by animals from river to land also occurs. On the Pacific coast of North America, salmonid fishes transport substantial amounts of nitrogen from ocean to river; the nitrogen is further transported to forests and agricultural fields by bears and other carnivores. In some areas of Alaska, the growth rate of Sitka spruce (*Picea sitchensis*) is tripled along the rivers where salmonids breed (Helfield and Naiman 2001). In one basin in southern California, about 20 % of the nitrogen in grape crops is derived from the ocean (Merz and Moyle 2006).

These examples show how the linkage of different ecosystems influences ecosystem functioning through bottom-up processes. These processes are relatively easy to imagine, but top-down processes also enhance the level of ecosystem functions. Coral reefs are among the major ecosystems in tropical and subtropical regions and have extraordinarily high productivity and high species richness. Similar to the foregoing examples, allochthonous input from terrestrial ecosystems is important indeed, but a top-down process also has a role in the persistence of coral reefs. Algae-feeding fish, such as the parrotfishes (*Scarus ceruleus*), reduce the coverage of corals by algae, thus reducing competition for light. If the abundance of algae-feeding fish were reduced, the system could exhibit a regime shift, from coral dominated to algae dominated (Edwards et al. 2010), leading to a large reduction in species diversity. As these fishes require both coral and mangrove ecosystems (habitat complementation) (Unsworth et al. 2008; Luo et al. 2009), mangrove ecosystems appear to have a positive indirect effect on the persistence of coral reefs. In the Caribbean Sea, coral reefs have persisted in areas where there are mangrove forests within 10 km (Edwards et al. 2010).

Recent theoretical studies have revealed that coupling of different ecosystems or habitats could stabilize the dynamics of mutual ecosystems through top-down forces (McCann et al. 2005; Rooney et al. 2006, 2008). Consider two hypothetical food webs in different ecosystems that have different temporal dynamics of primary productivity. This assumption is plausible, as examples are known in terrestrial and aquatic environments (Nakano and Murakami 2001; Gonzalez et al. 2009). When a top predator belonging to one ecosystem also consumes prey in another ecosystem at a moderate rate, the dynamics of the two ecosystems become stable (McCann et al. 2005) because the predator dynamics become stabilized. Hence, the ecosystem dynamics as a whole are stabilized. This stabilizing effect is pronounced when the linkage of the two ecosystems mediated by the top predator is weak, as is shown by the higher minimum densities of top predators in a weak coupling strength (Fig. 2.5, right panel). However, as the linkage becomes stronger, the two ecosystems behave as if they are a single ecosystem and the dynamics become destabilized again (Fig. 2.5, right panel). Although there are as yet no empirical demonstrations of this process, carnivorous mammals linking forest and river ecosystems, and carnivorous fish linking benthic and limnetic habitats, may have such functions.

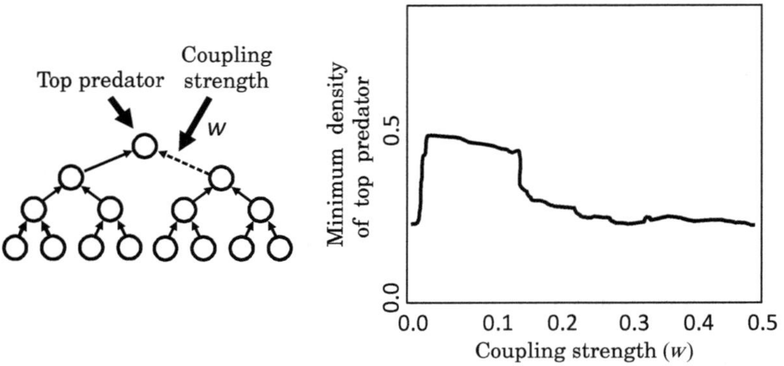

Fig. 2.5 Linkage of two food webs with a top predator and the consequences of linkage on the dynamic stability of the top predator. Note that the minimum population density of the predator increases only when the coupling strength of the two food webs is weak. (Modified from McCann et al. 2005)

Enhancement of Ecosystem Resilience Against Disturbance Through Differential Responses of Landscape Elements

In nature, some disturbances are small enough that a system can recover quickly to its original state, but others are so strong that it takes a long time to recover, or recovery may even be impossible. Resilience is one measure of stability, which is defined as the amount of disturbance that a system can absorb while remaining within the same basin of attraction or state (Holling 1973). Note that this definition is different from that of engineering resilience, that is, the rate at which a system returns to a single steady state (Pimm 1982; Tilman 1999). As is often discussed in regard to species diversity, ecosystem diversity or landscape heterogeneity can also buffer against disturbance, and it contributes to ecosystem resilience.

In semiarid regions such as those in North America, wildfire has been the major disturbance to ecosystems and has created a mosaic landscape with different successional stages. This landscape structure in turn determines the magnitude of wildfires (Turner and Romme 1994), because mature stands harbor abundant fuel, including dead trees and litter, whereas young stands have accumulated only small amounts of organic materials and function as barriers to fire spread (Romme and Despain 1989). This condition is a kind of self-organized system thought to have a high degree of resilience against disturbance (Turner et al. 2001). However, management in the northwestern United States in the late nineteenth century was aimed at suppressing fire and created an extensive homogeneous landscape consisting of mature forests. The large-scale wildfire that occurred in Yellowstone National Park in 1988 might have been the result of this homogeneous landscape structure (Romme and Despain 1989). Recent outbreaks of spruce budworm (*Choristoneura fumiferana*) in North American conifer forests appear to have occurred by a similar mechanism. This forest pest occasionally causes serious damage to spruce and firs and can spread extensively in mature forests. The homogeneous landscape created by conservative fire management might have induced severe outbreaks of this insect species (Swetnam and Lynch 1993).

Coral reefs often experience severe disturbances from typhoons, cyclones, and hurricanes. The degree of damage is not homogeneous in space, but some locations are more vulnerable than others, and the vulnerable locations differ in terms of their disturbance events. This variability may result from differences in topography, depth, direction of tidal currents, and successional stage among coral reefs (Connell et al. 1997). Colonization of highly disturbed sites by corals from less disturbed sites maintains the regional-scale resilience of coral reefs (Nystrom and Folke 2001).

The common mechanism underlying the resilience of forests and coral reefs is the dynamic interactions and interdependencies between subsystems. This mechanism is called "spatial resilience," the dynamic capacity to cope with disturbance and avoid thresholds of regime shifts at spatial scales larger than individual ecosystems (Nystrom and Folke 2001). The mechanism is basically the same as that of the "insurance hypothesis" (Yachi and Loreau 1999) of species diversity, whereby diversity of the responses of species within a functional group to environmental change stabilizes the system dynamics as a whole. However, if the composition of subsystems exceeds a certain threshold because of environmental homogenization through, for example, anthropogenic degradation or management, then disturbance will spread over the entire system, inducing a regime shift (Fig. 2.6).

Landscape Heterogeneity Does Not Always Enhance Ecosystem Resilience

Ecosystem diversity or landscape heterogeneity has a positive effect on ecosystem functioning and resilience. However, just as ecosystem diversity sometimes reduces species diversity through habitat fragmentation, ecosystem functioning and resilience could be degraded by a similar process. Examples are the increased vulnerability of trees along a forest edge to windthrow and insect outbreaks (Franklin and Forman 1987; Roland 1993). Because trees at the edge are likely to be exposed to strong winds and a relatively dry microclimate, the heterogeneous nature of a landscape composed of closed forests and open land can encourage extensive windthrow in a positive feedback manner. Forest damage by the western tent caterpillar (*Malacosoma californicum*) can spread in a similar way. In contrast to the spruce budworm, western tent caterpillar infestations, which are often found in fragmented forest landscapes (Roland 1993), may be caused by the enhancement of larval development by high temperatures or limitations imposed by fragmentation on the dispersal of natural enemies, including parasitoids and nuclear polyhedrosis virus.

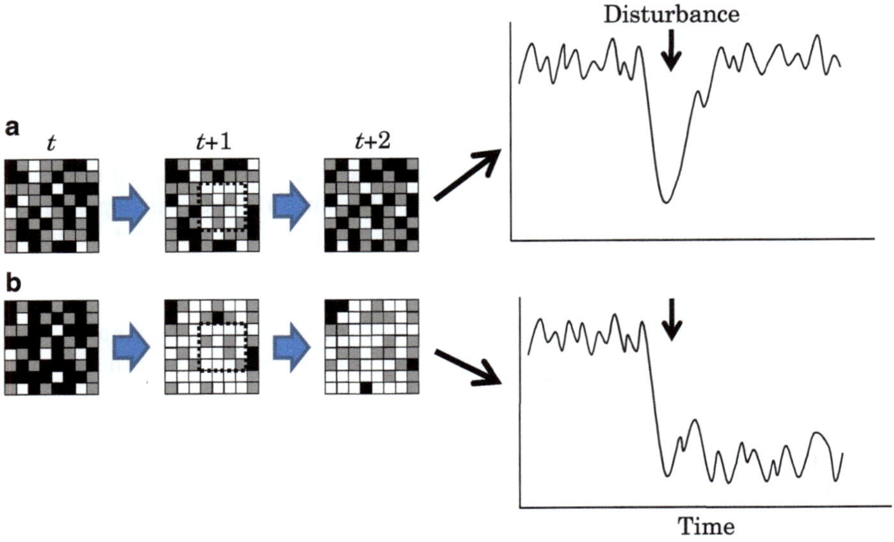

Fig. 2.6 Schematic representation of spatial resilience, as exemplified by landscapes composed of elements with different disturbance susceptibilities. After disturbance, the landscape recovers to the original state (*A*) or shifts to another state (*B*), depending on the initial landscape pattern. *Black*, susceptible patch; *gray*, nonsusceptible patch; *white*, empty patch. After disturbance (from t to $t + 1$), susceptible patches turn into empty patches while other patches remain unchanged. From $t + 1$ to $t + 2$, some of the nonsusceptible and empty patches turn into susceptible and nonsusceptible patches, respectively. *Dotted boxes* inside the "$t + 1$" panel indicate the area receiving disturbance

Thus, whether ecosystem diversity or landscape heterogeneity enhances the resilience of ecosystems depends on the nature of the disturbance processes, as well as the degree of fragmentation. Another important point to note is that the natural heterogeneity that has been created by interactions between abiotic and biotic environments has a long history, including the evolutionary history of the organisms within these environments; this natural heterogeneity appears to be quite different from anthropogenic landscape heterogeneity, especially in the case of recent human interference (Romme 2005; Strayer 2005). As a consequence, natural heterogeneity is likely to be more resilient than artificial heterogeneity in the face of disturbances. Human-created landscapes also tend to have higher contrasts between landscape components and to have sharp boundaries (Strayer 2005). For instance, the boundaries between forests and croplands are generally clearer than those between forests and natural grasslands. Natural landscape elements are thus exposed to discontinuously changing environments created by human activities along their edges, and this could cause the system dynamics to become unstable.

Ecosystem Services

Ecosystem services are defined by the Millennium Ecosystem Assessment as the benefits that people obtain from ecosystems. The MA distinguishes four broad benefits: supporting services, provisioning services, regulating services, and cultural services. It is now well recognized that ecosystem services are often in trade-off relationships, making it difficult to increase them simultaneously in a given ecosystem (Millennium Ecosystem Assessment 2005; Rodriguez et al. 2006). For instance, although the productivity of crops, timber, and biofuels is usually higher in homogeneous environments spread over vast areas, regulating services such as water purification, protection from soil erosion, improvement of pollination efficiency, and pest control appear to deteriorate in these environments, as these services are generally based on the complex interactions between biotic and abiotic components (Millennium Ecosystem Assessment 2005). A case study in Quebec, Canada, revealed that there were negative correlations between provisioning services (production of crops and pork) and all regulating services examined (Raudsepp-Hearne et al. 2010). It is obvious, therefore, that ecosystem diversity is necessary to maintain various types of ecosystem services.

Previous assessments of ecosystem services had serious drawbacks, however, because most of them have been based on land-use data. Using such data, it is difficult to quantitatively estimate ecosystem drivers such as nutrient fluxes and abundance of keystone species, and nonlinear responses of ecosystems cannot be estimated (Naidoo et al. 2008). Mechanistic understandings of the trade-offs between different ecosystem services are therefore required, not only for precise prediction but also to seek practical ways of reducing or mitigating trade-offs. Bennett et al. (2009) proposed that it is important to separate apparent trade-offs into direct causal interactions and indirect processes mediated by a common external driver (Fig. 2.7). For instance, continual input of fertilizers to cropland is necessary to maintain crop production, but it may result in eutrophication of rivers and coastal areas and thereby degrade various provisioning and regulating services. There are no direct trade-offs here, as increased crop growth or harvesting itself does not lead to increased nutrient input into rivers, but a common driver—addition of fertilizer—induces an apparent trade-off. An example of a direct trade-off can be seen in the afforestation of grasslands on the South American Pampas. Afforestation increases carbon sequestration, but it increases evapotranspiration as well, leading to a reduction in river water flow (Bennett et al. 2009; Fig. 2.7). Here, the two ecosystem services, carbon sequestration and water supply, are in a trade-off relationship. When the trade-off is indirect and has a common driver, it is possible to reduce the degree of the trade-off by manipulating ecosystems. For instance, as riparian forests function to reduce nutrient flow from croplands to river (Lovell and Sullivan 2006), creating riparian forests could mitigate the trade-offs in crop production and ecosystem services provided by river and coastal areas (Fig. 2.7). Thus, ecosystem diversity does not merely support different ecosystem services, but it also plays a role in reducing trade-offs between different ecosystem services.

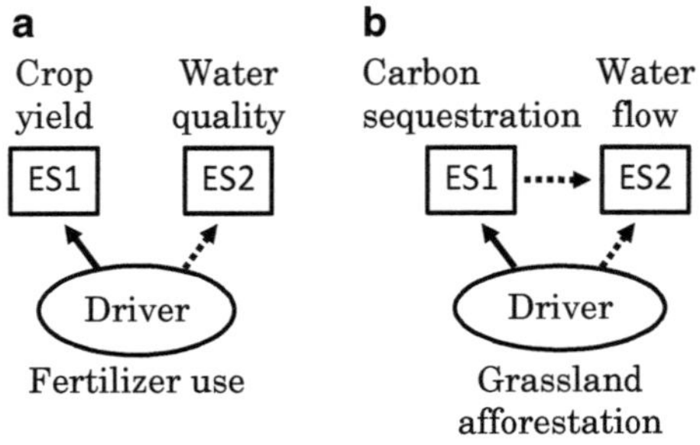

Fig. 2.7 Trade-off of ecosystem services can be caused by different responses to the same driver of change (**a**) or interactions between ecosystem services (**b**). *ES1* and *ES2* represent different ecosystem services: *dotted paths* represent negative effects, and *solid lines* are positive effects. (Modified from Bennett et al. 2009)

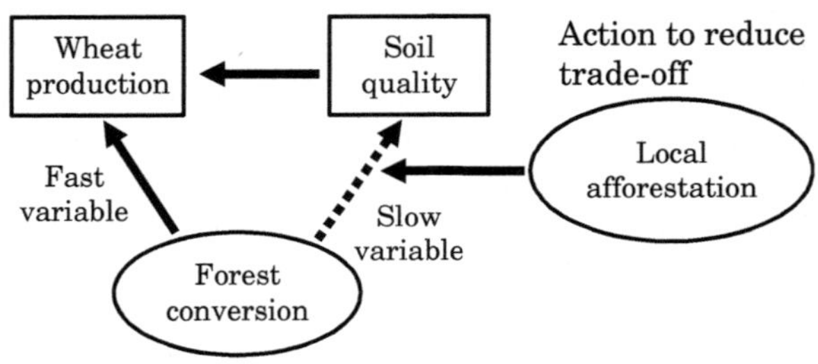

Fig. 2.8 The trade-off of ecosystem services can be resolved or ameliorated by the presence of another ecosystem. *Dotted paths* represent negative effects, and *solid lines* are positive effects

Most earlier studies also lacked temporal perspective on the trade-offs in a given ecosystem service; this is particularly important for provisioning services. Monocultures that use large amounts of fertilizers and pesticides may benefit from increased short-term crop production, but in the long run they are likely to be vulnerable to physical disturbance and epidemic disease (Hooper et al. 2005). Soil deterioration and changes in groundwater levels are slow processes; changes may emerge over time scales of several decades (Rodriguez et al. 2006) and are difficult to predict from land-use data. One way to reduce such temporal trade-offs between crop production and soil deterioration is to change land-use patterns before unwanted effects emerge. The land-use rotation that is often seen in agroforestry systems is an example (Lawrence and Schlesinger 2001; Diemont and Martin 2009); such rotations imply that landscape heterogeneity consisting of different stand-age classes is necessary for minimizing temporal trade-offs. Another way to reduce temporal trade-offs is to create a landscape element that enhances regulating services to maintain provisioning services in the long run (Fig. 2.8). In South Australia, extensive areas of eucalyptus forest have been converted to wheat fields, but the reduced evapotranspiration that resulted caused an increase in groundwater levels, resulting in high salinity levels on the ground surface (slow variable) and thereby lowering wheat production (fast variable): this is an example of temporal trade-offs in provisioning services. By considering topological and geological characteristics, it has been possible to identify effective locations for planting eucalyptus trees to reduce water levels (Farrington and Salama 1996). This case also suggests the importance of landscape heterogeneity or ecosystem diversity for providing sustainable ecosystem services.

Conclusions with Asian Perspectives

In this chapter, we have described the effects of ecosystem diversity or landscape heterogeneity on three properties: species diversity, ecosystem functioning, and ecosystem services. It seems obvious that different ecosystems harbor different biological communities and have different ecosystem characteristics, but we have tried to move beyond this basic statement by considering additive and nonadditive (or emergent) effects of ecosystem diversity on the enhancement and stability of these three properties. The logical structure is similar to the relationship between species diversity and ecosystem functioning. "Sampling effect" and "niche complementarity" (Tilman 1999) are types of additive effects that explain the positive relationship between species diversity and ecosystem functioning, whereas "facilitation effect" (Cardinale et al. 2002) refers to emergent effect. Not only does the mere presence of different ecosystems matter, but also the linkages or interactions between such ecosystems have beneficial effects on the three properties discussed here. By using this analogy, we are able to conceptualize the relationship between ecosystem diversity and species diversity or ecosystem functioning/services. We have also discussed how landscape heterogeneity sometimes has negative effects.

Finally, we provide here a brief perspective on how the views presented here on ecosystem diversity contribute to the understanding of biodiversity and ecosystem resilience in monsoon Asia. Monsoon Asia has a huge human population and harbors a rich biological diversity, from tropical to alpine. Moreover, monsoon Asia has many islands and several archipelagos with steep volcanic mountains, as well as broad areas of continental shelf and coral reefs. Because diverse ecosystems are embedded in relatively narrow areas, within which there are numerous types of human activity, the issue of how to conserve and manage ecosystem diversity is central to achieving a society in harmony with nature. We have emphasized here that ecosystem diversity plays a role in enhancing species diversity, ecosystem functioning, and ecosystem services, but it sometimes has negative consequences. These context dependencies appear to have emerged from the historical background of regional biota and the nature of landscape heterogeneity. This inference is supported by the rich biodiversity and high-level ecosystem functioning and services of Japan's traditional Satoyama landscape (Washitani 2001; Kobori and Primack 2003). However, whether a similar argument can be applied to other regions is not yet clear. For instance, agroforestry ecosystems in the tropics are considered sustainable and are occasionally highly biologically diverse (Bhagwat et al. 2008), but it is questionable whether such ecosystems can support endemic and endangered species comparable to natural forests in the tropics (Noble and Dirzo 1997; Garcia-Fernandez et al. 2003). Moreover, in the face of global climate change, systems that are currently resilient could become nonresilient with time. Therefore, rather than simply labeling ecosystem diversity and traditional management systems as important or sustainable, we need to accumulate scientific knowledge to predict the situations in which, and the lengths of time for which, a given composite system will be resilient in the face of local human activity and global climatic change.

Acknowledgments We thank T. Fukami, Y. Osada, and S. Nishijima for their comments on the earlier version of the manuscript. This study was partly supported by the Ministry of the Environment, Japan (S-9).

References

Becker CG, Fonseca CR, Haddad CFB, Batista RF, Prado PI (2007) Habitat split and the global decline of amphibians. Science 318:1775–1777

Begon M, Townsend CR, Harper JL (2006) Ecology: from individuals to ecosystems. Blackwell, Oxford

Bennett EM, Peterson GD, Gordon LJ (2009) Understanding relationships among multiple ecosystem services. Ecol Lett 12(12):1394–1404

Bhagwat SA, Willis KJ, Birks HJB, Whittaker RJ (2008) Agroforestry: a refuge for tropical biodiversity? Trends Ecol Evol 23(5):261–267

Cardinale BJ, Palmer MA, Collins SL (2002) Species diversity enhances ecosystem functioning through interspecific facilitation. Nature (Lond) 415:426–429

Connell JH, Hughes TP, Wallace CC (1997) A 30-year study of coral abundance, recruitment, and disturbance at several scales in space and time. Ecol Monogr 67(4):461–488

Croll DA, Maron JL, Estes JA, Danner EM, Byrd GV (2005) Introduced predators transform subarctic islands from grassland to tundra. Science 307:1959–1961

Darwin C (1872) The origin of species, 6th London edn. Thompson and Thomas, Chicago

Diemont SAW, Martin JF (2009) Lacandon Maya ecosystem management: sustainable design for subsistence and environmental restoration. Ecol Appl 19(1):254–266

Dunning JB, Danielson BJ, Pulliam HR (1992) Ecological processes that affect populations in complex landscapes. Oikos 65:169–175

Edwards HJ, Elliott IA, Pressey RL, Mumby PJ (2010) Incorporating ontogenetic dispersal, ecological processes and conservation zoning into reserve design. Biol Conserv 143(2):457–470

Elton CS (1958) The ecology of invasions by plants and animals. Methuen, London, p 18

Fahrig L (2003) Effects of habitat fragmentation on biodiversity. Annu Rev Ecol Evol Syst 34:487–515

Fahrig L, Baudry J, Brotons L, Burel FG, Crist TO, Fuller RJ, Sirami C, Siriwardena GM, Martin JL (2011) Functional landscape heterogeneity and animal biodiversity in agricultural landscapes. Ecol Lett 14(2):101–112

Farrington P, Salama RB (1996) Controlling dryland salinity by planting trees in the best hydrogeological setting. Land Degrad Dev 7(3):183–204

Fisher RA (1930) The genetical theory of natural selection. Clarendon, Oxford

Franklin JE, Forman RTT (1987) Creating landscape patterns by forest cutting: ecological consequences and principles. Landsc Ecol 1(1):5–18

Franlkin J (2010) Mapping species distributions: spatial inference and prediction. Cambridge University Press, Cambridge

Garcia-Fernandez C, Casado MA, Perez MR (2003) Benzoin gardens in North Sumatra, Indonesia: effects of management on tree diversity. Conserv Biol 17(3):829–836
Gaston KJ (1996) Biodiversity: a biology of numbers and difference. Blackwell, Oxford
Gonzalez A, Mouquet N, Loreau M (2009) Biodiversity as spatial insurance: the effects of habitat fragmentation and dispersal on ecosystem functioning. In: Naeem S, Bunker DE, Hector A, Loreau M, Perrings C (eds) Biodiversity, ecosystem functioning, and human wellbeing: an ecological and economic perspective. Oxford University Press, Oxford, pp 134–146
Hairston NG, Smith FE, Slobodkin LB (1960) Community structure, population control and competition. Am Nat 94:421–425
Helfield JM, Naiman RJ (2001) Effects of salmon-derived nitrogen on riparian forest growth and implications for stream productivity. Ecology 82(9):2403–2409
Holling CS (1973) Resilience and stability of ecological systems. Annu Rev Ecol Syst 4:1–23
Holling CS (1992) Cross-scale morphology, geometry, and dynamics of ecosystems. Ecol Monogr 62:447–502
Holt RD (2009) Bringing the Hutchinsonian niche into the 21th century: ecological and evolutionary perspectives. Proc Natl Acad Sci USA 106(2):19659–19665
Hooper DU, Chapin FS III, Ewel JJ, Hector A, Inchausti P et al (2005) Effects of biodiversity on ecosystem functioning: a consensus of current knowledge. Ecol Monogr 75:3–36
Johnson SP (1993) The Earth Summit: the United Nations Conference on Environment and Development (UNCED). Graham and Trotman, London
Katayama N, Saito D, Amano T, Miyashita T (2011) Effects of modern drainage systems on the spatial distribution of loach in rice ecosystems. Aquat Conserv Mar Freshw Ecosyst 21:146–154
Kato N, Yoshio M, Kobayashi R, Miyashita T (2010) Differential responses of two anuran species breeding in rice fields to landscape composition and spatial scale. Wetlands 30:1171–1179
Kobori H, Primack RB (2003) Participatory conservation approaches for Satoyama, the traditional forest and agricultural landscape of Japan. Ambio 32:307–311
Lawrence D, Schlesinger WH (2001) Changes in soil phosphorus during 200 years of shifting cultivation in Indonesia. Ecology 82:2769–2780
Lovell ST, Sullivan WC (2006) Environmental benefits of conservation buffers in the United States: evidence, promise, and open questions. Agric Ecosyst Environ 112(4):249–260
Luo JG, Serafy JE, Sponaugle S, Teare PB, Kieckbusch D (2009) Movement of gray snapper *Lutjanus griseus* among subtropical seagrass, mangrove, and coral reef habitats. Mar Ecol Prog Ser 388:255–269
May RM (1973) Complexity and stability in model ecosystems. Princeton University Press, Englewood Cliffs, NJ
McCann KS, Rasmussen JB, Umbanhowar J (2005) The dynamics of spatially coupled food webs. Ecol Lett 8(5):513–523
McGarigal K, Cushman SA, Neel MC, Ene E (2002) FRAGSTATS: Spatial Pattern Analysis Program for Categorical Maps. Computer software program produced by the authors at the University of Massachusetts, Amherst. http://www.umass.edu/landeco/research/fragstats/fragstats.html
Merz JE, Moyle PB (2006) Salmon, wildlife, and wine: marine-derived nutrients in human-dominated ecosystems of central California. Ecol Appl 16(3):999–1009
Millennium Ecosystem Assessment (2005) Millennium ecosystem assessment. Island Press, Washington, DC
Miyashita T, Chishiki Y, Takagi SR (2012) Landscape heterogeneity at multiple spatial scales enhances spider species richness in an agricultural landscape. Popul Ecol 54:573–581
Naeem S, Bunker D, Hector A, Loreau M, Perrings C (2009) Biodiversity, ecosystem functioning, and human wellbeing: an ecological and economic perspective. Oxford University Press, Oxford
Naidoo R, Balmford A, Costanza R, Fisher B, Green RE, Lehner B et al (2008) Global mapping of ecosystem services and conservation priorities. Proc Natl Acad Sci U S A 105:9495–9500

Nakano S, Murakami M (2001) Reciprocal subsidies: dynamic interdependence between terrestrial and aquatic food webs. Proc Natl Acad Sci USA 98:166–170
Noble IR, Dirzo R (1997) Forests as human-dominated ecosystems. Science 277:522–525
Nystrom M, Folke C (2001) Spatial resilience of coral reefs. Ecosystems 4(5):406–417
Oksanen L, Oksanen T (2000) The logic and realism of the hypothesis of exploitation ecosystems. Am Nat 155:703–723
Osawa S, Katsuno T (2001) Dispersal of the brown frogs *Rana japonica* and *R. ornativentris* in the forest of the Tama Hills. Curr Herpetol 20:1–10
Pickett SRA, Siriwardena GM (2011) The relationship between multi-scale habitat heterogeneity and farmland bird abundance. Ecography 34:955–969
Pimm SL (1982) Food webs. Chapman & Hall, London
Polis GA, Hurd SD (1996) Linking marine and terrestrial food webs: allochthonous input from the ocean supports high secondary productivity on small islands and coastal land communities. Am Nat 147:396–423
Polis GA, Power ME, Huxel GR (2004) Food webs at the landscape level. University of Chicago Press, Chicago
Raudsepp-Hearne C, Peterson GD, Bennett EM (2010) Ecosystem service bundles for analyzing tradeoffs in diverse landscapes. Proc Natl Acad Sci USA 107:5242–5247
Ray N, Adams J (2001) A GIS-based vegetation map of the world at the Last Glacial Maximum (25,000–15,000 BP). Internet Archaeology 11 (http://intarch.ac.uk/journal/issue11/rayadams_toc.html)
Rodriguez JP, Beard TD, Bennett EM, Cumming GS, Cork SJ, Agard J, Dobson AP, Peterson GD (2006) Trade-offs across space, time, and ecosystem services. Ecol Soc 11(1):28
Roland J (1993) Large-scale forest fragmentation increases the duration of tent caterpillar outbreak. Oecologia (Berl) 93:25–30
Romme WH, Despain DG (1989) Historical perspective on the Yellowstone fires of 1988. BioScience 39(10):695–699
Romme WH (2005) The importance of multiple spatial heterogeneity in wildland fire management and research. In: Lovett G, Jones CG, Turner MG, Weathers KC (eds) Ecosystem function in heterogeneous landscapes. Springer, New York, pp 353–356
Rooney N, McCann K, Gellner G, Moore JC (2006) Structural asymmetry and the stability of diverse food webs. Nature (Lond) 442:265–269
Rooney N, McCann KS, Moore JC (2008) A landscape theory for food web architecture. Ecol Lett 11(8):867–881
Sakai S, Yamaguchi N, Momose H, Higuchi H (2011) Seasonal shifts in foraging site and prey of grey-faced buzzards (*Butastur indicus*), breeding in Satoyama habitat of central Japan. Ornithol Sci 10:51–60
Strayer DL (2005) Challenges in understanding the functions of ecological heterogeneity. In: Lovett GM, Turner MG, Jones CG, Weathers KC (eds) Ecosystem function in heterogeneous landscapes. Springer, New York, pp 411–425
Swetnam TW, Lynch AM (1993) Multicentury, regional-scale patterns of western spruce budworm outbreaks. Ecol Monogr 63(4):399–424
Szabo P, Meszena G (2006) Spatial ecological hierarchies: coexistence on heterogeneous landscapes via scale niche diversification. Ecosystems 9:1009–1016
Tilman D (1999) The ecological consequences of changes in biodiversity: a search for general principles. Ecology 80:1455–1474
Turner MG, Romme WH (1994) Landscape dynamics in crown fire ecosystems. Landsc Ecol 9(1):59–77
Turner MG, Gardner RH, O'Neill RV (2001) Landscape ecology in theory and practice. Springer, New York
Unsworth RKF, De Leon PS, Garrard SL, Jompa J, Smith DJ, Bell JJ (2008) High connectivity of Indo-Pacific seagrass fish assemblages with mangrove and coral reef habitats. Mar Ecol Prog Ser 353:213–224

Uruma H, Nishijima S, Kobayashi R, Miyashita T (2012) Effectiveness of conservation-oriented agricultural practices on amphibians inhabiting Sado Island, Japan, with a consideration of spatial structure. Jpn J Conserv Ecol 17:155–164 (in Japanese)

Washitani I (2001) Traditional sustainable ecosystem 'Satoyama' and biodiversity crisis in Japan: conservation ecological perspective. Glob Environ Res 5:119–133

Yachi S, Loreau M (1999) Biodiversity and ecosystem functioning in a fluctuating environment: the insurance hypothesis. Proc Natl Acad Sci USA 96:1463–1468

Zipkin EF, Dewan A, Royle JA (2009) Impacts of forest fragmentation on species richness: a hierarchical approach to community modeling. J Appl Ecol 46:815–822

Part II
Status Quo of Biodiversity Monitoring

Chapter 3
Monitoring Breeding Bird Populations in Taiwan

Chie-Jen Ko, Meng-Wen Fan, Ruey-Shing Lin, Chien-Chung Cheng, and Pei-Fen Lee

Abstract The Taiwan Breeding Bird Survey (BBS Taiwan) is a national monitoring project with the aim of establishing a breeding bird population index for conservation. It was established as a citizen science project with the working group operating on a cross-organizational cooperation basis. From 2009 to 2012, BBS Taiwan has obtained data from more than 300 sampling sites and has reported on national population trends for more than 60 % of the diurnal breeding bird species in Taiwan. Descriptions of sampling design, survey methods, and current results are given in this chapter. With ongoing monitoring projects covering other taxa such as anurans, butterflies, bats, and reptiles, biodiversity observation networks in Taiwan are expected to provide critical information to aid conservation and climate change adaptation strategies, and also to increase public awareness of the importance of biodiversity.

Keywords Aichi biodiversity targets • Bird conservation • East Asia • Indicator • National biodiversity monitoring systems • Point-count method • Public engagement • Stratified random sampling • Subtropical climate

C.-J. Ko • P.-F. Lee (✉)
Institute of Ecology and Evolutionary Biology, National Taiwan University, Taipei, Taiwan
e-mail: leepf@ntu.edu.tw

M.-W. Fan • R.-S. Lin
Endemic Species Research Institute, Nantou, Taiwan

C.-C. Cheng
Chinese Wild Bird Federation, Taipei, Taiwan

S.-i. Nakano et al. (eds.), *Integrative Observations and Assessments*, Ecological Research Monographs, DOI 10.1007/978-4-431-54783-9_3, © Springer Japan 2014

Purpose

Nation-wide wildlife monitoring projects are vital for biodiversity conservation, especially as early warnings of environmental crises and for evaluating the efficiency of implemented conservation strategies. However, such projects have been very rare in the Asia-Pacific region, including Taiwan. Wild bird populations are among the few taxa for which large-scale surveys have been conducted in Taiwan. Dating back to 1974, the New Year Bird Count was the earliest attempt to collect nation-wide bird census data at a specific time of the year (Yen 1982), by encouraging birders to go out birding and upload their records in December and January. Such activities still continue to be implemented in several regions (e.g., Kenting National Park Headquarters 2012). Bird counts focusing on specific species such as migratory raptors (Yang et al. 2011), or in local regions such as Important Bird Areas, have been the main source of long-term biodiversity monitoring during the late twentieth century. These counts have been mostly organized by nongovernment organizations (NGOs). Public participation has hence played an important role in these early attempts of data gathering and biodiversity monitoring.

However, transforming observation and count activities into a well-functioning monitoring project presents difficulties. NGOs in Taiwan were enthusiastic to make contributions when it came to saving the environment, but they were unable to establish persuasive monitoring designs and proper analyses. Academic institutions have the ability to conduct sound monitoring surveys and can also produce numerous biodiversity reports, but such surveys were generally short lived since resources were bound to specific projects only specifically instituted by different agencies. Government institutions, on the other hand, hold the greatest potential to undertake long work tasks, but with limited resources the goal of running large-scale surveys was elusive at best. In short, it was soon recognized that to give birth to a monitoring project on a large spatial and temporal scale, a cross-organizational citizen science collaborating project would be vital.

Development History of the Taiwan Breeding Bird Survey

Despite the abundant examples of migratory species and nonbreeding season surveys, thoughts of conducting projects targeting breeding bird monitoring did not literally emerge until the onset of the twenty-first century. Largely inspired by the headline indicators for sustainable development in the UK, and the North American Breeding Bird Survey, the first suggestions of designing a comprehensive large-scale breeding bird survey in Taiwan were made in 2001. Suggestions were for a project aimed at establishing nation-wide indicators of ecological sustainable developments using breeding bird populations (Chiou et al. 2001). In the following year, to exchange information and gain further consensus on this new idea, a conference titled “Planning and Promoting the Breeding Bird Survey” was held by the Endemic Species Research Institute in Nantou County in Central Taiwan (Endemic Species Research Institute 2002). After the conference, related articles were published in both government reports (Hsu et al. 2003) and journals targeting Taiwanese birders (Hsu 2003a, b) to help raise public awareness. By then, most of the proposed sampling schemes, survey methods, and basic structure for the breeding bird survey were arranged. Furthermore, the idea that the survey should be designed as a citizen science project had also been adopted (Hsu et al. 2003).

In 2007, a project aimed at evaluating the status of biodiversity in Taiwan funded by the Forestry Bureau gave researchers at National Taiwan University a promising chance to bring the idea of a breeding bird survey into reality (Lee et al. 2007). After a final review of previous efforts and preliminary tests of survey methods in 2008, the Taiwan Breeding Bird Survey (BBS Taiwan) was launched in the breeding season of 2009, by National Taiwan University in collaboration with the Chinese Wild Bird Federation (CWBF). The Endemic Species Research Institute (ESRI) joined BBS Taiwan's working group in 2010 in joint partnership. BBS Taiwan entered its fourth year in 2012, and with its steady growth in site coverage and participants, it is now confidently planning ahead as far as its tenth anniversary year.

Approach

Cross-Organizational Collaboration

BBS Taiwan was designed as a citizen science project and implemented by joint partnerships with NGOs, government agencies, and academic institutions. Cross-organizational collaboration helped to overcome the three main challenges we face in Taiwan when establishing a large-scale monitoring project: (1) shortage of resources, (2) difficulties in persistence, and (3) lack of scientific involvement. Joint partnership within the BBS Taiwan working group includes the Chinese Wild Bird Federation, the Endemic Species Research Institute of the Council of Agriculture, and the Institute of Ecology and Evolutionary Biology of National Taiwan University (Table 3.1).

Survey Training Workshops

One of the major challenges in implementing citizen science projects is to ensure the quality of data. Because these data would be used to build national indicators aimed for providing important information for biodiversity conservation, it would be of the most important basal issues to be carefully addressed. Recognizing that data would be collected by amateurs who generally had not been through formal

Table 3.1 Roles of each organization in the Breeding Bird Survey (BBS) Taiwan joint partnership

Collaborating organizations	Description of role in collaborations
中華民國野鳥學會 CHINESE WILD BIRD FEDERATION	The Chinese Wild Bird Federation (CWBF) with associated Wild Bird Societies within each county is one of the largest NGOs in Taiwan. Holding large numbers of experienced birders as a rich human resource base, they run various bird surveys in areas such as Important Bird Areas, but generally run short in having resources to produce sound scientific reports.
	In the BBS Taiwan project, CWBF plays the role of recruiting and organizing volunteering surveyors, with the assistance of regional organizers working in each wild bird society.
農委會特有生物研究保育中心 TAIWAN ENDEMIC SPECIES RESEARCH INSTITUTE.COA	As the main government institute in Taiwan, which puts great effort in running surveillance surveys of species distribution data, the Endemic Species Research Institutes (ESRI) holds funds that are relatively stable annually, hence giving ESRI the ability to establish long-term research or monitoring projects, but still with limited resources to build large-scale surveys.
	ESRI stands as the central communicator and organizer between participants of BBS Taiwan and is responsible for arranging survey workshops and publications of survey reports.
ECOLOGY & EVOLUTIONARY BIOLOGY 台灣大學 生態學與演化生物學研究所 Institute of Ecology and Evolutionary Biology National Taiwan University	The Institute of Ecology and Evolutionary Biology of National Taiwan University (NTU) is an academic institution holding faculties that focus on bird survey methods and establishments of monitoring projects. As resources are usually bound to projects from different agencies, NTU faces the challenge of maintaining long-term projects and recruiting surveyors.
	NTU is responsible to the BBS Taiwan project for providing suggestions on survey methods, sampling schemes, and analysis of survey results.

training, the BBS Taiwan workgroup arranged workshops in each county of Taiwan to recruit surveyors and to train them in correct survey methods. The training courses included four essential sections:

1. Introduction to biodiversity monitoring and the goals of BBS Taiwan.
2. Step-by-step survey methods, with a strong emphasis on the concept of standard protocols and the importance of compliance.
3. Birding by ear: because point counts rely strongly on the ability to identify bird species by auditory cues, this section is set to learn and recognize the song and calls of breeding bird species, an exercise commonly neglected by birders in Taiwan.
4. Practice in the field: an exercise of BBS Taiwan survey methods in the field led by experienced surveyors.

Sampling Scheme

The central idea of BBS Taiwan's sampling scheme is to encompass a set of survey sites that would be representative of all major habitats in Taiwan. We employed stratified random sampling to select our sampling sites. The first step was to divide the island of Taiwan into 41 ecoregions (Su 1992), with a further division into three different elevation zones (0–1,000, 1,000–2,500, 2,500–4,000 m), hence resulting in 91 strata. The second step was to locate areas that are reachable by roads within each stratum. This criterion was added because many montane areas in Taiwan would be difficult and dangerous to be surveyed by volunteers if there was no access road. The final step was to randomly select sites within each stratum with numbers fitting the proportion of area of each stratum to cover the whole of Taiwan Island. Taiwan's main island has an area of about 36,000 km^2, and we set 5 % coverage as a first-stage goal. Consequently, with a survey site expected to cover around 4 km^2, we randomly selected 450 survey sites, which were provided to be adopted by surveyors (Fig. 3.1).

Survey Methods

From among the various bird survey methods, we decided to use a point-count method (Bibby et al. 2000). This approach was preferred over a transect method, which is also a common technique in breeding bird monitoring (e.g., UK breeding bird survey), mainly because of the difficulty in executing transect surveys in the rugged topography in the mountainous regions of Taiwan (Chiou et al. 2001). Distance sampling (Buckland et al. 2001) and time removal model (Farnsworth et al. 2002) were both implemented into the survey design to account for imperfect detection. Each BBS Taiwan site consists of 6–10 points, and is surveyed twice a year, which gives each site a minimum sampling effort of 12 times: this is the minimum sampling effort required to attain reliable data with a coefficient of variance less than 20 % in the breeding season within each site on Taiwan (Shiu and Lee 2003). The count duration of each point is 6 min, which suggests the likelihood of detecting at least 80 % of the species at each sites (Ding 1993; Shiu and Lee 2003). Surveys are performed within the breeding season between March and June, with sites at higher elevations surveyed in a later sequence. Surveys are conducted within 4 h of sunrise on each day, which is estimated to be the most efficient time period to detect birds in the breeding season (Ding 1993; Chiou et al. 2001).

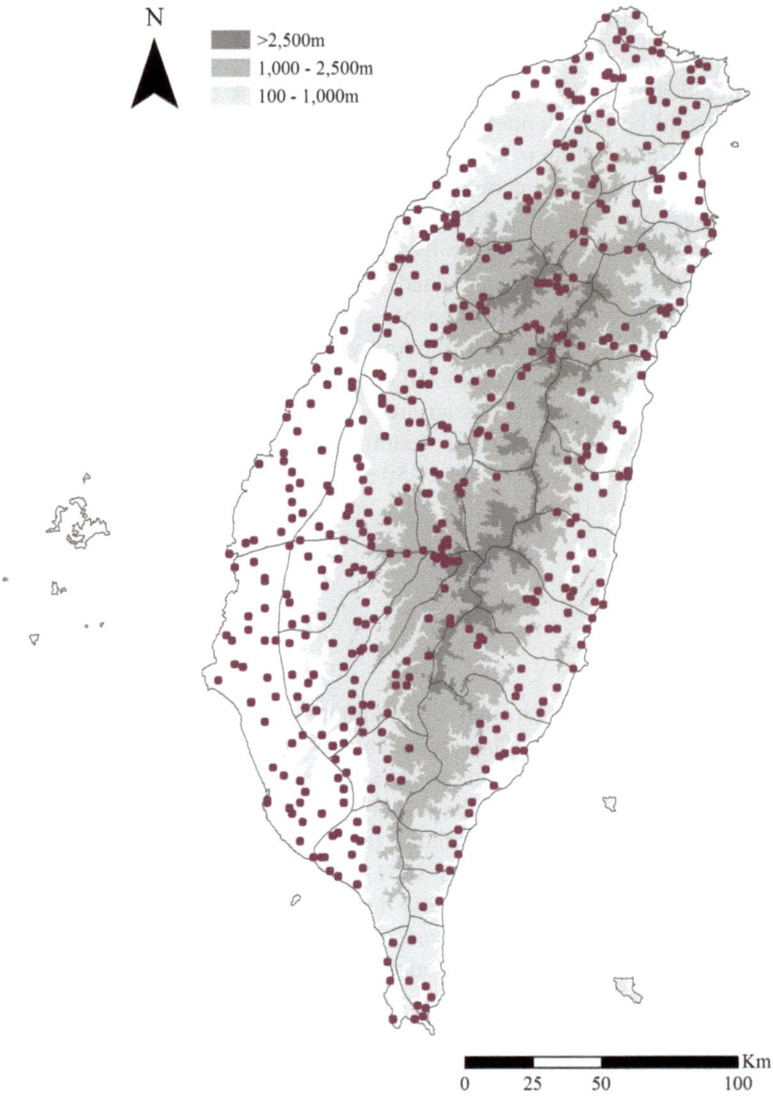

Fig. 3.1 Breeding Bird Survey (BBS) Taiwan sampling sites selected by stratified random sampling. Border lines are ecological climate zones that were delineated by Su (1992)

Analysis

Bird Population Indicators

Tracking trends of breeding bird populations is one of the main goals of BBS Taiwan. For the establishment of population indicators of Taiwan breeding birds, we used TRIM (TRends and Indices for Monitoring data), a software program using

Poisson regression to detect trends of abundance data (van Strien et al. 2004). Regression models with site and year effects included are established for each species in TRIM: count data can be weighted by sampling effort, accounting for covariates (such as habitat types), and missing data are also acceptable. According to these advantages, TRIM is widely used to analyze large-scale wildlife population monitoring data, including the Pan-European Common Bird Monitoring Scheme organized by the European Bird Census Council.

Target species of BBS Taiwan were common diurnal breeding birds. The breeding status of birds was determined by the checklist of the birds of Taiwan (Juan et al. 2012), which is updated annually by the Bird Record Committee of the CWBF. Both residents and summer visitors were included as breeding birds. However, several breeding bird species may not be appropriately covered by our sampling scheme and survey methods. Such species include (1) nocturnal birds such as owls and nightjars, (2) colony breeding birds such as egrets or swifts, and (3) breeding bird species that also have large wintering (or passage) populations present within BBS Taiwan's survey season (March to June), such as White Wagtail (*Motacilla alba*) and Little Ringed Plover (*Charadrius dubius*).

National indicators were established for bird species that have data coverage over 30 survey sites. Because geographic, climate, and land-use patterns are significantly different between the different regions of Taiwan (Shiu 2003), we also developed indicators for the North, West, and Eastern regions of Taiwan, which are expected to reveal population trends that are specific to each region.

Current Achievements

Since the launch of BBS Taiwan in 2009, both the number of surveyors and sites sampled have grown encouragingly. In 2012, survey participants have numbered more than 250 persons, and the number of sites sampled per year is currently over 300 sites, rapidly approaching 450 sites: the original target. Spatial distribution of sampled sites was generally homogeneous across Taiwan, covering all major habitats from sea level to nearly 4,000 m into the mountains (Fig. 3.2). Still, future efforts are anticipated to fill gaps, especially in the Northern Region and at certain elevation zones.

Building population indices for reporting the status of breeding birds is the most important goal of BBS Taiwan, and currently the population status of 63 species has been reported, which represents more than 60 % of diurnal land breeding birds on Taiwan. These species include 11 endemic species such as the Taiwan Barbet (*Megalaima nuchalis*) and the Formosan Magpie (*Urocissa caerulea*), and also 12 species listed in the Protected Species List of Taiwan by the Council of Agriculture of the Executive Yuan. IUCN Red List species, including the Styan's Bulbul (*Pycnonotus taivanus*), Taiwan Partridge (*Arborophila crudigularis*), and the Taiwan Hwamei (*Garrulax taewanus*), were also effectively monitored.

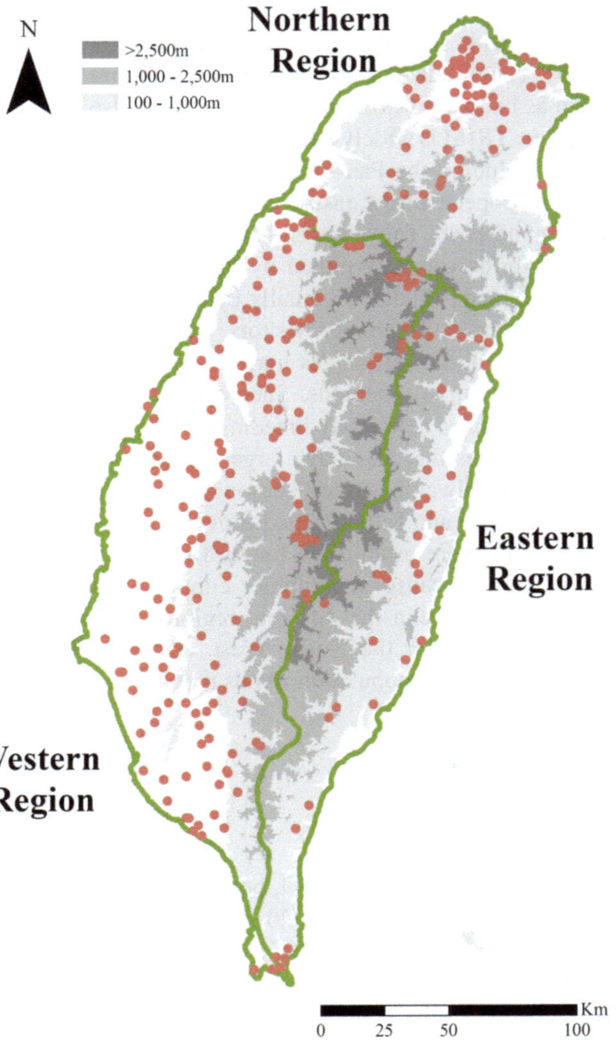

Fig. 3.2 Sites sampled by volunteers within 2009–2011. *Green line* delineates the borders of Northern, Eastern, and Western regions that were defined by Shiu (2003)

Table 3.2 summarizes significant population changes of monitored species since the initiation of BBS Taiwan. Significant trends are defined by log-linear regression models established by TRIM with slopes significantly different from zero ($p < 0.05$). Regression models with imputed indices were adopted as suggested by Pannekoek and van Strien (2005), with serial correlation and overdispersion taken into account. Full results and additional outputs of the BBS Taiwan working team can be found in annual reports published by the Endemic Species Research Institute (Lee et al. 2010b; Ko et al. 2012), PDF versions are also available on the BBS Taiwan website: http://bbstaiwan.tw/. A noteworthy finding of the current result is that forest species

Table 3.2 Breeding bird species with significant population changes between 2009 and 2011 in Taiwan

Species name	Scientific name	Number of sites	Index	LCL	UCL	Trend-class 2009–2011	Status in Taiwan
Crested Serpent-Eagle	*Spilornis cheela*	74	212	133	565	Strong increase*	Resident
Eurasian Moorhen	*Gallinula chloropus*	58	123	123	352	Strong increase*	Resident
Oriental Turtle-Dove	*Streptopelia orientalis*	58	−50	24	89	Steep decline*	Resident
Himalayan Cuckoo	*Cuculus saturatus*	65	−46	33	83	Steep decline*	Summer visitor
Lesser Coucal	*Centropus bengalensis*	51	−35	42	95	Moderate decline*	Resident
Pacific Swallow	*Hirundo tahitica*	143	36	102	176	Moderate increase*	Resident
Striated Swallow	*Cecropis striolata*	77	−31	46	99	Moderate decline*	Resident
Styan's Bulbul	*Pycnonotus taivanus*	28	24	103	148	Moderate increase*	Resident
Black Bulbul	*Hypsipetes leucocephalus*	195	28	109	149	Moderate increase**	Resident
Long-tailed Shrike	*Lanius schach*	31	−49	19	100	Moderate decline*	Resident
Black-naped Monarch	*Hypothymis azurea*	168	23	101	148	Moderate increase*	Resident
Taiwan Scimitar-Babbler	*Pomatorhinus musicus*	196	46	121	173	Strong increase**	Resident
Gray-cheeked Fulvetta	*Alcippe morrisonia*	161	37	109	167	Moderate increase**	Resident
Vinous-throated Parrotbill	*Paradoxornis webbianus*	42	−63	16	71	Steep decline**	Resident
Green-backed Tit	*Parus monticolus*	40	−41	41	82	Steep decline**	Resident
Fire-breasted Flowerpecker	*Dicaeum ignipectum*	34	−47	28	89	Steep decline*	Resident
Javan Myna	*Acridotheres javanicus*	111	57	113	207	Strong increase*	Introduced species
Large-billed Crow	*Corvus macrorhynchos*	63	125	121	360	Strong increase*	Resident

Significance: $*p < 0.05$, $**p < 0.01$

Index is change of population index between 2011 and 2009, with 2009 set to 100. LCL and UCL, lower and upper confidence intervals of the regression line, respectively and nonforest species appear to behave oppositely in their direction of population trends. Of the eight species that declined significantly, 75 % were grassland or farmland species, including Oriental Turtle-Dove (*Streptopelia orientalis*), Himalayan Cuckoo (*Cuculus saturatus*), Lesser Coucal (*Centropus bengalensis*), Striated Swallow (*Cecropis striolata*), Long-tailed Shrike (*Lanius schach*), and Vinous-throated Parrotbill (*Paradoxornis webbianus*). In contrast, only 30 % of the ten increasing species prefer grassland or farmland habitats. As the decline of farmland birds is of major concern in Europe and the UK (Benton et al. 2003), whether the current finding in Taiwan implicates a similar situation of farmland biodiversity loss may require additional evidence. Nevertheless, as the status of agro-ecosystems in East Asia is generally unknown (but see Amano 2009), data obtained by BBS Taiwan provide a promising chance to fulfill this gap.

Future Perspectives

Current achievements of BBS Taiwan bear very important and exciting information for Taiwan and the East Asian region. BBS Taiwan presents the possibility of implementing biodiversity monitoring successfully on a national scale, in a region where similar projects were relatively rare. Data collected by BBS Taiwan also have the potential to be used for large-scale research, including the validation of species distribution modeling, gap analysis for evaluation of conservation areas, reevaluation of conservation status of breeding birds in Taiwan, role of landscape heterogeneity with species richness (Miyashita et al., in press), and many more establishing issues. Still, much work lies ahead; most fundamentally in the effort to encourage surveyors to repeat surveys for as long as possible, and also the recruitment and training of new incoming participants. The former task will rely on improving information sharing between project organizers and survey volunteers; whereas the latter has the important goal to maintain data quality. Collaborations with other government agencies and NGOs that express a strong interest in biodiversity conservation are in the pipeline, with the insight that these efforts would raise public awareness for greater biodiversity conservation.

Based on the successful experience of BBS Taiwan, the idea of building a bird monitoring system to census birds in a more comprehensive way has recently been suggested. The system will be composed of several monitoring systems focusing on common breeding, nocturnal breeding, wetland, endangered, wintering migratory, and alien birds, respectively. The obvious way to overcome the foreseeable difficulties is expanding and strengthening the partnerships among NGOs, government agencies, and academic institutions as well as encouraging more volunteers to become involved in well-designed citizen science projects. Meanwhile, projects aimed at monitoring bird population demographics through constant-effort banding (Lin 2012) and the occurrences of alien invasive birds (Fang et al. 2011) have been conducted since 2009 and 2010, respectively. Other projects, such as monitoring of migratory landbirds and wetland birds, are in the preliminary stage of testing field methods.

In addition to birds, increasing numbers of taxa are surveyed on a nation-wide scale in Taiwan, such as anurans, butterflies, bats, and reptiles (Lee et al. 2010a; Deng et al. 2012). The stages of project development differ among taxa. The anuran monitoring network initiated in 2003 is now the most important source regarding the status of amphibians in Taiwan (Yang and Gong 2012). Projects targeting butterflies and bats are focused on designing standardized survey methods and implementing educational workshops to create public interest in these species (Lee et al. 2010a). The reptile project was originally formed by a group of enthusiastic people who were concerned about the negative influences of roadkill and proactively reported their records through Facebook during 2011. Gradually, systematic survey methods have been developed, and hundreds of volunteers continually reported roadkill records with detailed information (Deng et al. 2012). One thing in common among these monitoring projects is that they share the same belief that, to successfully achieve our goals of effective biodiversity monitoring, the need to incorporate the concept of citizen science into survey designs is crucial.

Overall, biodiversity monitoring workgroups in Taiwan are putting great effort into the establishment of even greater national biodiversity indicators in the near future. These indicators, which are essential for biodiversity conservation and climate change adaptation in Taiwan, are expected to trigger large-scale monitoring on more taxa and in more countries within the Asia-Pacific region.

Acknowledgments We thank all participants of BBS Taiwan, including regional organizers, workshop speakers, and survey volunteers. We are also grateful to Tatsuya Amano, Mark B. Wilkie, and Michihiko K. Saito for providing valuable comments to improve the manuscript. The BBS Taiwan workgroup thanks the Forestry Bureau, Council of Agriculture, and the Biodiversity Research Center, National Taiwan University, for funding the project.

References

Amano T (2009) Conserving bird species in Japanese farmland: past achievements and future challenges. Biol Conserv 142:1913–1921

Benton TG, Vickery JA, Wilson JD (2003) Farmland biodiversity: is habitat heterogeneity the key? Trends Ecol Evol 18:182–188

Bibby CJ, Burgess ND, Hill DA, Mustoe SH (2000) Bird census techniques, 2nd edn. Academic, London

Buckland ST, Anderson DR, Burnham KP, Laake JL, Borchers DL, Thomas L (2001) Introduction to distance sampling. Oxford University Press, New York

Chiou C-R, Lee P-F, Chang C-R, Shiu H-J, Chen I-C, Wu T-Y, Lee Y-C, Chen Y-R, Yang T-T (2001) Establishing indicators for ecological sustainability in Taiwan: a monitoring system using avian bird census data. Environmental Protection Administration, Taipei (in Chinese, English abstract)

Deng D, Mai G-S, Hsu C-H, Chuang T-R, Lin T-E, Lin H-H, Shao K-T, Lemmens R, Kraak M-J (2012) Using social media for collaborative species identification and occurrence: issues, methods, and tools. First ACM SIGSPATIAL international workshop on crowdsourced and volunteered geographic information. ACM SIGSPATIAL GEOCROWD, CALIFORNIA

Ding, T-S (1993) Avian community ecology of mature forests on Mt. Yushan. Master Thesis. National Taiwan University, Taipei (in Chinese, English abstract)

Endemic Species Research Institute (2002) Conference proceedings on the development and implementation of the Taiwan breeding bird survey. Endemic Species Research Institute, Nantou (in Chinese)

Fang W, Lin Y-S, Lin C-Y, Fan M-W (2011) Alien birds with potential threat to Taiwan. Endemic Species Research Institute, Nantou (in Chinese)

Farnsworth GL, Pollock KH, Nichols JD, Simons TR, Hines JE, Sauer JR (2002) A removal model for estimating detection probabilities from point-count surveys. Auk 119:414–425

Hsu F-H (2003a) Survey methods for breeding bird survey in Taiwan (I). Feathers 16:6–9 (in Chinese)

Hsu F-H (2003b) Survey methods for breeding bird survey in Taiwan (II). Feathers 16:12–15 (in Chinese)

Hsu F-H, Yao C-T, Lin R-S (2003) Study of the development and implementation of breeding bird survey in Taiwan (II). Endemic Species Research Institute, Nantou (in Chinese)

Juan C-S, Ding T-S, Wu J, Lin R-S, Yang Y-H, Tsai Y-J, Pan C-Y (2012) The 2012 CWBF checklist of the birds of Taiwan. Chinese Wild Bird Federation, Taipei (in Chinese)

Kenting National Park Headquarters (2012) The Kenting National Park new year bird count survey report 2012. Kenting National Park Headquarters, Pingtung (in Chinese)

Ko C-J, Fan M-W, Hsieh C-F, Huang G-W, Chang T-Y, Yo C-H, Yeh C-W, Lin R-S, Cheng C-C, Lee P-F (2012) The Taiwan breeding bird survey 2011. Endemic Species Research Institute, Nantou (in Chinese, English abstract)

Lee L-L, Chao J-T, Yang Y-T, Lee P-F, Cheng H-C, Hsu C-L (2010a) Establishing national biodiversity indicators and monitoring programs of selected taxa (III). Forestry Bureau, Council of Agriculture, Taipei (in Chinese)

Lee L-L, Huang C-L, Lin I-C (2007) 2007 Status report of biodiversity in Taiwan. Forestry Bureau, Council of Agriculture, Taipei (in Chinese)

Lee P-F, Ko C-J, Huang G-W, Kao W-S, Wu T-Y, Lin H-S, Chen W-J, Lin R-S, Fan M-W, Hsieh C-F, Yu W-D (2010b) The Taiwan breeding bird survey 2009–2010. Endemic Species Research Institute, Nantou (in Chinese)

Lin R-S (2012) Manual to monitoring for avian productivity and survival in Taiwan (MAPS Taiwan). Endemic Species Research Institute, Nantou (in Chinese)

Miyashita T, Amano T, Yamakita T (in press) Effects of ecosystem diversity on species richness and ecosystem functioning and services: a general conceptualization. Biodiversity Observation Network in the Asia-Pacific Region, Tokyo

Pannekoek J, van Strien AJ (2005) TRIM 3 Manual (TRends & Indices for Monitoring data). Statistics Netherlands, Voorburg

Shiu H-J (2003) Spatial and seasonal variations in avian assemblages in Taiwan. PhD dissertation. National Taiwan University, Taipei (in Chinese, English abstract)

Shiu H-J, Lee P-F (2003) Assessing avian point-count duration and sample size using species accumulation functions. Zool Stud 42:357–367

Su HJ (1992) A geographical data organization system for the botanical inventory of Taiwan. Bot Inst Acad Sin Monogr Ser 12:23–36

van Strien A, Pannekoek J, Hagemeijer W, Verstael T (2004) A log-linear Poisson regression method to analyse bird monitoring data. In: Anselin A (ed) Bird numbers 1995. Proceedings of the international conference and 13th meeting of the European bird census council. Pärnu, Estoni. Bird Census News 13(2000):33–39

Yang C-H, Yang M-Y, Hung F-L, Hsieh W-K (2011) Investigation on spring and fall populations of migrating raptors at the Kenting National Park in 2011. Kenting National Park Headquarters, Pingtung (in Chinese)
Yang Y-J, Gong W-B (2012) Ecological monitoring of Taiwan anuran by using investigative data of volunteers. Chin Biosci 54:65 (in Chinese)
Yen C-W (1982) The Taiwan new year bird count. Taiwan Tourism Bureau, Taipei (in Chinese)

Chapter 4
Current Status of the National Coral Database in Japan: Dataset Development, Applications, and Future Directions

Hiroya Yamano, Chuki Hongo, Kaoru Sugihara, Yumiko Yara, Yushin Nakao, and Masahiko Fujii

Abstract Corals are facing rapid environmental changes including global-scale climate changes and local-scale anthropogenic influences. Japan covers a wide latitudinal range, stretching from subtropical to temperate areas, and latitudinal limits of coral reefs and coral distributions are observed around the Japanese islands. Further, at some islands, land development is causing sedimentation of coral reefs via the rivers. The variation in environmental parameters in Japan provides one of the best opportunities to examine the relationships between changes in coral reef status and environmental variables.

In this chapter, we review the current status of the national coral reef database of Japan and present the applications and possible future directions of the database. The database was developed by collecting information from published reports and scientific papers in a survey spanning the period from the 1930s to the present. Currently, coral coverage and species data originate from ~10,000 and ~1,000 sites, respectively.

H. Yamano (✉) • K. Sugihara • Y. Yara
Center for Environmental Biology and Ecosystem Studies, National Institute for Environmental Studies, 16-2 Onogawa, Tsukuba, Ibaraki 305-8506, Japan
e-mail: hyamano@nies.go.jp

C. Hongo
Center for Environmental Biology and Ecosystem Studies, National Institute for Environmental Studies, 16-2 Onogawa, Tsukuba, Ibaraki 305-8506, Japan

Department of Physics and Earth Sciences, University of the Ryukyus, 1 Senbaru, Nishihara, Okinawa 903-0213, Japan

Y. Nakao
Overseas Environmental Cooperation Center, 7th Floor, Shibakoen Annex Bldg., 3-1-8, Shibakoen, Minato-ku, Tokyo 105-0011, Japan

Graduate School of Environmental Science, Hokkaido University, N10W5, Kita-ku, Sapporo, Hokkaido 060-0810, Japan

M. Fujii
Faculty of Environmental Earth Science, Hokkaido University, N10W5, Kita-ku, Sapporo, Hokkaido 060-0810, Japan

S.-i. Nakano et al. (eds.), *Integrative Observations and Assessments*, Ecological Research Monographs, DOI 10.1007/978-4-431-54783-9_4,

The databases enable the detection of decreases in coral coverage caused by coral bleaching from anomalously high sea surface temperatures in subtropical areas, as well as poleward range expansions of coral distributions in temperate areas. In addition to these specific applications, by integrating environmental variables and their future projections, the database could play a central role in the integration of observations, statistical modeling, and future projections to understand coral dynamics and to establish proper management protocols for coral reef ecosystems.

Keywords BISMaL • Coral • Coverage • Environmental change • Japan • Monitoring • OBIS • Species

Introduction

Corals play a fundamental role in primary production and habitat formation for numerous other species in tropical to temperate marine areas. However, corals are facing rapid environmental changes including global-scale climate changes and local-scale anthropogenic influences (e.g., Hoegh-Guldberg et al. 2007; Carpenter et al. 2008). Changes to their status resulting from environmental changes could cause fundamental modifications of coastal ecosystems (Pratchett et al. 2009).

Databases based on quantitative surveys have contributed significantly to quantitative assessments of coral decline in the past few decades (Gardner et al. 2003; Bellwood et al. 2004; Bruno and Selig 2007; Osborne et al. 2011; Sweatman et al. 2011; Magdaong et al. this volume). Such databases have been developed for the Caribbean Sea, the Great Barrier Reef, and the Philippines. In addition, a database by Bruno and Selig (2007) covers the Indo-Pacific region, but there are limited data on some countries, such as Japan. Hence, it is necessary to supplement this database with contributions from more regions and countries. In this chapter, we review the current status of the national coral reef database of Japan and present its applications and possible future directions.

Setting

Japan covers a wide latitudinal range, from subtropical to temperate areas (Fig. 4.1). The Japanese islands form an almost continuous chain, and corals are distributed in a stepping-stone sequence along a latitudinal gradient, because of the Kuroshio and Tsushima warm currents, which appear to transport coral larvae from south to north (Veron and Minchin 1992). Coral reef geomorphology and coral communities show remarkable latitudinal gradients in response to decreases in sea surface temperature (SST) (Veron and Minchin 1992; Kayanne et al. 2004). Well-developed coral reefs have formed among the Ryukyu Islands of the southern part of Japan (from

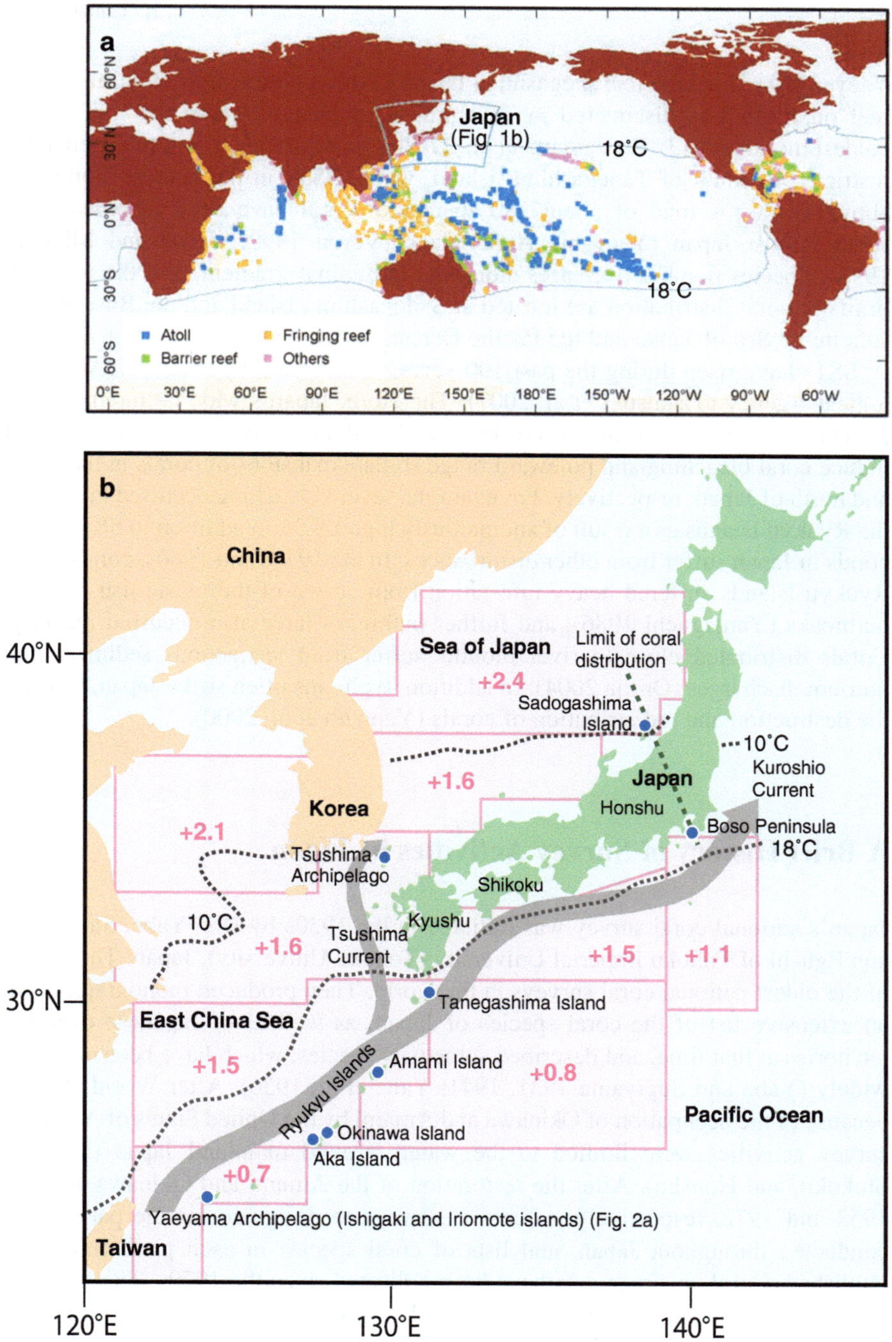

Fig. 4.1 Japan and its surrounding environment. **a** Distribution of corals and coral reefs of the world. Data from ReefBase. Japan is located at the northern limit of coral and coral reef distributions. **b** Currents and sea surface temperature (SST) isolines in the coldest month around Japan. The northern limits of coral and coral reef distributions correspond generally to the 10 ° and 18 °C isotherms, respectively. *Numbers in rectangles* indicate increases in winter SST (°C) over the past 100 years (Takatsuki et al. 2007). This figure indicates the locations of the places mentioned in this report

Yaeyama Archipelago to Tanegashima Island). Although the highest latitude coral reef on Earth was discovered in Tsushima Archipelago, where the SST in the coldest month is ~13 °C (Yamano et al. 2012b), coral reef distribution is generally restricted to south of Tanegashima Island, where SSTs in the coldest month are above 18 °C. A total of about 380 species of coral have been recorded from southernmost Japan (Yaeyama Archipelago: Veron 1992; Veron and Minchin 1992). Species number decreases along the latitudinal gradient, and the northern limits of coral distribution are located at Sadogashima Island and the Boso Peninsula in the Sea of Japan and the Pacific Ocean, respectively.

SSTs have risen during the past 100 years, as indicated by the Japan Meteorological Agency (Takatsuki et al. 2007). Therefore, Japan's wide latitudinal range provides a unique opportunity for examining the effects of rising SSTs, which could induce coral bleaching and poleward range shifts/expansions of corals in the south and north of Japan, respectively. For example, severe bleaching occurred in 1998 in the Ryukyu Islands as a result of anomalously high SSTs. In addition to SST limits, corals in Japan suffer from other disturbances. In the 1970s and 1980s, corals in the Ryukyu Islands suffered heavy infestation from crown-of-thorns starfish (COTS) outbreaks (Yamaguchi 1986), and further outbreaks have also occurred recently. Corals distributed close to river mouths suffer from terrigenous sediment and nutrient discharges (Omija 2004). In addition, typhoons often strike Japan, causing the destruction and redistribution of corals (Yamano et al. 2000).

A Brief History of Survey Activities in Japan

Japan's national coral survey was initiated in the 1930s by Drs. Yabe, Sugiyama, and Eguchi of Tohoku Imperial University (Tohoku University), Japan. This is one of the oldest national coral surveys in the world. They produced monographs with an extensive list of the coral species of Japan, as well as of Japanese overseas territories at that time, and described some new species, which have been accepted widely (Yabe and Sugiyama 1931, 1941; Yabe et al. 1936). After World War II, because of the occupation of Okinawa and Amami by the United States of America, survey activities were limited to the waters around mainland Japan (Kyushu, Shikoku, and Honshu). After the restoration of the Amami and Okinawa areas in 1953 and 1972, respectively, extensive surveys to designate marine parks were conducted throughout Japan, and lists of coral species in each prefecture were published. Coral coverage data have been collected since the 1970s. The Ministry of the Environment of Japan (formerly the Environment Agency) set up the National Survey of the Natural Environment to determine the status of land, surface water, and coastal areas (including coral reefs) throughout the country

(Tsuzawa 1990; Akasaka et al., this volume). The survey has been conducted about every 5 years since 1973.

National coral survey activities were enhanced by the establishment of the International Coral Reef Initiative (ICRI), which was announced at the First Conference of the Parties of the Convention on Biological Diversity (CBD-COP1) in 1994, to conserve coral reefs and related ecosystems. In the ICRI framework, the Global Coral Reef Monitoring Network (GCRMN) was established and published reports on the status of coral reefs in 1998, 2000, 2002, 2004, and 2008 (Wilkinson 1998, 2000, 2002, 2004, 2008). GCRMN also launched a global-scale database for coral reefs (ReefBase; http://www.reefbase.org/). In Japan, the Ministry of the Environment established the International Coral Reef Research and Monitoring Center on Ishigaki Island in 2000. The center serves as a focal point for coral reef conservation in Southeast Asia, and the information gathered on the status of coral reefs has contributed to the GCRMN reports. The Ministry of the Environment further developed the Monitoring Sites 1,000 Project to achieve the long-term monitoring of ecosystems in Japan. Japanese corals have been monitored extensively under this project since 2003. In 2004, when the 10th International Coral Reef Symposium was held in Okinawa, the book *Coral Reefs of Japan* (Ministry of the Environment and Japanese Coral Reef Society 2004) was published and distributed to provide a general overview of corals and coral reefs in Japan.

Although most surveys have been field based, remote sensing techniques have been used to map coral reefs since the 1990s. The first national coral distribution map, published in 1996, was based on field surveys and the interpretation of aerial photographs. In response to the availability of satellite images, the most recent map was generated by ALOS AVNIR2 satellite images in 2010. All these works have been converted into geographic information system (GIS) data and may be downloaded at websites operated by the Ministry of the Environment (e.g., Japan Integrated Biodiversity Information System, http://www.biodic.go.jp/english/J-IBIS.html; International Coral Reef Research and Monitoring Center, http://www.coremoc.go.jp/).

In addition to these government-based surveys, there are several other survey activities. Reef Check was launched during the first International Year of the Reef (1997) as an international, voluntary, community-based coral reef monitoring protocol (Hodgson 1999). In Japan, Reef Check Japan (http://reefcheck.jp/) and the Okinawa Reef Check & Research Group (http://reefcheck.net/) conduct annual surveys. Public awareness activities were further enhanced during the second International Year of the Reef (2008), and another citizen monitoring program, the Sango (Coral) Map Project (http://www.sangomap.jp/), was launched. The data collected helped validate the ALOS AVNIR2-derived map in 2010 (Namizaki et al., in press).

Local- to regional-scale monitoring has also been conducted actively since the 1970s by local organizations and researchers. These surveys were reviewed by

Kimura (2004). One outstanding example is a monitoring program organized by Okinawa Prefecture (Okinawa Prefecture 2010). In the program, permanent monitoring sites were established on three islands of Okinawa Prefecture (18 sites at Okinawa Island, 2 at Aka Island, and 4 at Ishigaki Island), and coral species and coverage in 2 × 2 m quadrats, along with fine sediment accumulation, have been surveyed annually since 1995.

National Coral Reef Database in Japan

The significant amount of coral surveys in Japan provides the opportunity to construct a national coral reef database, which is subdivided into a coral coverage database and a coral species database.

Coral Coverage Database

Reef Check has collected data in Japan since 1997, and these data were used by Bruno and Selig (2007), but the sample number was not sufficient to reveal national-scale changes. Since 2003, the Monitoring Sites 1,000 Project has collected coral coverage data throughout the regions with corals from subtropical to temperate areas, but these data have not been published. The first national coral coverage database was generated by Nakao et al. (2009), who collected coral coverage data from ~10,000 locations from the 1970s to 2007 based on 43 reports and on data provided by the Monitoring Sites 1,000 Project and Reef Check (Fig. 4.2a). All these data were converted into GIS data, including locations (latitude and longitude), the attributes of coral coverage, and additional information that was available (e.g., the occurrence of COTS). When the coordinates of the sample sites were not available, these were obtained from maps that indicated sampling locations.

The data showed significant variation in characteristics, which raised the importance of quality control of the data (Nakao et al. 2009). Only 2.5 % of the sampling points were measured by objective methods such as line transect or quadrat methods. Many data were semiquantitative, with ranges of coral coverage (e.g., 0–24 %). In such cases, the original range and the intermediate value of coral coverage (e.g., a coverage of 0–24 % would have an intermediate value of 12 %) were given in the database, and the intermediate value was used to detect trends in coral coverage. When reviewing the sampling sites by decade, there were 143 locations in the 1970s, 717 locations in the 1980s, 5,737 locations in the 1990s, and 2,675 locations in the 2000s. The locations in the 1990s accounted for more than half of all sites. Regional variation was also found in the number of survey sites. For example, the Yaeyama Archipelago had 4,817 survey sites, which accounted for nearly half the total number of sites.

a

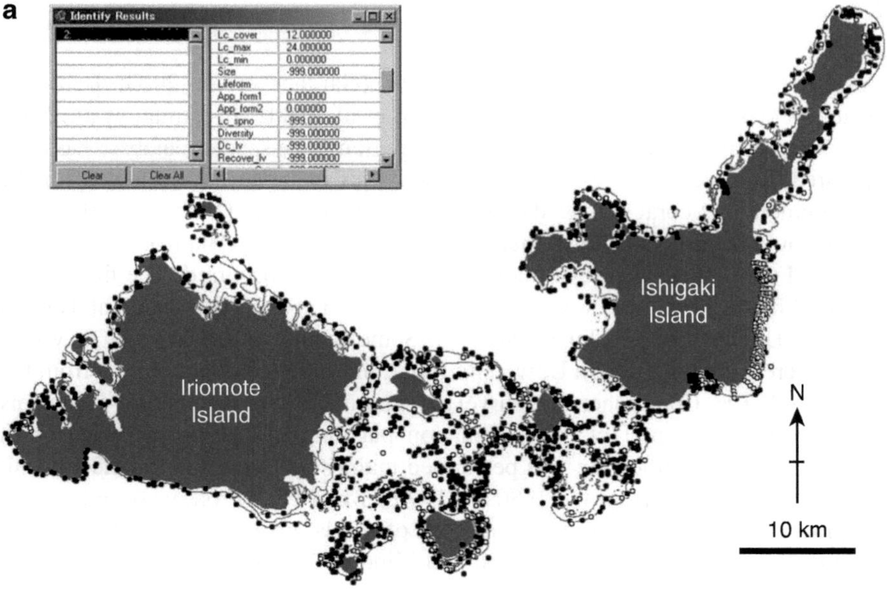

b

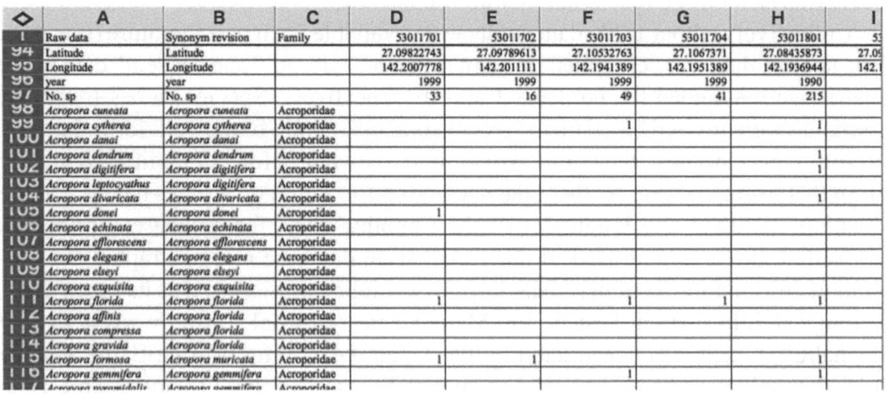

◇	A	B	C	D	E	F	G	H	I
1	Raw data	Synonym revision	Family	53011701	53011702	53011703	53011704	53011801	53
94	Latitude	Latitude		27.09822743	27.09789613	27.10532763	27.1067371	27.08435873	27.0
95	Longitude	Longitude		142.2007778	142.2011111	142.1941389	142.1951389	142.1936944	142.
96	year	year		1999	1999	1999	1999	1990	
97	No. sp	No. sp		33	16	49	41	215	
98	*Acropora cuneata*	*Acropora cuneata*	Acroporidae						
99	*Acropora cytherea*	*Acropora cytherea*	Acroporidae			1		1	
100	*Acropora danai*	*Acropora danai*	Acroporidae						
101	*Acropora dendrum*	*Acropora dendrum*	Acroporidae					1	
102	*Acropora digitifera*	*Acropora digitifera*	Acroporidae					1	
103	*Acropora leptocyathus*	*Acropora digitifera*	Acroporidae						
104	*Acropora divaricata*	*Acropora divaricata*	Acroporidae					1	
105	*Acropora donei*	*Acropora donei*	Acroporidae	1					
106	*Acropora echinata*	*Acropora echinata*	Acroporidae						
107	*Acropora efflorescens*	*Acropora efflorescens*	Acroporidae						
108	*Acropora elegans*	*Acropora elegans*	Acroporidae						
109	*Acropora elseyi*	*Acropora elseyi*	Acroporidae						
110	*Acropora exquisita*	*Acropora exquisita*	Acroporidae						
111	*Acropora florida*	*Acropora florida*	Acroporidae	1		1	1	1	
112	*Acropora affinis*	*Acropora florida*	Acroporidae						
113	*Acropora compressa*	*Acropora florida*	Acroporidae						
114	*Acropora gravida*	*Acropora florida*	Acroporidae						
115	*Acropora formosa*	*Acropora muricata*	Acroporidae	1	1			1	
116	*Acropora gemmifera*	*Acropora gemmifera*	Acroporidae			1		1	

Fig. 4.2 Examples from the database. **a** Survey locations for coral coverage (*closed dots*) and coral species (*open dots*) around Yaeyama Archipelago. *Inset table* shows example attributes of the sites, including not only coral coverage but also survey period (year, month, day, and time if available), occurrence of other organisms (e.g., crown-of-thorns starfish, COTS), physical environment (e.g., water depth and sedimentation), and the reference (Nakao et al. 2009). **b** Partial list of coral occurrence in the species database shows the presence/absence of each species (with synonyms, if available) at each survey location

In addition to this national database, Omija et al. (2000), Kinjo et al. (2005), and Hongo and Yamano (2013) collected coral coverage and species data from surveys performed at Okinawa Island between 1995 and 2009. All surveys were conducted in a consistent manner, with a permanent quadrat (2 × 2 m) placed on the reef. Thus, analysis based on consistent sampling standards was possible.

Coral Species Database

Despite emerging coral coverage databases (Gardner et al. 2003; Bellwood et al. 2004; Bruno and Selig 2007; Osborne et al. 2011; Sweatman et al. 2011; Magdaong et al., this volume), no long-term, region-wide species database has been prepared. Long-term, large-scale survey activities in Japan have allowed for the development of a coral species database for mainland Japan (Yamano et al. 2011), which could be integrated into a national database including data from the Ryukyu Islands. Four distinct periods of coral exploration have occurred. The first period occurred in the 1930s, and was undertaken by Drs. Yabe and Sugiyama of Tohoku Imperial University (Tohoku University). Some of the specimens that they collected were stored in Tohoku University Museum. The second period was from the 1960s to 1970s. Comprehensive surveys of marine environments and organisms were conducted as part of the designation planning of marine protected areas (MPAs). Coral identification was performed mainly by Drs. Utinomi and Eguchi. The third period was from the 1980s to 1990s, when Dr. Veron, who revised and integrated the recent taxonomic framework of corals, visited Japan and examined coral distribution (Veron 1992; Veron and Minchin 1992). The last period covers the late 1990s to the present, when Japanese researchers have conducted field surveys based on the recent taxonomic framework.

Collectively, about 1,000 data points were available from past literature (Fig. 4.2). The species data were compiled into a matrix, and "present" and "absent" data in the matrix were assigned the numbers 1 and 0, respectively. Species names from past literature were revised based on the recent coral taxonomic framework formulated by Veron (1992), because the coral taxonomic framework before 1990 differed greatly from that in current use. As these data contain information on species names and the locations (latitude and longitude) of the sampling sites, the database format was converted into a format used in the global-scale database Oceanographic Biogeographic Information System (OBIS, http://www.iobis.org/), as well as its Japan node, Biological Information System for Marine Life (BISMaL; http://www.godac.jp/bismal/e/), which was constructed by the Japan Agency for Marine-Earth Science and Technology (JAMSTEC) and is operated by the Global Oceanographic Data Center of JAMSTEC (Yamamoto et al. 2012). Therefore, the coral species database is ready to be incorporated into global-scale species databases.

Applications of the Database

Trends in Coral Coverage

Figure 4.3 shows the trends in coral coverage among the Ryukyu Islands and Okinawa Island based on Nakao et al. (2009) and Hongo and Yamano (2013), respectively. Coverage among the Ryukyu Islands was maintained at about 20 % in

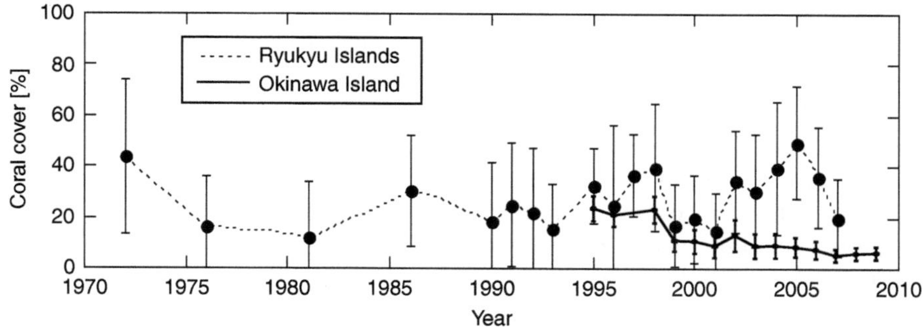

Fig. 4.3 Trends in coral coverage for the Ryukyu Islands (Nakao et al. 2009; Fujii and Yamano 2010) and Okinawa Island (Hongo and Yamano 2013). *Vertical bars* indicate standard deviation

the 1970s and 1980s. At the beginning of the 1990s, however, it rapidly increased and had reached 40 % by the mid-1990s. After that, it began to decline again, and by the end of the 1990s, coverage had fallen to 20 %. Gradually, from the end of the 1990s to the middle of the first decade of the twenty-first century, coverage again increased to more than 40 % before falling sharply again in 2007. Coral coverage at Okinawa Island, however, showed different trends. Although the coverage was about 25 % before 1998, it decreased dramatically to less than 10 % after 1999, and has shown no recovery since then. A general correspondence between disturbance and coral coverage may be found. The low coral coverage in the 1970s and 1980s could be explained by COTS outbreaks and terrestrial runoff, and the decrease in 1998 followed a severe mass bleaching event that occurred throughout the world (Wilkinson 2000).

Although coral coverage in the Caribbean Sea has declined from ~50 % to ~10 % during the past 35 years (Gardner et al. 2003), that among the Ryukyu Islands has shown cycles of increases and decreases between 20 % and 40 %, which might indicate that Japanese corals have high resilience to disturbance. However, caution should be taken to avoid overgeneralization, because trends may differ among regions, as shown (Fig. 4.3) by the different trends at Okinawa Island.

Species-Level Responses to Rising SST

Japan spans a wide latitudinal range, and thus it is expected that responses to rising SSTs could show different characteristics between subtropical (Ryukyu Islands) and temperate (mainland Japan) areas. Anomalously high SSTs can cause coral bleaching in subtropical areas, although overall increases in SST may lead to poleward range expansions of warm-water coral species into temperate areas. As shown in Fig. 4.3, coral coverage significantly decreased because of bleaching in 1998, but the study of species-level responses has been limited to specific sites (Loya et al. 2001; van Woesik et al. 2011). Based on the coral species database,

Omija et al. (2000) identified species that were resistant and vulnerable to bleaching in 1998. Further, Hongo and Yamano (2013) calculated the abundance of each species as the number of sites in which a species occurred divided by the total number of sites, and showed species-specific trajectories from 1995 to 2009 at a regional scale. They revealed that at Okinawa Island, the abundance of *Acropora digitifera*, a key species for coral reef formation, decreased significantly after 1998, and has not yet recovered, whereas massive-type *Porites* species have shown stability. This observation strongly suggests that the "winners" and "losers" of bleaching events (Loya et al. 2001; van Woesik et al. 2011) at specific sites can be examined over a large spatial scale.

Warm-water coral species have recently been documented in some temperate areas of Japan (Nojima and Okamoto 2008; Nomura et al. 2008), raising awareness of the need for national-scale surveys to reveal the extent and speed of range expansions. Based on the coral occurrence records for four periods (1930s, 1960s–1970s, 1980s–1990s, and the present) in eight temperate regions in the coral species database, Yamano et al. (2011) revealed that four species (*Acropora hyacinthus*, *Acropora muricata*, *Acropora solitaryensis*, and *Pavona decussata*) have expanded their ranges into temperate areas at speeds reaching up to 14 km/year. Thus, temperate areas may serve as refugia for tropical corals in an era of global warming, whereas corals in tropical areas suffer declines because of rising SSTs. The range expansions were shown to change temperate coastal biodiversity, as obligate coral-dwelling crabs also extended their ranges as their host corals moved north (Yamano et al. 2012a).

Future Directions

The development of a long-term, large-scale database of coral records is essential to understand baseline coral status and to examine the causes of any changes that occur (Gardner et al. 2003). The database presented here contributes significantly to these goals. To the best of our knowledge, the coral species database is the first long-term, large-scale species database in the world, and as such has allowed us to reveal poleward range expansions of tropical reef corals into temperate areas (Yamano et al. 2011). The national databases on coral coverage and species will be used further to examine the causes and effects of such disturbances over large spatial extents. To enhance use of the database, some future issues are presented.

Quality Control

Quality control and characterization are critical issues for the database. For coral coverage data, as Sweatman and Syms (2011, p. 661) emphasized, "combining data collected at different spatial scales (quadrats and transects in the past mixed with more recent whole-reef averages from manta tows) are likely to overestimate decline, because whole-reef averages will very rarely reach the high cover values that can occur at the quadrat scale." Analysis of the data considering the survey method could be needed; this, in turn, means that systematic monitoring would be needed to detect changes. Based on the construction process of the coral coverage database, Nakao et al. (2009) proposed unifying the formats of monitoring methods and of the obtained data and minimizing any bias in regions and years for future monitoring. In addition, the survey methods would need to be compared to quantify differences among methods. Then, multi-site, field-based monitoring with a unified format (e.g., the Monitoring Sites 1,000 Project and Reef Check), combined with mapping by satellite data that are subject to rigorous evaluation of their classification accuracy (Scopélitis et al. 2010; Yamano 2013), may be encouraged for continuous monitoring.

Quality control is also critical for coral species data. Corals are one of the most difficult taxa to identify to the species level. They have few taxonomically significant morphological characteristics, because such characters frequently show polymorphism, intraspecific variation, and phenotypic plasticity in response to environmental factors (Veron 2000). Further, their occurrence records may be affected significantly by survey method. For example, snorkeling cannot detect species that occur at depths greater than 10 m. To overcome these issues, Yamano et al. (2011) used a rigorous quality control method for species occurrence records from ten regions, where species that were abundant in the regions of interest were selected to avoid the possibility that past surveys had missed their occurrence; species distributed typically at depths of less than 10 m were selected, as some past surveys used snorkeling or observations through a water glass from a boat; and species with distinct growth forms that could be identified in situ were selected, because in some previous reports, observers identified species in situ without collecting specimens. Similar assessments would be needed for the species database.

In addition, for future surveys and monitoring, the establishment of a solid baseline for coral identification is urgently needed. For this purpose, integration between morphological and phylogenetic works would be of primary importance (Fukami et al. 2010), because emerging phylogenetic techniques provide new insights into coral taxonomy (Fukami et al. 2008). Based on the baseline data obtained, it would be possible to publish identification guidelines and hold training workshops to enhance coral identification skills and produce more coral surveyors and researchers.

Expanding the Database

The database may be expanded to include other coral-related organisms. Corals play a fundamental role in primary production and habitat formation for numerous other species in tropical and subtropical areas, and thus changes in coral community structure (e.g., losses from bleaching and new settlement by poleward range expansions) will affect other coral-related organisms (Pratchett et al. 2009; Yamano et al. 2012a). Corals also compete with other benthic organisms such as algae, which also play a fundamental role in coastal ecosystems. Bruno et al. (2009) expanded their coral coverage database (Bruno and Selig 2007) to include algae cover, and found that only 4 % of the reefs examined showed coral-dominant to algae-dominant phase shifts (see Hughes 1994). Expanding the databases to consider ecosystem processes and functions would provide a holistic understanding of coastal ecosystems and biodiversity dynamics in a changing world.

Global-scale databases of coral reefs and marine biodiversity have been developed (e.g., ReefBase and OBIS) and have contributed significantly to large-scale assessments of coral status and biodiversity (Tittensor et al. 2010). The existing databases, including those presented in this chapter, will be integrated and incorporated into these global-scale databases.

Further Uses of the Database

One effective way to use the database is to apply statistical modeling to correlate environmental factors with distribution, which has been shown to be a powerful tool for explaining and predicting the spatial distribution of organisms (Tittensor et al. 2010). As demonstrated in Fig. 4.3, trajectories in coral coverage differ among regions and/or islands. Furthermore, even on the same island, coral distribution should show variation in response to environmental factors, as represented by the zonal distribution of corals (Done 1983). Recent advances in data mining and reanalysis techniques have made it possible to map factors on a global scale and to classify coral reef environments based on these factors (Shimazaki et al. 2006; Freeman et al. 2012). Sediment and nutrient discharge from agricultural activities may affect river and coastal ecosystems (Fabricius 2005), and making the land–sea connection is strongly encouraged to enable the establishment of integrated coastal reserve planning (Stoms et al. 2005). Because detailed national land use/land cover data are available in Japan (Akasaka et al., this volume), statistical modeling that integrates marine and terrestrial variables would help explain and predict the spatial distributions of corals.

Evaluations of future coral habitats using SST projections derived from climate models have shown higher bleaching frequency and continuous poleward range expansions in Japan during the twenty-first century (Yara et al. 2009, 2011) (Fig. 4.4). Yara et al. (2012) added ocean acidification data to the SST projections and reported a more pessimistic result that corals would be suppressed by ocean acidification even if poleward range expansions could occur. These results need to be validated by observations. Because such projections could be applied to the twentieth century and evaluated based on empirical data, the database could be refined and used for validating future projections.

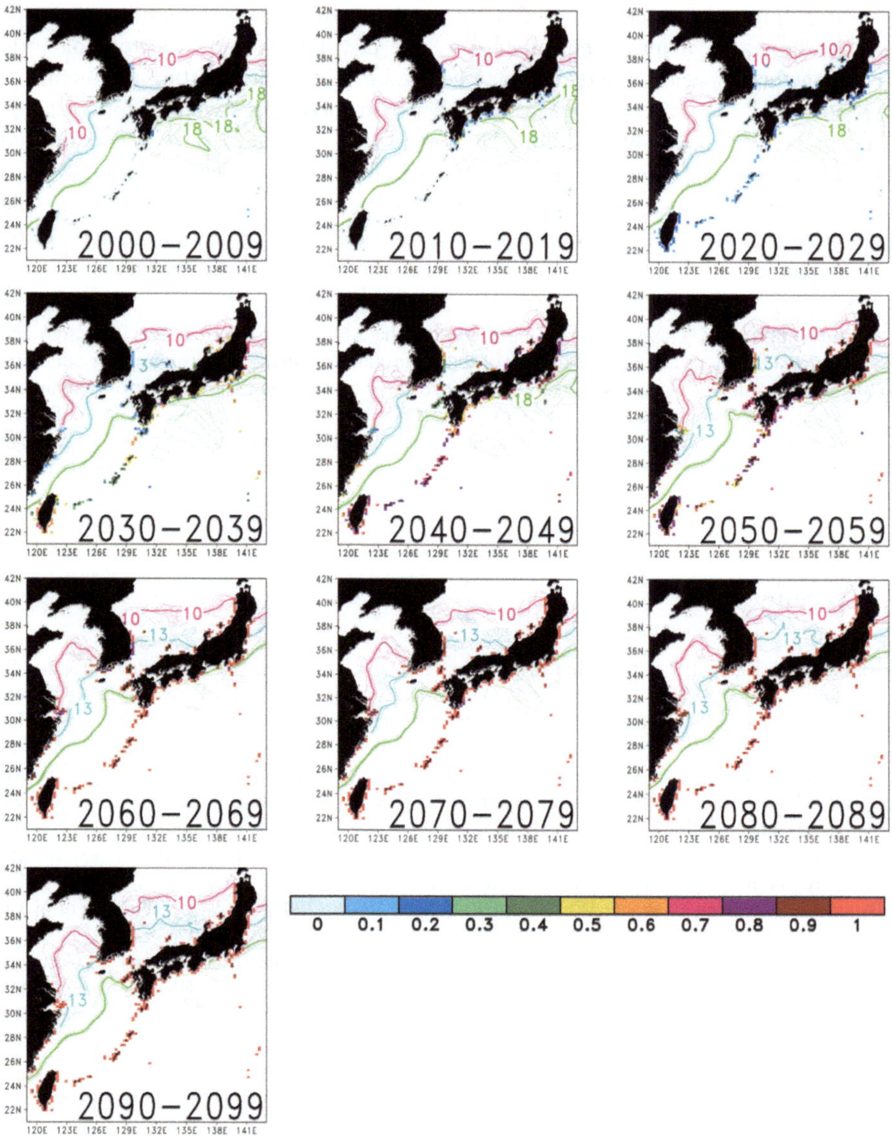

Fig. 4.4 Projected northern limits of coral distribution (10 °C isotherm in the coldest month) and distribution of the subtropical coral community (18 °C isotherm in the coldest month) that contribute to reef building. In the Ryukyu Islands, the probabilities (frequency) of high water temperatures with the potential to induce severe bleaching or death are shown. The projections are based on projected monthly mean SSTs by a high-resolution climate projection model for 2000–2099. *Color bar* indicates the frequency of high water temperatures in each decade in the Ryukyu Islands. (Adapted from Yara et al. 2009)

Modeling and projections based on the database will contribute significantly to the establishment of management strategies. The distribution and trajectory of coral coverage, and species abundance, may be used to reevaluate existing MPAs and to propose new MPAs based on recent advances in modeling development for MPA selection (Possingham et al. 2000). These data may also be used to establish a Red List according to the International Union for Conservation of Nature extinction risk assessment criteria (Carpenter et al. 2008). The database will play a central role in the integration of observations, statistical modeling, and future projections to understand the dynamics and to establish proper management of coral reef ecosystems.

Acknowledgments This study is a contribution to the S-9 Project (Environment Research and Technology Development Fund, Ministry of the Environment, Japan), Grant-in-Aid for Scientific Research on Innovative Areas "Coral reef science for symbiosis and coexistence of human and ecosystem under combined stresses" (No. 20121004, Ministry of Education, Culture, Sports, Science and Technology [MEXT], Japan), the Program for Risk Information on Climate Change (MEXT), and the Biodiversity Priority Program (National Institute for Environmental Studies).

References

Akasaka M, Takenaka A, Ishihama F, Kadoya T, Ogawa M, Osawa T, Yamakita T, Tagane S, Ishii R, Nagai S, Taki H, Akasaka T, Oguma H, Suzuki T, Yamano H (this volume) Development of a national land use/cover dataset to estimate biodiversity and ecosystem services. In: Nakano S, Yahara T, Nakashizuka T (eds) The biodiversity observation network in the Asia-Pacific region: integrative observations and assessments of Asian biodiversity. Springer, Tokyo

Bellwood DR, Hughes TP, Folke C, Nyström M (2004) Confronting coral reef crisis. Nature (Lond) 429:827–833

Bruno JF, Selig ER (2007) Regional decline of coral cover in the Indo-Pacific: timing, extent, and subregional comparisons. Plos ONE 2(8):e711

Bruno JF, Sweatman H, Precht WF, Selig ER, Schutte VGW (2009) Assessing evidence of phase shifts from coral to macroalgal dominance on coral reefs. Ecology 90:1478–1484

Carpenter KE, Abrar M, Aeby G, Aronson RB, Banks S, Bruckner A, Chiriboga A, Cortés J, Delbeek JC, DeVantier L, Edgar GJ, Edwards AJ, Fenner D, Guzmán HM, Hoeksema BW, Hodgson G, Johan O, Licuanan WY, Livingstone SR, Lovell ER, Moore JA, Obura DO, Ochavillo D, Polidoro BA, Precht WF, Quibilan MC, Reboton C, Richards ZT, Rogers AD, Sanciangco J, Sheppard A, Sheppard C, Smith J, Stuart S, Turak E, Veron JEN, Wallace C, Weil E, Wood E (2008) One-third of reef-building corals face elevated extinction risk from climate change and local impacts. Science 321:560–563

Done TJ (1983) Coral zonation: its nature and significance. In: Barnes DJ (ed) Perspectives in coral reefs. Australian Institute of Marine Science, Townsville, pp 107–147

Fabricius KE (2005) Effect of terrestrial runoff on the ecology of corals and coral reefs: review and synthesis. Mar Pollut Bull 50:125–146

Freeman LA, Miller AJ, Norris RD, Smith JE (2012) Classification of remote Pacific coral reefs by physical oceanographic environment. J Geophys Res 117, C02007

Fujii M, Yamano H (2010) Response to Omori (2010) Comment on "Development and analysis of a coral cover database in seas close to Japan". J Jpn Coral Reef Soc 12:53–57

Fukami H, Tachikawa H, Suzuki G, Nagata S, Sugihara K (2010) Current status and problems with the identification and taxonomy of zooxanthellate scleractinian corals in Japan. J Jpn Coral Reef Soc 12:17–31

Fukami H, Chen CA, Budd AF, Collins A, Wallace C, Chuang Y-Y, Chen C, Dai C-F, Iwao K, Sheppard C, Knowlton N (2008) Mitochondrial and nuclear genes suggest that stony corals are monophyletic but most families of stony corals are not (Order Scleractinia, Class Anthozoa, Phylum Cnidaria). PLoS One 3:e3222

Gardner TA, Côté IM, Gill JA, Grant A, Watkinson AR (2003) Long-term region-wide declines in Caribbean corals. Science 301:958–960

Hodgson G (1999) A global assessment of human effects on coral reefs. Mar Pollut Bull 38:345–355

Hoegh-Guldberg O, Mumby PJ, Hooten AJ, Steneck RS, Greenfield P, Gomez E, Harvell CD, Sale PF, Edwards AJ, Caldeira K, Knowlton N, Eakin CM, Iglesias-Prieto R, Muthiga N, Bradbury RH, Dubi A, Hatziolos ME (2007) Coral reefs under rapid climate change and ocean acidification. Science 318:1737–1742

Hongo C, Yamano H (2013) Species-specific responses of corals to bleaching events on anthropogenically turbid reefs on Okinawa Island, Japan, over a 15-year period (1995–2009). PLoS ONE 8:e60952

Hughes TP (1994) Catastrophes, phase shifts, and large-scale degradation of a Caribbean coral reef. Science 265:1547–1551

Kayanne H, Hongo C, Yamano H (2004) Coral reef landforms in Japan. In: Ministry of the environment and Japanese Coral Reef Society (eds) Coral reefs of Japan. Ministry of the Environment, Tokyo, pp 14–19

Kimura T (2004) Coral monitoring in Japan. In: Ministry of the environment and Japanese Coral Reef Society (eds) Coral Reefs of Japan. Ministry of the Environment, Tokyo, pp 78–88

Kinjo K, Higa E, Uehara M (2005) Transition of sediment of red soil on seabed and its impact on corals. Ann Rep Okinawa Pref Inst Health Environ 39:63–74

Loya Y, Sakai K, Yamazato K, Nakano Y, Sambali H, van Woesik R (2001) Coral bleaching: the winners and the losers. Ecol Lett 4:122–131

Magdaong E, Yamano H, Fujii M (this volume) Development of a large-scale, long-term coral cover and disturbance database in the Philippines. In: Nakano S, Yahara T, Nakashizuka T (eds) The biodiversity observation network in the Asia-Pacific region: integrative observations and assessments of Asian biodiversity. Springer, Tokyo

Ministry of the Environment and Japanese Coral Reef Society (2004) Coral reefs of Japan. Ministry of the Environment, Tokyo

Nakao Y, Yamano H, Fujii M, Yamanaka Y (2009) Development and analysis of a coral cover database in seas close to Japan. J Jpn Coral Reef Soc 11:109–129

Namizaki N, Yamano H, Suzuki R, Oohori K, Onaga H, Kishimoto T, Sagawa T, Machida Y, Yasumura, S, Satoh T, Shigiya T, Shibata T, Tsuchikawa M, Miyamoto Y, Harukawa K, Hirate K, Furuse K, Hokoyama K, Yamanaka Y, Wagatsuma T (in press) Potential of citizen monitoring programs for marine areas—two-year activities of Sango (Coral) Map Project. Galaxea J Coral Reef Stud

Nojima S, Okamoto M (2008) Poleward range expansion and bleaching of corals. Nippon Suisan Gakaishi 74:884–888

Nomura K, Uchida H, Fukuda T (2008) Progressive faunal inventory of hermatypic corals in Kushimoto, Wakayama Prefecture, Japan. Nanki-Seibutsu 50:191–200

Okinawa Prefecture (2010) Report of fixed-point observation of red soil sedimentation in 2009. Okinawa Prefecture

Omija T (2004) Coral monitoring in Japan. In: Ministry of the environment and Japanese Coral Reef Society (eds) Coral reefs of Japan. Ministry of the Environment, Tokyo, pp 64–68

Omija T, Nakasone K, Mitsumoto H, Uehara M, Oshiro T (2000) Comparison of sediment-resistance of corals with bleaching-resistance. Ann Rep Okinawa Pref Inst Health Environ 34:69–76

Osborne K, Dolman AM, Burgess SC, Johns KA (2011) Disturbance and the dynamics of coral cover on the Great Barrier Reef (1995–2009). Plos One 6:e17516s

Possingham HP, Ball IR, Andelman S (2000) Mathematical methods for identifying representative reserve networks. In: Burgman M, Ferson S (eds) Quantitative methods for conservation biology. Springer, New York, pp 291–305

Pratchett MS, Wilson SK, Graham NAJ, Munday PL, Jones GP, Polunin NVC (2009) Coral bleaching and consequences for motile reef organisms: past, present and uncertain future effects. In: van Oppen JH, Lough JM (eds) Coral bleaching: patterns, processes, causes and consequences. Springer, Berlin, pp 139–158

Scopélitis J, Andrefouet S, Phinn S, Arroyo L, Dalleau M, Cros A, Chabanet P (2010) The next step in shallow coral reef monitoring: combining remote sensing and in situ approaches. Mar Pollut Bull 60:1956–1968

Shimazaki H, Yamano H, Yokoki H, Yamaguchi T, Chikamori M, Tamura M, Kayanne H (2006) Global mapping of factors controlling reef-island formation and maintenance. In: Proceedings of the 10th international coral reef symposium. Okinawa, pp 1577–1584

Stoms DM, Davis FW, Andelman SJ, Carr MH, Gaines SD, Halpern BS, Hoenicke R, Leibowitz SG, Leydecker A, Madin EMP, Tallis H, Warner RR (2005) Integrated coastal reserve planning: making the land-sea connection. Front Ecol Environ 3:429–436

Sweatman H, Syms C (2011) Assessing loss of coral cover on the Great Barrier Reef: a response to Hughes et al. (2011). Coral Reefs 30:661–664

Sweatman H, Delean S, Syms C (2011) Assessing loss of coral cover on Australia's Great Barrier Reef over two decades with implications for longer-term trends. Coral Reefs 30:521–531

Takatsuki Y, Kuragano T, Shiga T, Bungi Y, Inoue H, Fujiwara H, Ariyoshi M (2007) Long-term trends in sea surface temperature adjacent to Japan. Sokko-Jiho 74:S33–S87

Tittensor DP, Mora C, Jetz W, Lotze HK, Ricard D, Berghe EV, Worm B (2010) Global patterns and predictors of marine biodiversity across taxa. Nature (Lond) 466:1098–1101

Tsuzawa M (1990) National Surveys on the Natural Environment ("Green Census"). J Geogr 99:64–71

van Woesik R, Sakai K, Ganase A, Loya Y (2011) Revisiting the winners and the losers at a decade after coral bleaching. Mar Ecol Prog Ser 434:67–76

Veron JEN (1992) Hermatypic corals of Japan. Australian Institute of Marine Science, Townsville

Veron JEN (2000) Corals of the world. Australian Institute of Marine Science, Townsville

Veron JEN, Minchin PR (1992) Correlations between sea surface temperature, circulation patterns and the distribution of hermatypic corals of Japan. Continent Shelf Res 12:835–857

Wilkinson CR (1998) Status of coral reefs of the world: 1998. Australian Institute of Marine Science, Queensland

Wilkinson CR (2000) Status of coral reefs of the world: 2000. Australian Institute of Marine Science, Queensland

Wilkinson CR (2002) Status of coral reefs of the world: 2002. Australian Institute of Marine Science, Queensland

Wilkinson CR (2004) Status of coral reefs of the world: 2004. Australian Institute of Marine Science, Queensland

Wilkinson CR (2008) Status of coral reefs of the world: 2008. Australian Institute of Marine Science, Queensland

Yabe H, Sugiyama T (1931) Reef corals found in the Japanese seas. Sci Rep Tohoku Imp Univ Sec Ser (Geol) 16:143–167

Yabe H, Sugiyama T (1941) Recent reef-building corals from Japan and the South Sea Islands under the Japanese mandate. II. Sci Rep Tohoku Imp Univ Sec Ser (Geol) Spec 2:67–91

Yabe H, Sugiyama T, Eguchi M (1936) Recent reef-building corals from Japan and the South Sea Islands under the Japanese mandate. I. Sci Rep Tohoku Imp Univ Sec Ser (Geol) Spec 1:1–66

Yamaguchi M (1986) *Acanthaster planci* infestations of reefs and coral assemblages in Japan: a retrospective analysis of control efforts. Coral Reefs 5:23–30

Yamamoto H, et al (2012) BISMaL, a marine biological database online at the global oceanographic data center of JAMSTEC. In: Nakano S, Yahara T, Nakashizuka T (eds) The biodiversity observation network in the Asia-Pacific Region: Toward future development of monitoring. Springer, Tokyo, pp 247–256

Yamano H (2013) Multispectral applications. In: Goodman JA, Purkis SJ, Phinn SR (eds) Coral reef remote sensing: a guide for multi-level sensing, mapping and assessment. Springer, Dordrecht, pp 51–78

Yamano H, Kayanne H, Yonekura N, Kudo K (2000) 21-year changes of backreef coral distribution: causes and significance. J Coast Res 16:99–110

Yamano H, Sugihara K, Nomura K (2011) Rapid poleward range expansion of tropical reef corals in response to rising sea surface temperatures. Geophys Res Lett 38, L04601

Yamano H, Sugihara K, Goto K, Kazama T, Yokoyama K, Okuno J (2012a) Ranges of obligate coral-dwelling crabs extend northward as their hosts move north. Coral Reefs 31:663

Yamano H, Sugihara K, Watanabe T, Shimamura M, Hyeong K (2012b) Coral reefs at 34°N, Japan: exploring the end of environmental gradients. Geology 40:835–838

Yara Y, Fujii M, Yamanaka Y, Okada N, Yamano H, Oshima K (2009) Projected effects of global warming on coral reefs in seas close to Japan. J Jpn Coral Reef Soc 11:131–140

Yara Y, Oshima K, Fujii M, Yamano H, Yamanaka Y, Okada N (2011) Projection and uncertainty of the poleward range expansion of coral habitats in response to sea surface temperature warming: a multiple climate model study. Galaxea J Coral Reef Stud 13:11–20

Yara Y, Vogt M, Fujii M, Yamano H, Hauri C, Steinacher M, Gruber N, Yamanaka Y (2012) Ocean acidification limits temperature-induced poleward expansion of coral habitats around Japan. Biogeosciences 9:4955–4968

Chapter 5
Development of a Large-Scale, Long-Term Coral Cover and Disturbance Database in the Philippines

Evangeline T. Magdaong, Hiroya Yamano, and Masahiko Fujii

Abstract The Philippines is located in the tropics and contains highly diverse coral reefs, as it lies within the Coral Triangle, which is the global center of marine biodiversity. However, this biodiversity "hotspot" has experienced reef degradation driven by natural disturbances and intense anthropogenic pressures. Consequently, coral researchers are focusing on promoting awareness of local reefs through the conservation, management, and assessment of reefs to monitor coral health. The status of coral reefs in the Philippines has been reported periodically since surveys were initiated in the 1970s. This study presents the preliminary steps to develop a large-scale, long-term coral cover database in the Philippines. The research was conducted through personal communications and electronic search of coral studies by different sectors (government, nongovernmental organizations, and reef scientists) conducted all over the Philippines. The collated data consist of living hard coral cover percentages from 2,349 reef benthic surveys conducted between 1978 and 2010. These data were analyzed to assess the spatial and temporal variability of Philippine reefs and to determine recent trends in coral cover. Overall, the results revealed an increase in areas with low levels (<25 %) of coral cover and a larger reduction in areas with high levels (>75 %) of coral cover. The results indicated that the Visayan Sea area is at high risk, with the highest ratio of sites with low levels of cover across time periods. Based on a review of the disturbances driving the reef decline, anthropogenic impacts are the most prevalent threat to Philippine reefs.

E.T. Magdaong (✉)
Graduate School of Environmental Science, Hokkaido University, N10W5, Kita-ku, Sapporo, Hokkaido 060-0810, Japan
e-mail: emagdaong@ees.hokudai.ac.jp

H. Yamano
Center for Environmental Biology and Ecosystem Studies, National Institute for Environmental Studies, 16-2 Onogawa, Tsukuba, Ibaraki 305-8506, Japan

M. Fujii
Faculty of Environmental Earth Science, Hokkaido University, N10W5, Kita-ku, Sapporo, Hokkaido 060-0810, Japan

S.-i. Nakano et al. (eds.), *Integrative Observations and Assessments*, Ecological Research Monographs, DOI 10.1007/978-4-431-54783-9_5,

Keywords Anthropogenic impacts • Coral reefs • Database • Disturbance • Living hard coral cover • Philippines

Introduction

The Philippines lies in the tropical Indo-West Pacific (4°N–21°N and 116°E–127°E), and consists of more than 7,100 islands (Fig. 5.1). It has a total coastline of 36,289 km (NAMRIA, cited by BFAR 2009), which includes an estimated 25,819

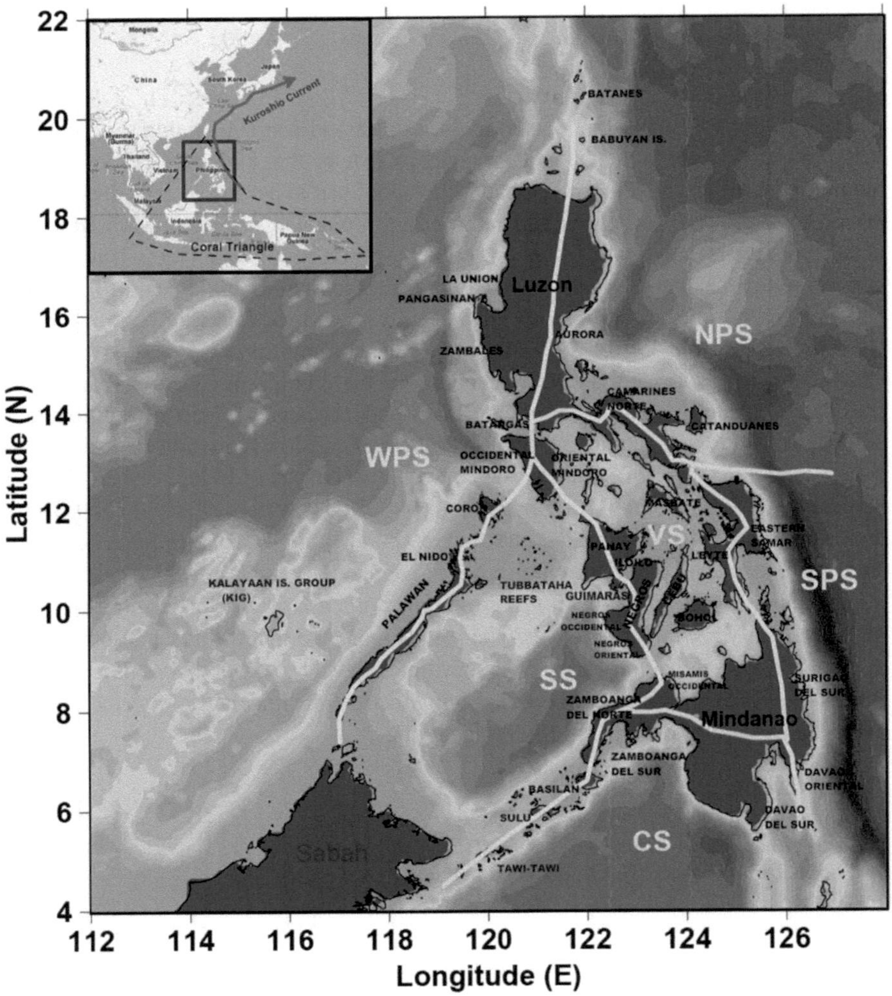

Fig. 5.1 Map of the study site and the classification of biogeographic regions modified from PhilReefs (2003). *WPS* Western Philippine Sea, *NPS* Northern Philippine Sea, *SPS* Southern Philippine Sea, *VS* Visayan Sea, *SS* Sulu Sea, *CS* Celebes Sea

(Burke et al. 2002; Tun et al. 2008) to 27,000 km^2 (White and Cruz-Trinidad 1998; BFAR 2009) of commonly fringing reefs. Atoll reefs appear near Tubbataha, Sulu Island, the northernmost islands of Babuyan and Batanes, and in the southeastern and western Negros and Guimaras; barrier reefs appear near Palawan and Tawi-tawi, and a double barrier can be found in Danajon Bank, Bohol. (Alcala et al. 1987).

The Philippines is located in the Coral Triangle, the center of marine biodiversity, and thus is home to highly diverse coral and fish species (Carpenter and Springer 2005). Approximately 468 species of scleractinian coral (Licuanan and Capili 2004; Veron and Fenner 2000), 915 reef fish species (Burke et al. 2002), 820 species of macrobenthic algae (Trono 1999), 16 seagrass species (Fortes and Santos 2004), and 44 mangrove species (Licuanan et al. 2011; Spalding et al. 2010) can be found in the Philippines (see Nañola et al. 2006). The islands may be the main source of coral larvae in the northwestern Pacific, as they are located in the upstream area of the Kuroshio Current. However, this center of endemism has been identified as highly at risk (Roberts et al. 2002) and has the highest level of vulnerable and threatened coral species (Carpenter et al. 2008). In a recent Reefs at Risk assessment (Burke et al. 2011), the Philippines was listed as highly vulnerable to reef degradation, with very low adaptive capacity. A recent status report from the Philippines demonstrated a steady decline in excellent reefs (with >75 % coral cover) and an increase in poor reefs (with <25 % coral cover) (Nañola et al. 2006). Consequently, research efforts have been geared toward protecting and rehabilitating reefs, and understanding reef health by developing indices to monitor coral health (Yap 1997).

Studies have shown that marine ecosystems have been greatly affected by anthropogenic and natural stresses (Burke et al. 2002; Carpenter et al. 2008). Coral cover is a critical measure of habitat loss and degradation (Bruno and Selig 2007), and because hard corals form the structural framework of the reef, the percentage cover of living hard corals is used as an index of reef condition in most studies (Sweatman et al. 2011). Licuanan and Gomez (2000) suggested that hard coral cover provides a better measure for comparison across sites than does living coral cover, because soft corals are not abundant in offshore reef areas. Long-term coral cover assessments have revealed that in many major reef systems worldwide, coral declines in response to disturbances; these areas include the Caribbean (Gardner et al. 2003), the Great Barrier Reef (De'ath et al. 2012), the Indo-Pacific (Bruno and Selig 2007), and Japan (Nakao et al. 2009).

Threats to Philippine reefs come primarily from human-related factors caused by increasing population; the most significant disturbances are associated with fishing, particularly overfishing and destructive fishing (Gomez 1997; PhilReefs 2003).

We conducted an initial analysis of collated data based on the existing methods used in various reports to clarify the recent trends in coral cover. We began by collating living hard coral cover from various studies to create a database, using more publications and a longer period than the research conducted by Bruno and Selig (2007).

We also collated data about disturbances that may have affected the coral reef status. This chapter presents our progress in creating a long-term and wide-scale database of living hard coral cover and disturbances in the Philippines.

Methods

Coral Cover Data

The state of Philippine coral reefs has been periodically reported since 1979, with the first systematic survey initiated in 1976 (see Gomez and Alcala 1979). We compiled percentage living scleractinian cover from quantitative surveys obtained from scientific and gray literature available online as well as through personal communications with nongovernmental organizations such as Reef Check and Coral Cay Conservation. Electronic searches consisted of browsing reef-related websites such as ReefBase and OneOcean.org, search engines such as Google Scholar, and the online journals of *Coral Reefs*, *Marine Pollution Bulletin*, and *Proceedings of the International Coral Reef Symposium*. The results presented here are preliminary findings; meta-analysis cannot be performed at this time because a number of studies published in the 1980s were not accessible online, either because they were published in Philippine journals or because no electronic copy is available for download. The data were summarized first as overall annual means, and second as subset of biogeographic regional means; these were then further classified as quartile categories to assess the condition of the reefs.

For overall means, all data points were pooled per year and averaged regardless of the depth or the method used in the surveys. A subset was made by grouping the surveyed sites per region following the six biogeographic classifications used in PhilReefs (2003), adapted from Aliño and Gomez (1994): (1) North Philippine Sea (NPS); (2) South Philippine Sea (SPS); (3) Visayan Sea (VS); (4) South China Sea, herein referred to as West Philippine Sea (WPS); (5) Sulu Sea (SS); and (6) Celebes Sea (CS) (Fig. 5.1). Biogeographic regional means were also averaged similarly to overall annual means. The surveys were further classified according to quartile categories, as suggested by Gomez and Alcala (1979) to determine the condition of the reefs. Although other indices can be used in combination with other benthic life categories, this initial assessment was limited to using living hard coral cover (LHCC). For quartile classification, the sites with repeated measurements were averaged and then classified based on coral cover percentages: <25 % as Poor; 25–49 % as Fair; 50–75 % as Good; and >75 % as Excellent. For each time period, the quartile categories were presented as percentages of the total sites, for instance, 161 of 372 sites fell into the Fair category, which included about 43 % of all sites.

Aside from annual means, we also examined LHCC trends based on the disturbance events defined by Bruno and Selig (2007) for the Indo-Pacific, which coincided with strong El Niño Southern Oscillation (ENSO) events (1983, 1998, and 2010) during the following time periods: 1978–1983, 1984–1996, 1997–2004, and 2005–2010.

Disturbance Data

The historical timeline of coral disturbances was constructed based partly on the study of Melbourne-Thomas et al. (2011) in the western Philippines. We reviewed threats to the reef from published reports, mostly included in coral status reports in the Philippines (Gomez and Alcala 1979; Gomez et al. 1981; Gomez et al. 1994; Licuanan and Gomez 2000; Nañola et al. 2006). For anthropogenic disturbances, we conducted a thorough literature search using ReefBase to determine when activities were initiated, increased, decreased, or ceased. Supplementary information was also obtained from the surveys conducted by Reef Check and interviews with reef scientists.

Results

Coral Cover

A total of 2,349 LHCC surveys were collated from studies conducted at 571 sites, 156 cities/municipalities, and 38 provinces between 1978 and 2010. To date, the inventory includes 1,239 more transects than the report on Philippines coral cover from the 1968 to 2004 Indo-Pacific study conducted by Bruno and Selig (2007). In the quartile category analyses, sites with repeated measurements were averaged, reducing the 2,349 surveys to 1,270 sites. The survey depths ranged from 2 to 12 m, and coral cover was estimated using various methods such as snorkeling and scuba diving, line and point intercepts, phototransects, and video transects (see Appendix). Most of the data collected were recorded in the 2000s, and data from this decade accounted for 71 % of the data, including the Reef Check monitoring that began in 1997. The greatest number of data points was from 2006, with 325 transects (13.8 % of the total); no coral cover data were recorded in 1987. Table 5.1 lists the data sources and corresponding number of data points gathered per decade.

The LHCC annual means ranged from 18.7 % in 1985 to 57.5 % in 2009, with an overall average of 33.8 % (Fig. 5.2), a slight increase from the reported LHCC of 32.3 % in the 1990s (Licuanan and Gomez 2000). Based on the annual means, the country's coral cover through the years appears to be stable, although the number of data points increased in the early 1990s, which may be attributed either to an increase in data collection or to the availability of online literature from this time (Fig. 5.2). This information also concurred with biogeographic annual means,

Table 5.1 Sources of substrate data

Source of data	1978–1989	1990–1999	2000–2010	Total
PhilReefs Series				**999**
PhilReefs (2003)	13	282	191	486
PhilReefs (2005)	1	63	314	378
PhilReefs (2008)	0	6	121	127
Reboton (2002)	3	5	0	8
Reef Check	0	30	653	**683**
ReefBase				**121**
Aliño et al. (1985)	2	0	0	2
Arquiza (1999)	0	27	8	35
De Jesus et al. (2001)	0	25	0	25
Hodgson and Dixon (2000)	1	0	0	1
Palaganas et al. (1985)	10	0	0	10
PRRCFI (2004)	0	0	22	22
Reboton and Calumpong (2000)	0	1	1	2
Ross et al. (2000)	0	4	2	6
Sotto et al. (2001)	0	11	0	11
UNESCO (2002)	0	7	0	7
Coastal Conservation Education Foundation				**300**
Christie et al. (2002)	4	8	0	12
Christie (2005)	0	0	7	7
White et al. (1999)	5	10	0	15
White et al. (2002)	10	13	25	48
White et al. (2004)	3	1	0	4
White et al. (2005)	0	22	20	42
White et al. (2006)	0	0	23	23
White et al. (2007)	8	26	45	79
White et al. (2008)	13	19	38	70
Coral Cay Conservation				**23**
Bochove et al. (2006)	0	0	15	15
Daly and Bochove (2009)	0	0	2	2
Solandt et al. (2003)	0	3	3	6
OneOcean.org				**89**
UPVFI (2005)	0	0	1	1
UPVFI (2007)	0	0	21	21
USAID/DA-BFAR FISH Project (2010)	0	0	62	62
Valle et al. (2000)	0	5	0	5
Boquiren et al. (2010)	0	0	8	8
De Guzman (2004)	0	0	8	8
Dygico (2006)	0	4	7	11
Raymundo et al. (2007)	0	0	2	2
Reef Check MPA Report (2006)	0	0	2	2
Russ and Alcala (1998)	2	7	0	9
Shuman et al. (2005)	0	0	4	4

(continued)

Table 5.1 (continued)

Source of data	1978–1989	1990–1999	2000–2010	Total
The Louis Berger Group Inc. (2008)	0	0	2	2
Vergara (2009)	0	0	63	63
White (1984)	25	0	0	25
Total	**100**	**579**	**1,670**	**2,349**

The English in this document has been checked by at least two professional editors, both native speakers of English. For a certificate, please see: http://www.textcheck.com/certificate/GOJB7J

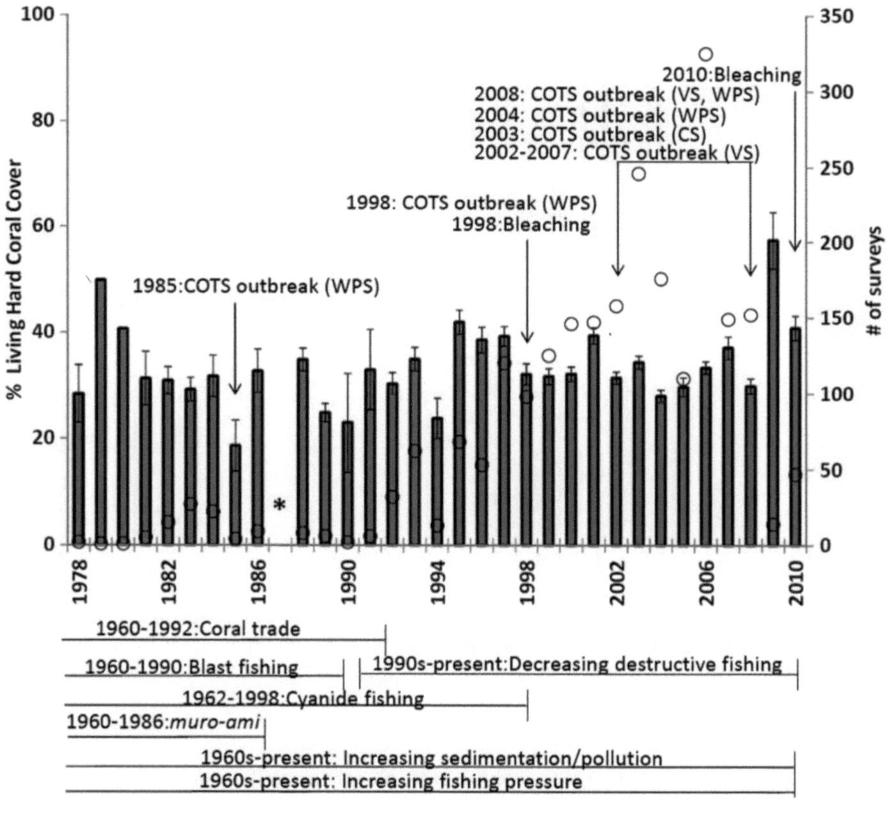

Fig. 5.2 Annual mean hard coral cover (%) from 1978 to 2010 and the historical timeline of major disturbances to Philippine reefs (except 1987 because of lack of data). *Error bars* denote one standard error of the mean. *Open circles* are the number of surveys per year (*right axis*)

wherein all regions were represented in the 1990s up to recent years, particularly NPS, SPS, and CS, for which 1980s data were lacking (Fig. 5.3). Among the regions, VS was the most surveyed, with 1,078 data points (~46 % of the total), followed by WPS with 566 (24 %); NPS was the least surveyed area with 50 (~2 %) data points.

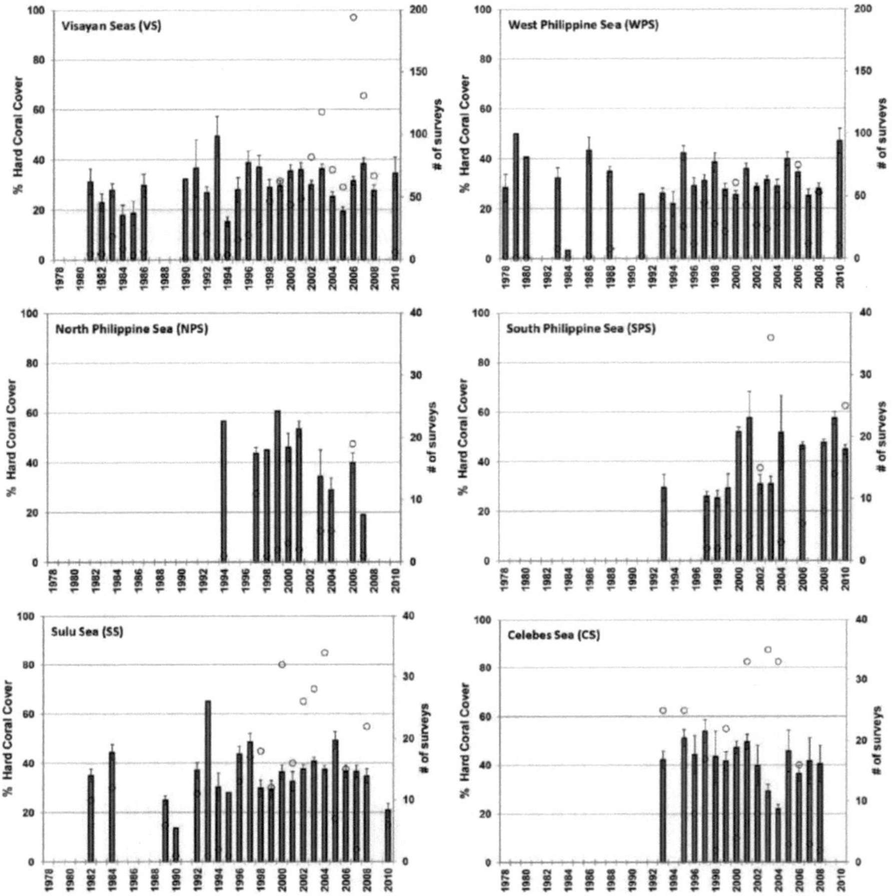

Fig. 5.3 Mean hard coral cover (%) from 1978 to 2010 in various biogeographic regions. *Open circles* denote the total number of surveys per year. Note the different scale in the number of surveys (right *y*-axis) in VS and WPS from the other regions

A similar trend was also observed based on the quartile categories. Overall, a large proportion of the sites (40–70 %) fell into the Fair category across time periods (Fig. 5.4). Poor sites steadily increased from 23.5 % in 1984–1996 to 39.8 % in 2005–2010; Good sites increased gradually from 13.8 % in 1997–2004 to 16.4 % in 2005–2010. A small percentage of sites fell into the Excellent category: 2.3 % and 2.9 % in 1984–1996 and 1997–2004, respectively, and the percentage of these sites declined significantly to 0.5 % in the period 2005–2010. Reef condition varied per biogeographic region across time periods, but the majority of the reefs was classified into the Fair category, except in VS (Fig. 5.5). In the VS region, the percentage of reefs in the Excellent condition increased from 1.9 % to 3.4 % between the periods 1984–1996 and 1997–2004, concomitant with a slight

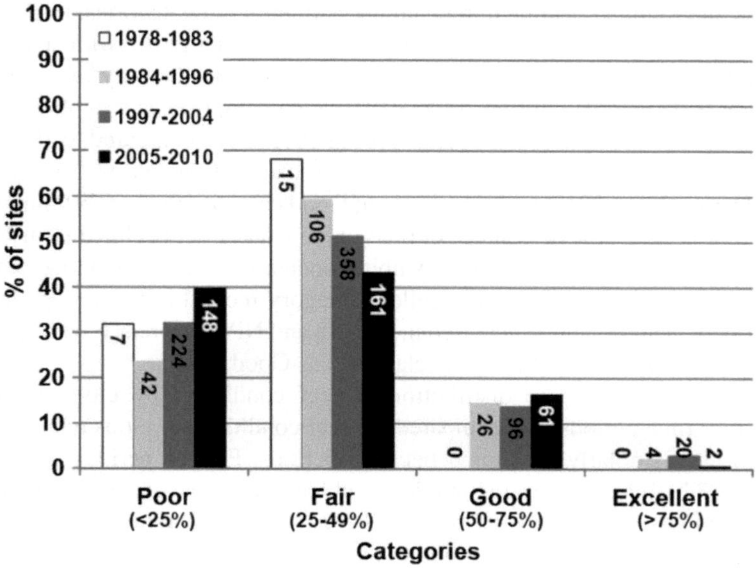

Fig. 5.4 Overall reef condition based on the quartile categories over four time periods. Values denote the number of sites covered per period

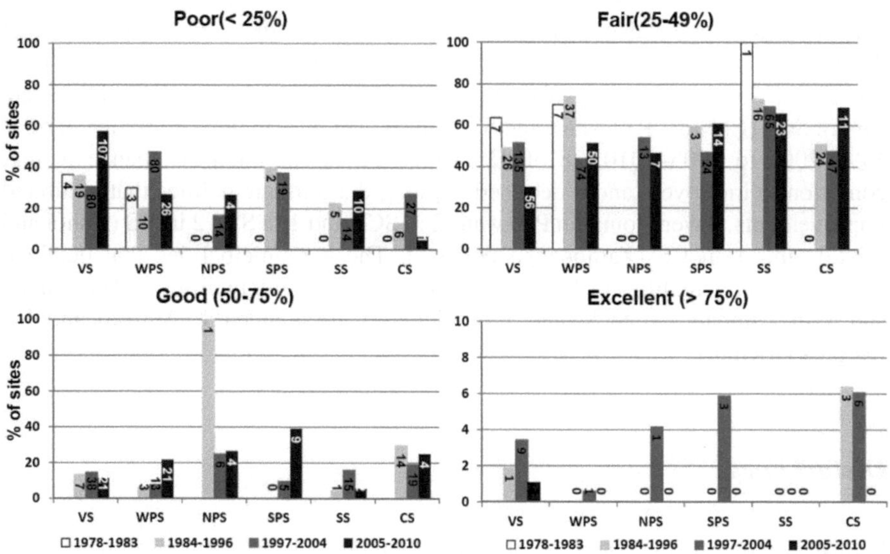

Fig. 5.5 Coral condition based on the quartile categories per biogeographic region over four time periods. Values denote the number of surveys covered

increase and minimal decline in reefs in the Fair and Poor categories, respectively. However, Poor reefs dominated at 58 % during 2005–2010, with a threefold reduction in reefs in the Excellent category. The VS was the only region to have Excellent reefs recorded during 2005–2010, which comprised 0.5 % of all sites. In contrast, WPS exhibited improvements during 2005–2010 compared to 1984–1996 and 1997–2004; in this area, reefs in the Poor category increased from 20 % to 47 %. During 2005–2010, the percentage of Good category reefs in WPS increased from 7.7 % to 22 %; the percentage of reefs classified as Poor declined from 48 % to 27 %. Similarly, both SPS and CS exhibited increases in Fair and Good category sites during 2005–2010, but no Excellent category reefs. In the same period, the ratio of Poor category reefs increased in both SS and NPS, whereas fewer sites in SS and slightly more sites in NPS were classified as Good.

Figure 5.5 presents the distribution of reef conditions by city/municipality during four time periods. We compared the reef conditions of cities/municipalities that had been similarly monitored between periods. For the periods from 1978–1983 to 1984–1996, the sites in both Poor and Fair conditions remained the same in 9 municipalities. From 1984–1996 to 1997–2004, of the 42 municipalities that had been similarly monitored, reef conditions did not change in 23 municipalities, improved in 12 municipalities, and declined in 7 municipalities. Of the 12 municipalities that improved, 9 were located in provinces of the VS region, namely Negros Oriental (4 municipalities), Bohol (2 municipalities), Cebu (1 municipality), Misamis Occidental (1 municipality), and Siquijor (1 municipality). Another 2 municipalities from SCS (Pangasinan and Zambales) and 1 from SS (Nueva Valencia, Guimaras) also exhibited improvement. Of the 7 municipalities that declined from 1997 to 2004, 3 were located in the CS region (Davao del Norte, Davao del Sur, and Saranggani); the rest were distributed in the VS (Bantayan, Cebu), SCS (El Nido, Palawan), and SS (Culasi, Antique) regions. From the periods 1997–2004 to 2005–2010, 28 of 42 municipalities exhibited no change in reef condition, 6 improved, and 8 declined. Among the 6 municipalities that exhibited improvements, 3 were found in Palawan (2 in SCS and 1 in SS), 2 in VS (Bohol and Leyte), and 1 in CS (Zamboanga del Sur). The 8 municipalities that declined included 3 municipalities in Tawi-tawi (SS region), 2 municipalities in Pangasinan, and 1 in Zambales in SCS, and 1 municipality each in NPS (Aurora) and VS (Negros Oriental).

Disturbances

Figure 5.2 presents the historical timeline of major reef disturbances in the Philippines from 1960 to 2010. The main causes of reef destruction during the 1970s to 1980s appear to be siltation or sedimentation resulting from deforestation and land development, followed by illegal and destructive fishing and coral harvesting for trade and construction (Gomez et al. 1981; Gomez and Alcala 1979). Similar threats were also predominant at varying scales in the 1990s, including blast fishing, cyanide fishing, siltation leading to pollution, overfishing, and massive bleaching from 1998 to 1999 (Licuanan and Gomez 2000). Reefs at Risk (Burke et al. 2002) estimated that more than 80 % of the reefs were threatened by overfishing and that 98 % of the reefs were at risk due to local human induced pressures. Indicators included coastal development (mining of sand and coral, dredging, land filling, coastal construction, and discharge of sewage), marine-based pollution (pollution from ports, oil spills and leakage, ballast and bilge discharge, and dumping), sedimentation (erosion potential), overfishing, and destructive fishing. No updated reports on issues and threats were published in the 2000s, with the exceptions of Tun et al. (2008) and Burke et al. (2011). Tun et al. (2008) identified an increase in coastal development and marine-based pollution and medium to high threats from overfishing and destructive fishing, and reported that sedimentation had changed from a medium to a low threat in the Philippines. The local integrated threats (combined impacts of overfishing and destructive fishing, marine-based pollution and damage, coastal development, and watershed-based pollution) are reportedly the most severe threats to Philippines reefs, particularly overfishing and destructive fishing, because a large number of people are dependent on the reefs (Burke et al. 2011).

Destructive Fishing and Overfishing

Various destructive fishing techniques have been used through the years, and destructive fishing practices have remained the major disturbance affecting the condition of the reefs in the country. *Muro-ami* is a Japanese fishing technique that was introduced to the Philippines before the Second World War, whereby swimmers pound the corals to drive fishes into nets. *Kayakas* is a modified form of *muro-ami* in which bamboo is used instead of rocks to frighten the fish; it is typically conducted in shallower reefs than *muro-ami* (Carpenter and Alcala 1977; Gomez et al. 1987). Commercial *muro-ami* was the leading fish-producing sector from 1965 to 1970, and *kayakas* has been regarded as equally productive (Carpenter and Alcala 1977). However, studies have shown that both operations resulted in reductions to coral cover through coral breakage, and induced overfishing because of the slow recovery of damaged corals and replenishment of fish stocks in frequently fished areas (see review by Gomez et al. 1987). Alcala and Gomez (1985) reported on four reefs that had been damaged by destructive fishing. Sumilon Island, Cebu was protected in 1970 until it was opened to fishing in 1984 and subsequently subjected to occasional *muro-ami* and dynamiting. The Selinog reef in Bohol Sea and Hulao-hulao in Panay Gulf also underwent occasional *muro-ami*, but at Apo Island, Negros Oriental, the activity ceased in 1983 and was banned in 1986 (Aliño et al. 2004). Blast fishing and sedimentation are the two main causes of coral destruction in Southeast Asia (Yap and Gomez 1985). Blast fishing is a destructive and wasteful method that damages the coral habitat, fish stocks, and nontarget marine organisms (Alcala 2000; Alcala and Gomez 1987). It started in the

1930s and became widely used after the Second World War because of the availability of ammunition. In Bolinao, Pangasinan, from late 1987 to mid-1989, the recorded frequency of blast fishing was ten blasts per hour (McManus et al. 1992). The incidence of blast fishing was observed to decrease or cease in the 1990s; this has been attributed to increased environmental awareness and the enforcement of laws, but was actually largely because of the depleted fishery stocks (Alcala 2000). Alcala (2000) reviewed the occurrence of blast fishing in the Philippines until the 1990s. Blast fishing was conducted in the following areas: northern Philippines (Babuyan Island group), western Philippines (Palawan except Puerto Princesa), Central Visayas (Danajon reef, Bohol and Mactan Island in 1997), Mindanao (Basilan and Zamboanga Peninsula, eastern Mindanao, Ozamis, Misamis Occidental), Sulu Archipelago (Turtle Islands), Sulu Sea (Tubbataha reef from 1991 to 1995). Areas with decreased activity or that were free of blast fishing included Negros Oriental (since 1980), western Visayas (Sagay, Negros Occidental), northwestern Luzon (Masinloc, Zambales and Bolinao, Pangasinan), western Philippines (Puerto Princesa, Palawan), and northern Mindanao (Camiguin Island, Dapitan, Zamboanga del Norte; Misamis Oriental, and Lopez Jaena, Misamis Occidental). Anecdotal evidence suggests that dynamite fishing still persists to the present, but at a localized level.

Cyanide fishing became popular in the marine aquarium fish trade and live reef fish food trade. The use of cyanide began in Batangas and Central Visayas in 1962, when the Philippines dominated ornamental fish exports (Rubec 1987, cited by Rubec et al. 2000). Sodium cyanide (NaCN) is a toxic chemical that is used to stun hard-to-catch fish which cannot be caught using nets. Approximately 70 % of aquarium fish caught in the Philippines have been reported to contain cyanide, resulting in high rates of delayed mortality among exported ornamental fish (Ochavillo et al. 2004). The strong demand for live fish for food in Hong Kong, Taiwan, and China brought about the live reef food fish trade (LRFFT), which also involved using cyanide rather than hook-and-line to catch fish (Licuanan and Gomez 2000). LRFFT collection areas included Coron Island and the Calamianes Group of Islands in northern Palawan, Balabac in Palawan, Surigao, Catanduanes, Camarines, and the Polillo Islands in Quezon (Mamauag 2004). Aquarium fish have been collected in Zambales, Ilocos region, Bohol, Leyte, Samar, Dipolog, Negros Oriental, Zamboanga, Siquijor, Polillo, Palawan, Bicol region, Masbate, Mindoro, Marinduque, Masbate, and Batangas (Albaladejo and Corpuz 1982). These highly selective fisheries significantly diminished the density of targeted species at collection sites in the Palawan and Cebu-Bohol areas (Ochavillo et al. 2004). Rubec et al. (2000) reported that the exposure of corals to low concentrations of cyanide impairs photosynthesis and causes a loss of zooxanthellae, resulting in coral mortality. By 1998, the percentage of ornamental fish testing positive for cyanide was reportedly reduced to 20 % (Rubec et al. 2000).

Overfishing is a pervasive problem linked to the prevalence of destructive fishing in the Philippines. According to Pauly et al. (1989), the overfishing problem in the Philippines can be defined as Malthusian overfishing, whereby fishermen opt to fish destructively as fishery resources are depleted by increasing numbers of fishers in the community. The decline in species richness and the absence of target fish species in the Visayas region is proof over a short ecological time scale of an ecosystem being degraded by overfishing and overexploitation (Nañola et al. 2010).

Coral Trade

The Philippines had been a major international supplier of stony corals for at least three decades (Mulliken and Nash 1993); Wood and Wells (1988) traced the country's trade dominance back to the 1950s. More than 13,000 tons of corals were estimated to have been exported for aquarium, ornamental, and surgical purposes (Mulliken and Nash 1993). In response to the overexploitation and poaching of corals, the government issued a ban on the harvest and commercial export of corals in 1977 and a Presidential Decree 1698 in 1980 (Gomez et al. 1994; Wood and Wells 1988; Ross 1984). However, commercial trading did not cease until the early 1990s, and its documented peak was between 1986 and 1989 (Mulliken and Nash 1993). This overexploitation reduced the abundance of harvested corals at collection sites (Ross 1984). Presently, coral harvesting is still prevalent in the form of poaching. Poaching is a persistent problem in areas such as Palawan, which has experienced high levels of foreign intrusion since the early 1970s. From 1995 to 2002 alone, 675 foreign fishers were caught illegally harvesting corals, giant clams, turtles, wrasses, and groupers, among other taxa (Benavente-Villena and Pido 2004).

Crown-of-Thorns Starfish (COTS)

The first observation of *Acanthasther* predation was in 1976 at a reef near Sumilon Island (Alcala 1976) COTS outbreaks were also observed in the 1980s in Alcoy and Bantayan Island in Cebu (Yap and Gomez, 1985). COTS feed on fast-growing corals and are responsible for the mass mortality of coral species in the Indo-Pacific (Miller 2002). The mechanism that triggers COTS outbreaks is still unknown (Burke et al. 2011), but siltation through nutrient runoff (Brodie et al. 2005; Fabricius et al. 2010) and the overharvesting of herbivores that prey on starfish (Dulvy et al. 2004) may contribute to the problem. Nañola et al. (2004) reported starfish outbreaks in 1985 (El Nido, Palawan), 1998 (Batangas), 2003 (Davao Gulf), and 2004 (El Nido, Palawan). We used Reef Check's COTS observations and plotted the annual abundance of COTS on Philippine reefs. The mean abundance of COTS per 100 m^2 per site was categorized as either low (<0.3), medium (0.30–0.75), or high (>0.75). The CRC Reef Research Center suggested that healthy reefs can sustain 30 COTS per hectare (10,000 m^2) or 0.30 per 100 m^2 of COTS; a density above this limit is considered an "active outbreak" (CRC Reef Research Center 2003). Here, we considered the "high" category to represent a

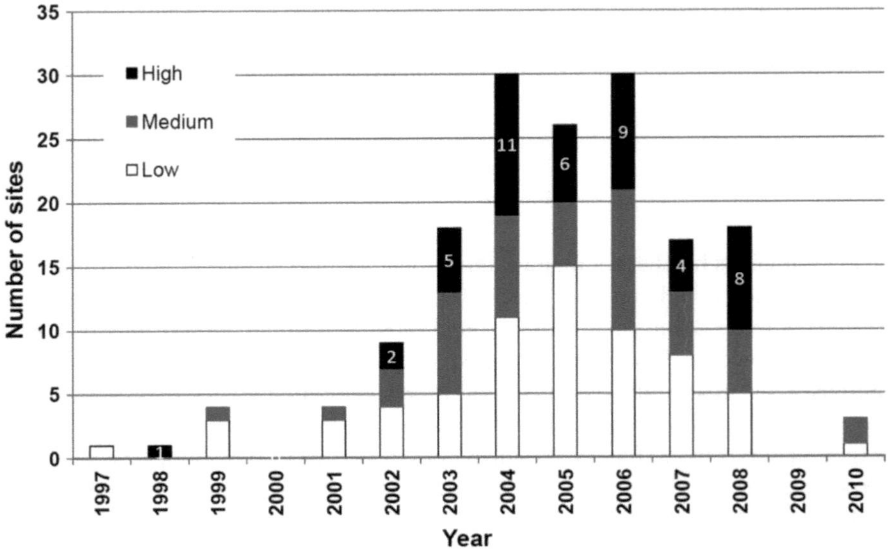

Fig. 5.6 Frequency of crown-of-thorns starfish (COTS), *Acanthaster planci* outbreaks from 1997 to 2010. Values denote the occurrence of high densities of COTS observed in the surveys conducted by Reef Check

COTS outbreak and identified the years with the greatest impacts from COTS (Fig. 5.6). The largest starfish outbreaks occurred in 2004 and 2006, and Cebu was consistently observed to have high densities of COTS between 2002 and 2008. Other areas with severe outbreaks included Palawan (1998, 2008), Bohol (2005, 2007), Negros Oriental (2002), Negros Occidental (2005), and Leyte (2008).

Terrestrial Influences

Reefs located near streams and rivers are most likely to be affected by episodes of perturbation because of high levels of exposure to suspended solids. Increased siltation from ashfall and volcanic mudflows from the Mt. Pinatubo eruption in 1991 substantially affected coastal areas in Zambales, killing approximately 40 % of corals near Capones Island (Pajaro et al. 1993). From 1997 to 1999, sedimentation caused by suspended solids from the river was the major threat reported in Sipalay, Negros Occidental (Beger and Harborn 2000). Siltation caused by man-made activities also poses a deleterious threat (Gomez et al. 1994). In Bacuit Bay, Palawan, reductions in coral cover were highly correlated with siltation caused by logging activities in 1986 (Hodgson 1989). High levels of mine tailing discharge caused siltation stress and reduced diversity in coral communities on reefs located 9 km from the outfall in Toledo, Cebu (Aliño 1984).

Anomalously High Sea Surface Temperatures

Coral bleaching was unknown in the Philippines until the global mass bleaching event caused by ENSO in 1998. Extensive elevations in sea surface temperature (SST) related to ENSO were also reported in 1983–1984 and 1987–1988 (reviewed by Arceo et al. 2001). However, only a single bleaching observation was reported in the Philippines, in June 1983 in Pangasinan (Yap et al. 1992), and one comprehensive study (Arceo et al. 2001) focused on the 1998 mass coral bleaching in the Philippines. Arceo et al. (2001) reported that the elevated temperatures from April to November followed a southward movement, with initial bleaching reports from the northwestern part of the Philippines in June 1998, and proliferated all over the country in August. They also found species of *Acropora* to be most susceptible to bleaching, with coral mortality rates reaching almost 50 %. Since then, bleaching has been monitored, and approximately 47 reports of observed bleaching were made from May to September 2010 in the Philippines (http://phcoralbleaching.crowdmap.com/). According to the World Meteorological Organization (WMO), 2010 had one of the warmest global temperatures ever recorded, similar to those in 1998 and 2005 (WMO 2010). In contrast to the 1998 bleaching, the 2010 bleaching was reported in various regions in the country, but the bleaching was observed to be patchy (W. Campos, personal communication).

Tropical Cyclones

The Philippine Area of Responsibility (PAR) is affected by an average of 19 tropical cyclones annually (Quibilan and Aliño 2006). These storms may not only kill or damage corals; they may directly or indirectly alter the physical and biological environment, affecting species recovery (Connell et al. 1997). On September 2, 1984, Typhoon Ike (local name, Nitang), with sustained winds of 220 kmh^{-1}, devastated the pristine reefs of Pescador Island, Cebu. The living coral cover on one reef was reduced from 52 % before the storm to 4 % 2 weeks after the typhoon (Alcala et al. 1986) and recovered to 45 % after 4 years (Alcala and Gomez 1990). Alcala and Gomez (1990) estimated that it would take 5 years for a typhoon-damaged reef to recover to a classification as Good. Anecdotal accounts from scuba divers referred to uprooted and overturned branching and foliose corals after Tropical Storm Washi (local name, Sendong) hit Negros Oriental on December 17, 2011. The storm devastated the reefs at Apo Island, and the sanctuary was closed to tourism activity to ensure recovery. Similarly, dead corals smothered with silt were observed on Poblacion reef in Dauin, Negros Oriental, which was attributed to heavy flooding brought by Tropical Storm Washi (J. Tanangonan, personal communication).

Discussion and Conclusions

Since the first national state of the reefs report, Philippines reefs have been considered to be influenced primarily by man-induced disturbances. The greater number of reefs in Poor and Fair categories in the 1976 survey (Gomez et al. 1981) and high percentage of reefs in the Fair category among the present data suggest that the trends are consistent. The historical timeline in Fig. 5.2 shows the predominant impact of anthropogenic activities through the years. Overfishing and destructive fishing still remain the most serious of the local threats to reefs (Burke et al. 2002, 2011). An estimated coastal population of 41 million (as of 2007) lives within 30 km of a reef and depends heavily on reef fisheries (Burke et al. 2011). The combined effects of local impacts and climate change currently pose a significant threat. Thermal stress associated with ENSO was evident from 1996 to 2006 (Peñaflor et al. 2009). The mass bleaching event in 1998 resulted in a slight decline in average coral cover from 39 % in 1997 to 32 %, which was maintained for 3 years before recovering in 2001 to 39 % (Fig. 5.2). Along with overfishing, destructive fishing, and thermal stress from climate change, the past decade has also seen reefs threatened by siltation and pollution associated with sewage, mining, mariculture, land clearing, deforestation, and agricultural fertilizer runoff, among other damage. From 2002 to 2008, COTS outbreaks were reported in various areas; these were particularly high in 2004 and 2006, but the most significant impact was in 2004 when the mean overall coral cover was at its lowest during the 2000s (Fig. 5.2).

The overall annual means revealed that, on a nation-wide scale, coral cover appears to be stable, with no major decline through the years. However, if the Poor and Excellent coral cover categories are used as a measure of habitat degradation, they may indicate that Philippine reefs have experienced habitat loss, particularly from 2005 to 2010, when there was a fivefold reduction in the Excellent category and an increasing trend toward the Poor category, a trend that has continued since 1984–1996. This inverse trend between the Excellent and Poor categories can also be observed in the VS region. The overall trends may have been influenced by trends in the well-surveyed VS region. The results may be skewed toward well-sampled areas and are dependent on the available data (Bruno and Selig 2007; Nañola et al. 2006; Nakao et al. 2009; Omori 2010), but the overall trends are consistent with previous reports on different areas in the 1990s (Licuanan and Gomez 2000) and in 2000–2004 (Nañola et al. 2006).

As the number of Excellent reefs has declined recently, trends in the Poor and Good categories may provide a better comparison across regions. From 2005 to 2010, the ratio of sites in the Poor category increased in the VS and SS regions while those in the Good category decreased. The WPS, SPS, and CS regions, in contrast, had positive growth in the percentage of sites classified as Good and significant declines in reefs classified as Poor. For NPS, both the Poor and Good categories increased, but the increase was more substantial in the Poor category

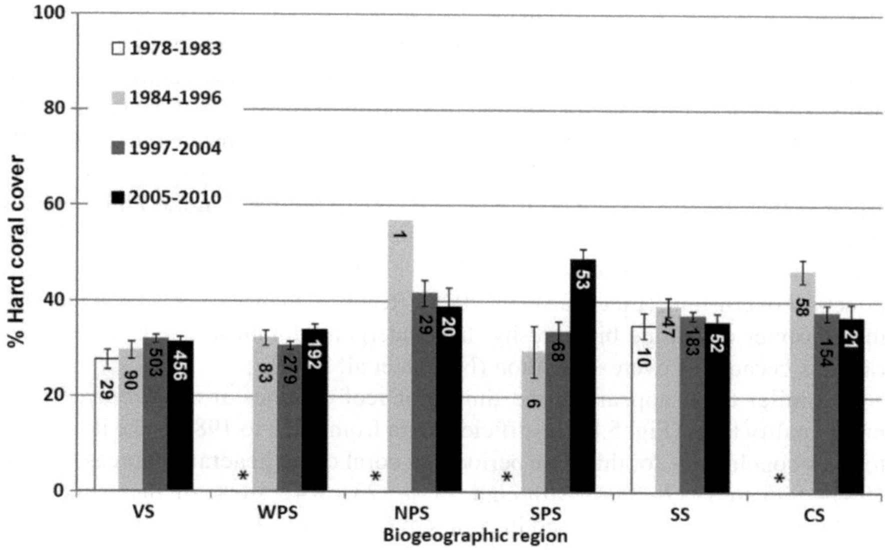

Fig. 5.7 Mean hard coral cover (%) in six biogeographic regions over four time periods. *Error bars* denote one standard error of the mean. *Asterisk* indicates no data

than in the Good category. The dominance of reefs classified as Poor and the decline of all other categories (Fair, Good, and Excellent) from 2005 to 2010 suggest that VS is at high risk. Reefs are expected to be in better condition in the WPS, SPS, and CS regions, particularly in SPS, where no reefs were classified as Poor in 2005–2010. This finding is comparable with the biogeographic regional means, which reveal that coral cover dropped in VS, SS, and NPS while WPS and SPS have improved in recent years (Fig. 5.7).

Because most of the surveys were undertaken in Marine Protected Areas (MPAs), the observed increase in coral cover may be linked to improvements in MPAs in the surveyed regions. The 1998 Republic Act 8,550 states that all coastal municipalities should set aside 15 % of their coastal areas as fish sanctuaries (Licuanan and Gomez 2000). The number of marine reserves, sanctuaries, and parks in the country grew from 565 in 2000 to 1,169 in 2007, concomitant with the observed expansion of the reserves from 10 ha to 11–100 ha (Arceo et al. 2008). However, these changes are still insufficient to ensure that more reefs can be categorized as Excellent; in 2005–2010, five biogeographic regions had no reefs classified as Excellent. Furthermore, only 30 % of MPAs were effectively managed (Alcala et al. 2008; Arceo et al. 2008). The VS region has the most marine reserves; the numbers increased exponentially from 1992 to 2005, and grew more slowly thereafter (Alcala et al. 2008). The timing of this slow increase in the number of marine reserves coincides with the noted increase in the percentage of reefs classified as Poor in 2005–2010 in the VS region. The rapid increase in MPAs until 2005 may have been reflected in increases in observed coral cover in 1997–2004, which, after the staggered increase in the number of MPAs and continuous degradation of reefs outside MPAs, led to a decline in observed coral cover in 2005–2010. Alcala et al. (2008) noted that fewer MPAs were established in 2006–2008 (40 marine reserves) than in 2000–2002 (177 marine reserves). Nañola et al. (2010) reported that the small number of effectively managed reefs in the Philippines is inadequate to override the high rate of overexploitation in VS (Weeks et al. 2009), which may need to protect at least 30 % of its coastal area to compensate for the effects of overfishing (Licuanan et al. 2008). Recent evidence has revealed that VS, the epicenter of marine biodiversity, has undergone diminished reef fish species richness because of overexploitation (Nañola et al. 2010).

A similar trend appears in the timing of reef declines at the surveyed city/municipality level (Fig. 5.8). Insufficient data from 1978 to 1983 make it difficult to draw conclusions for this time period, but coral cover generally increased from 1984–1996 to 1997–2004. Although 55 % (23 of 42 sites) of the reefs were stable, 28 % (12 sites) increased in coral cover and only 16 % (7 sites) decreased in coral cover. From 1997–2004 to 2005–2010, stable reefs increased to 67 %, but there was a higher proportion of declines (19 %) compared to improved reefs (14 %) and a substantial reduction in improved reefs, down from the previous 28 % to 14 %.

The results revealed that trends on the national and biogeographic levels may not be the same for all sites, so in future analyses it may be worthwhile to investigate individual reef sites using time series data to clarify general trends based on monitoring site. Preliminary findings revealed that the VS region is at high risk, which warrants immediate concern and conservation efforts. WPS and SPS have improved slightly, but in recent years no reefs have been classified as Excellent. Proposed conservation efforts entail improving the management of MPAs and increasing the number of no-take marine reserves. It is probably still important to educate coastal communities and local government about the importance of effectively managing the reefs to their benefit.

The risks to Philippine reefs are not limited to anthropogenic stress, but also include considerable damage from acute and chronic disturbances such as typhoons and wave action. The classification of reefs in uninhabited areas (such as the Kalayaan Island Group) as Poor is mainly the result of typhoons and monsoon conditions (Quibilan and Aliño 2006). Identifying the significant factors affecting the coral distribution in a particular area would be helpful to determine which areas require the most protection from human activities. At present, the selection of MPAs relies on "opportunistic" consideration, which requires scientific basis in designing the reserves (Licuanan et al. 2011). To this end, future research will involve sourcing additional data points from unpublished data, particularly from the 1980s, to enable long-term coral cover assessment. The analysis of coral cover change in the Philippines is a preliminary step toward an extensive analysis of coral cover on a regional scale, which may have implications for the potential role of Philippine reefs in ensuring coral biodiversity in Taiwan and Japan via the Kuroshio Current.

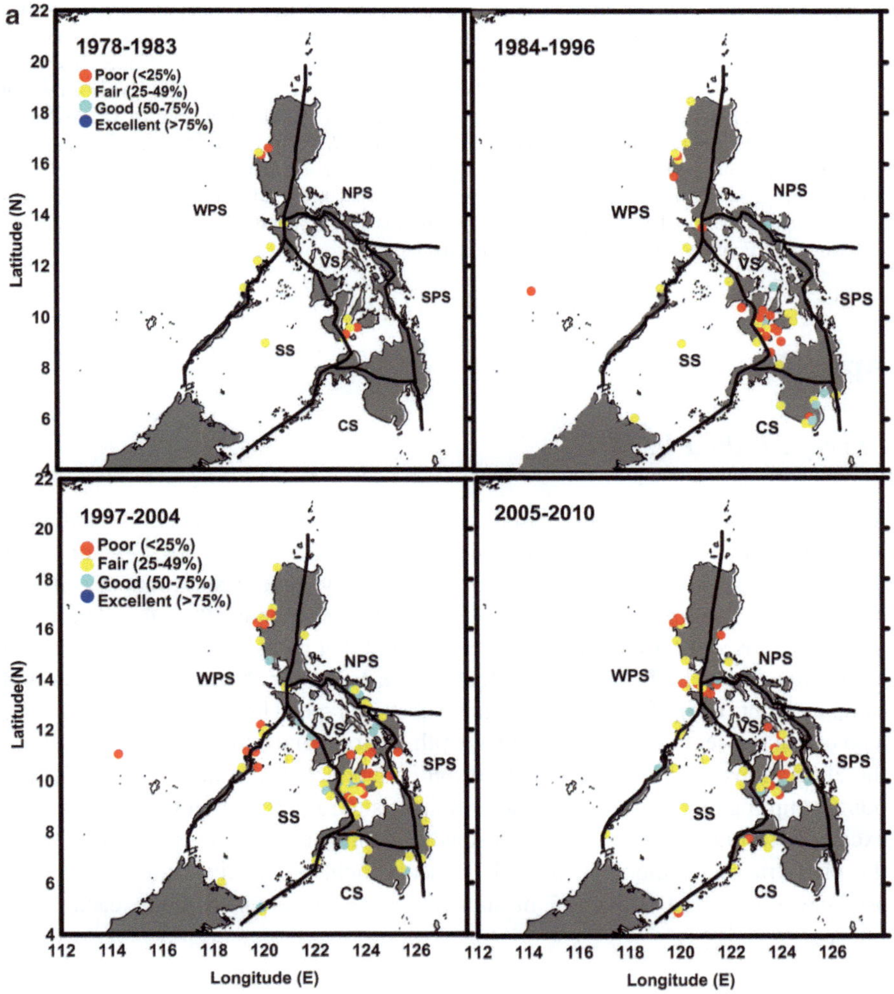

b

Change in condition	1978-1983 to 1984-1996	1984-1996 to 1997-2004	1997-2004 to 2005-2010
Stable	9	23	28
Improved	0	12	6
Decline	0	7	8
Total	9	42	42

Fig. 5.8 (**a**) Spatial distribution of reef condition based on quartile categories per city/municipality over four time periods. (**b**) Change in reef condition of city/municipality monitored between the periods

Acknowledgments We extend our gratitude to all who contributed to make this study possible. Special thanks to Reef Check, especially Director Gregor Hodgson and Ms. Jenny Mihaly for providing data. Thanks also to Coral Cay Conservation, PhilReefs (Marine Science Institute), Dr. Alan White, Dr. Cleto Nañola, Mr. Lambert Meñez, and Mr. Mark Vergara, for their prompt responses to our inquiries and for providing copies of their publications, and to Dr. Wilfredo Campos and Dr. Angel Alcala for granting us interviews. We also thank Dr. Jean Tanangonan, Dr. Rene Abesamis, Ms. Emily Layos, Mr. Melchor Deocadez, and the Hokkaido University Sustainable Low-Carbon Society Project for their assistance and support. This paper is a contribution to the S-9 Project and is supported by the Environment Research and Technology Development Fund, Ministry of the Environment, Japan.

Appendix

Survey Method Used

Most of the collated data were based on the line intercept transect (LIT) method to assess benthic communities. A transect line, typically a 10-, 20-, 30-, or 50-m measuring tape, is placed (randomly or permanently) on the reef along a constant depth contour and growth forms are estimated in situ under the tape. The point intercept transect (PIT) method is similar, except that measurements are conducted at specific intervals. For instance, Reef Check lays four 20-m transect lines and samples points every 50 cm to determine substrate type. The video transect method also uses a transect line placed along a depth contour, and a video is recorded along the entire transect. Substrate identification is conducted in a laboratory using the point sampling technique. The same principle is used in the phototransect method, except that a digital camera is used instead of a video camera. Vergara (2009) used the phototransect method in the 63 transects included in this study. A total of 50 images covering 0.4×0.5 m substrate areas were taken along each 50-m transect.

Data Sources

PhilReefs Series

Coral Reef Information Network of the Philippines (Philreefs) (2003) Philippine coral reefs through time (2003): workshop proceedings. Second of the atlas of Philippine coral reefs series. Coral Reef Information Network of the Philippines (PhilReefs), University of the Philippines Marine Science Institute, Quezon City

Coral Reef Information Network of the Philippines (Philreefs) (2005) Reefs through time: 2004 biennial report on the status of Philippine coral reefs. Coral Reef Information Network of the Philippines (PhilReefs), and the Marine Science Institute, University of the Philippines, Diliman, Quezon City

Coral reef information network of the Phillipine (PhilReefs) (2008) Reefs through time: initiating the state of the coast report. Coral reef information network of the Philippines (PhilReefs), MPA support network, Marine Environment and Resources Foundation Inc and the Marine Science Institute. University of the Philippines, Diliman, Quezon City

Reboton C (2002) Apo Island. In: Aliño P, Miclat E, Nañola C Jr, Roa-Quiaoit H, Campos R (eds) Atlas of Philippine coral reefs. Philippine coral reef information (Philreefs). Goodwill Trading Co., Inc., Quezon City, pp 132–136

ReefBase

Aliño P, Banzon V, Yap H, Gomez E, Morales V, Bayoneto R (1985) Recovery and recolonization on a damaged back reef area at Cangaluyan is (Northern Philippines). Proceedings of the 5th international Coral Reef Congress, vol 4, pp 279–284

Arquiza Y (1999) Rhythm of the sea: coastal environmental profile of San Vicente, Palawan. Coastal Resource Management Project, Cebu City

De Jesus E, Diamante-Fabunan D, Nañola C, White A, Cabangon H (2001) Coastal environmental profile of the Sarangani Bay Area Mindanao, Philippines. Coastal Resource Management Project, Cebu City

Hodgson G, Dixon J (2000) El Nido revisited: ecotourism, logging and fisheries. In: Cesar H (ed) Collected essays on the economics of coral reefs. CORDIO, Kalmar University, Sweden

Palaganas V, Sy J, Aliño P (1985) Coral communities of the Tubbataha Atolls. Proceedings of the 5th international coral reef congress, vol 6, pp 237–242

Philippine Reef and Rainforest Conservation Foundation, Inc. (2004) Danjugan Island Marine Reserve and Sanctuaries (DIMRS) Reef Monitoring Program. In: UNEP (ed) Monitoring coral reefs for better management schemes. UNEP EAS/RCU Small Grants Fund Programme, Bangkok, Thailand

Reboton C, Calumpong H (2000) How much damage do divers and snorkelers do to corals? Proceedings of the 9th international coral reef symposium, vol 2, pp 787–790

Ross M, White A, Sitoy A, Menguito T (2000) Experience from improving management of an "Urban" marine protected area: Gilutongan marine sanctuary, Philippines. Proceedings of the 9th international coral reef symposium, vol 2, pp 641–648

Sotto F, Gatus J, Ross M, Portigo M, Freire F (2001) Coastal environmental profile of Olango Island, Cebu, Philippines. Coastal Resource Management Project, Cebu City

UNESCO (2002) An ecological assessment of Ulugan Bay, Palawan, Philippines. CSI info N° 12, UNESCO, Paris

Coastal Conservation Education Foundation

Christie P (2005) Observed and perceived environmental impacts of marine protected areas in two Southeast Asia sites. Ocean Coast Manag 48:252–270

Christie P, White A, Deguit E (2002) Starting point or solution? Community-based marine protected areas in the Philippines. J Environ Manag 66:441–454

White A, Christie P, Apurado J, Meneses A, Ovenden M, Tesch S, White E (2002) Summary field report: coral reef monitoring in Cebu, Negros and Siquijor, Philippines, March 23–31, 2002. The Coastal Conservation and Education Foundation, Inc. and Coastal Resource Management Project, Cebu City

White A, Christie P, Meneses A, Maypa A, Stockwell B, Tesch S, White E, Mueller T (2004) Summary field report: coral reef monitoring expedition to Tubbataha Reefs National Marine Park, Sulu Sea, Philippines, April 3–11, 2004. The Coastal Conservation and Education Foundation, Inc. and Coastal Resource Management Project, Cebu City

White A, Maypa A, Tesch S, Stockwell B, Meneses A, White E, Mueller T (2005) Summary field report: coral reef monitoring expedition to Mabini and Tingloy, Batangas, Philippines, March 19–27, 2005. The Coastal Conservation and Education Foundation, Inc. and the Fisheries Improved for Sustainable Harvest (FISH) Project, Cebu City

White A, Maypa A, Tesch S, Meneses A, Stockwell B, White E, Martinez R (2006) Summary field report: coral reef monitoring expedition to Calamianes Islands, Palawan, Philippines, April 8–16, 2006. The Coastal Conservation and Education Foundation, Inc. and the Fisheries Improved for Sustainable Harvest (FISH) Project, Cebu City

White A, Meneses A, Tesch S, Maypa A, Stockwell B, White E, Martinez R (2007) Summary field report: coral reef monitoring in Bohol, Philippines, March 26–April 3, 2007. The Coastal Conservation and Education Foundation, Inc. and Fisheries Improved for Sustainable Harvest Project, Cebu City

White A, Maypa A, Tesch S, Diaz R, Martinez R, White E (2008) Summary field report: coral reef monitoring expedition to Tubbataha Reefs Natural Park, Sulu Sea, Philippines, March 26–April 1, 2008. The Coastal Conservation and Education Foundation, Inc., Cebu City

OneOcean.org

UPVFI (2005) Baseline assessment of existing or potential MPA's in FISH Project focal areas: Tawi-tawi. Technical Report, University of the Philippines-Visayas, Miag-ao, Iloilo

UPVFI (2007) Monitoring and baseline assessment of marine protected areas in the Fish Project focal areas: Tawi-tawi. Technical Report, University of the Philippines-Visayas, Miag-ao, Iloilo

USAID/DA-BFAR FISH Project (2010) Consolidated report: monitoring of the capture fisheries and marine protected areas (reef habitats) in the FISH Project's focal areas: Coron Bay, Danajon Bank, Lanuza Bay and Tawi-tawi. FISH Project Final Report, Tetra Tech EM Inc., Ortigas Center, Pasig City, p 143

Valle IS, Cristobal MCB, White AT, Deguit E (2000) Coastal environmental profile of the Malalag Bay Area, Davao del Sur, Philippines. Coastal Resource Management Project, Cebu City

Coral Cay Conservation

Bochove J, Wood O, Holman K, Head C, Raines P (2006) Coral reef resource assessment and management recommendations, Polillo Islands, Philippines. Coral Conservation, London

Daly R, Bochove JWV (2009) Napantao fish sanctuary report 2009. Coral Cay Conservation, Southern Leyte

Solandt J-L, Harding SP, Beger M, Dacles TP, Raines PS (2003) Effects of 1998 bleaching event on a large Pavona clavus colony in a Philippine marine protected area. Silliman J 44:19–32

Other Studies

Boquiren R, Di Carlo G, Quibilan M (2010) Climate change vulnerability assessment of the Verde Island passage, Philippines. Technical report. Conservation International, Arlington, Virginia, USA

De Guzman A (2004) A fishery in transition: impact of a community marine reserve on a coastal fishery in Northern Mindanao, Philippines. Research Report 2004 RR6. EEPSEA, Singapore

Dygico M (2006) Tubbataha reefs a marine protected area that works, a case study on the Philippines. WWF-Philippines, Quezon City

Raymundo L, Maypa A, Gomez E, Pablina C (2007) Can dynamite-blasted reefs recover? A novel, low-tech approach to stimulating natural recovery in fish and coral populations. Mar Pollut Bull 54:1009–1019

Reef Check MPA Report (2006) Hambongan marine sanctuary management report. Marine Aquarium Council, Reef Check, Conservation and Community Investment Forum

Russ G, Alcala A (1998) Natural fishing experiments in marine reserves 1983–1993: community and trophic responses. Coral Reefs 17:383–397

Shuman C, Hodgson G, Ambrose R (2005) Population impacts of collecting sea anemones and anemonefish for the marine aquarium trade in the Philippines. Coral Reefs 24:564–573

The Louis Berger Group Inc. (2008) Regional Impacts Projects (RIP) R-04 Tawi-tawi Bridge/Road Partnership Environmental Assessment Report. USAID's Growth with Equity in Mindanao (GEM-3) Program, Manila

Vergara M (2009) Coral community structure of the Bolinao reef system, Northwestern Philippines. M.S. thesis, University of the Philippines, Quezon City
White A (1984) Marine parks and reserves: management for Philippine, Indonesian and Malaysian coastal reef environments. Ph.D. dissertation, University of Hawaii

References

Albaladejo V, Corpuz V (1982) A market study of the aquarium fish industry of the Philippines: an assessment of the growth and the mechanisms of the trade. In: Gomez E, Birkeland C, Buddemeier R, Johannes R, Marsh J Jr, Tsuda R (eds) Proceedings of the 4th international coral reef symposium, vol 1, pp 75–81
Alcala A (1976) Population densities of the Crown-of-Thorns Starfish in two Philippine reefs. Silliman J 23:279–287
Alcala A (2000) Blast fishing in the Philippines, with notes on two destructive fishing activities. Silliman J 41:26–47
Alcala A, Gomez E (1985) Fish yields of coral reefs in Central Philippines. Proceedings of the 5th international coral reef congress, vol 5, pp 521–524
Alcala A, Gomez E (1987) Dynamiting coral reefs for fish: a resource destructive fishing method. In: Salvat B (ed) Human impacts on coral reefs: facts and recommendations. Antenne Museum E.P.H.E, French Polynesia
Alcala L, Gomez E (1990) Recovery of a reef from typhoon damage at Pescador Island, Cebu, Central Visayas, Philippines. In: Proceedings of the first ASEAN symposium on Southeast Asia marine science and environment protection. UNEP Reg Seas Rep Stud 116:105–109
Alcala A, Gomez E, Alcala L, Luchavez T (1986) Notes on the recovery of the coral reef at Pescador Island off Moalboal, Cebu, Philippines from typhoon damage. Silliman J 33:24–30
Alcala A, Gomez E, Yap H (1987) Philippine coral reefs: status and human responses to changes. Resour Manag Optim 4:297–340
Alcala A, Bucol A, Nillos-Kleiven P (2008) Directory of marine reserves in the Visayas Philippines. Foundation for the Philippine Environment and Silliman University-Angelo King Center for Research and Environmental Management, Dumaguete City
Aliño P (1984) The effect of mine tailings on the structure of coral communities at Toledo, Cebu. In: Koh L, Hew C (eds) Proceedings of 3rd symposium on our environment. National University of Singapore, Singapore
Aliño P, Gomez E (1994) Philippine coral reef conservation: its significance to the South China Sea. In: Yamamoto K, Ishijima S, Sakihara S, Taira H, Shimabukuro Z, Teruya F, Nishihira F (eds) Development and conservation in the Asia-Pacific region. Proceedings of regional conference east–west center association, pp 222–229
Aliño P, Nañola C, Campos W, Hilomen V, Uychiaoco A, Mamauag S (2004) Philippine coral reef fisheries: diversity in adversity. In: DA-BFAR (ed) In turbulent seas: the status of Philippine marine fisheries. Coastal Resource Management Project, Cebu City
Arceo H, Quibilan M, Aliño P, Lim G, Licuanan W (2001) Coral bleaching in the Philippines reefs: coincident evidences with mesoscale thermal anomalies. Bull Mar Sci 69:579–593
Arceo H, Aliño P, Gonzales R (2008) Where are we now with marine protected areas? In: Coral Reef Information Network of the Philippines (PhilReefs) (ed) Reefs through time 2008: initiating the state of the coasts reports. Coral Reef Information Network of the Philippines (PhilReefs), MPA Support Network, Marine Environment and Resources Foundation, Inc. and the Marine Science Institute, University of the Philippines, Diliman, Quezon City
Beger M, Harborn A (2000) Southern Negros development programme. Summary report to the municipality of Sipalay. Coral Conservation Cay LTD, UK

Benavente-Villena A, Pido M (2004) Poaching in Philippine marine waters: intrusion of Chinese fishing vessels in Palawan waters. In: DA-BFAR (ed) In turbulent seas: the status of Philippine marine fisheries. Coastal Resource Management Project, Cebu City

BFAR (2009) Philippine Fisheries Profile 2009. www.bfar.da.gov.ph. Accessed 31 Mar

Brodie J, Fabricius K, De'ath G, Okaji K (2005) Are increased nutrient inputs responsible for more outbreaks of crown-of-thorns starfish? An appraisal of the evidence. Mar Pollut Bull 51:266–278

Bruno J, Selig E (2007) Regional decline of coral cover in the Indo-Pacific: timing, extent, and subregional comparisons. PLoS One 2:e711

Burke L, Selig L, Spalding M (2002) Reefs at risk in South-East Asia. The World Resource Institute, London

Burke L, Reytar K, Spalding M, Perry A (2011) Reefs at risk revisited. World Resources Institute, Washington, DC

Carpenter K, Alcala A (1977) Philippine coral reef fisheries resources. Part II. Muro-ami and Kayakas reef fisheries, benefit or bane? Philipp J Fish 15:217–235

Carpenter K, Springer V (2005) The center of the center of marine shore fish biodiversity: the Philippine Islands. Environ Biol Fishes 72:467–480

Carpenter K, Abrar M, Aeby G, Aronson R, Banks S, Bruckner A, Chiriboga A, Cortés J, Delbeek J, DeVantier L, Edgar G, Edwards A, Fenner D, Guzmán H, Hoeksema B, Hodgson G, Johan O, Licuanan W, Livingstone S, Lovell E, Moore J, Obura D, Ochavillo D, Polidoro B, Precht W, Quibilan M, Reboton C, Richards Z, Rogers A, Sanciangco J, Sheppard A, Sheppard C, Smith J, Stuart S, Turak E, Veron J, Wallace C, Weil E, Wood E (2008) One-third of reef-building corals face elevated extinction risk from climate change and local impacts. Science 321:560–563

Connell J, Hughes T, Wallace C (1997) A 30-year study of coral abundance recruitment, and disturbance at several scales in space and time. Ecol Monogr 67:461–488

Coral Reef Information Network of the Philippines (Philreefs) (2003) Philippine coral reefs through time (2003): workshop proceedings. Second of the atlas of Philippine coral reefs series. Coral Reef Information Network of the Philippines (PhilReefs), University of the Philippines Marine Science Institute, Quezon City

CRC Reef Research Center (2003) Crown-of thorns on the Great Barrier Reef. http://www.reef.crc.org.au/publications/brochures/COTS_web_Nov2003.pdf. Accessed 4 April

De'ath G, Fabricius K, Sweatman H, Puotinen M (2012) The 27-year decline of coral cover on the Great Barrier Reef and its causes. Proc Natl Acad Sci 109(44):17995–17999. doi:10.1073/pnas.1208909109

Dulvy N, Freckleton R, Polunin N (2004) Coral reef cascades and the indirect effects of predator removal by exploitation. Ecol Lett 7:410–416

Fabricius K, Okaji K, De'ath G (2010) Three lines of evidence to link outbreaks of the crown-of-thorns seastar *Acanthaster planci* to the release of larval food limitation. Coral Reefs 29:593–605

Fortes M, Santos K (2004) Seagrass ecosystem of the Philippines: status, problems and management directions. In: DA-BFAR (ed) In turbulent seas: the status of Philippine marine fisheries. Coastal Resource Management Project, Cebu City

Gardner T, Cotê I, Gill J, Grant A, Watkinson A (2003) Long-term region-wide declines in Caribbean corals. Science 301:958–960

Gomez E (1997) Reef management in developing countries: a case study in the Philippines. Coral Reefs 16:S3–S8

Gomez E, Alcala A (1979) Status of Philippine coral reefs 1978. In: Proceedings of the international symposium on the marine biogeography and evolution in the southern hemisphere. DSIR Information Series, vol 137, pp 645–661

Gomez E, Alcala A, San Diego A (1981) Status of Philippines coral reefs—1981. Proceedings of the 4th international coral reef symposium, vol 1, pp 275–282

Gomez E, Alcala A, Yap H (1987) Other fishing methods destructive to corals. In: Salvat B (ed) Human impacts on coral reefs: facts and recommendations. Antenne Museum, French Polynesia

Gomez E, Aliño P, Yap H, Licuanan W (1994) A review of the status of Philippine reefs. Mar Pollut Bull 29:62–68
Hodgson G. (1989) The effects of sedimentation on Indo-Pacific reef corals. Ph.D. dissertation, University of Hawaii, Manoa, Hawaii
Licuanan W, Capili E (2004) New records of stony corals from the Philippines previously known from the peripheral areas of Indo-Pacific. Raffles Bull Zool 52:285–288
Licuanan W, Gomez E (2000) Philippine coral reefs, reef fishes, and associated fisheries: status and recommendations to improve their management. Global Coral Reef Monitoring Network (GCRMN) Report
Licuanan W, Mamauag S, Gonzales R, Aliño P (2008) The minimum sizes of fish sanctuaries and fishing effort reductions needed to achieve sustainable coastal fisheries in Calauag and Tayabas Bays. Philipp Agric Scientist 91:51–59
Licuanan W, Medina M, Luzon K, Samson M, Nañola C, Rollon R, Rolleda M (2011) Priority reef areas in the Pacific coast of the Philippines for marine protected area deployment. Philipp Agric Scientist 94:384–400
Mamauag S (2004) The live reef food fish trade in the Philippines. In: DA-BFAR (ed) In turbulent seas: the status of Philippine marine fisheries. Coastal Resource Management Project of the Department of Environment and Natural Resources, Cebu City
McManus J, Nañola C, Reyes R, Kesner K (1992) Resource ecology of the Bolinao coral reef system. ICLARM Stud Rev 22:117
Melbourne-Thomas J, Johnson C, Aliño P, Geronimo R, Villanoy C, Gurney G (2011) A multi-scale biophysical model to inform regional management of coral reefs in the western Philippines and South China Sea. Environ Model Softw 26:66–82
Miller I (2002) Crown-of-thorns and other coral predators. In: Best B, Pomeroy R, Balboa C (eds) Implications for coral reef management and policy: relevant findings from the 9th International coral reef symposium. USAID, Washington, DC
Mulliken T, Nash S (1993) The recent trade in Philippine corals. Traffic Bull 13:97–105
Nakao Y, Yamano H, Fujii M, Yamanaka Y (2009) Development and analysis of a coral cover database in seas close to Japan. J Jpn Coral Reef Soc 11:109–209 (in Japanese with English abstract)
Nañola C, Alcala A, Aliño P, Arceo H, Campos W, Gomez E, Licuanan W, Quibilan M, Uychiaoco A, White A (2004) Status report on coral reefs in the Philippines. Poster presented at the 10th ICRS in Okinawa, Japan. 28 June to 2nd July 2004
Nañola C, Aliño P, Arceo H, Licuanan W, Uychiaoco A, Quibilan M, Campos W, Alcala A, White A, Gomez E (2006) Status report on coral reefs of the Philippines, 2004. Proceedings of the 10th international coral reef symposium, pp 1055–1061
Nañola C, Aliño P, Carpenter K (2010) Exploitation-related reef fish species richness depletion in the epicenter of marine biodiversity. Environ Biol Fishes 90:405–420
Ochavillo D, Hodgson G, Shuman C, Ruz R (2004) Status of the Philippine marine aquarium fish trade. In: DA-BFAR (ed) In turbulent seas: the status of Philippine marine fisheries. Coastal Resource Management Project, Cebu City
Omori M (2010) Comment on "Development and analysis of a coral cover database in seas close to Japan". J Jpn Coral Reef Soc 12:49–51 (in Japanese with English abstract)
Pajaro M, Aliño P, Bermas N, Atrigenio M (1993) The impact of Mt. Pinatubo's eruption on the coral reef benthic communities in Zambales. Proceedings of the 7th international coral reef symposium, p 309
Pauly D, Silvestre G, Smith I (1989) On development fisheries and dynamite: a brief review of tropical fisheries management. Nat Resour Model 3:307–329
Peñaflor E, Skirving W, Strong A, Heron S, David L (2009) Sea-surface temperature and thermal stress in the coral triangle over the past two decades. Coral Reefs 28:841–850
Quibilan M, Aliño P (2006) Coral community structure of western Philipine reefs I: spatial patterns. Proceedings of the 10th international coral reef symposium, pp 341–350

Roberts C, McClean C, Veron J, Hawkins J, Allen G, McAllister D, Mittermeier C, Schueler F, Spalding M, Wells F, Vynne C, Werner T (2002) Marine biodiversity hotspots and conservation priorities for tropical reefs. Science 295:1280–1284

Ross M (1984) A quantitative study of the stony coral fishery in Cebu, Philippines. Mar Ecol Prog Ser 5:75–91

Rubec P, Cruz F, Pratt V, Oellers R, Lallo F (2000) Cyanide-free, net-caught fish for the marine aquarium trade. Live Reef Fish Inf Bull 7:28–34

Spalding M, Kainuma M, Collins L (2010) World atlas of mangroves. EarthScan, Washington, DC

Sweatman H, Delean S, Syms C (2011) Assessing loss of coral cover on Australia's Great Barrier reef over two decades, with implications for longer-term trends. Coral Reefs 30:521–531

Trono G (1999) Diversity of the seaweed flora of the Philippines and its utilization. Hydrobiologia 398/399:1–6

Tun K, Chou L, Yeemin T, Phongsuwan N, Amri A, Ho N, Sour K, Long N, Nañola C, Lane D, Tuti Y (2008) Status of the coral reefs in Southeast Asia. In: Wilkinson C (ed) Status of coral reefs of the world: 2008. Global Coral Reef Monitoring Network and Reef and Rainforest Research Center, Townsville

Veron J, Fenner D (2000) Corals (Zooxanthellate Scleractinia) of the Calamianes Islands, Palawan Province, Philippines. RAP Bull Biol Assess 17:24–26

Weeks R, Russ G, Alcala A, White A (2009) Effectiveness of marine protected areas in the Philippines for biodiversity conservation. Conserv Biol 24(2):531–540. doi:10.1111/j.1523-1739.2009.01340.x

White A, Cruz-Trinidad A (1998) The values of Philippine coastal resources: why protection and management are critical. Coastal Resource Management Project, Cebu City

Wood E, Wells S (1988) The Marine Curio trade conservation issues. A report for the Marine Conservation Society, Herefordshire

World Meteorological Organization (2010) 2010 in the top three warmest years, 2001–2010 warmest 10-year period. http://www.wmo.int/pages/mediacentre/press_releases/pr_904_en.htm. Accessed 20 April

Yap H (1997) Assessment of the present health of coral reefs in the Philippines. In: Grigg R, Birkeland C (eds) Status of coral reefs in the Pacific. UNIHI Sea Grant CP-98-01

Yap H, Gomez E (1985) Coral reef degradation and pollution in East Asian Seas region. In: Dahl A, Carew-Reid J (eds) Environment and resources in the South Pacific: a regional approach, vol 69, UNEP Reg Seas Rep Stud., pp 185–207

Yap H, Aliño P, Gomez E (1992) Trends in growth and mortality of three coral species (Anthozoa: Scleractinia), including effects of transplantation. Mar Ecol Prog Ser 83:91–101

Chapter 6
Zooxanthellate Scleractinian Corals of Jeju Island, Republic of Korea

Kaoru Sugihara, Hiroya Yamano, Kwang-Sik Choi, and Kiseong Hyeong

Abstract We report on seven zooxanthellate scleractinian corals (hermatypic or reef-building corals) found at the Jeju Island, off the southern coast of Korea. The Jeju Island, located at one of the northern limits of coral distribution in the East China Sea, has experienced significant sea surface temperature rises in the past century. In this chapter, we describe the systematic characteristics and identification keys of these seven species. This study thus provides a baseline not only for understanding coral community structure at the latitudinal limits of distribution, but also for examining ecosystem modification resulting from global warming.

Keywords Coral • Distribution • Identification • Jeju • Korea • Scleractinia • Taxonomy • Zooxanthellate

Introduction

Zooxanthellate scleractinian corals (hermatypic or reef-building corals; hereafter called "corals") are distinguished from azooxanthellate corals (ahermatypic corals) by their symbiotic algae (*Symbiodinium* spp.). With the aid of these symbiotic algae, they precipitate calcium carbonate ($CaCO_3$) much faster than azooxanthellate

K. Sugihara (✉) • H. Yamano
Center for Environmental Biology and Ecosystem Studies, National Institute for Environmental Studies, 16-2 Onogawa, Tsukuba, Ibaraki 305-8506, Japan
e-mail: sugihara.kaoru@nies.go.jp

K.-S. Choi
School of Marine Biomedical Science, Jeju National University, 102 Jejudaehakno, Jeju 690-756, Republic of Korea

K. Hyeong
Deep-Sea and Seabed Resources Research Department, Korea Institute of Ocean Science and Technology, Ansan, P.O. Box 29, Seoul 425-600, Republic of Korea

S.-i. Nakano et al. (eds.), *Integrative Observations and Assessments*, Ecological Research Monographs, DOI 10.1007/978-4-431-54783-9_6, © Springer Japan 2014

corals to create skeletons and three-dimensional structures. Therefore, they play a fundamental role in primary production and habitat formation for numerous other marine organisms.

Coral distribution is not limited to tropical and subtropical areas. Corals are distributed also in temperate areas, although they do not normally form coral reefs there (Veron and Minchin 1992; Kleypas et al. 1999; Yamano et al. 2012b). In regions affected by poleward warm currents, such as Korea and Japan, corals are distributed at higher latitudes than in other regions. Remarkable declines in coral diversity have been observed along a latitudinal gradient around Korea and Japan in response to northward-decreasing sea surface temperatures (SSTs) (Fig. 6.1a). SSTs have risen by 1.2 °–1.3 °C during the past 100 years around Korea and Japan (Takatsuki et al. 2007), in response to which poleward range expansions of corals have been documented, by a series of field surveys, at speeds up to 14 km/year during the past 80 years on temperate Japanese coasts (Yamano et al. 2011). Organisms associated with corals have also shown range expansions following the coral expansions (Yamano et al. 2012a). Accordingly, corals on the Korean coasts may also have experienced range extensions, as Kim et al. (2009) detected a recent range expansion of the seagrass species *Halophila nipponica* in Korea.

The Jeju Island is located approximately 90 km south of the Korean Peninsula. The island is ellipsoidal in shape, with the long and short axes measuring 74 and 32 km, respectively (Fig. 6.1). The Jeju Island formed from basaltic and trachytic volcanism from the Pliocene to Quaternary periods. At the center of the island, the volcanic Mt. Halla rises 1,950 m above sea level. Basaltic and trachytic lava flows and pyroclastic rocks cover most of the island's land surface, but the volcano is no longer active (Won 1976). The Jeju Island displays a temperate oceanic climate. The southern coast of the island is influenced by the Yellow Sea Warm Current (YSWC) (Fig. 6.1). The current system, however, is complex, and two origins have been proposed. The first is that the YSWC is a branch of the Tsushima Warm Current, and the second is that the YSWC is a branch of a current from Taiwan (Lie et al. 2000; Ichikawa and Beardsley 2002). SSTs on the south coast of Jeju Island ranged from 15 °C in February to 27 °C in August (Lie et al. 2000; Abu Affan and Lee 2004).

Although soft corals and azooxanthellate corals have been studied widely in Korea (Song 1991; Hwang and Song 2009), information on zooxanthellate corals has been scarce. At the Jeju Island, zooxanthellate corals have been commonly found only around Seogwipo on the southern coast. To date, Song (1991, 2004) and Song et al. (2003) have reported three zooxanthellate coral species (*Montipora trabeculata*, *Alveopora japonica*, and *Psammocora profundacella*), and Park and Choi (2001) also reported six zooxanthellate coral species [*Alveopora japonica*, *Alveopora* sp., *Montastrea* sp. (?), *Oulastrea* sp. (?), and two *Acropora* sp.], although the organism recorded in this study as *Montastrea* sp. may in fact have been *Zoanthus* sp. Recently, Denis et al. (2013) reported *A. japonica* beds at Biyangdo, located to the northwest of Jeju Island.

In this chapter, corals of Jeju Island are described systematically with identification keys, and the species described by previous studies were reexamined.

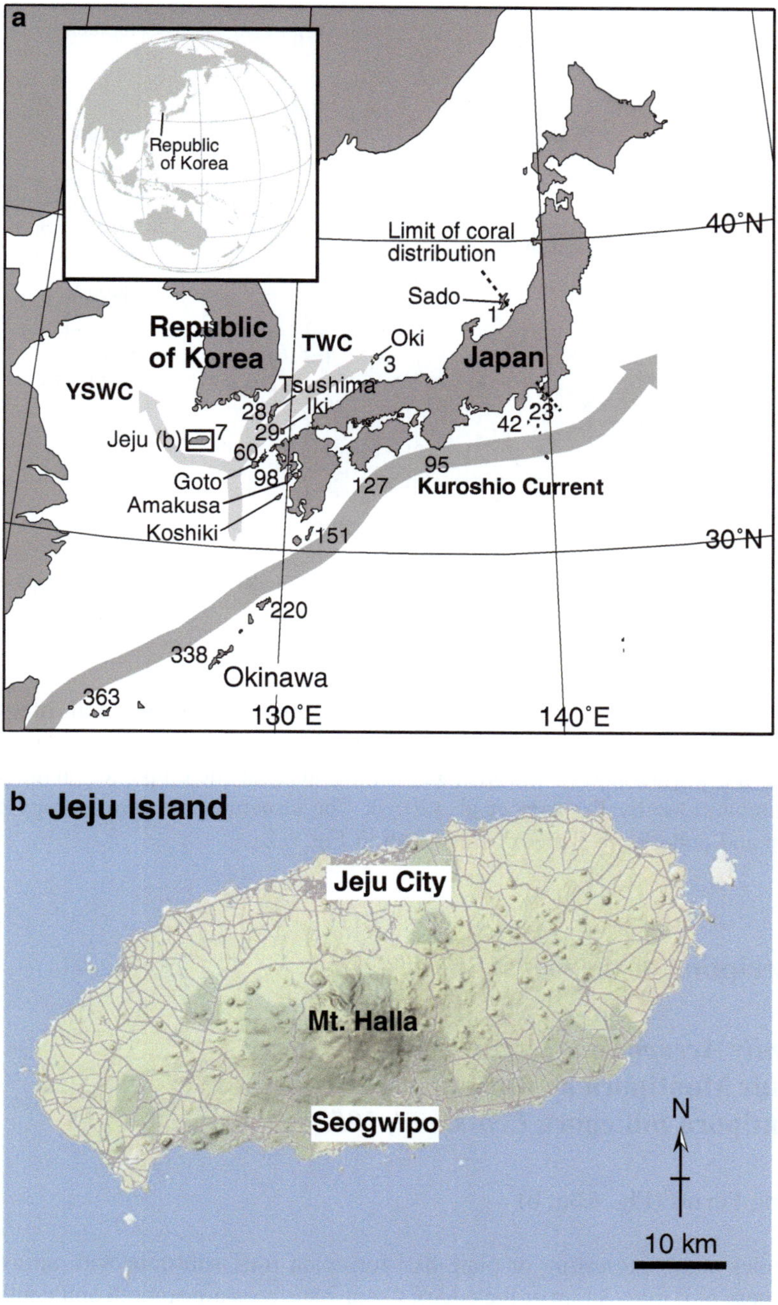

Fig. 6.1 (**a**) The Republic of Korea and a schematic figure of the current system that indicates the locations of the places referred to in this report. *Numbers* are numbers of coral species based on Veron and Minchin (1992; Amakusa Islands), Nomura et al. (1994; Oki Islands), and Honma and Kitami (1978; Sado Island), with updates by the first author (see Table 6.1). *YSWC* Yellow Sea Warm Current, *TWC* Tsushima Warm Current. (**b**) Close-up view of Jeju Island (Google Map image)

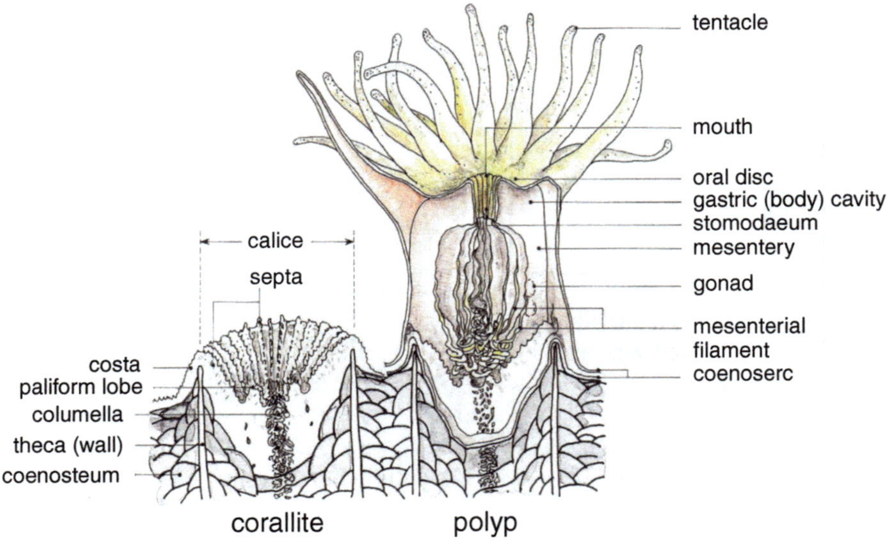

Fig. 6.2 Structure and terminology of the polyp and underlying skeleton. (Adapted from Nishihira and Veron 1995)

This study was accomplished through field observations of coral life forms (Fig. 6.1b) and examination of skeletal morphology based on the recent taxonomic framework formulated by Veron et al. (1996) and subsequent updates on the family Psammocoridae by Benzoni et al. (2010). The structure and terminology of the polyp and underlying skeleton are shown in Fig. 6.2.

Description of the Coral Species

Family Acroporidae Verrill, 1902
Genus Montipora *de Blainville, 1830*
Montipora millepora *Crossland, 1952*

Living Forms (Fig. 6.3a, b)

Colonies show encrusting or plate-like forms on rock substrate with an average thickness of 6 mm (3 mm at the edge). Colony surfaces are smooth and commonly include small nodular mounds with maximum dimensions of 5 mm in height and 10 mm in width (Fig. 6.3a). Colony edges are blunt, usually bent downward over the edge of the substrate, but upward at one corner of the colony edge. Colonies frequently reach 1 m in diameter. When bristle worms live in the skeleton, colony

Fig. 6.3 *Montipora millepora.* (**a**) Live colony with typical nodular mounds and a foliose edge free from the rock substrate. (**b**) Surface of a live colony with a large number of chimneys arising from the growth of bristle worms. (**c**) Corallum with a smooth surface. (**d**) Small corallites evenly distributed on the corallum surface. *Bar* 2 mm. (Photographs **a** and **b** provided by Chang Ho Lee)

shapes are submassive and contain large numbers of chimneys inside which worms reside (Fig. 6.3b). Colonies appear light green, dark green, or brown color in most parts except for the edges, which appear white.

Skeletal Features (Fig. 6.3c, d)

Corallites are small (usually <0.5 mm) and distributed uniformly on the colony surface more than a diameter apart from each other. Corallites have well-developed primary septa and a few septa in the second cycle. No columella exists. Some corallites are immersed in the coenosteum, but most are raised over the coenosteum surface as conical or roughly cylindrical mounds. The coenosteum has a reticulate shape with fine and pointed spinules on the surface.

Habitat

This species is commonly distributed on rock substrate at depths of 5–10 m in semi-exposed environments affected by wave action.

Distribution Around Korea

This species is common but inconspicuous in the Indo-Pacific region (Veron 2000).
Jeju Island: Song (1991, 2004) (as *M. trabeculata*), Park and Choi (2001, p. 59) (as *Acropora* sp.), Song et al. (2003) (as *M. trabeculata*); Iki Island: Yamano et al. (2004) (as *Montipora mollis*); Goto Archipelago: Sugihara et al. (2009); Amakusa Islands: Nishihira and Veron (1995).

Taxonomic Note

Montipora millepora is primarily distinguished from *Montipora efflorescens* Bernard 1897, which is the senior synonym of *M. trabeculata* Bernard 1897, by its colony forms, smaller corallites, and the absence of thecal papillae (Veron and Wallace 1984). Colony forms of *M. millepora* are encrusting or plate-like with small nodular mounds with maximum dimensions of 5 mm in height and 10 mm in width, whereas those of *M. efflorescens* are massive with a surface of irregular mounds developing into globular protuberances with maximum dimensions of 15 mm in height and 12 mm in width. Corallites of *M. efflorescens* are relatively larger (0.6–0.7 mm diameter) than those of *M. millepora* (<5 mm) and have well-developed thecal papillae covered with elaborated spicules.

Family Poritidae Gray, 1842
Genus Alveopora *de Blainville, 1830*
Alveopora japonica *Eguchi, 1965*

Living Forms (Fig. 6.4a, b)

Colonies are submassive, massive, or hemispherical. Most colonies measure 2–3 cm in diameter, but occasionally can grow to more than 10 cm in diameter. Colonies appear larger than their skeleton sizes because polyps with 12 tentacles are extended during the day. Polyp colors are usually pale brown or green. The tips of the polyp tentacles are white or cream colored and may be pointed or knob like in shape.

Skeletal Features (Fig. 6.4c, d)

Corallites are subrounded or polygonal, ranging from 2 to 5 mm in diameter. Corallite walls are thin, perforated, and pointed. Dentations are developed on the wall tops. Septa are not platy and consist of a vertical row of spines. The septa number 12; 6 of the primary septa are distinct and longer than the remaining secondary septa that lie between them. Columellae are rudimentary and may be formed by the well-developed primary septa at the centers of corallites.

Fig. 6.4 *Alveopora japonica*. (**a**) Live colony with long polyps extended during the day. (**b**) Polyp tentacles with white and knob-like tips. (**c**) Small corallum with hemispherical growth form. (**d**) Corallites having porous walls and aligned spiny septa. *Bar* 2 mm

Habitat

Colonies of *A. japonica* are found at depths of 5–25 m in turbid environments protected from wave action. They are sometimes densely distributed over a small area, probably because they are a hermaphroditic brooding species that releases larvae, which settle within several hours (Harii et al. 2001).

Distribution Around Korea

This species is endemic around Korea and Japan (Veron 2000). Large populations are found around Seogwipo and Biyangdo, Jeju Island.

Jeju Island: Song (1991, 2004), Park and Choi (2001, p. 57), Park and Choi (2001, p. 58) (as *Alveopora* sp. and *Acropora* sp.), Song et al. (2003) and Denis et al. (2013); Oki Islands: Nomura et al. (1994); Tsushima Archipelago: Eguchi (1973); Iki Islands: Eguchi and Fukuda (1972), Yamano et al. (2004); Goto Archipelago: Sugihara et al. (2009); Amakusa Islands: Nishihira and Veron (1995).

Fig. 6.5 *Psammocora albopicta*. (**a**) Live colony with a hillocky surface. (**b**) Live colony with a smooth appearance arising from the short and densely distributed tentacles on the colony surface. *Holes* indicate the outlets of siphons of boring bivalves. (**c**) Small corallites evenly spaced on the corallum surface. (**d**) Corallites surrounded by enclosed petaloid costosepta. *White arrows* indicate columellae made of styliform processes. *Bar* 2 mm. (Photograph **a** provided by Jin-Soo Kim)

Taxonomic Note

This species falls phylogenetically in the cluster of the family Acroporidae (Fukami et al. 2008), although the current family is Poritidae (Veron 2000). Rigorous examination based on traditional morphology and taxonomic treatment following the International Code of Zoological Nomenclature (ICZN) would be needed to move this species from the Poritidae to the Acroporidae as per Dai and Horng (2009a).

Family Psammocoridae Chevalier and Beauvais, 1987
Genus Psammocora *Dana 1846*
Psammocora albopicta *Benzoni, 2006*

Living Forms (Fig. 6.5a, b)

Colonies are mainly encrusting, following the surface of the underlying substrate. Knobby or nodular mounds up to several centimeters in size can be scattered on the colony surface. The colony diameter ranges from several centimeters to more than 1 m. Colony edges may be slightly detached from the bottom substrate. The colony surface is usually as smooth as velvet, occasionally undulating, or hillocky. The outlets of the siphons of boring bivalves may be scattered on the surface. Color ranges from yellowish green to dark brown. Tentacles with rounded tips are short and even and distributed densely on colony surfaces. Tentacles extend during the daytime.

Skeletal Features (Fig. 6.5c, d)

The corallum surface is basically smooth. Short and discontinuous collines may be irregularly developed on the corallum surface. Corallites are very small (approximately 1 mm in diameter), shallow, and evenly distributed over the corallum. The columella is made of a single styliform process and surrounded by enclosed petaloid or lamellar-shaped septa. Granular ornamentations are well developed on the upper septal margins, giving the corallum a hirsute appearance. The shape of the septa and development of septal ornamentation show a high degree of variation even within different parts of the same colony. Seven to nine septa reach the central fossa, and three to four of the septa are not fused with other septa.

Habitat

This species is commonly found on rocky foreshores deeper than 10 m with a regular tidal current.

Taxonomic Note

Psammocora albopicta is a new species that was described recently by Benzoni (2006). The same species found around Japan has been once identified as *Psammocora superficialis* Gardiner, 1898 in previous literature (Nishihira and Veron 1995). Benzoni et al. (2010) synonymized *P. superficialis* with *P. profundacella* Gardiner, 1898. This species is distinguished from *P. profundacella* and other species in the same genus by its smaller calice and columella, indistinct collines, and the number of corallites per square centimeter on the corallum surface (Benzoni et al. 2007).

Distribution Around Korea

This species is presently known from the Arabian Gulf, the Gulf of Aden, and Malaysia, Indonesia, Japan, and southeast Australia (Benzoni 2006).

Jeju Island: this study; Goto Archipelago: K. Sugihara (personal observation); Amakusa Islands: Nishihira and Veron (1995) (p. 199, middle photograph as *P. superficialis*).

Fig. 6.6 *Psammocora profundacella*. (**a**) Live colony showing oral discs of a different color from the other living parts. (**b**) Live colony with the polyp tentacles extended during daytime. (**c**) Corallum with an irregularly ridged surface. *Black arrows* indicate collines discontinuously developed on the corallum surface. (**d**) Lamellar and regularly spaced costosepta form a compact mesh. *Bar* 2 mm. (Photograph **b** provided by Chang Ho Lee)

Psammocora profundacella Gardiner, 1898

Living Forms (Fig. 6.6a, b)

Colonies are encrusting or submassive. The size of the colony rarely exceeds 15–20 cm in diameter in tropical and subtropical regions (Benzoni et al. 2010), but usually reaches 30 cm in temperate regions around Jeju and Japan. The edges of colonies are thin, undulated, and rarely detached from the bottom substrate. The colony surface shows a rugged appearance from the well-developed ridges or collines. Small nodular mounds are scattered on the colony surface. Tentacles extend during the day. The tentacles are usually short and slightly tapered to a rounded tip. The polyp color varies from light brown to dark green. The oral disc at the center of the corallite can have a different color from other living parts.

Skeletal Features (Fig. 6.6c, d)

Collines with either rounded or sharp upper edges are usually short and discontinuous on the surfaces of the corallum (exoskeleton of a coral colony). Corallites are single or a few corallites are arranged in a series between collines. The calice ranges in diameter from 2 to 3 mm. The columella is often reduced to a single styliform process within the calice. Septa can be petaloid or lamellar in shape and have well-developed granulations at their upper margins. The septa frequently fuse together, and the primary and secondary septa reach the calicular fossa. The septa-costae are regularly spaced, forming a compact mesh. The wall is formed basically by small rods or bars connecting opposed faces of adjacent septa, and forms boundaries between adjacent corallites.

Habitat

This species is commonly found on shallow and exposed rocky foreshores as well as in reef environments protected from wave action. Song (1991) collected the specimen from a water depth of 5 m.

Distribution Around Korea

This species is widespread throughout the Indo-Pacific (Veron 2000).

Jeju Island: Song (1991, 2004); Oki Islands: Nomura et al. (1994); Tsushima Archipelago: Eguchi (1973); Goto Archipelago: Eguchi (1971); Amakusa Islands: K. Sugihara (personal observation).

Taxonomic Note

Stefani et al. (2008) and Benzoni et al. (2010) synonymized *P. superficialis* Gardiner, 1898 with this species based on both phylogenetic and morphometric analyses.

Family Faviidae Gregory, 1900
Genus **Oulastrea** *Milne Edwards and Haime, 1848*
Oulastrea crispata *(Lamarck, 1816)*

Living Forms (Fig. 6.7a, b)

This species forms plocoid, encrusting, or submassive colonies. Colonies usually grow to only a few centimeters across, but sometimes reach sizes exceeding 10 cm.

Fig. 6.7 *Oulastrea crispata*. (**a**) Live colony encrusting bottom substrate. (**b**) Upper septal margins on the corallite walls appear to be white even when the polyps are extended. (**c**) Corallum showing the brown skeleton. (**d**) Corallites are subrounded and closely spaced; septa appear to be rough because of the well-developed spicules and granules on their surfaces. *Bar* 2 mm. (Photographs **a** and **b** provided by Chang Ho Lee)

This species is known as "zebra coral" because of its distinctive dark brown or black skeleton with white upper margins to the radiating septa. Pale-colored tentacles sometimes extend during the day.

Skeletal Features (Fig. 6.7c, d)

The corallum skeletons are dark brown or black in color. Corallites are subrounded and rounded in shape and are of uniform size ranging in diameter from 4 to 7 mm. They are closely compacted or widely spaced, usually with an alternation of long and short septa. Septa are usually thin and slope uniformly to the weakly developed columella. The septa number 30–40, about half of which reach the columella. Septal teeth and granules are well developed on the septal margins and on their faces, respectively. The paliform lobes common in other faviid corals are rarely developed. A high degree of latitudinal variation in skeletal morphology can be found in the West Pacific (Chen et al. 2011).

Habitat

This species usually grows on rocky substrate in the infralittoral zone and sometimes on foreshores at depths of 10–20 m.

Taxonomic Note

Fukami et al. (2008) indicated that this species was not phylogenetically close to species in the Faviidae but rather to those in the Fungiidae. However, taxonomic treatments following ICZN are needed to assign this species to the Fungiidae, although Dai and Horng (2009b) included this species in the Fungiidae based only on the phylogenetic results of Fukami et al. (2008).

Distribution Around Korea

This species is widespread throughout the Indo-West Pacific (Veron 2000). The highest latitude occurrence record is from Sado Island, Japan (38 °N, 138.5 °E) (Honma and Kitami 1978). It is known as a species that is tolerant to low SSTs, as it is associated with the stress-tolerant *Symbiodinium* clade D (Chen et al. 2003; Lien et al. 2007).

Jeju Island: Park and Choi (2001, p. 58); Oki Islands: Nomura et al. (1994); Tsushima Archipelago: Sugihara and Yamano (2004), Sugihara et al. (2009); Iki Islands: Yamano et al. (2004); Amakusa Islands: Nishihira and Veron (1995), Chen et al. (2011)

Family Mussidae Ortmann, 1890
Genus Acanthastrea *Milne Edwards and Haime, 1848*
Acanthastrea *cf.* lordhowensis *Veron and Pichon, 1982*

Living Forms (Fig. 6.8a)

Colonies are encrusting to submassive in shape and sometimes grow to approximately 20 cm across. Living colonies are fleshy and colorful in appearance. The polyps extending during the day are gray, red, purple, and green. The oral disc can also have a different color from the other living parts. Concentric wrinkles surrounding the oral discs may be recognized on the polyp surfaces.

Habitat

Acanthastrea lordhowensis appears to be distributed in shallow and sheltered environments.

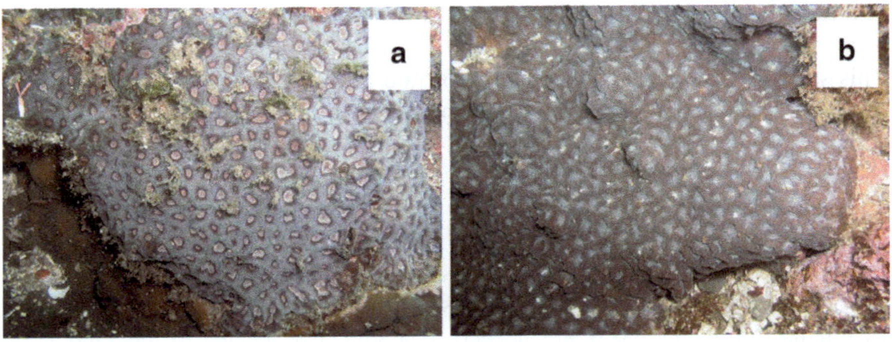

Fig. 6.8 (**a**) Live colony of *Acanthastrea* cf. *lordhowensis* showing fleshy polyps with concentrated wrinkles within the calices. (**b**) Live colony of *Favites* ? sp. with thick irregular walls. (Photographs provided by Chang Ho Lee)

Family Faviidae Gregory, 1900
Genus Favites *Link, 1807*
Favites *? sp.*

Living Forms (Fig. 6.8b)

Colonies are cerioid with encrusting and submassive forms. The colony may have rugged surfaces because of its thick walls with irregular upper margins.

Discussion and Conclusion

Although two corals could not be identified to species level because of lack of colony specimens, seven coral species were found at the Jeju Island, Republic of Korea. We found and described three species (*Psammocora albopicta*, *Acanthastrea* cf. *lordhowensis*, and *Favites* ? sp.) that have not been described in previous studies. A remarkable decrease in species numbers has been noted in Japan along a latitudinal gradient (Veron and Minchin 1992; Sugihara et al. 2009; Table 6.1, Fig. 6.9). Although a need remains for the detailed examination of two species (*A.* cf. *lordhowensis* and *Favites* ? sp.), of which only photographs were available, all seven species are expected to be found in the Iki-Tsushima region, and three of the seven (*A. japonica*, *P. profundacella*, and *O. crispata*) were found in the Oki Islands (Nomura et al. 1994; Sugihara et al. 2009; Table 6.1). The species composition of Jeju Island thus sits between those of the Iki-Tsushima region and the Oki Islands (Table 6.1, Fig. 6.9). The corals in the Jeju Island and the Oki Islands are distributed sparsely and do not form large communities. These features display a marginal feature at the end of the latitudinal gradient.

Table 6.1 Coral species list of Japanese sites (Goto Archipelago, Iki Islands, Tsushima Archipelago, and Oki Islands; see Fig. 6.1) close to Korea and Jeju Island

Family	Species	Goto Archipelago	Iki Islands	Tsushima Archipelago	Jeju Island	Oki Islands
Acroporidae	*Acropora hyacinthus*	+				
	Acropora japonica	+		+		
	Acropora muricata	+				
	Acropora pruinosa	+	+	+		
	Acropora solitaryensis	+		+		
	Acropora cf. *valida*	+				
	Montipora cf. *hispida*	+				
	Montipora informis	+	+			
	Montipora millepora	+			+	
	Montipora turgescens	+		+		
Agariciidae	*Leptoseris mycetoseroides*	+	+			
	Pavona decussata	+				
	Pavona explanulata	+				
Astrocoeniidae	*Stylocoeniella guentheri*	+	+	+		
Dendrophylliidae	*Turbinaria mesenterina*	+				
	Turbinaria peltata	+	+	+		
Faviidae	*Caulastrea tumida*	+	+	+		
	Cyphastrea japonica	+	+	+		
	Cyphastrea serailia	+	+	+		
	Favia speciosa	+	+	+		
	Favia aff. *veroni*	+				
	Favites abdita	+	+	+		
	Favites aff. *flexuosa*	+				
	Favites pentagona	+				
	Favites? sp.				+	
	Goniastrea aspera	+				
	Goniastrea australensis	+	+	+		
	Goniastrea deformis	+	+			
	Oulastrea crispata	+	+	+	+	+
	Oulophyllia crispa	+				
	Platygyra contorta	+	+	+		
	Plesiastrea versipora	+	+	+		
	Leptastrea pruinosa	+	+	+		
	Leptastrea purpurea?	+				
	Montastraea curta	+				
	Montastraea valenciennesi	+				
Fungiidae	*Lithophyllon undulatum*	+	+	+		
	Cycloseris explanulata	+				
	Cycloseris cyclolites			+		
Merulinidae	*Hydnophora bonsai*	+				
	Hydnophora exesa	+	+	+		

(continued)

Table 6.1 (continued)

Family	Species	Goto Archipelago	Iki Islands	Tsushima Archipelago	Jeju Island	Oki Islands
Mussidae	*Acanthastrea hemprichii*	+				
	Acanthastrea hillae	+				
	Acanthastrea lordhowensis	+	+	+		
	Acanthastrea cf. *lordhowensis*				+	
	Blastomussa wellsi	+				
	Cynarina lacrymalis	+	+	+		
	Lobophyllia robusta	+				
	Micromussa amakusensis	+				
Pectiniidae	*Echinophyllia aspera*	+	+	+		
	Echinophyllia cf. *orpheensis*	+	+	+		
	Mycedium elephantotus	+				
	Oxypora lacera	+				
	Pectinia ayleni	+	+	+		
Pocilloporidae	*Pocillopora damicornis*	+				
Poritidae	*Alveopora japonica*	+	+	+	+	+
	Goniopora djiboutiensis	+	+			
	Goniopora lobata	+	+	+		
	Porites heronensis	+	+	+		
Psammocoridae	*Psamocora profundacella*	+	+	+	+	+
	Psammocora albopicta	+			+	
Siderastreidae	*Coscinaraea monile*	+	+	+		
Trachyphilliidae	*Euphyllia ancora*	+				
Genus number		35	25	25	6	3
Species number		60	29	29	7	3

The Japanese species list was updated based on recent surveys by the first author (K. Sugihara) with reexamination of past published records: Eguchi (1971, 1973; Eguchi and Fukuda (1972; Nomura et al. (1994; Sugihara and Yamano (2004; Yamano et al. (2004; Sugihara et al. (2009)+, presence of the species

Tropical and subtropical corals are being degraded by environmental changes, including global-scale climate changes and local-scale anthropogenic influences (Hoegh-Guldberg et al. 2007; Carpenter et al. 2008). SST increases are a major cause of coral bleaching in tropical and subtropical areas, and temperate areas could serve as refugia for corals in an era of global warming as a result of poleward range expansion of their habitats (Yamano et al. 2011) (Fig. 6.9). The consequences of the changes to coastal ecosystems and biodiversity (Pratchett et al. 2009; Yamano et al. 2012a) are of primary concern. This report thus provides baselines not only to further the understanding of the distribution of corals at the latitudinal limits of

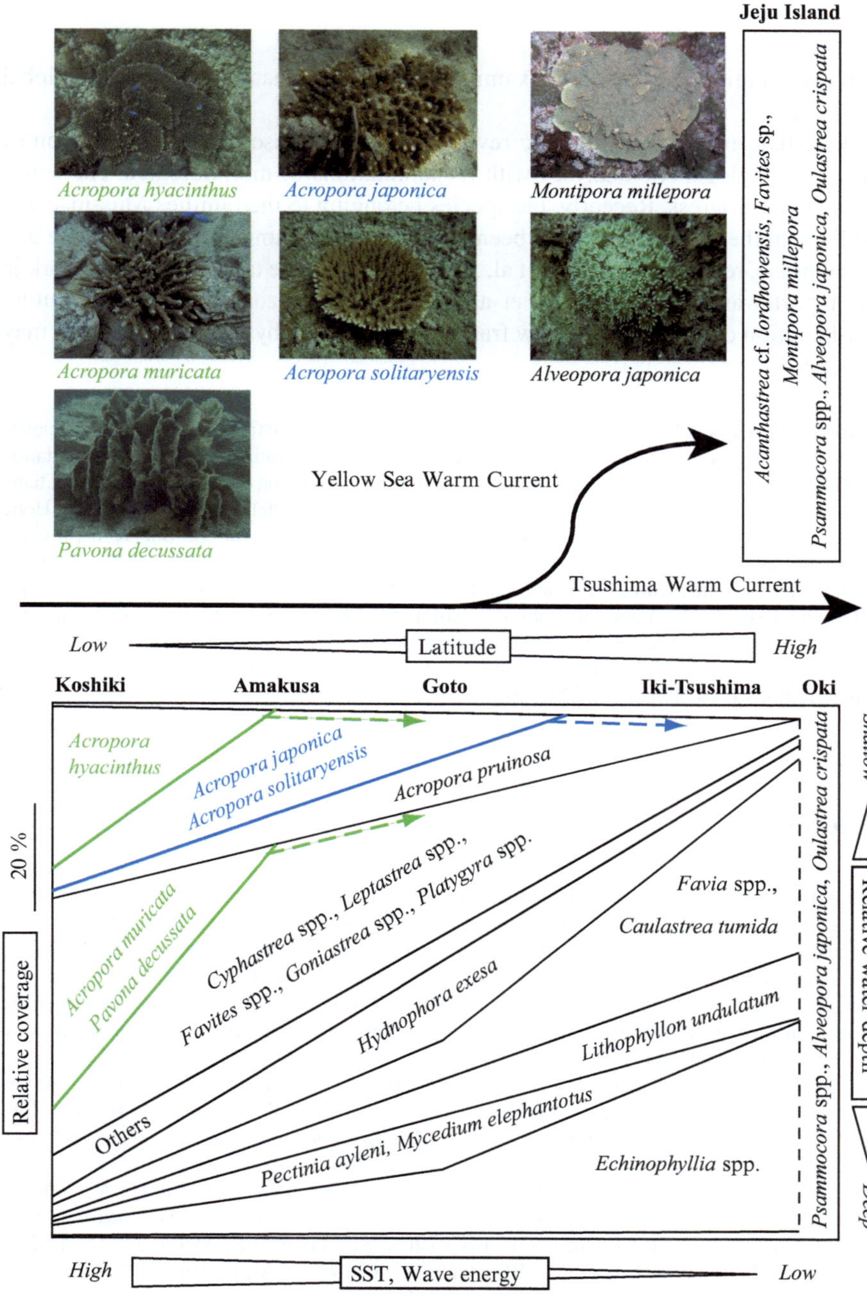

Fig. 6.9 Distribution of major coral species in Japanese sites close to Korea (Sugihara et al. 2009; Fig. 6.1) and Jeju Island (this study) along a latitudinal gradient with schematic current patterns. Jeju Island is located downstream of Yellow Sea Warm Current, a branch of Tsushima Warm Current (Fig. 6.1), and the species composition is between those of the Iki-Tsushima region and the Oki Islands. *Photographs* show representative coral species of the Koshiki Islands, the Amakusa-Goto region, and Jeju Island, from *left* to *right*. Species named in *green* and *blue* showed recent poleward range expansions in response to rising sea surface temperature (*SST*); *colored dashed arrows* indicate the expansions (Yamano et al. 2011)

their distribution but also to examine community changes caused by global warming.

Note that extensive taxonomic revisions of corals based on the integration of emerging phylogenetic analyses with reexaminations of morphological characteristics are in progress. Recently, the species belonging to the families Mussidae and Faviidae in the Indo-Pacific have been revised into the families Lobophyliidae and Merulinidae, respectively (Budd et al. 2012). Although the taxonomic framework in this report was based on Veron et al. (1996) and Benzoni et al. (2010), future consideration of the emerging new framework proposed by Budd et al. (2012) may be needed.

Acknowledgments This study was supported by the Japan Society for the Promotion of Science (JSPS) and the Korea Science and Engineering Foundation/National Research Foundation (KOSEF/NRF) under the Japan–Korea Basic Scientific Cooperation Program. We thank Chang Ho Lee, Sung-Hwan Cho, and Michiyo Shimamura for their field support. Hyun-Ki Hong translated Korean literature on corals into English. Chang Ho Lee and Jin-Soo Kim provided some of the photographs of living corals. This study is a contribution to a collaborative study on corals of the Tsushima/Korea Strait and Jeju Island between the Korea Institute of Ocean Science and Technology (KIOST) and the National Institute for Environmental Studies (NIES), the S-9 Project of Environment Research and Technology Development Fund, Ministry of the Environment, Japan, and KIOST/Ministry of Land, Transport and Maritime Affairs (MLTM) climate change projects (F64310404 and PM57090).

References

Abu Affan M, Lee J-B (2004) Seasonal characteristics of phytoplankton dynamics and environmental factors in the coast of Mara-do and U-do, Jeju Island, Korea. Algae 19:235–245

Benzoni F (2006) *Psammocora albopicta* sp. nov., a new species of Scleractinian coral from the Indo-West Pacific (Scleractinia; Siderastreidae). Zootaxa 1358:49–57

Benzoni F, Stefani F, Stolarski J, Pichon M, Mitta G, Galli P (2007) Debating phylogenetic relationships of the scleractinian *Psammocora*: molecular and morphological evidences. Contrib Zool 76:35–54

Benzoni F, Stefani F, Pichon M, Galli P (2010) The name game: morpho-molecular species boundaries in the genus *Psammocora* (Cnidaria, Scleractinia). Zool J Linn Soc 160:421–456

Budd AF, Fukami H, Smith ND, Knowlton N (2012) Taxonomic classification of the reef coral family Mussidae (Cnidaria: Anthozoa: Scleractinia). Zool J Linn Soc 166:465–529

Carpenter KE, Abrar M, Aeby G, Aronson RB, Banks S, Bruckner A, Chiriboga A, Cortés J, Delbeek JC, DeVantier L, Edgar GJ, Edwards AJ, Fenner D, Guzmán HM, Hoeksema BW, Hodgson G, Johan O, Licuanan WY, Livingstone SR, Lovell ER, Moore JA, Obura DO, Ochavillo D, Polidoro BA, Precht WF, Quibilan MC, Reboton C, Richards ZT, Rogers AD, Sanciangco J, Sheppard A, Sheppard C, Smith J, Stuart S, Turak E, Veron JEN, Wallace C, Weil E, Wood E (2008) One-third of reef-building corals face elevated extinction risk from climate change and local impacts. Science 321:560–563

Chen CA, Lam KK, Nakano Y, Tsai WS (2003) A stable association of the stress-tolerant zooxanthellae, *Symbiodinium* clade D, with the low-temperature-tolerant coral, *Oulastrea crispata* (Scleractinia: Faviidae) in subtropical non-reefal coral communities. Zool Stud 42:545–550

Chen KS, Hsieh HJ, Keshavmurthy S, Leung JKL, Lien IT, Nakano Y, Plathong S, Huang H, Chen CA (2011) Latitudinal gradient of morphological variations in zebra coral *Oulastrea crispata* (Scleractinia: Faviidae) in the West Pacific. Zool Stud 50:43–52

Dai CF, Horng S (2009a) Scleractinia fauna of Taiwan. I. The complex group. National Taiwan University, Taipei

Dai CF, Horng S (2009b) Scleractinia fauna of Taiwan. II. The robust group. National Taiwan University, Taipei

Denis V, Chen CA, Song JI, Woo S (2013) *Alveopora japonica* beds thriving under kelp. Coral Reefs 32:503

Eguchi M (1971) Scleractinia of candidate sites for marine parks in Nagasaki Prefecture. Report of Marine Parks Center, Nagasaki Prefecture, pp 19–31

Eguchi M (1973) Coral fauna of the Tsushima Islands. Nagasaki Prefecture. Report of Marine Parks Center, Nagasaki Prefecture, pp 45–56

Eguchi M, Fukuda T (1972) Invertebrata (chiefly coral fauna) of Marine Park of Iki Island, Nagasaki Prefecture. Report of Marine Parks Center, Nagasaki Prefecture, pp 45–58

Fukami H, Chen CA, Budd AF, Collins A, Wallace C, Chuang Y-Y, Chen C, Dai C-F, Iwao K, Sheppard C, Knowlton N (2008) Mitochondrial and nuclear genes suggest that stony corals are monophyletic but most families of stony corals are not (Order Scleractinia, Class Anthozoa, Phylum Cnidaria). PloS One 3:e3222

Harii S, Omori M, Yamakawa H, Koike Y (2001) Sexual reproduction and larval settlement of the zooxanthellate coral Alveopora japonica Eguchi ath high latitude

Hoegh-Guldberg O, Mumby PJ, Hooten AJ, Steneck RS, Greenfield P, Gomez E, Harvell CD, Sale PF, Edwards AJ, Caldeira K, Knowlton N, Eakin CM, Iglesias-Prieto R, Muthiga N, Bradbury RH, Dubi A, Hatziolos ME (2007) Coral reefs under rapid climate change and ocean acidification. Science 318:1737–1742

Honma Y, Kitami T (1978) Fauna and flora in the waters adjacent to the Sado Marine Biological Station. Niigata University, Annual Report of the Sado Marine Biological Station, Niigata, vol 8, pp 7–81

Hwang SJ, Song JI (2009) Sexual reproduction of soft coral, *Scleronephthya gracillimum* (Alcyonacea: Nephtheidae) based on long-term collection from Jejudo Island, Korea. Galaxea J Coral Reef Stud 11:155–167

Ichikawa H, Beardsley RC (2002) The current system in the Yellow and East China seas. J Oceanogr 58:77–92

Kim JB, Park J-I, Jung C-S, Lee P-Y, Lee K-S (2009) Distributional range extension of the seagrass *Halophila nipponica* into coastal waters off the Korean peninsula. Aquat Bot 90:269–272

Kleypas JA, McManus JW, Menez LAB (1999) Environmental limits to coral reef development: where do we draw the line? Am Zool 39:146–159

Lie H-J, Cho C-H, Lee J-H, Lee S, Tang Y (2000) Seasonal variation of the Cheju Warm Current in the northern East China Sea. J Oceanogr 56:197–211

Lien YT, Nakano Y, Plathong S, Fukami H, Wang JT, Chen CA (2007) Occurrence of the putatively heat-tolerant *Symbiodinium* phylotype D in high-latitudinal outlying coral communities. Coral Reefs 26:35–44

Nishihira M, Veron JEN (1995) Hermatypic corals of Japan. Kaiyu-sha, Tokyo

Nomura K, Kajimura M, Uchida H (1994) The reef-building corals in the Oki Islands. Mar Parks J 106:7–11

Park H-S, Choi S-S (2001) Guide book to marine life of Korea. Pungdeung, Seoul

Pratchett MS, Wilson SK, Graham NAJ, Munday PL, Jones GP, Polunin NVC (2009) Coral bleaching and consequences for motile reef organisms: past, present and uncertain future effects. In: van Oppen JH, Lough JM (eds) Coral bleaching: patterns, processes, causes and consequences. Springer, Berlin, pp 139–158

Song J-I (1991) The systematic study on the Korean Anthozoa 12. Order Scleractinia. Kor J Syst Zool 7:127–150

Song J-I (2004) Illustrated encyclopedia of fauna and flora of Korea, vol 39. Anthozoa. Daehan, Seoul

Song J-I, Kim J-H, Choi K-S (2003) Coral in coast of Seogwipo. Korea Environmental Industry Technology Institute, Seoul

Stefani F, Benzoni F, Pichon M, Mitta G, Galli P (2008) Genetic and morphometric evidence for unresolved species boundaries in the coral genus *Psammocora* (Cnidaria; Scleractinia). Hydrobiologia 596:153–172

Sugihara K, Yamano H (2004) Tsushima Archipelago. In: Ministry of the Environment and Japanese Coral Reef Society (ed) Coral reefs of Japan. Ministry of the Environment, Tokyo, pp 245–247

Sugihara K, Sonoda N, Imafuku T, Nagata S, Ibusuki T, Yamano H (2009) Latitudinal changes in hermatypic coral communities from west Kyushu to Oki Islands in Japan. J Jpn Coral Reef Soc 11:51–67

Takatsuki Y, Kuragano T, Shiga T, Bungi Y, Inoue H, Fujiwara H, Ariyoshi M (2007) Long-term trends in sea surface temperature adjacent to Japan. Sokko-Jiho 74:S33–S87

Veron JEN, Minchin PR (1992) Correlations between sea surface temperature, circulation patterns and the distribution of hermatypic corals of Japan. Cont Shelf Res 12:835–857

Veron JEN, Wallace C (1984) Scleractinia of eastern Australia. Part V. Family Acroporidae. AIMS, Townsville

Veron JEN, Odorico DM, Chen CA, Miller DJ (1996) Reassessing evolutionary relationships of scleractinian corals. Coral Reefs 15:1–9

Veron JEN (2000) Coral of the world. Volumes 1-3. Townsville, Queensland: Australian Institute of Marine Science

Won CK (1976) Study of petro-chemistry of volcanic rocks in Jeju Island. J Geol Soc Korea 12:207–226

Yamano H, Sugihara K, Nakai T, Yamagawa O (2004) Iki Islands. In: Ministry of the Environment and Japanese Coral Reef Society (ed) Coral reefs of Japan. Ministry of the Environment, Tokyo, pp 242–244

Yamano H, Sugihara K, Nomura K (2011) Rapid poleward range expansion of tropical reef corals in response to rising sea surface temperatures. Geophys Res Lett 38:L04601

Yamano H, Sugihara K, Goto K, Kazama T, Yokoyama K, Okuno J (2012a) Ranges of obligate coral-dwelling crabs extend northward as their hosts move north. Coral Reefs 31:663

Yamano H, Sugihara K, Watanabe T, Shimamura M, Hyeong K (2012b) Coral reefs at 34°N, Japan: exploring the end of environmental gradients. Geology 40:835–838

Chapter 7
Landscape Mosaicness in the Ocean: Its Significance for Biodiversity Patterns in Benthic Organisms and Fish

Takehisa Yamakita and Tadashi Miyashita

Abstract Although the importance of landscape mosaicness for biodiversity has been repeatedly stressed in terrestrial ecosystems, the significance of its application in the ocean has rarely been addressed. This application appears to be particularly important in East and Southeast Asia because the coastal landscape is complicated by the presence of many islands and peninsulas. Here we review basic differences in environments between marine and terrestrial areas. We then explore how the presence of multiple landscape elements enhances the diversity of marine organisms. The additive effect of multiple landscape elements on the biodiversity of benthic organisms and fish appears to be widespread. Nonadditive or emergent effects of multiple landscape elements have been found for groups of organisms exhibiting ontogenetic or daily habitat changes, particularly for fish in tropical coastal areas. Processes of determining the biodiversity of fish may be hierarchically organized by different factors at different spatial scales. We also report advanced research on the effect of the terrestrial landscape on fish in coastal ecosystems. We propose that it is important to consider spatial and temporal scales as well as land–ocean relationships. Because landscape analysis of marine areas is still at an early stage, theories and techniques developed in terrestrial ecosystems will help us understand general principles underlying the generation of biodiversity in the ocean.

Keywords Coastal ecosystem • Coral reef • East and Southeast Asia • Life history • Mangrove • Marine biodiversity • Seagrass • Seascape mosaicness • Spatial scale

---

T. Yamakita (✉)
Marine Biodiversity Research Program, Institute of Biogeosciences, Japan Agency for Marine-Earth Science and Technology, 2-15, Natsushima-cho, Yokosuka-city, Kanagawa 237-0061, Japan
e-mail: yamakitat@jamstec.go.jp

T. Miyashita
Department of Ecosystem Studies, The University of Tokyo, 1-1-1 Yayoi, Bukyo-ku, Tokyo 113-8657, Japan

S.-i. Nakano et al. (eds.), *Integrative Observations and Assessments*, Ecological Research Monographs, DOI 10.1007/978-4-431-54783-9_7, © Springer Japan 2014

Introduction

Consideration of interactions among organisms and environments at broad spatial scales is gaining importance for both scientific and social reasons. The Aichi Target, the goal of global conservation of biodiversity until 2020, proposed the establishment of protected areas that will help stop the loss of biodiversity (Convention for Biological Diversity CBD 2011). However, the method for estimating biodiversity at broad spatial scales while considering the landscape structure is a central issue. The SATOYAMA initiative, which aims to achieve sustainable human development harmonized with nature, was also adopted. In this initiative, human-mediated mosaic landscapes, typically found in traditional rural landscapes in Japan, are considered to enhance biodiversity (Kato 2001). Revealing the mechanisms for maintenance of biodiversity in heterogeneous landscapes is necessary for effective management of such human-mediated landscapes. The role of landscape mosaicness in biodiversity has been addressed before the adoption of such international targets. There are two primary processes by which biodiversity is enhanced in heterogeneous landscapes, namely, the additive effect and the nonadditive or emergent effect (Fig. 7.1) (Miyashita 2013). The combination of multiple ecosystems will increase the number of species in them because of different species compositions in different landscape elements. The combination of two ecosystems may also increase the evenness of species because the dominant species in a particular ecosystem will not dominate in another ecosystem. These effects can be considered to be additive effects derived from the combination of different ecosystems and are known in various organisms such as birds and butterflies (Kerr et al. 2001; St.-Louis et al. 2009). The emergent effect, "habitat complementation," is also found in species using different landscape elements in different life cycle stages or for different purposes such as nest sites and feeding sites (Dunning et al. 1992). The emergent effect may also be found in organisms that utilize various prey from different ecosystems, although the consumers themselves do not move, such as salmonid fish in mountain streams consuming both aquatic and terrestrial insects (Nakano and Murakami 2001). Not only composition (component) diversity but also configuration complexity influences biodiversity. The edge effect is one of the best known examples of landscape configuration that influences biodiversity both positively and negatively (Ries et al. 2004).

A limited number of studies have concerned the concept of "landscape" in marine ecosystems. More broadly, some researchers state that there are few important contributions of marine science to general ecology in recent years (Stergiou et al. 2005), partly because of the lack of theoretical considerations and the presence of communication barriers between marine ecologists and other ecologists (Webb 2012). However, these researchers have also pointed out that general papers can be published, regardless of whether the research is on terrestrial or marine areas. For example, the intermediate disturbance hypothesis proposed by Connell (1978) is well known as a unified model explaining biodiversity in both terrestrial and coastal ecosystems. Comparison of different

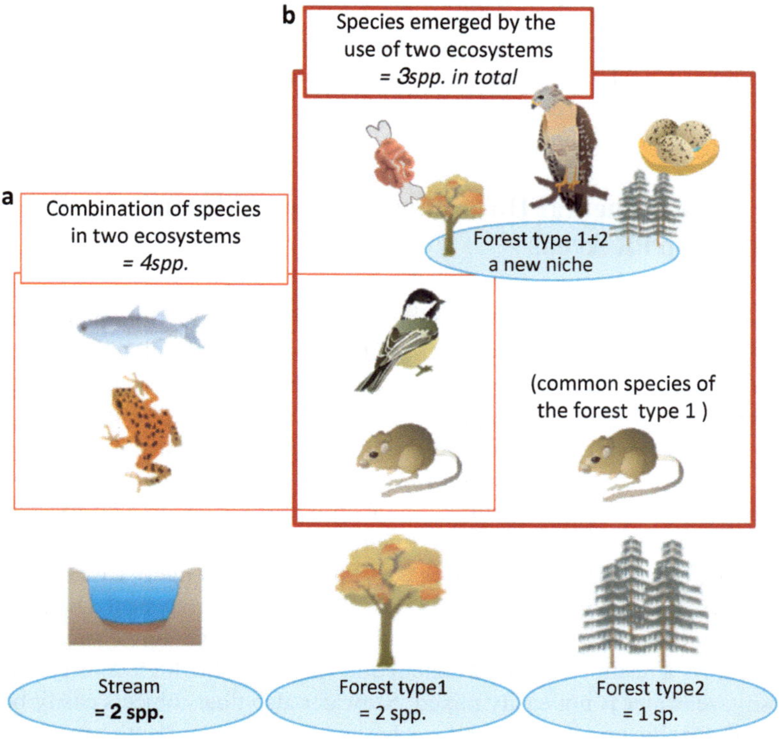

Fig. 7.1 Conceptual image of species diversity created by the mosaicness of the landscape. (**a**) Example of the additive effect of two landscape elements. Combination of forest and stream species increased the number of species in this landscape. (**b**) Synergetic effect of the combination of two landscape elements. In spite of the lack of the additive effect (presence of a common species), a new niche emerged for higher tropic levels by the combination of forests suitable for nesting and forests used as a feeding ground. [Images courtesy of the Integration and Application Network, University of Maryland Center for Environmental Science (ian.umces.edu/symbols/)]

systems using the same principles will help us discover the fundamental mechanisms for structuring and functioning ecosystems.

In this chapter, we first show the general differences in environments and the characteristics of organisms between terrestrial and marine areas. Following this, we summarize biodiversity research conducted in marine areas and discuss whether the techniques developed in terrestrial research can be applied to marine landscapes or whether new analytical tools applicable to marine landscapes are needed.

East and Southeast Asian countries harbor numerous islands and peninsulas. For example, Indonesia, the Philippines, and Japan are island countries, whereas Malaysia, Thailand, and South Korea are peninsular countries. These areas have inherently complex landscapes, including coastlines. In addition, the coastal areas in Southeast Asia are called the "Coral Triangle," and biodiversity in these areas is the highest in the world. Various organisms depend on coral reefs, and research therein is progressing. Therefore, in the latter half of this chapter, we propose suitable methods for analyzing the relationship between landscape complexity and biodiversity, referring to previous studies conducted in tropical areas.

Basic Environmental Differences Between Marine and Terrestrial Areas

Can landscape ecology be used in the ocean? Before entering into details, we will clarify the differences between the characteristics of terrestrial and marine areas. Marine areas cover 70 % of the earth surface, whereas terrestrial areas cover 30 %, more than 30 % of which is occupied by forests. Elevation of terrestrial areas ranges from 0 to 8,848 m, with an average of 840 m. In contrast, the depth of marine areas ranges from 0 to 10,920 m, with an average of 3,729 m. Terrestrial areas show complexity in geographic structures, and their surface is covered by various types of organic matter. In marine areas, relatively homogeneous environments are widespread, and inorganic substances cover the bottom in many cases. Moreover, seawater, a medium in marine areas, has the following physical characteristics compared with air: high specific heat results in fewer temperature changes in water over time. As a result of this heat conduction characteristic, high density, and high viscosity, seawater is not easily mixed. Seawater also floats objects easily because of its high specific gravity. Sunlight reaches only to a shallow depth because of high absorbance. Consequently, in marine areas, the stratification and gradient of physical environments prevail at smaller spatial and temporal scales compared with those in terrestrial areas. Interspecific interactions between organisms are relatively rare, and physical factors are typically more important for determining abundance and species richness of organisms in marine areas.

Because of the aforementioned environmental characteristics, many organisms in the ocean have a floating life form, particularly in larval stages. More than 80 % of benthic animals in shallow water are estimated to have a floating life form (Thorson 1964; Goshima 2003; Kinlan and Gaines 2003). This larval life stage is often a small-sized planktonic life form that enables long-distance dispersal. The number of larvae produced is large and they have high mortality rates, resulting in high variation in spatiotemporal dynamics of populations. Omnivorous species are dominant, particularly in secondary consumers, which is probably an adaptation to the high degree of fluctuation in prey population size (Andersson et al. 2007). Many taxa of various sizes exist in marine areas. This taxonomic diversity ranges from protozoa to invertebrates and is usually higher than that in terrestrial areas. Even plants are composed of several phyla, a characteristic that is rarely seen in terrestrial environments. In contrast, species diversity is generally higher in terrestrial areas. Among marine areas, species diversity is relatively high in coastal areas. In fish, species richness recorded in coastal areas is ten times higher compared with that recorded in other areas (Gray 1997; Angel 1993; Franklin 1993). These characteristics of environments and organisms are important for considering the relationship

Table 7.1 Difference in mosaicness between biodiversity in the ocean and on land (also see Carr et al. 2003)

		Ocean	Land
Landscape mosaicness			
	Dominant surface cover	Inorganic materials (sediment dominated)	Organic materials (forest dominated)
	Environment heterogeneity	Homogeneous	Heterogeneous
	Seasonality	Low	High
	Physical stress	Vertically structured	Dependent on geography and seasons
	Dispersal	Directed by currents	Dependent on geography and seasons
	Species distribution	Mainly caused by environment	Species interaction also plays important role
	Landscape constructed by organisms	Coral reef, seagrass/algae, some shells	Forests, grasslands
Biodiversity			
	Major floating fauna	Fishes, plankton (include larvae)	Birds, insects, bacteria
	Major benthic fauna	Shells, Polychaeta, etc.	Larvae, soil animals such as mites, earthworms
	Phylogenic diversity	High (35 phyla)	Low (11 phyla)
	Species diversity	Low (higher in coastal areas)	High (more than 400,000 beetles)
	Feeding characteristics	Dominant omnivores	Herbivores and carnivores (many specialists)

between landscape and biodiversity in the ocean. In Table 7.1, we have summarized how these characteristics are related to the landscape mosaicness–species diversity relationship.

Here we should define the word landscape. Landscape ecology emphasizes the interactions between spatial patterns and ecological processes, that is, the causes and consequences of spatial heterogeneity across a range of scales (Turner 1989). The landscapes on which we focus here are defined as the areas spatially heterogeneous in at least one factor of interest (Forman 1995; Turner et al. 2001). In some reports, landscape is considered to be a mosaic of ecosystems or habitats. Here we consider the elements that comprise mosaicness as landscape elements, irrespective of the spatial size or characteristics of the system. In terrestrial landscapes, mixtures of landscape elements such as forests and crop fields are considered as a landscape. In marine landscapes (also called "seascapes"), some researchers consider a seagrass bed as a landscape because of the mixture of sand and vegetation. Other researchers consider a coastal area, which is a combination of seagrass beds, coral reefs, and mangroves, as a landscape. This difference is based on the spatial scale and resolution of habitat classification. Here we refer to both types of classifications as examples of mosaic landscapes.

Effect of Mosaic Landscape on Life Forms of Benthic Marine Organisms

In terrestrial ecosystems, studies have revealed the effects of multiple landscape elements on organisms with low mobility. Plants and insects are examples comparable with benthic marine organisms. Although the mobility of plants is generally lower than that of benthic animals, passive transportation such as that by wind, birds, and insects is predominant in the propagule stage, which is similar to that seen in benthic organisms. After plant seeds are dispersed to a particular site, their ability to establish there is totally dependent on the local environment, which is also true for benthic organisms.

Are benthic organisms also affected by mosaic landscapes through changes in life history? Most benthic organisms change their life form after settlement on the ground (here we use "settlement on the ground" to describe "landing" to distinguish it from "establishment"). Although there is a difference in the length of periods among species, benthic organisms have a planktonic larval stage, with the exception of some species, such as some jellyfishes and snails, that are born directly. After landing on a substrate, benthic organisms spend a low-mobility period on an adhered object, inside bottom sediment, or on the substrata. Dispersal of benthic organisms at the planktonic larval stage is mainly passive. Factors influencing the transportation of larvae at large spatial scales include current flow at a scale of 1 km to tens of kilometers, whereas those affecting that transportation of larvae at the local scale include shore types or the coastline structure. After settlement, because the habitat where the larva can grow is restricted, the success of establishment is influenced by the local landscape structure. However, the larva itself is known to judge whether it is in an appropriate settlement location (Pineda 2000; Fukaya et al. 2010). Larvae actively perform up-and-down movements during their travel, and they also "lick" the surface of the landing location to decide whether they will remain on the location at the time of landing. As shown earlier, benthic organisms are affected by large-scale landscapes through physical processes. They are also affected by small-scale landscapes through interactions of local physical factors and their biological characteristics. Similar landscape effects derived from different spatial and temporal scales are observed in terrestrial ecosystems, for example, wind dispersal in plants and small insects. It is thus likely that knowledge from terrestrial research is widely applicable to benthic organisms.

Here we show an example of the distributions of benthos in a heterogeneous tidal flat (Fig. 7.2; Tsuchiya and Kurihara 1976). As shown in the upper half of the figure, landscape elements can be classified as sand, sandy mud, and mud. Alternatively, they can be classified as sand and mud. In any case, a combination of different landscape elements increased the number of species at the landscape level; this can serve as the simplest example of the additive effect of landscape mosaicness, which is probably widespread in marine areas.

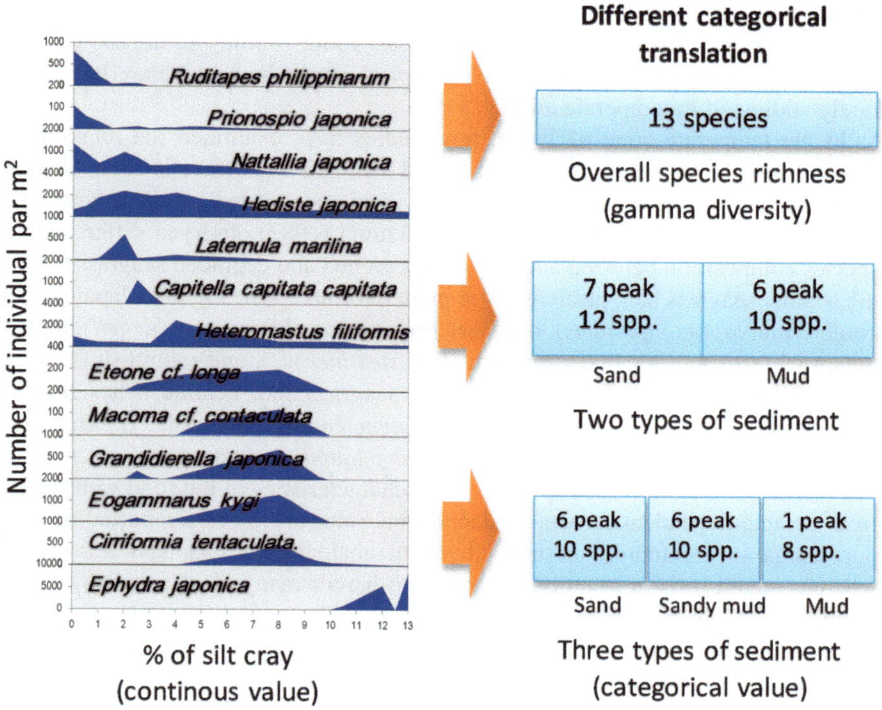

Fig. 7.2 Distribution of species in Gamou tidal flat. The number of species increases depending on categorization. (Modified from Tsuchiya and Kurihara 1976)

Compared with these additive effects of landscape elements on species richness, nonadditive effects of landscape elements are rarely reported in benthos (Minello et al. 1994). Several studies have shown that the colonization rate of little neck clams is higher at the edge of a seagrass bed than inside the bed because of larval accumulation by the physical setting (Bologna and Heck 2000; Heck and Valentine 2006). This observation suggests the importance of landscape heterogeneity; however, whether the little neck clam requires both seagrass and sandy areas for population persistence, a necessary condition for the nonadditive effect, remains unclear. To our knowledge, no study has shown that multiple landscape elements are essential for the completion of life cycles in benthic organisms.

Importance of Landscape Mosaicness for Temperate Coastal Fish

In terrestrial areas, the effects of the landscape on various organisms such as birds and butterflies are well known (Marini et al. 2009). In coastal areas, seagrass, algal beds, and coral reefs are known to have important functions as nursery areas of fish in general. The linkage between these ecosystems is expected to affect fish productivity and diversity (Nagelkerken et al. 2000; Mumby et al. 2004). Such topics have been recently advanced in tropical areas; however, they have been rarely addressed in temperate areas.

In the temperate zone of Japan, few studies have examined the relationship between large-scale landscape structures and fish. However, there is circumstantial evidence that commercially important individual species depend on seagrass and algal beds (Reviewed in Terawaki 1997). Azuma (1982) observed differences in species composition between a healthy seagrass bed and degraded seagrass bed. In the healthy seagrass bed, piscivorous or planktivorous species such as Japanese sea perch (*Lateolabrax japonicus*), black sea bream (*Acanthopagrus schlegeli*), red sea bream (*Pagrus major*), black rockfish (*Sebastes inermis*), and rabbitfish (*Siganus fuscescens*) are abundant. In the degraded seagrass bed, benthic fishes such as Richardson dragonets (*Repomucenus valenciennei*), poryfish (*Leiognathus nuchalis*), and a type of goby (*Acentrogobius pflaumii*) increased with decrease in piscivorous or planktivorous species. Thus, characteristics of the food web shifted from pelagic dependent to benthic dependent. Saito (1979) also reported that the pure sargasso community harbors abundant abalone, *Batillus*, and sea urchin, whereas mixed beds of seagrass and sargasso harbor many black rockfish, wrasse (*Halichoeres* spp.), black sea bream, spottybelly greenlings (*Hexagrammos agrammus*), and sea cucumber.

These examples suggest that mixtures of two types of landscape elements, namely, the difference in the status of seagrass beds as well as the difference in the bed-forming species, enhance species richness of fish at the landscape level, implying additive effects of landscape heterogeneity.

Some fish species appear to require different landscape elements to complete their life cycle, suggesting a nonadditive effect of landscape combinations on abundance and species richness. Kikuchi (1982) classified utilization patterns of seagrass beds by fish on the basis of seasonality and dependency (Table 7.2). For example, seasonal residents that utilize seagrass beds as breeding, nursery, or foraging sites may require different landscape elements to complete their life cycle.

The black rockfish is a well-known species that uses multiple landscape elements (Shoji 2009; Terawaki 1997). Newborn larval fish (~5 mm) spend a floating life stage offshore from February to April. After developing the ability to swim, they move into a bay and begin to inhabit a seagrass bed. When they become large-sized individuals, they shift their feeding habit from plankton to benthos and shift their home range to deeper waters. In the Tokyo Bay, they move to seaweed beds (sargasso beds) around June, and they supposedly move to rocky areas outside the bay after reaching maturity in December. This is an example of complementary usage of multiple landscape elements.

At a shorter time scale, daily or tidal-based cycles of using different landscape elements are well known (Sogard and Able 1994; Shoji and Tanaka 2005; Unsworth et al. 2008). For instance, carnivorous gray snapper (*Lutjanus griseus*) uses mangrove and other habitats at night and seagrass beds in the daytime, which was revealed by telemetry (Luo et al. 2009).

Table 7.2 Categorization of life history and seagrass dependency of fish (Kikuchi 1982)

1. Year-round resident species
White spotted pygmy filefish (*Rudarius ercodes*), redfin velvet fish (*Hypodytes rubripinnis*)
Sculpin (*Pseudblennius cottoides*), small gobies
Seaweed pipefish (*Syngnathus schlegeli*), seahorse (*Hippocampus coronatus*)
2. Seasonal resident species
Black rockfish (*Sebastes inermis*), Japanese surfperch (*Ditrema temminck*), black sea bream (*Acanthopagrus schlegeli*)
Japanese seaperch (*Lateolabrax japonicus*), red sea bream (*Chrysophrys major*), common Japanese conger (*Conger myriaster*)
Fat greenling (*Hexagrammos otakii*), spottybelly greenlings (*Hexagrammos agrammus*), multicolorfin rainbowfish (*Halichoeres poecilopterus*)
Squid, octopus, blue crabs, sea cucumbers
3. Intermediate between 2 and 4
Bigfin reef squid (*Sepioteuthis lessoniana*)
Crocodile flathead (*Inegocia guttata*)
Japanese silver-biddy (*Gerres japonicus*), ponyfish (*Nuchequula nuchalis*)
Blue swimming crab (*Portunus pelagicus*, *Portunus trituberculatus*), shore swimming crab (*Charybdis japonica*)
4. Occasional visitor species
Grass puffer (*Takifugu niphobles*), blue crabs
5. Incidentally visiting species
Other species

In conclusion, there are two different time scales at which different landscape elements are utilized by fish. However, in either case, the extent to which landscape heterogeneity enhances abundance and species richness of temperate fish communities is yet to be demonstrated (Wedding et al. 2008).

Importance of Mosaic Landscapes Focusing on Tropical Coastal Areas

An increasing number of studies have considered landscape factors influencing abundance and species richness of tropical fish communities; however, most of these studies are quite recent (Boström et al. 2011). Unsworth et al. (2008) showed the importance of the combination of multiple landscape elements on the diversity of fish in Wakatobi Marine National Park (MNP) of Indonesia. They pointed out that species richness and the biomass of fish in seagrass beds near mangrove forests were more than twice those in seagrass beds located far from mangrove forests (Fig. 7.3). In addition, they showed differences in species composition. Plankton feeders or vegetation feeders are abundant in seagrass beds located near mangrove forests; however, predatory and omnivorous fishes are dominant in seagrass beds near coral reefs, which indicates that compositional diversity of landscape elements is important for sustaining diverse fish communities.

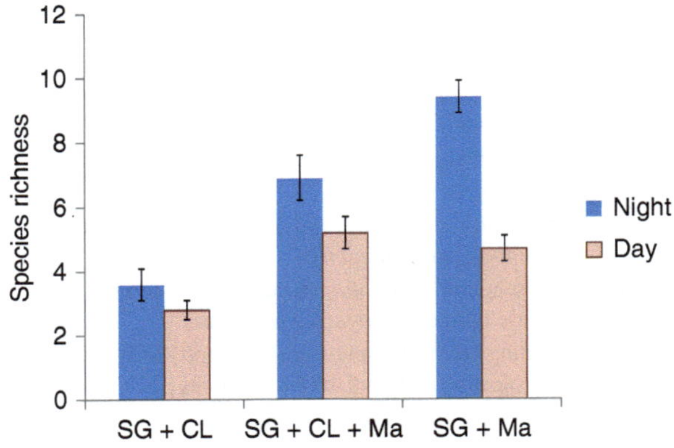

Fig. 7.3 Number of species in each combination of landscape elements (*SG*, seagrass; *CL*, coral leaf; *Ma*, mangrove). Data were extracted from Unsworth et al. (2008)

Another study conducted in coral reefs in the Caribbean Sea showed somewhat different results. Grober-Dunsmore et al. (2007, 2008) computed various landscape indices within a 100-m circle for each of 14 reefs and compared them with the distribution of fish. The indices were as follows: mean patch size, reef size, patch–edge ratio, patch diversity and richness, habitat diversity and richness, area of each habitat (bedrock, deep, algal plain, pavement, reef, sand, seagrass), and rugosity (a measure of roughness of the bottom surface). Stepwise multiple regression revealed that seagrass area had a positive effect on average species richness; however, habitat diversity was not effective for explaining either species richness or the biomass of fish. Thus, the effect of landscape was the result of the area of an individual landscape element and not the diversity of landscape elements.

The contradictory results of the two aforementioned studies (one showed a significant landscape effect, whereas the other did not) can be explained by the differences in (1) regional ecological characteristics, (2) classification categories of landscape elements, and (3) spatial scales considered. For example, mangrove forests appear to have little function as nursery areas for juvenile fish inhabiting coral reefs or seagrass beds in the Indian–Pacific area. The method of categorization may also be involved because coarse resolution could mask the effects of the fine-scale landscape structure (Kendall et al. 2011). However, the most plausible reason is the mismatch of spatial scales of observations (Mellin et al. 2007). When the spatial range (extent) of observations is changed, it becomes more difficult to accurately capture the overall dynamics of the focal system. When spatial resolution is changed, the landscape indices also change depending on the pixel size (Yamakita et al. 2011; Yamakita and Nakaoka 2011). Therefore, it is important to explicitly recognize appropriate spatial scales.

Pittman et al. (2004) explicitly considered the interactions of landscape elements at different spatial scales that may determine species richness and abundance of fish in the East Coast of Australia. Using a combination of aerial photographs, satellite images, sonar, and field surveys, they classified the landscape into 11 classes, including bottom types (coarse sand, sandy-mud, and muddy sand), type of species that form landscape elements (mangrove, oyster reef), species of seagrass, and disturbance locations. Based on the landscape map thus created, they computed landscape indices at two different spatial levels, namely, whole-landscape level and patch level. To detect the interactions of environmental factors at different spatial scales, they performed path analysis. As a result, patch diversity had an indirect positive effect on species richness and the density of juvenile fish through the percent coverage of seagrass beds in a tidal flat. Although causal structures consisting of factors from different spatial scales are occasionally complex, recently developed techniques such as Bayesian networks allow us to treat complicated nonparametric causal relationships. It is important to remember that for performing such analyses, researchers must carefully establish experimental designs with sufficient sample sizes before conducting field surveys.

Future Landscape Research in Marine Areas

From this review (summarized in Table 7.3), it can be said that the techniques developed in landscape ecology in terrestrial ecosystems are essentially applicable to ocean ecosystems, particularly for benthic organisms and fish. However, this does not necessarily mean that exactly the same principles apply to both terrestrial and ocean ecosystems. To provide further insight into future studies in the ocean, we focus on spatial and temporal scales and causal structures of multiple factors that may influence biodiversity in heterogeneous ocean landscapes.

First, spatial and temporal scales of various processes could differ between terrestrial and oceanic areas because the scale of dispersal or movement is not always the same as that in terrestrial species. Few studies have examined the correspondence between the dispersal range of benthic organisms and the extent of the background landscape. In addition, the distance and direction of dispersal may differ depending on the location because of ocean currents. Although fish behavior from an ethological standpoint is relatively well studied, how far each fish species can move to use different landscapes or to what extent each fish is dependent on each landscape element remains largely unknown. Some researchers have also stated spatial scales at which interspecific interactions occur are broader in the ocean (Ian Perry and Ommer 2003; Azovsky 2000). Thus, quantification of spatial usage by marine organisms in the landscape context is important. This understanding could bind the gap between terrestrial and ocean ecosystems, where different classifications of landscape elements have been used. In terrestrial organisms such as butterflies, there is an abundance of information on the ecology of the movements of individual species, which enables biologically meaningful classifications of landscape elements and weighing of each landscape element to estimate dispersal capability in a landscape (Kuroe et al. 2011). If such measurements or estimations are also possible in marine areas, definitions of landscape elements become more ecologically meaningful. Before accumulation of detailed

Table 7.3 Summary of the importance of mosaicness considered in this review

Types of target species	Target groups and phenomenon	Related reference cited in the text	Element of the landscape	Spatial size of the landscape	Proof of diversity-mosaicness relationships
Benthic organisms					
	Occurrence of larval settlement of barnacle species	Pineda (2000)	Combination of different types of substrate	Meter to kilometer: a scale including different types of bottom	Not investigated
	Species number and density of benthos in the tidal flat	Tsuchiya and Kurihara (1976)	Different percentage of mud	5-km-wide tidal flat from back side marsh to foreshore	Not investigated
	Density of a shell settlement at the edge of the landscape	Heck and Valentine (2006)	Edge length of a seagrass bed	Meter to kilometer: size of a seagrass bed	Not investigated
Temperate fishes					
	Species number and characteristics of fishes in two types of seagrass	Azuma (1982)	Healthy/ degraded seagrass	Meter to kilometer: size of a seagrass bed	Not investigated
	Species number and characteristics of fishes in sargasso bed	Saito (1979)	Sargasso bed	Meter to kilometer: size of combination of different types of the beds	Not investigated
	Seasonal species number and characteristics of fishes	Kikuchi (1982)	Seagrass and other components in different seasons	0 to 50 km: depends on types of fishes	Not investigated
		Shoji (2009)			Not investigated

	Life stage-dependent use of landscape by black rockfish		Seagrass, sargasso, and rocky area	50 km :half of Tokyo Bay, which contains three types of landscape	
Tropical fishes					
	Frequency of use on each landscape by gray snapper	Luo et al. (2009)	Seagrass, mangrove, and coral reef	5 km: the daily movement range of the fishes at this study site	Not investigated
	Fish biomass and species number in a seagrass bed	Unsworth et al. (2008)	Seagrass with mangrove/ coral reef	Kilometer scale: combination of three types of landscape	Yes
	Fish biomass and species number in a mixed landscape	Grober-Dunsmore et al. (2008)	26 habitat classes including rugosity of coral reefs	Different levels of landscape parameters from 100 to 1 km	No
	Fish biomass and species number in a mixed landscape	Pittman et al. (2004)	Seagrass, mangrove, and coral reefs	Three levels of landscape. Mosaicness in 300-m habitat type and within patch	Yes, but indirectly

Tropical research is advanced in diversity-landscape relationships (see Mellin et al. 2009 for more examples) information, however, it seems useful to classify species based on the spatial scale of movements and to identify the relevant landscape elements. As a rough approximation, birds and fish are both mobile organisms, whereas plants and benthos are rather sedentary, except at the very early stages. This sort of analogy may be useful, at least at the early phase of landscape studies. In addition to these views on organisms, spatial and temporal scales at which an environmental factor acts on organisms also differ between the land and ocean (Hatcher et al. 1987; Bengtsson et al. 2003). A relationship between magnitude of disturbance and landscape structure in coral reefs has been shown in recent years (Mumby and Steneck 2008); however, the temporal dynamics of landscape structures have not been well studied. Measuring the temporal dynamics of landscape indices is important for understanding landscape–organism relationships in the ocean.

Second, it may not be possible to achieve an accurate biodiversity assessment of coastal areas at the landscape level because external factors from terrestrial and offshore environments are likely to be involved. In tropical Asia, coastal landscapes are under high human pressure for land reclamation for aquaculture ponds, ports, and roads (Pas Ong and Lebel 2000). Even in the case of the traditional use of mangrove forests (Nip and Wong 2010), increasing human populations accelerate overexploitive logging. In addition, the importance of the inflow from terrestrial areas has been demonstrated (Klein et al. 2012; Schlünz and Schneider 2000). In Japan, some forested areas have been historically protected by law for fisheries. Although the close linkage between forests and coastal areas has been empirically suggested by local knowledge, solid scientific evidence has been provided only very recently by statistical analysis using data of terrestrial land use and coastal landscapes (Yamakita et al. 2014; Klein et al. 2012). We propose new perspectives for integrated analysis of the biodiversity in coastal landscapes, with consideration of terrestrial and offshore factors (Figs. 7.4, 7.5). Here, both terrestrial and offshore factors affect local species richness and abundance in a hierarchical manner through landscape or local factors. It is important to note that not all variables are directly measurable, such as the amount of water flow and water quality from rivers as well as local food supply. In these cases, indirect measures may be useful, including terrestrial land use, human population density, and offshore water temperatures estimated by satellite data. With the use of these variables and the causal structure as described here, we can predict how land-use alteration and global warming affect biodiversity in coastal landscapes. We are still at a very early stage of the application of landscape analysis to the ocean (Klein et al. 2012; Webb 2012). The combination of an a priori approach by statistical analysis using large databases and an a posteriori approach by experiments are promising for the future progress of the study of landscape–biodiversity relationships in the ocean.

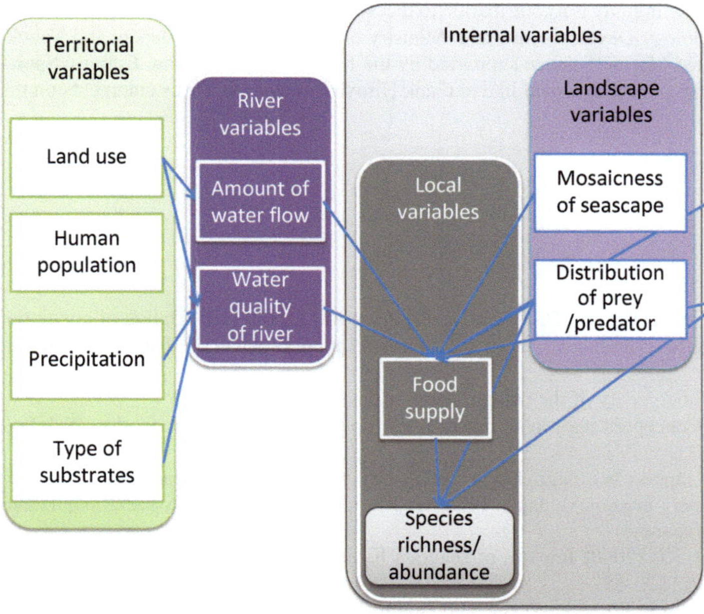

Fig. 7.4 Conceptual image of integrated analysis of biodiversity in coastal landscapes, including terrestrial and offshore factors (focused on food availability)

Fig. 7.5 Seagrass beds and mangrove forests in Trang, Thailand. (Photograph by Nakaoka Lab, Akkeshi Marine Station, Hokkaido University)

Acknowledgments This study is a contribution to the S-9 projects supported by the Environment Research and Technology Development Fund, Ministry of the Environment, Japan, and Tohoku Ecosystem-Associated Marine Science supported by the Ministry of Education, Culture, Sports, Science & Technology. We also would like to thank Hiroya Yamano giving us comments on this article.

References

Andersson J, Byström P, Claessen D, Persson L, De Roos AM (2007) Stabilization of population fluctuations due to cannibalism promotes resource polymorphism in fish. Am Nat 169 (6):820–829

Angel MV (1993) Biodiversity of the pelagic ocean. Conserv Biol 7:760–772

Azovsky AI (2000) Concept of scale in marine ecology: linking the words or the worlds. Web Ecol 1(1):28–34

Azuma M (1982) Seagrass bed degradation and fishery products. In: Seagrass bed (especially eelgrass) and fishery organisms. Japan Fisheries Resource Conservation Association, Tokyo, pp 106–149 (in Japanese)

Bologna PAX, Heck KL (2000) Impacts of seagrass habitat architecture on bivalve settlement. Estuar Coast 23(4):449–457

Bengtsson J, Angelstam P, Elmqvist T, Emanuelsson U, Folke C, Ihse M, Moberg F, Nyström M (2003) Reserves, resilience and dynamic landscapes. AMBIO 32:389–396

Boström C, Pittman SJ, Simenstad C, Kneib RT (2011) Seascape ecology of coastal biogenic habitats: advances, gaps, and challenges. Mar Ecol Prog Ser 427:191–217

Carr MH, Neigel JE, Estes JA, Andelman SJ, Warner RR, Largier JL (2003) Comparing marine and terrestrial ecosystems: implications for the design of coastal marine reserves. Ecol Appl 13:90–107

Convention for Biological Diversity (CBD) (2011) Strategic Plan for Biodiversity 2011–2020 (COP 10 Decision X/2; including the Aichi Targets). http://www.cbd.int/decision/cop/?id=12268. Accessed 3 Dec 2011

Connell JH (1978) Diversity in tropical rain forests and coral reefs. Science 199:1302–1310

Dunning JB, Danielson BJ, Pulliam HR (1992) Ecological processes that affect populations in complex landscapes. Oikos 65:169–175

Forman RTT (1995) Some general principles of landscape and regional ecology. Landsc Ecol 10:133–142

Franklin JF (1993) Preserving biodiversity: species, ecosystems, or landscapes? Ecol Appl 3:202–205

Fukaya K, Okuda T, Nakaoka M, Hori M, Noda T (2010) Seasonality in the strength and spatial scale of processes determining intertidal barnacle population growth. J Anim Ecol 79:1270–1279

Goshima S (2003) Relationships between benthic ecology and fishery. Tokai University Press, Tokyo, pp 369–406 (in Japanese)

Gray JS (1997) Marine biodiversity: patterns, threats and conservation needs. Biodivers Conserv 6:153–175

Grober-Dunsmore R, Frazer TK, Lindberg WJ, Beets J (2007) Reef fish and habitat relationships in a Caribbean seascape: the importance of reef context. Coral Reefs 26:201–216

Grober-Dunsmore R, Frazer TK, Beets JP, Lindberg WJ, Zwick P, Funicelli NA (2008) Influence of landscape structure on reef fish assemblages. Landsc Ecol 23:37–53

Hatcher BG, Imberger J, Smith SV (1987) Scaling analysis of coral reef systems: an approach to problems of scale. Coral Reefs 5:171–181

Heck KL, Valentine JF (2006) Plant–herbivore interactions in seagrass meadows. J Exp Mar Biol Ecol 330:420–436

Ian Perry R, Ommer RE (2003) Scale issues in marine ecosystems and human interactions. Fish Oceanogr 12:513–522

Kato M (2001) SATOYAMA and Biodiversity Conservation: SATOYAMA as Important Insect Habitats. Glob Environ Res Eng Ed 5:135–150

Kendall MS, Miller TJ, Pittman SJ (2011) Patterns of scale-dependency and the influence of map resolution on the seascape ecology of reef fish. Mar Ecol Prog Ser 427:259–274

Kerr JT, Southwood TRE, Cihlar J (2001) Remotely sensed habitat diversity predicts butterfly species richness and community similarity in Canada. Proc Natl Acad Sci USA 98:11365

Kikuchi T (1982) Fish community in a seagrass bed. In: Report of seagrass subdivision. Study on fishery environment. Japan Fisheries Resource Conservation Association, Tokyo, pp 49–105 (in Japanese)

Kinlan BP, Gaines SD (2003) Propagule dispersal in marine and terrestrial environments: a community perspective. Ecology 84:2007–2020

Klein CJ, Jupiter SD, Selig ER, Watts ME, Halpern BS, Kamal M, Roelfsema C, Possingham HP (2012) Forest conservation delivers highly variable coral reef conservation outcomes. Ecol Appl 22:1246–1256

Kuroe M, Yamaguchi N, Kadoya T, Miyashita T (2011) Matrix heterogeneity affects population size of the harvest mice: Bayesian estimation of matrix resistance and model validation. Oikos 12:271–279

Luo J, Serafy JE, Sponaugle S, Teare PB, Kieckbusch D (2009) Movement of gray snapper *Lutjanus griseus* among subtropical seagrass, mangrove, and coral reef habitats. Mar Ecol Prog 380:255–269

Marini L, Fontana P, Battisti A, Gaston KJ (2009) Agricultural management, vegetation traits and landscape drive orthopteran and butterfly diversity in a grassland forest mosaic: a multi scale approach. Insect Conserv Divers 2:213–220

Mellin C, Andrefouët S, Ponton D (2007) Spatial predictability of juvenile fish species richness and abundance in a coral reef environment. Coral Reefs 26:895–907

Mellin C, Andréfouët S, Kulbicki M, Dalleau M, Vigliola L (2009) Remote sensing and fish–habitat relationships in coral reef ecosystems: review and pathways for systematic multi-scale hierarchical research. Mar Pollut Bull 58:11–19

Minello TJ, Zimmerman RJ, Medina R (1994) The importance of edge for natant macrofauna in a created salt marsh. Wetlands 14:184–198

Miyashita T (2013) The significance of ecosystem diversity on species richness and ecosystem sustainability: a general conceptualization. In: Nakano S, Yahara T, Nakashizuka T (eds) Integrative observations and assessments, Ecological research monographs. Springer, Tokyo, pp 29–47

Mumby PJ, Steneck RS (2008) Coral reef management and conservation in light of rapidly evolving ecological paradigms. Trends Ecol Evol 23:555–563

Mumby PJ, Edwards AJ, Arias González JE, Lindeman KC, Blackwell PG, Gall A, Gorczynska MI, Harborne AR, Pescod CL, Renken H (2004) Mangroves enhance the biomass of coral reef fish communities in the Caribbean. Nature (Lond) 427:533–536

Nagelkerken I, Van der Velde G, Gorissen MW, Meijer GJ, Van't Hof T, Den Hartog C (2000) Importance of mangroves, seagrass beds and the shallow coral reef as a nursery for important coral reef fishes, using a visual census technique. Estuar Coast Shelf Sci 51:31–44

Nakano S, Murakami M (2001) Reciprocal subsidies: dynamic interdependence between terrestrial and aquatic food webs. Proc Natl Acad Sci USA 98:166

Nip THM, Wong CK (2010) Juvenile fish assemblages in mangrove and non-mangrove soft-shore habitats in eastern Hong Kong. Zool Stud 49:760–778

Pas Ong S, Lebel L (2000) Political transformation and the environment in Southeast Asia. Environment 42:8–19

Pineda J (2000) Linking larval settlement to larval transport: assumptions, potentials, and pitfalls. Oceanogr East Pac 1:84–105
Pittman SJ, McAlpine CA, Pittman KM (2004) Linking fish and prawns to their environment: a hierarchical landscape approach. Mar Ecol Prog 283:233–254
Ries L, Fletcher RJ Jr, Battin J, Sisk TD (2004) Ecological responses to habitat edges: mechanisms, models, and variability explained. Annu Rev Ecol Evol Syst 35:491–522
Saito Y (1979) Basic of marine forest construction technics. Fish Technics Manag 171:77–82 (in Japanese)
Schlünz B, Schneider RR (2000) Transport of terrestrial organic carbon to the oceans by rivers: re-estimating flux-and burial rates. Int J Earth Sci 88(4):599–606
Shoji J (2009) Fishes and seagrass/algal bed. Seizando, Tokyo (in Japanese)
Shoji J, Tanaka M (2005) Daily ration and prey size of juvenile piscivore Japanese Spanish mackerel. J Fish Biol 67:1107–1118
Sogard SM, Able KW (1994) Diel variation in immigration of fishes and decapod crustaceans to artificial seagrass habitat. Estuar Coast 17:622–630
Stergiou KI, Browman HI, Cole JJ, Halley JM, Paine RT, Raffaelli D, Solan M, Webb TJ, Stenseth NC, Mysterud A (2005) Bridging the gap between aquatic and terrestrial ecology. Mar Ecol Prog Ser 304:271–307
St.-Louis V, Pidgeon AM, Clayton MK, Locke BA, Bash D, Radeloff VC (2009) Satellite image texture and a vegetation index predict avian biodiversity in the Chihuahuan Desert of New Mexico. Ecography 32:468–480
Terawaki T (1997) Function of seagrass bed. In: Research review Series 4: function of seagrass/algal bed (ed) The Council for the Promotion of experiment and research on fisheries industry. pp 82–10 (in Japanese)
Thorson G (1964) Light as an ecological factor in the dispersal and settlement of larvae of marine bottom invertebrates. Ophelia 1:167–208
Tsuchiya M, Kurihara Y (1976) Studies on the distribution of the macrobenthos with special reference to the movement of the fine particles at Gamo Lagoon, Miyagi Prefecture. Physiol Ecol Jpn 17:145–151 (in Japanese)
Turner MG (1989) Landscape ecology: the effect of pattern on process. Annu Rev Ecol Syst 20:171–197
Turner MG, Gardner RH, O'Neill RV (2001) Landscape ecology in theory and practice: pattern and process. Springer, New York
Unsworth RKF, De León PS, Garrard SL, Jompa J, Smith DJ, Bell JJ (2008) High connectivity of Indo-Pacific seagrass fish assemblages with mangrove and coral reef habitats. Mar Ecol Prog 353:213–224
Webb TJ (2012) Marine and terrestrial ecology: unifying concepts, revealing differences. Trends Ecol Evol 27(10):535–541. doi:10.1016/j.tree.2012.06.002
Wedding LM, Friedlander AM, McGranaghan M, Yost RS, Monaco ME (2008) Using bathymetric lidar to define nearshore benthic habitat complexity: implications for management of reef fish assemblages in Hawaii. Remote Sens Environ 112:4159–4165
Yamakita T, Nakaoka M (2011) Importance of considering grain and extent for the analysis on spatial dynamics: perspectives from comparison between theory and empirical example on seagrass bed dynamics in Tokyo Bay. Procedia Soc Behav Sci 21:177–183
Yamakita T, Watanabe K, Nakaoka M (2011) Asynchronous local dynamics contributes to stability of a seagrass bed in Tokyo Bay. Ecography 34:519–528
Yamakita T, Taki H, Kimiko O (2014) Quantitative effects of terrestrial and oceanic factors on the nationwide distribution of seagrass and algal beds at two different basin scales. Submitted

Part III
Historical Changes in Biodiversity and Challenges for Biodiversity Conservation

Chapter 8
Long-Term and Spatial Variation in the Diversity of Littoral Benthic Macroinvertebrate Fauna in Lake Biwa, Japan

Jun-ya Shibata, Zin'ichi Karube, Yoichirou Sakai, Tomohiro Takeyama, Ichiro Tayasu, Shigeo Yachi, Shin-ichi Nakano, and Noboru Okuda

Abstract Freshwater ecosystems provide ecosystem services essential to human well-being, such as provisioning of water and fishery resources, but they are the most vulnerable to human disturbances and concerns have been raised about their loss of biodiversity. The ancient Lake Biwa is one of valuable Asian freshwater ecosystem because of its high biodiversity and endemism but it has been exposed to various severe anthropogenic environmental changes during the past half century. Although we have great concern about the biodiversity loss in this lake, it remains highly unknown what kind of human disturbances cause deterioration of the ecosystem. Elucidation of environmental pressures for the biodiversity loss is important to illuminate subjects for conservation of biodiversity. We particularly focus on a littoral benthic macroinvertebrate fauna because its habitats, located in the interface between aquatic and terrestrial ecosystems, are those most vulnerable to human activities. We found that littoral benthic macroinvertebrate diversity was largely affected by pH, temperature, phytoplankton biomass, benthic microalgae biomass, and coverage of submerged plants. These environmental variables are largely associated with environmental problems that occurred in the Lake Biwa: eutrophication, warming, and massive submerged plant expansion. This finding suggests that past environmental problems caused serious impacts on the biodiversity of Lake Biwa.

Keywords Anthropogenic pressure • Eutrophication • Lake Biwa • Littoral environment • Submerged vegetation • Water temperature • Zoobenthos fauna

---

J.-y. Shibata (✉) • Y. Sakai • I. Tayasu • S. Yachi • S.-i. Nakano • N. Okuda
Center for Ecological Research, Kyoto University, Hirano 2-509-3, Shiga 520-2113, Japan
e-mail: shiba@ecology.kyoto-u.ac.jp

Z. Karube
National Institute for Environmental Studies, Onogawa 16-2, Tsukuba
Ibaraki 305-8506, Japan

T. Takeyama
Laboratory of Animal Sociology, Department of Biology and Geosciences, Graduate School of Science, Osaka City University, Sugimoto 3-3-138, Sumiyoshi-ku, Osaka 558-8585, Japan

S.-i. Nakano et al. (eds.), *Integrative Observations and Assessments*, Ecological Research Monographs, DOI 10.1007/978-4-431-54783-9_8, © Springer Japan 2014

Introduction

Lake Biwa dates back approximately 4 million years and is one of the largest representative ancient lakes in Asia. The lake contains a high biodiversity of about 1,700 aquatic species including 61 endemics (Nishino 2012). Since the 1960s, however, the Lake Biwa ecosystem has suffered deterioration of water quality and habitat loss for native aquatic organisms as a result of human activities within its watershed, such as human population increase, land development, and agricultural modernization (Rossiter 2000). Biodiversity loss has become increasingly evident in this lake (Nakanishi and Sekino 1996). For example, the abundance of several common endemic plankton species has substantially decreased in Lake Biwa over the past 40 years (Ichise et al. 2004). A long-term analysis has also demonstrated that eutrophication and increased water temperatures have led to drastic changes in pelagic plankton communities (Hsieh et al. 2010).

However, it remains unknown which types of human disturbances and resultant environmental changes in the lake ecosystem affect the diversity of benthic macroinvertebrates inhabiting littoral waters. Littoral lake habitats serve as the interface between aquatic and terrestrial ecosystems and are thus among those most vulnerable to human disturbances, especially at a catchment scale (Karube et al. 2010). It is crucial to understand the processes and mechanisms causing the loss of diversity of littoral benthic macroinvertebrate fauna during the past half century, as these taxa account for the majority of endemism in Lake Biwa (Nishino and Watanabe 2000). Some long-term monitoring data exist for littoral zoobenthic fauna in this lake, but most data are highly scattered or unpublished. Compilation and analysis of these archival data will help to trace the historical changes in the lake under recent human disturbances and to identify the driving factors of the drastic biodiversity loss in Lake Biwa.

In this study, we first analyzed existing historical data on littoral macroinvertebrate fauna in Lake Biwa in relationship to coastal environmental changes resulting from human disturbances since the 1960s. Using multivariate statistics and ordination methods, we then focused on spatial variation in local littoral macroinvertebrate communities in the current Lake Biwa and examined which environmental characteristics of their microhabitats affect community composition.

Methods

Archival Data

We used archival data on littoral benthic macroinvertebrate fauna in Lake Biwa to examine historical changes in these taxa since the 1960s when environmental alterations began to appear. The archival data were provided by the Shiga

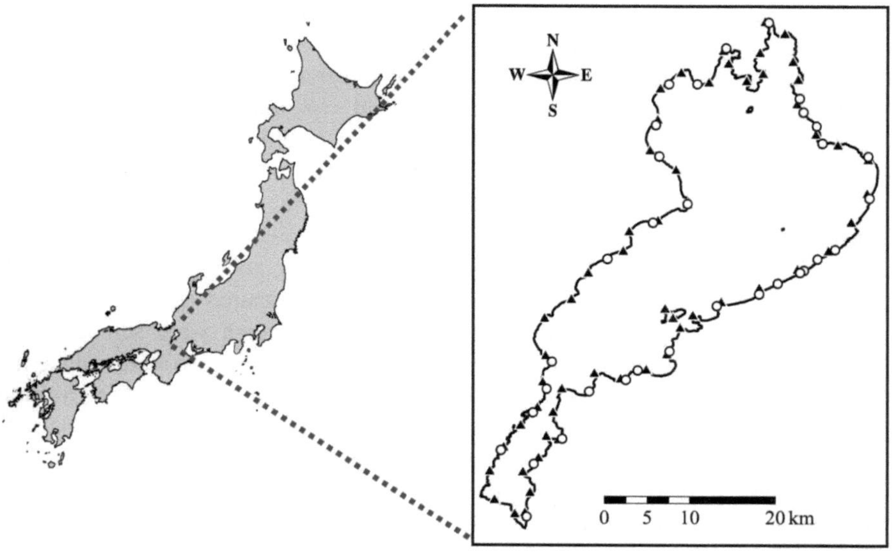

Fig. 8.1 Map of Lake Biwa. *Open circles* show observation sites in the 2006 survey; *closed triangles* show sampling sites in the historical monitoring from 1969 to 2003 by Shiga Prefectural Fisheries Institute

Prefectural Fishery Experimental Station. Monitoring surveys were conducted at 55 sites within littoral waters during the summers (July to August) of 1965, 1995, and 2003 (Fig. 8.1). At each site, 15 × 15 cm littoral sediment samples were collected from six depths (1–7 m) using an Ekman-Berge grab sampler, and benthic macroinvertebrates were sieved using a 1-mm-mesh sieve. Samples were then sorted and classified to the species level whenever possible, although some were identified to genus or a higher taxonomic group. Details of these monitoring surveys have been documented as annual reports elsewhere (Shiga Prefectural Fishery Experimental Station 1972, 1998, 2005).

Environmental Assessment of Littoral Habitats

To assess the habitat quality for littoral benthic macroinvertebrates in Lake Biwa, we conducted a field survey in July 2006. Benthic macroinvertebrates were collected from 33 sites in littoral waters shallower than 1 m in depth, using a 475-μm-mesh Surber net sampler with a 30 × 30 cm quadrat and three replicate samples (Fig. 8.1). At each sampling site, we also monitored environmental characteristics of the microhabitat in terms of water chemistry, physical habitat structure, food quality and quantity, and predation pressure, to examine how local benthic macroinvertebrate communities are colonized along environmental gradients of the littoral habitats.

For water chemistry, we measured the water temperature, pH, turbidity, and dissolved oxygen concentration using a multi-profiler (U-20; Horiba, Japan). In terms of physical habitat structure, we scored substratum size based on the dominant substrate types (rock = 1, pebble = 2, sand = 3, mud = 4) and coverage of submerged and emergent plants in rank order (no vegetation = 0, <25 % coverage = 1, 25–50 % = 2, 50–75 % = 3, and >75 % = 4).

As an indicator of food quality and quantity, we measured the abundance of particulate organic carbon (POC) and epilithic organic matter (EOM), which serve as the basal food sources for filter-feeding and grazing zoobenthos, respectively (Karube et al. 2010). The abundance of EOM was measured as the chlorophyll *a* concentration (mg cm^{-2}) of epilithic microalgae, which were scraped from 6 × 6 cm areas on each of five boulders at each site and then mixed to reduce the effect of spatial variation in algal biomass at a microhabitat scale. For POC, littoral water samples were collected from each site and filtered on precombusted (450 °C for 2 h) glass-fiber filters (GF/F, pore size ~0.7 μm; Whatman) after sieving through 150-μm mesh to remove coarse particles. The carbon content of particulate samples (POC) was measured using an elemental analyzer (EA1108; Fisons, Italy; EA1112; Thermo Fisher, Germany) at the Center for Ecological Research, Kyoto University. For details of the sampling methods, see Sakai et al. (2013).

To collect data on the density of littoral fishes, which are considered potential predators of benthic macroinvertebrates, we enclosed a given area of littoral waters at each sampling site with a 1-cm-mesh beach seine net and collected all fishes within the enclosure by towing the net toward the shoreline. Fish density was calculated as the number of fishes per unit area of the enclosure.

Statistical Analysis

We analyzed historical changes in littoral benthic macroinvertebrate communities in Lake Biwa using archival data from 1965, 1995, and 2003 (Shiga Prefectural Fishery Experimental Station 1972, 1998, 2005). We calculated the Shannon–Wiener *H'* as an index of littoral zoobenthic diversity for each depth at each site in each of the 3 years, and this value represented α-diversity. Interannual differences in α-diversity were tested using one-way analysis of variance (ANOVA) followed by a Sheffé's post hoc test. We also tested interannual differences in macroinvertebrate abundance using a likelihood ratio test (LRT) for a generalized linear model (GLM) with negative binomial distribution. We set the significance level of whole pairwise tests at 0.05 by applying a sequential Bonferroni adjustment (Rice 1989). To examine how the community composition of benthic macroinvertebrates varies along environmental gradients of littoral habitat, we performed redundancy analysis (RDA) and Monte Carlo permutation analysis of variance with the "vegan" package of R statistics software (R Development Core Team 2012). For these analyses, data on spatial patterns were processed in relationship to both benthic macroinvertebrate community composition and environmental variables. Before the RDA, community composition data were standardized using the Hellinger method (Legendre and Gallagher 2001).

Results

Long-Term Variation in Littoral Zoobenthic Communities

In total, 25 benthic macroinvertebrate taxa were recorded during the three monitoring surveys between 1969 and 2003 (Table 8.1). Oligochaetes and chironomid larvae always dominated the littoral zoobenthic communities, accounting for 57.9–84.6 % (in abundance) of the entire community, respectively, although chironomid abundance was exceptionally low in 2003. For seven of ten taxa for which the relative abundance was more than 1 %, we observed significant historical changes in their abundance. From 1969 to 1995, some taxa exhibited a significant increase in abundance (Planorbidae spp. and Corbiculidae spp.), whereas the abundance of other species substantially decreased (*Radix onchia*, Oligochaeta spp., and Gammaridae spp.; 1). In contrast, from 1995 to 2003, two annelid taxa significantly increased in abundance (Oligochaeta spp. and Pharyngobdellida spp.), whereas chironomids and two gastropods (Planorbidae spp. and Lymnaeidae spp.) showed marked decreases in abundance (Table 8.1).

The littoral zoobenthic communities exhibited long-term variation in α-diversity (Fig. 8.2). These long-term patterns varied among monitoring depths (Table 8.2). At monitoring depths of 1 and 2 m, α-diversity tended to steadily decrease as time elapsed. In contrast, the α-diversity of zoobenthic communities at a depth of 3 m significantly increased from 1969 to 1995. In waters deeper than 3 m, no clear historical changes were observed across the three monitoring years.

Spatial Variation in Littoral Zoobenthic Communities

In 2006, 25 macroinvertebrate taxa were found in the littoral waters of Lake Biwa. Although Oligochaeta and Chironomidae were among the most dominant within the littoral zoobenthos, community composition exhibited large spatial variation among monitoring sites along environmental gradients of littoral habitats (Tables 8.3, 8.4). According to the RDA, five of ten environmental variables were selected as significant factors explaining a large portion of the spatial variation in littoral zoobenthic community composition (Table 8.5). Two annelid taxa, the most dominant Oligochaeta spp. (Oli) and the common Pharyngobdellidae spp. (Pha), tended to appear at sites with high coverage of submerged vegetation, whereas the second most dominant Chironomidae spp. (Chi) preferred sites with

Table 8.1 List of benthic macroinvertebrate taxa found at the Lake Biwa littoral zone in 1969, 1995, and 2003

Taxon	1969	1995	2003
Annelida			
Oligochaeta			
Oligochaeta spp.	257.9^{ab} (0.0–835.6; 46.68 %)	148.4^{ac} (0.0–1,111.1; 33.55 %)	657.0^{bc} (59.3–2,029.6; 79.57 %)
Hirudinae			
Pharyngobdellida			
Pharyngobdellida spp.	17.6 (0.0–177.8; 3.19 %)	6.4^{a} (0.0–71.1; 1.45 %)	18.8^{a} (0.0–81.5; 2.28 %)
Mollusca			
Gastropoda			
Sorbeoconcha			
Pleuroceridae			
Pleuroceridae spp.	60.9 (0.0–222.2; 11.02 %)	51.0 (0.0–177.8; 11.53 %)	43.8 (0.0–124.4; 5.31 %)
Viviparidae			
Viviparidae spp.	16.9 (0.0–497.8; 3.05 %)	23.6 (0.0–222.2; 5.34 %)	10.8 (0.0–133.3; 1.31 %)
Valvatidae			
Biwakovalvata biwaensis			0.4 (0.0–7.4; 0.05 %)
Pulmonata			
Planorbidae			
Planorbidae spp.	6.0^{a} (0.0–186.7; 1.08 %)	39.5^{ab} (0.0–251.9; 8.94 %)	2.8^{b} (0.0–14.8; 0.34 %)
Ancyliae			
Laevapex nipponica		0.2 (0.0–7.4; 0.04 %)	0.0 (0.0–0.0; 0.00 %)
Physidae			
Physa acuta		0.5 (0.0–14.8; 0.12 %)	0.0 (0.0–8.9; 0.01 %)
Lymnaeidae			
Lymnaeidae spp.	3.4 (0.0–155.6; 0.61 %)	10.9^{a} (0.0–111.1; 2.47 %)	1.6^{a} (0.0–8.9; 0.20 %)
Radix onchia	4.7^{a} (0.0–133.3; 0.85 %)	0.0^{a} (0.0–0.0; 0.00 %)	1.2 (0.0–22.2; 0.14 %)

Bivalvia			
Unionoida			
Unionidae			
Unio spp.	6.1 (0.0–97.8; 1.11 %)	8.4 (0.0–66.7; 1.9 %)	3.2 (0.0–14.8; 0.39 %)
Anodonta woodiana		0.4 (0.0–11.1; 0.09 %)	0.0 (0.0–0.0; 0.00 %)
Anodonta calipygos		0.2 (0.0–8.9; 0.04 %)	0.0 (0.0–0.0; 0.00 %)
Cristaria plicata clessini	0.2 (0.0–8.9; 0.03 %)	0.0 (0.0–0.0; 0.00 %)	0.0 (0.0–0.0; 0.00 %)
Lanceolaria oxyrhyncha	0.5 (0.0–7.4; 0.08 %)	0.2 (0.0–7.4; 0.04 %)	0.0 (0.0–0.0; 0.00 %)
Veneroida			
Corbiculidae			
Corbiculidae spp.	17.8[a] (0.0–231.1; 3.21 %)	41.7[a] (0.0–755.6; 9.42 %)	41.9 (0.0–302.2; 5.08 %)
Arthropoda			
Insecta			
Diptera			
Chironomidae			
Chironomidae spp.	136.5[a] (0.0–822.2; 24.7 %)	107.6[b] (0.0–681.5; 24.34 %)	41.6[ab] (0.0–133.3; 5.04 %)
Tipulidae			
Tipulidae spp.	0.5 (0.0–26.7; 0.10 %)		
Ephemeroptera			
Ephemeroptera spp.	0.4 (0.0–11.1; 0.07 %)	0.6 (0.0–14.8; 0.15 %)	0.9 (0.0–17.8; 0.11 %)
Trichoptera			
Trichoptera spp.	1.2 (0.0–29.6; 0.22 %)	0.6 (0.0–11.1; 0.13 %)	1.6 (0.0–14.8; 0.19 %)
Coleoptera			
Psephenidae			
Psephenidae spp.	0.2 (0.0–11.1; 0.04 %)		
Odonata			
Gomphidae			
Gomphidae spp.	0.2 (0.0–7.4; 0.03 %)		

(continued)

Table 8.1 (continued)

Taxon	1969	1995	2003
Malacostraca			
Amphipoda			
Gammaridae			
Gammaridae spp.	19.5^{ab} (0.0–435.6; 3.52 %)	1.0^{a} (0.0–17.8; 0.23 %)	0.0^{b} (0.0–7.4; 0.01 %)
Decapoda			
Palaemonidae			
Palaemon paucidens	1.3 (0.0–22.2; 0.23 %)	1.0 (0.0–11.1; 0.21 %)	0.0 (0.0–0.0; 0.00 %)
Isopoda			
Corallanidae			
Tachea chinensis	1.0 (0.0–17.8; 0.19 %)		
Total	552.6 ind./m^2	442.2 ind./m^2	825.7 ind./m^2

Letters beside subtotal of each class and phylum level show results of pairwise test of likelihood ratio test (LRT) for a generalized linear model (GLM) with sequential Bonferroni adjustment. The pairs of years with same letter significantly differed (significance level < 0.05)
Mean values of individual density (ind. m^{-2}) are shown with range of the individual density and percentage of total number in parentheses

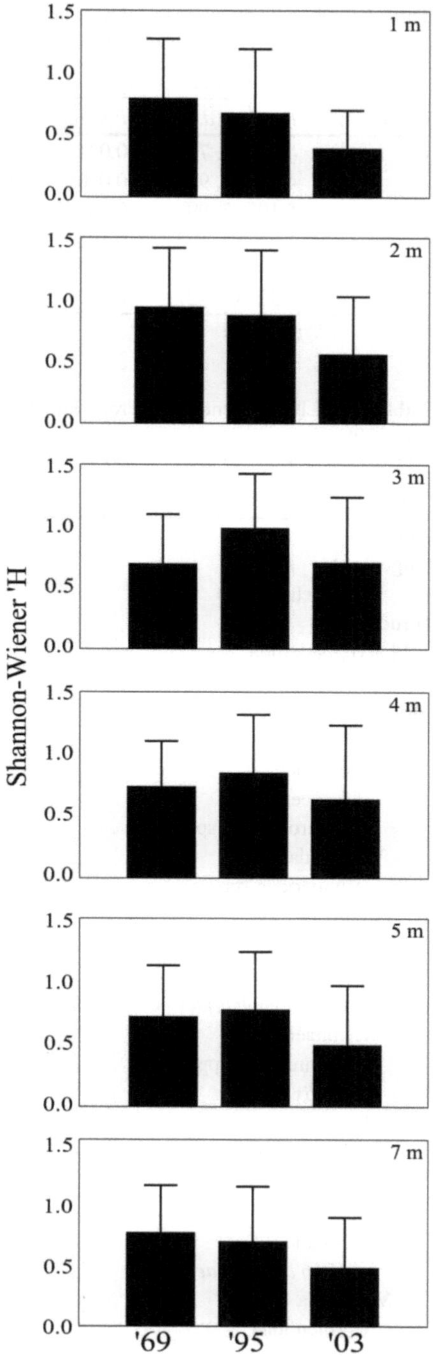

Fig. 8.2 Historical changes in the biodiversity of the littoral benthos from 1969 to 2003 at each water depth. *Error bars* show standard deviations

Table 8.2 Results of one-way analysis of variance (ANOVA) and post hoc test for comparison of α-diversity among years in each depth

	One-way ANOVA		P value in Sheffé's post hoc test		
Depth (m)	*F* value (*d.f.*)	*P* value	1969 vs. 1995	1995 vs. 1903	1969 vs. 1903
1	4.097 (2, 72)	0.021*	0.602	0.138	0.021*
2	4.314 (2, 90)	0.016*	0.855	0.072	0.020*
3	5.105 (2, 99)	0.008**	0.017*	0.068	0.997
4	1.643 (2, 103)	0.198	0.557	0.216	0.702
5	2.843 (2, 96)	0.063	0.880	0.070	0.178
7	2.339 (2, 72)	0.104	0.786	0.299	0.105

*$p < 0.05$, **$p < 0.01$

Table 8.3 List of benthic macroinvertebrate taxa found at the Lake Biwa littoral zone in July 2006

Taxon	Individual density (ind. m^{-2})	Abbreviation
Annelida		
Oligochaeta		
Oligochaeta spp.	465.0 (0.0–2,025.9)	Oli
Hirudinae		
Pharyngobdellida		
Pharyngobdellida spp.	64.6 (0.0–403.7)	Pha
Mollusca		
Gastropoda		
Sorbeoconcha		
Pleuroceridae		
Pleuroceridae spp.	50.5 (0.0–414.8)	Ple
Viviparidae		
Viviparidae spp.	8.4 (0.0–151.9)	Viv
Pulmonata		
Planorbidae		
Planorbidae spp.	4.5 (0.0–111.1)	Pla
Choanomphalodes perstriatulus	1.8 (0.0–37.0)	Cho
Lymnaeidae		
Lymnaeidae spp.	1.0 (0.0–14.8)	Lym
Radix onchia	2.0 (0.0–40.7)	Rad
Physidae		
Physa acuta	1.1 (0.0–11.1)	Phy
Bivalvia		
Unionoida		
Unionidae		
Unio douglasiae biwae	0.6 (0.0–11.1)	Uni
Veneroida		
Corbiculidae		
Corbiculidae spp.	56.6 (0.0–481.5)	Cor
Mytiloida		
Mytilidae		
Limnoperna fortunei	1.5 (0.0–48.1)	Lim

(continued)

Table 8.3 (continued)

Taxon	Individual density (ind. m^{-2})	Abbreviation
Arthropoda		
Insecta		
Diptera		
Chironomidae		
Chironomidae spp.	595 (0.0–3,829.6)	Chi
Ephemeroptera		
Ephemeroptera spp.	16.4 (0.0–251.9)	Eph
Plecoptera		
Plecoptera spp.	0.6 (0.0–18.5)	Plec
Trichoptera		
Trichoptera spp.	36.1 (0.0–340.7)	Tri
Psychomyiidae		
Psychomyiidae spp.	4.2 (0.0–74.1)	Psy
Coleoptera		
Psephenidae		
Eubrianax ramicornis	0.1 (0.0–3.7)	Eub
Odonata		
Odonata spp.	0.1 (0.0–3.7)	Odo
Arthropoda		
Malacostraca		
Amphipoda		
Gammaridae		
Gammaridae spp.	79.9 (0.0–1,366.7)	Gam
Cprophiidae		
Cprophiidae spp.	27.6 (0.0–633.3)	Cpr
Decapoda		
Decapoda spp.	0.2 (0.0–3.7)	Dec
Palaemonidae		
Palaemon paucidens	1.0 (0.0–11.1)	Pal
Macrobrachium nipponense	0.2 (0.0–3.7)	Mac
Isopoda		
Asellidae		
Asellus hilgendorfi	3.4 (0.0–44.4)	Ase

Mean values of individual density (ind. m^{-2}) are shown with range of the individual density in parentheses

low coverage of submerged vegetation (Fig. 8.3). The filter-feeding *Corbicula* (Cor) and water pollution-tolerant snails (Ple) were more abundant in eutrophic waters with high POC concentration. In contrast, the abundance of some grazing taxa, such as gammarids (Gam), Ephemeroptera (Eph), and planorbid snails (Cho), was higher in littoral habitats rich in epilithic algae (Fig. 8.3). Temperature and pH were also primary environmental factors affecting zoobenthic communities (Table 8.5, Fig. 8.3).

Table 8.4 Summary of microhabitat environments in the littoral ecosystem of Lake Biwa in July 2006

	Median	Minimum	Maximum
pH	6.89	5.52	8.08
Dissolved oxygen concentration (mg l^{-1})	9.71	5.68	13.17
Turbidity (NTU)	1.40	0.66	5.69
Water temperature (°C)	25.4	22.4	28.7
Suspended particulate organic matters (mg C l^{-1})	6.15	3.42	12.90
Epilithic microalgae abundance (mg Chl-a cm^{-2})	2.04	0.40	9.83
Submerged vegetation coverage	1.0	0.0	4.0
Emergent vegetation coverage	0.0	0.0	2.0
Fish density (ind. m^{-2})	0.022	0.000	0.369
Grain size	2.5	1.0	4.0

Table 8.5 Result of the stepwise variable selection in redundancy analysis for relationship between the benthos community structures and microhabitat environments in the littoral ecosystem of the Lake Biwa

	Redundancy analysis (RDA) scores		
	Axis 1	Axis 2	*P* value
pH	−0.269	−0.205	0.040
Dissolved oxygen concentration	–	–	–
Turbidity	–	–	–
Water temperature	−0.473	−0.143	0.002
Suspended particulate organic matter	−0.001	0.542	0.008
Epilithic microalgae abundance	0.055	−0.721	0.021
Submerged vegetation coverage	0.607	0.092	0.008
Emergent vegetation coverage	–	–	–
Fish density	–	–	–
Grain size	–	–	–
Eigenvalue	0.0460	0.0259	
Proportion of explained variance	0.143	0.081	

Summary of the first two RDA axes and *P* values of each selected variables in permutation ANOVA (1,000 permutations)

Discussion

Our long-term analysis based on archival data revealed that the community composition and biodiversity of benthic macroinvertebrates substantially varied in the littoral waters of Lake Biwa during the past half century. Particularly in shallower waters, the α-diversity of the zoobenthic communities exhibited significant, steady decreases as time elapsed. Most common taxa showed significant changes in abundance, although the temporal patterns differed greatly among these taxa. Our results support a commonly accepted, but previously untested, notion that biodiversity has declined in Lake Biwa since the 1960s when environmental alterations caused by anthropogenic disturbances became apparent (Rossiter 2000).

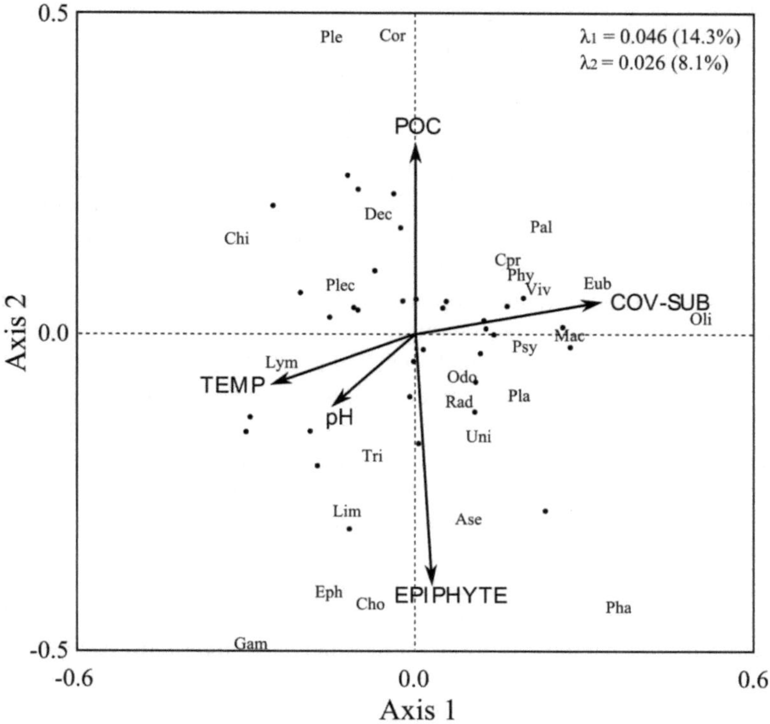

Fig. 8.3 Ordination diagram with the first two redundancy analysis (RDA) axes for the littoral benthic macroinvertebrate community structures from 2006. Plots show sites; abbreviations indicate 25 taxa defined in Table 8.1. *Arrows* show effects of environmental variables selected in the RDA variable selection on community structure: COV-SUB, coverage of submerged vegetation; TEMP, water temperature; pH, pH in water column; POC, concentration of suspended particulate organic carbon; EPIPHYTE, chlorophyll *a* abundance of epiphytes on the littoral bottom

Within large basins in Lake Biwa, the littoral benthic macroinvertebrates also exhibited large spatial variation in community composition along environmental gradients of littoral habitat. The RDA revealed that five environmental variables, namely, submerged vegetation coverage, POC, EOM abundance, water temperature, and pH, all of which are considered critical for the habitat quality of littoral zoobenthos, accounted for a large fraction of the spatial variation in community composition. These variables are closely related to environmental problems that have appeared in the lake during the past half century. For example, a massive expansion of submerged vegetation has occurred in the shallow south basin during recent years (Hamabata and Kobayashi 2002). Furthermore, as is common in lakes worldwide, the surface water temperature of Lake Biwa has been gradually increasing as a result of global warming during the past half century (Endoh et al. 1999; Hayami and Fujiwara 1999; Hsieh et al. 2010). Lake eutrophication has increased POC concentrations (as a proxy for phytoplankton biomass; Hsieh et al. 2010) as well as pH (as a proxy for phytoplankton photosynthetic activity; Brönmark and Hansson 2005), but has negatively affected EOM abundance through reduced light availability from increased turbidity (Vadeboncoeur et al. 2001, 2003; Kagatsume 2012).

In Lake Biwa, eutrophication progressed in the 1960s and peaked during the 1970s. However, temporal changes in the distribution of eutrophic and oligotrophic taxa along an eutrophication–oligotrophication gradient (the POC-EPIPHYTE axis in the RDA bi-plots) became unclear between the two earlier monitoring years, 1969 and 1995, when eutrophication occurred. This inconsistency may have resulted from a periodic mismatch between the incidence and monitoring of eutrophication, because reoligotrophication has gradually progressed since the mid-1980s as a result of increasing efforts by local governments to reduce nutrient loading to the Lake Biwa watershed (Hsieh et al. 2010; Kagatsume 2012).

In Lake Biwa, surface water temperature, which is an important environmental determinant of littoral zoobenthic fauna, has increased at a rate of 0.028 °C/year over the past half century, with an accelerating trend since 1990 (Hsieh et al. 2010). However, warm-water taxa, such as chironomids and lymnaeids, have decreased, rather than increased, during this warming period. Assuming a temperature increase rate of 0.028 °C/year, we estimate that lake water temperature increased by approximately 1 °C from 1969 to 2003. This long-term change in water temperature is small relative to the range (6.3 °C) of water temperatures measured in a variety of the current littoral habitats. For the warm-water taxa, long-term changes in their abundance might be attributed to other non-climate factors. However, we must take precautions against ongoing climate changes and better understand the effects of global warming on the population dynamics of littoral benthic macroinvertebrates and thus biodiversity as a whole in the future.

In the south basin of Lake Biwa, submerged macrophytes began to drastically increase in biomass in 1995. This phenomenon is well known as a catastrophic shift or regime shift between two stable states of phytoplankton-dominated turbid waters and macrophyte-dominated clear waters (Scheffer et al. 2001). During this period, two annelid taxa, which tend to dominate submerged vegetation habitats (Table 8.5, Fig. 8.3), markedly increased in abundance, whereas chironomids and lymnaeids, which tend to dominate areas with less submerged vegetation habitat (Table 8.5, Fig. 8.3), significantly decreased in abundance. This consistency between historical changes in abundance associated with the expansion of submerged vegetation and the patterns expected from distributional changes along submerged vegetation gradients in the spatial analysis support that the regime shift to a macrophyte-dominant system has had a large effect on littoral zoobenthic diversity in Lake Biwa. In this ecosystem, the regime shift is believed to have occurred in 1995 and was promoted by severe drought in 1994. The resultant drawdown facilitated macrophyte growth through increased light penetration to the lake bottom in the shallow south basin (Hamabata and Kobayashi 2002; Nishino et al. 2012). The lake water level can also be artificially manipulated by the government to prevent flooding (Nakamura et al. 2012). Ongoing climate changes and corresponding water regime management have strong potential to affect the community composition and diversity of benthic macroinvertebrates in the littoral waters of Lake Biwa.

Using archival data, we demonstrated that the diversity of benthic macroinvertebrates has actually declined within the littoral waters of Lake Biwa during the past half century. In addition, several environmental characteristics closely related to past environmental problems (eutrophication, warming, and regime shift) have strongly affected the composition of the littoral benthic macroinvertebrate community. To conserve biodiversity of littoral zoobenthos in this lake, the anthropogenic drivers causing environmental deterioration in local habitats should be identified, as mitigating environmental stressors is easier and more effective on a catchment scale than on a global scale (e.g., water warming). The most likely cause of freshwater environmental deterioration is human land use within catchments (Agarwal et al. 2005; Light and Marchetti 2007), which in turn affects material inputs to littoral waters through tributary rivers (Sakai et al. 2013) and subsequent trophic flows within littoral benthic communities. Land-use changes may also alter habitat quality for littoral zoobenthos in terms of physicochemical environments (Karube et al. 2010). At present, we have limited knowledge of the underlying mechanisms by which specific land uses alter the environments of littoral habitats and how such environmental alterations can cause community compositional changes. Path analysis is a powerful and promising tool to disentangle the complexity of causal relationships among human-derived forces and pressures, environmental states, resultant impacts on biota, and biological responses within a watershed, including terrestrial, river, and lake ecosystems (Sakai et al. 2013). This future challenge will be useful for the restoration of littoral habitats and the conservation of lake biodiversity.

Acknowledgments This study was supported by The Environment Research and Technology Development Fund (S-9) of the Ministry of the Environment, Japan and by the JSPS Grant-in Aid (No.24370010). We greatly appreciate the Shiga Prefectural Fishery Institute for providing monitoring data of the Lake Biwa benthos fauna. The present study was conducted using Cooperative Research Facilities (EA-IRMS) of Center for Ecological Research, Kyoto University.

References

Agarwal DK, Silander J, Gelfand AE, Dewar RE, Mickelson JG (2005) Tropical deforestation in Madagascar: analysis using hierarchical, spatially explicit, Bayesian regression models. Ecol Modell 185:105–131

Brönmark C, Hansson L (2005) The biology of lakes and ponds. Oxford University Press, New York

Endoh S, Yamashita S, Kawakami M, Okumura Y (1999) Recent warming of Lake Biwa water (in Japanese). Jpn J Limnol 60:223–228

Hamabata E, Kobayashi Y (2002) Present status of submerged macrophyte growth in Lake Biwa: recent recovery following a summer decline in the water level. Lake Reserv Res Manag 7:331–338

Hayami Y, Fujiwara T (1999) Recent warming of the deep water in Lake Biwa (in Japanese). Res Ocean 8:197–202

Hsieh CH, Ishikawa K, Sakai Y, Ishikawa T, Ichise S, Yamamoto Y, Kuo TC, Park HD, Yamamura N, Kumagai M (2010) Phytoplankton community reorganization driven by eutrophication and warming in Lake Biwa. Aquat Sci 72:467–483

Ichise S, Morita T, Wakabayashi T, Kusuoka Y, Nishino M (2004) Disappearance and its past distribution of the endemic protista *Difflugia biwae* Kawamur, 1918 in Lake Biwa, central Japan (in Japanese). Rep Shiga Pref Inst Public Health Environ Sci 39:57–63

Kagatsume T (2012) Water conservation policy of Shiga prefectural government. In: Kawanabe H, Nishino M, Maehata M (eds) Lake Biwa: interactions between nature and people. Springer, New York, pp 423–427

Karube Z, Sakai Y, Takeyama T, Okuda N, Kohzu A, Yoshimizu C, Nagata T, Tayasu I (2010) Carbon and nitrogen stable isotope ratios of macroinvertebrates in the littoral zone of Lake Biwa as indicators of anthropogenic activities in the watershed. Ecol Res 25:847–855

Legendre P, Gallagher E (2001) Ecologically meaningful transformations for ordination of species data. Oecologia (Berl) 129:271–280

Light T, Marchetti MP (2007) Distinguishing between invasions and habitat changes as drivers of diversity loss among California's freshwater fishes. Conserv Biol 21:434–446

Nakamura M, Ogino Y, Akiyama M, Moriyasu K (2012) Evolving history of Lake Biwa and Yodo River basin management. In: Kawanabe H, Nishino M, Maehata M (eds) Lake Biwa: interactions between nature and people. Springer, New York, pp 371–418

Nakanishi M, Sekino T (1996) Recent drastic changes in Lake Biwa bio-communities, with special attention to exploitation of the littoral zone. GeoJournal 40:63–67

Nishino M (2012) Biodiversity of Lake Biwa. In: Kawanabe H, Nishino M, Maehata M (eds) Lake Biwa: interactions between nature and people. Springer, New York, pp 31–153

Nishino M, Watanabe NC (2000) Evolution and endemism in Lake Biwa, with special reference to its gastropod mollusc fauna. Adv Ecol Res 31:151–180

Nishino M, Azuma Y, Tatumi M, Kaneko Y (2012) Ecological changes in the littoral of Lake Biwa and its adjacent areas. In: Kawanabe H, Nishino M, Maehata M (eds) Lake Biwa: interactions between nature and people. Springer, New York, pp 157–160

R Development Core Team (2012) R: a language and environment for statistical computing. R Foundation for Statistical Computing, Vienna. http://www.R-project.org/

Rossiter A (2000) Lake Biwa as a topical ancient lake. Adv Ecol Res 31:571–598

Rice R (1989) Analyzing tables of statistical tests. Evolution 43:223–225

Sakai Y, Karube Z, Takeyama T, Kohzu A, Yoshimizu C, Nagata T, Tayasu I, Okuda N (2013) Seasonal and site-specific variability in terrigenous particulate organic carbon concentration in near-shore waters of Lake Biwa, Japan. Limnology 14:167–177

Scheffer M, Carpenter S, Foley J, Folke C, Walker B (2001) Catastrophic shifts in ecosystems. Nature (Lond) 413:591–596

Shiga Prefectural Fishery Experimental Station (1972) Reports of the survey of littoral zone in Lake Biwa in 1969 (in Japanese). Shiga Prefectural Fishery Experimental Station, Shiga

Shiga Prefectural Fishery Experimental Station (1998) Reports of the survey of littoral zone in Lake Biwa in 1995 (in Japanese). Shiga Prefectural Fishery Experimental Station, Shiga

Shiga Prefectural Fishery Experimental Station (2005) Reports of the survey of littoral zone in Lake Biwa in 2002–2003 (in Japanese). Shiga Prefectural Fishery Experimental Station, Shiga

Vadeboncoeur Y, Lodge DM, Carpenter SR (2001) Whole-lake fertilization effects on distribution of primary production between benthic and pelagic habitats. Ecology 82:1065–1077

Vadeboncoeur Y, Jeppesen E, Vander Zanden MJ, Schierup H, Christoffersen K, Lodge DM (2003) From Greenland to green lakes: cultural eutrophication and the loss of benthic pathways in lakes. Limonol Oceanogr 48:1408–1418

Chapter 9
Research on the Biodiversity of the Seas Surrounding Malaysia

Zulfigar Yasin, Shau-Hwai Aileen TAN, and Yoshihisa Shirayama

Abstract This chapter integrates the findings of reviews of the marine biodiversity of Malaysian waters, providing a perspective on what is known and the major scientific gaps. Although Malaysia's seas are located at the world's most diverse region, the number of studies on marine biodiversity is considered relatively limited as the result of lack of funding and experts for taxonomic work. Deep-sea habitats also have not been explored, and species-rich taxonomic groups, especially of smaller organisms, remain poorly studied. In this chapter, we discuss how collaboration with international institutions is necessary to fill the knowledge gaps and improve the productivity and discovery of marine biodiversity in Malaysian waters.

Keywords Coral reef • Malaysian waters • Marine biodiversity • Matahari Expedition • NaGISA • SEAFDEC

Introduction

The Seas Surrounding Malaysia

Malaysia's coastline, stretching along the Malay Peninsula, Sabah, and Sarawak, bounds much of the southern part of the South China Sea and is one of the world's most diverse and productive seas. The country today has a 200-nautical-mile Exclusive Economic Zone (EEZ), making Malaysia's seas larger than its land territory. Malaysia is therefore very much a maritime nation with a long maritime

Z. Yasin • S.-H.A. TAN (✉)
School of Biological Sciences, Universiti Sains Malaysia, 11800 Penang, Malaysia
e-mail: aileen@usm.my

Y. Shirayama
Japan Agency for Marine-Earth Science and Technology (JAMSTEC), 2-15 Natsushima Cho, Yokosuka City, Kanagawa 237-0061, Japan

S.-i. Nakano et al. (eds.), *Integrative Observations and Assessments*, Ecological Research Monographs, DOI 10.1007/978-4-431-54783-9_9, © Springer Japan 2014

history strongly shaped by its geographic affinity to the sea. From the myriad species that inhabit the waters and coastal habitats, to the rich economic resources, Malaysia's seas are an ecosystem to be treasured and sustained.

Malaysia, both the Peninsular and Sabah-Sarawak, is surrounded by four seas. The Andaman Sea is found northwest of Peninsular Malaysia, with the Straits of Malacca and Johor along the west and south of the peninsula. The South China Sea lies off the east coast of Peninsular Malaysia and also the west coast of Malaysian Borneo, making up the state of Sabah and Sarawak. Sabah is bordered by the Sulu Sea to the north and the Sulawesi Sea to the south. Figure 9.1 shows the seas surrounding Malaysia.

It is now generally acknowledged that both the land and the marine areas around the region, which encompasses Malaysia, are one of the most biologically diverse areas in terms on the planet.

Malaysian Policy on Biodiversity

The Convention on Biological Diversity was adopted in Rio de Janeiro in June 1992 and had called for adoption of national strategies to conserve biodiversity. Following that, Malaysia developed the National Policy on Biological Diversity, which was officially adopted in 1998 under the Ministry of Science Technology and Innovation (MOSTI). The vision and policy statement of the document are as below:

- *Vision:* To transform Malaysia into a world center of excellence in conservation, research, and utilization of tropical biological diversity by the year 2020.
- *Policy statement:* To conserve Malaysia's biological diversity and to ensure that its components are utilized in a sustainable manner for the continued progress and socioeconomic development of the nation.

The overall framework of the Policy contains 15 strategies (Table 9.1) and 86 action plans, ranging from improving scientific knowledge and research to integrating biodiversity considerations into planning, as well as institutional and legislative development.

Studies on the Biodiversity of the Region

There have been four noteworthy programs to study the biodiversity of the seas around Malaysia. The programs, the scope of their research, and the areas covered are given in Table 9.2.

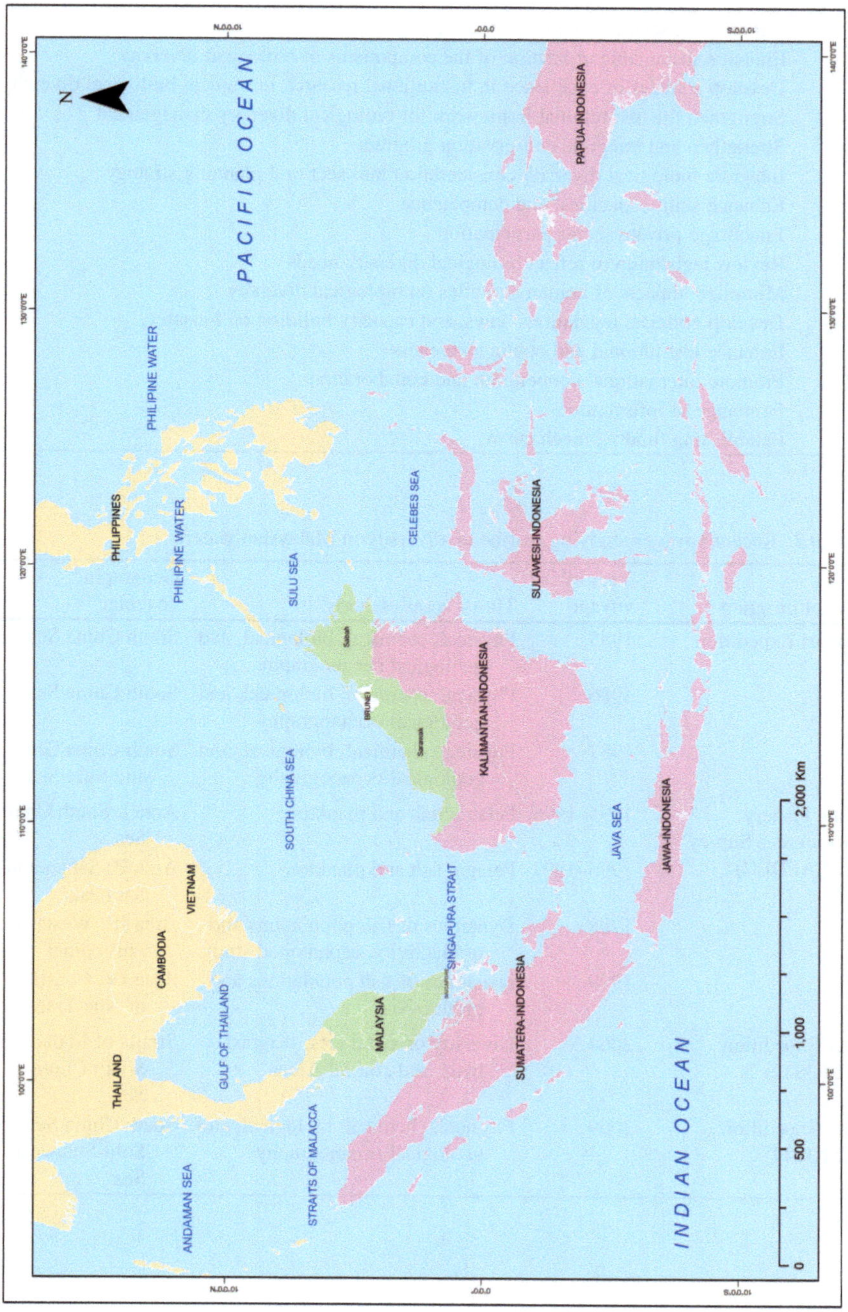

Fig. 9.1 The seas surrounding Malaysia

Table 9.1 Strategies in the Malaysian Policy on Biodiversity

Strategy	Component
I	Improve the scientific knowledge base
II	Enhance sustainable utilization of the components of biological diversity
III	Develop a center of excellence in institutional research in tropical biological diversity
IV	Strengthen the institutional framework for biological diversity management
V	Strengthen and integrate conservation program
VI	Integrate biological diversity consideration into sect oral planning strategy
VII	Enhance skill, capacities, and competence
VIII	Encourage private sector participation
IX	Review legislation to reflect biological diversity needs
X	Minimize impacts of human activities on biological diversity
XI	Develop policies, regulations, laws, and capacity building on biosafety
XII	Enhance institutional and public awareness
XIII	Promote international cooperation and collaboration
XIV	Exchange of information
XV	Establishing funding mechanism

Table 9.2 Recent voyages studying marine biodiversity in Malaysian waters

Name of program	Year of voyage	Thrust areas of study	Geographic coverage
Matahari Expedition	1985	Physical, chemical, biological, and geological oceanography	South China Sea
	1986	Physical, chemical, biological, and geological oceanography	South China Sea
	1987	Physical, chemical, biological, and geological oceanography	South China Sea and Sulu Sea
Marine Fishery Resources Survey (SEAFDEC)	1995–1996	Pelagic fish and plankton	Area I: South China Sea
	1996–1997	Pelagic fish and plankton	Area II: Sabah and Sarawak
	1998	Dynamics of fish populations and productivity, cephalopod study	Area III: Western Philippines
	1999	Dynamics of fish populations and productivity	Area IV: Coastline of Nha Trang
Roses Expedition (MOSTI)	2004	Diversity on coral refs, seagrass beds, and coastal Areas	Straits of Malacca, South China Sea
Prime Expedition (MOSTI)	2009	Physical, chemical, biological, and geological oceanography	South China Sea, Sulu-Sulawesi Sea

Matahari Expedition

The Matahari Expedition was initiated to determine the nature and distribution of living resources in the Exclusive Economic Zone (EEZ) of Malaysia. Within the EEZ is an area of 312 km of the seas adjacent to Malaysia's coast, with an estimation of 332,673 km^2 or four times her land area of the seas around her. The Matahari Expedition was a 5-year program to survey the offshore waters of Malaysia in the South China Sea and the Sulu Sea.

The expedition started in 1985, covering an area of 8,000 km off the coast of Terengganu. In 1986, 12,000 km^2 was covered off the coast of Pahang and east Johor. In 1987 and 1989, the expedition moved to the offshore waters off Sarawak and Sabah, respectively. All the Matahari Expeditions were conducted aboard the research vessel *Kagoshima Maru*, belonging to the Faculty of Fisheries, Kagoshima University. Figure 9.2 shows the study areas of all the Matahari expeditions.

The Malaysian expedition team, headed by Universiti Pertanian Malaysia, was composed of 31 scientists, students, technicians, and laboratory assistants. Japanese experts under the auspices of the Japan International Cooperation Agency (JICA) were also involved in the expedition. The scientific disciplines covered a wide variety of fields of marine science including physical, chemical, biological, and geological oceanography, fisheries, and marine pollution.

Marine Fishery Resources Survey (SEAFDEC)

SEAFDEC (Southeast Asian Fisheries Development Corporation) is a regional cooperative program for the Southeast Asian nations. Its members consist of Brunei, Cambodia, Indonesia, Japan, Malaysia, Philippines, Singapore, Thailand, and Vietnam. SEAFDEC has carried out a large research program in the water of its member countries in the South China Sea for many years.

Perceiving the vital importance and the deficiency of fundamental information and data for fisheries planning and management, the SEAFDEC had launched an Inter-Departmental Collaborative Research Programme in 1995. The main aim of the research program was to evaluate the fisheries resources and their associated parameters that stretched from the Gulf of Thailand, the South China Sea, and the Philippines seas, as a basal objective in the development of sustainable fisheries. Five main fields of scientific endeavor were studied among the 30 or 40 scientists in each cruise. The cruises were conducted using the Training & Research Vessel M.V. *SEAFDEC*.

The program was divided into four defined survey areas. Table 9.3 shows the information on the different cruises. Figure 9.3 shows the study area of the survey done by SEAFDEC.

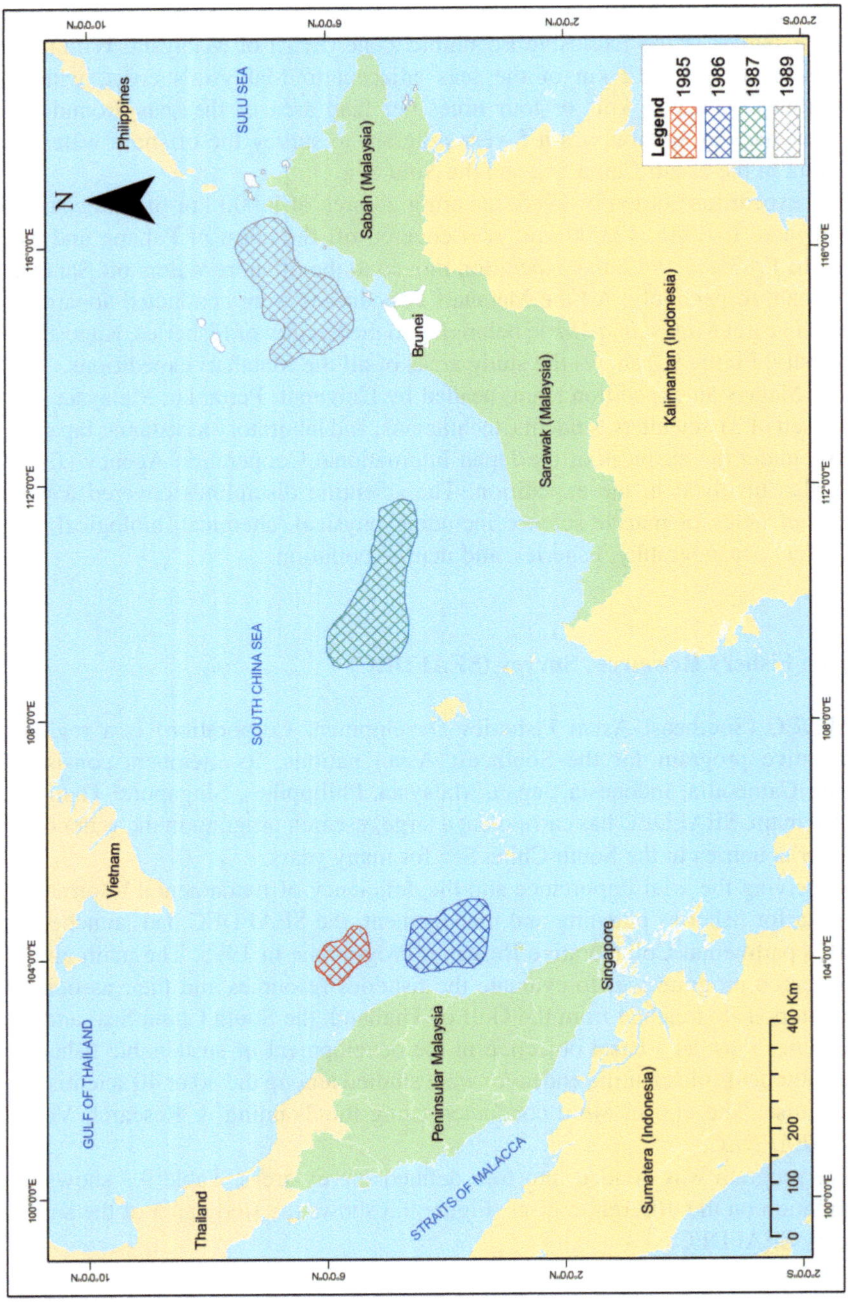

Fig. 9.2 Study areas covered during the Matahari Expedition in 1985, 1986, 1987, and 1989

Table 9.3 Areas surveyed during the Marine Fishery Resources Survey by SEAFDEC

Area	Year	Geographic coverage
I	1995–1996	Gulf of Thailand and the East Coast of Peninsular Malaysia
II	1996–1997	Sabah, Sarawak, and Brunei
III	1998	Western Philippines
IV	1999	Coastline of Nha Trang, Vietnam

The ROSES Expedition 2004

The Research on the Seas and Islands of Malaysia (ROSES) program is an ambitious project that aimed to explore the richness of Malaysia's marine realm for the Science of Biotechnology and Biodiversity. A scientific voyage sponsored by the Ministry of Science, Technology and Innovation and led by Universiti Sains Malaysia, the marine voyage started in June 2004. The program involved 11 Malaysian institutions active in marine science including universities, government agencies, the private sector, and the Malaysian Navy.

On board the research vessel, the *Allied Commander*, were 30 marine scientists supported by more than 52 other staff in various laboratories around the country. The scientific disciplines ranged from satellite-based remote sensing, marine biotechnology, environmental science, coral reef studies, marine taxonomy, and reference collection to remote monitoring of the oceans. In all, 38 reefs were investigated from the Straits of Malacca to the extreme boundaries of Malaysia's South China Sea. It was the longest and largest voyage ever conducted for the nation, covering a distance of more than 7,000 km. Figure 9.4 shows the survey route during the ROSES expedition.

The Prime Marine Scientific Expedition 2009

The Prime Marine Scientific Expedition 2009 was organized by the Ministry of Science, Technology and Innovation of Malaysia, through the National Oceanography Directorate in close cooperation with the National Hydrographic Centre, Royal Malaysian Navy. The expedition, which began on 18 June 2009 and finished on 2 August 2009, covered the Malaysian territorial waters of the South China Sea and the Sulu-Sulawesi Seas, using survey vessel KD *PERANTAU* of the Royal Malaysian Navy. Figure 9.5 shows the survey route during the expedition.

It was envisioned to be the catalyst for more sustainable oceanographic and marine scientific research among Malaysian marine scientists. This 46-day marine scientific expedition, an integrated program that involved 56 researchers and students from various government agencies and marine science and oceanography institutes from the local universities, encompassed all major disciplines including biological, geological, chemical, and physical oceanography.

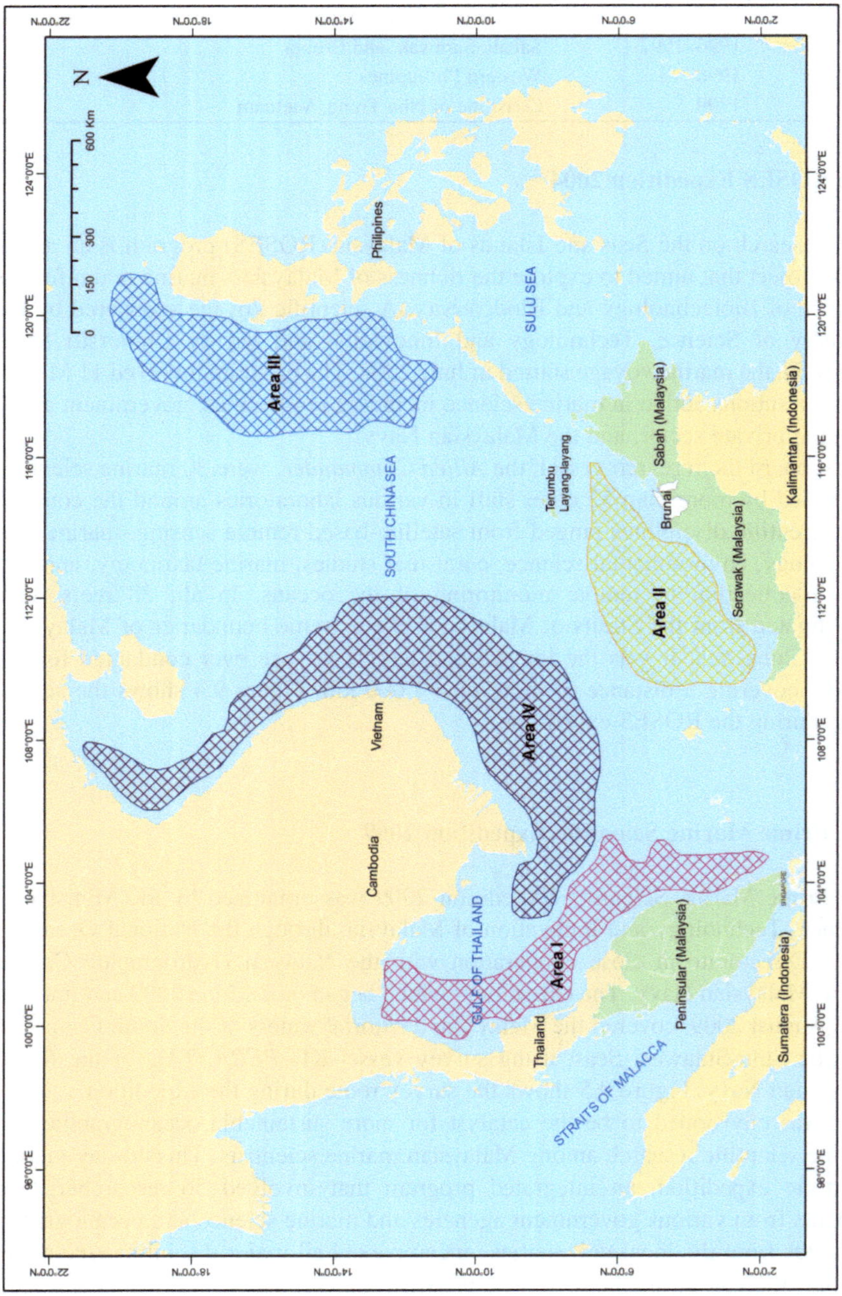

Fig. 9.3 Survey areas covered by SEAFDEC (Southeast Asian Fisheries Development Corporation) from 1995 to 1999

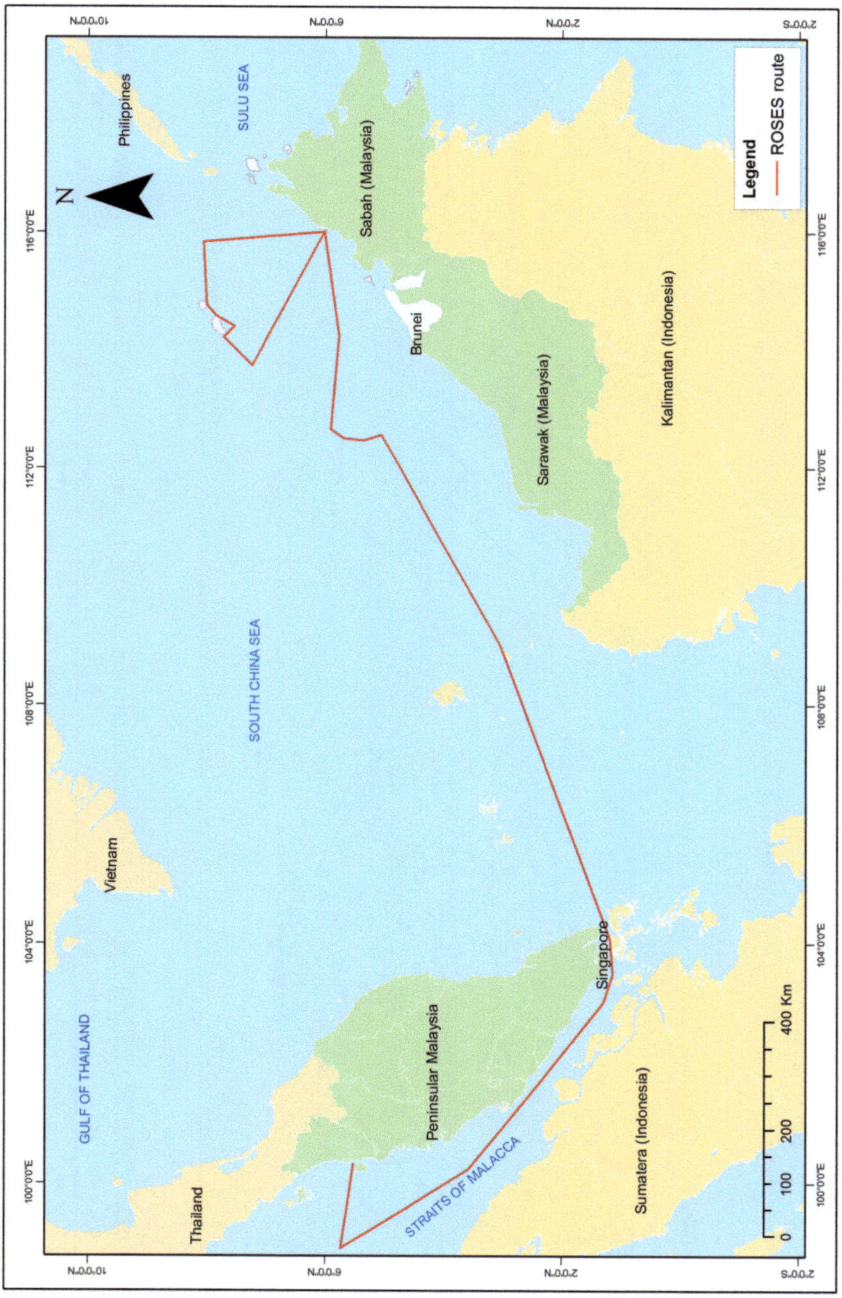

Fig. 9.4 Survey route during the ROSES Expedition in 2004

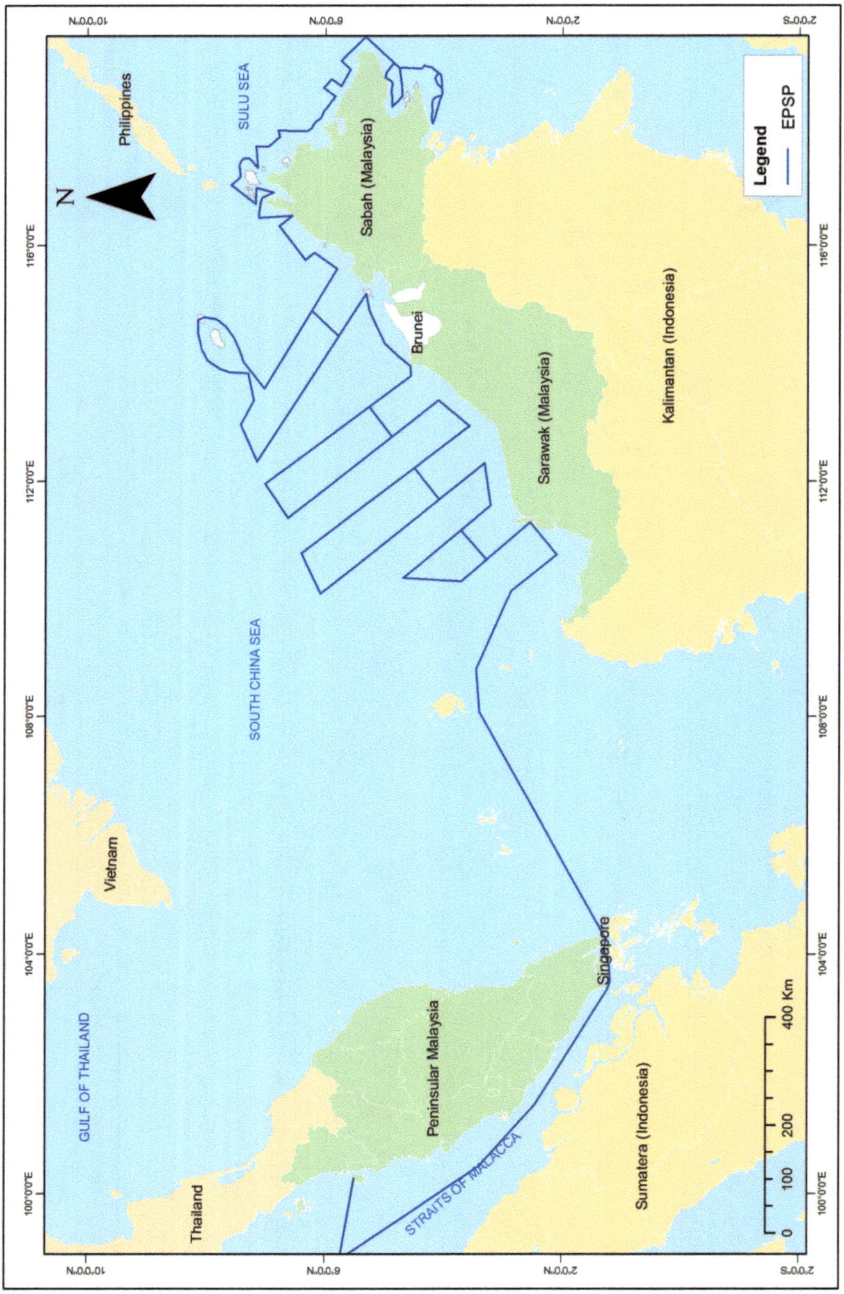

Fig. 9.5 Survey route during the Prime Marine Scientific Expedition in 2009

Collaborative Research on Marine Biodiversity Studies

Although located in the Earth's richest genetic resource, there is little or very limited taxonomic capability among Malaysian scientists. Malaysia had relied on the major world centers of expertise during the early colonial days. Malaysia is now faced with a taxonomic crisis, especially in the pursuit of national programs for food security, conservation, and sustainable use of natural resources and biodiversity.

Following that, Malaysian has been seeking international assistance through networking and collaborative research to establish a team of trained taxonomists in marine biodiversity. Below are some of the organizations with which Malaysian scientists have been working closely during past decades:

- Natural Geography In Shore Areas (NaGISA) is one of the first projects to participate in the Census of Marine Life (CoML)
- ASEAN-Australian Project on Marine Science
- Japanese Society for the Promotion of Science (JSPS)
- IOC-Westpac Projects, e.g., Invasive Species, HAB
- The Malaysian Antarctic Research Programmed (MARP)
- Coral Triangle Initiative (CTI)
- ASEAN Centre for Biodiversity (ACB)

Through this networking, collaborative research involving field sampling, taxonomy workshops, and seminars were conducted. Emphasis was on capacity building of young taxonomists in the field of marine biodiversity.

Methods

The published and unpublished data from many institutions in Malaysia were brought together to review what was known about marine biodiversity in this region. Marine biodiversity in Malaysia had been studied during the various expeditions but most of the studies were done on the coastal areas. Figures 9.6, 9.7, 9.8, 9.9, 9.10, and 9.11 show the study areas where marine biodiversity research was conducted.

In this chapter, we investigated the collective knowledge assembled by the various institutions and summarized their research resources, depository areas, and issues on the study of marine biodiversity in Malaysia. Future directions of marine biodiversity study in Malaysia are also discussed here.

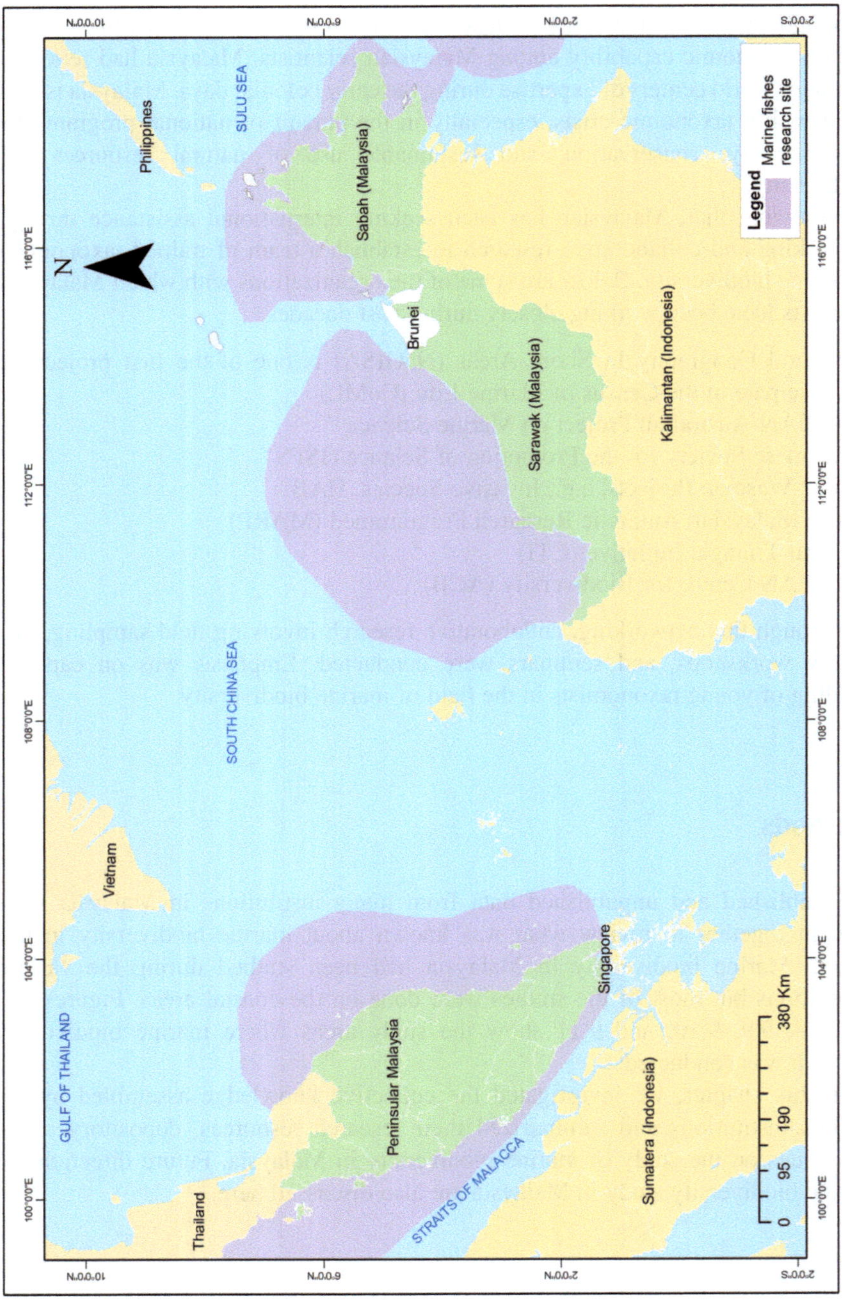

Fig. 9.6 Research sites of marine fishes in Malaysia

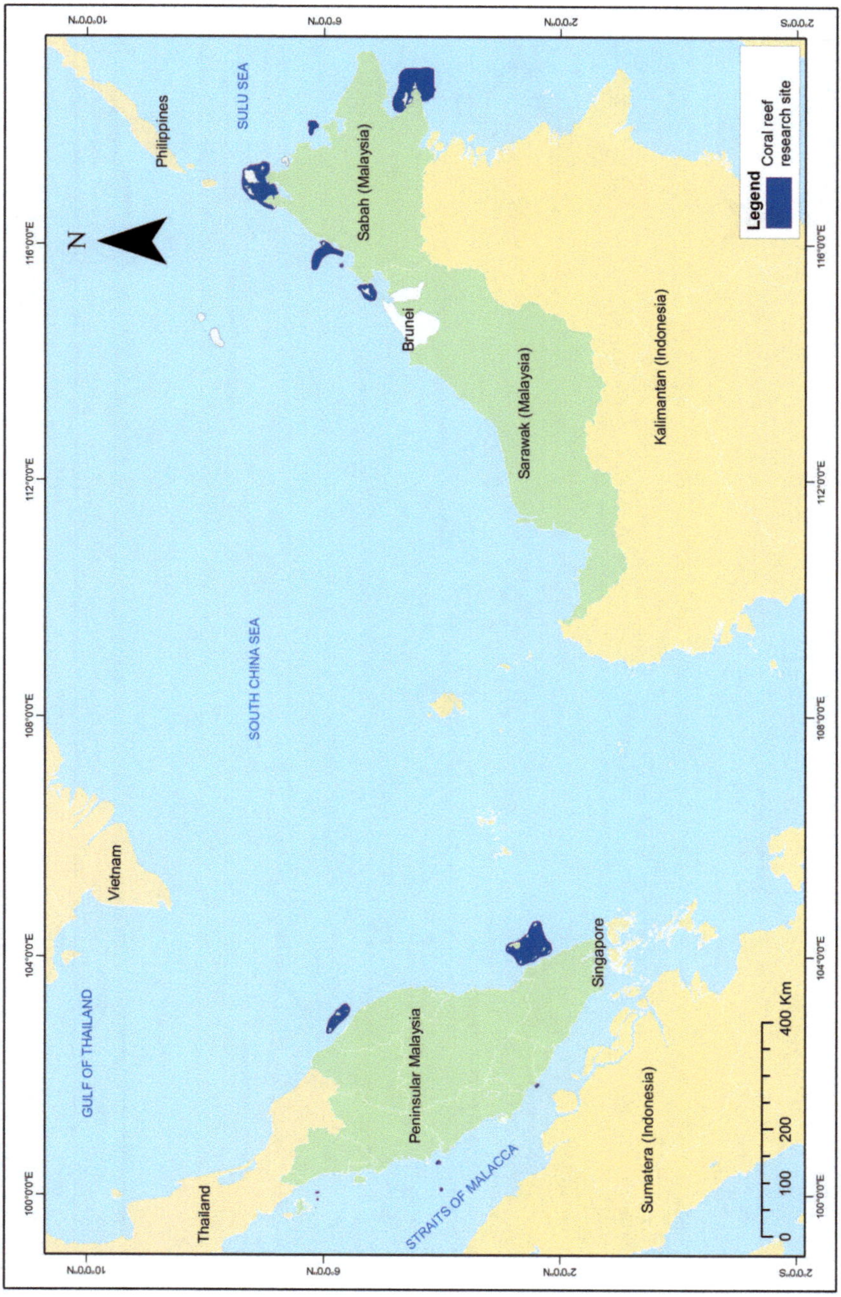

Fig. 9.7 Research sites of coral reefs in Malaysia

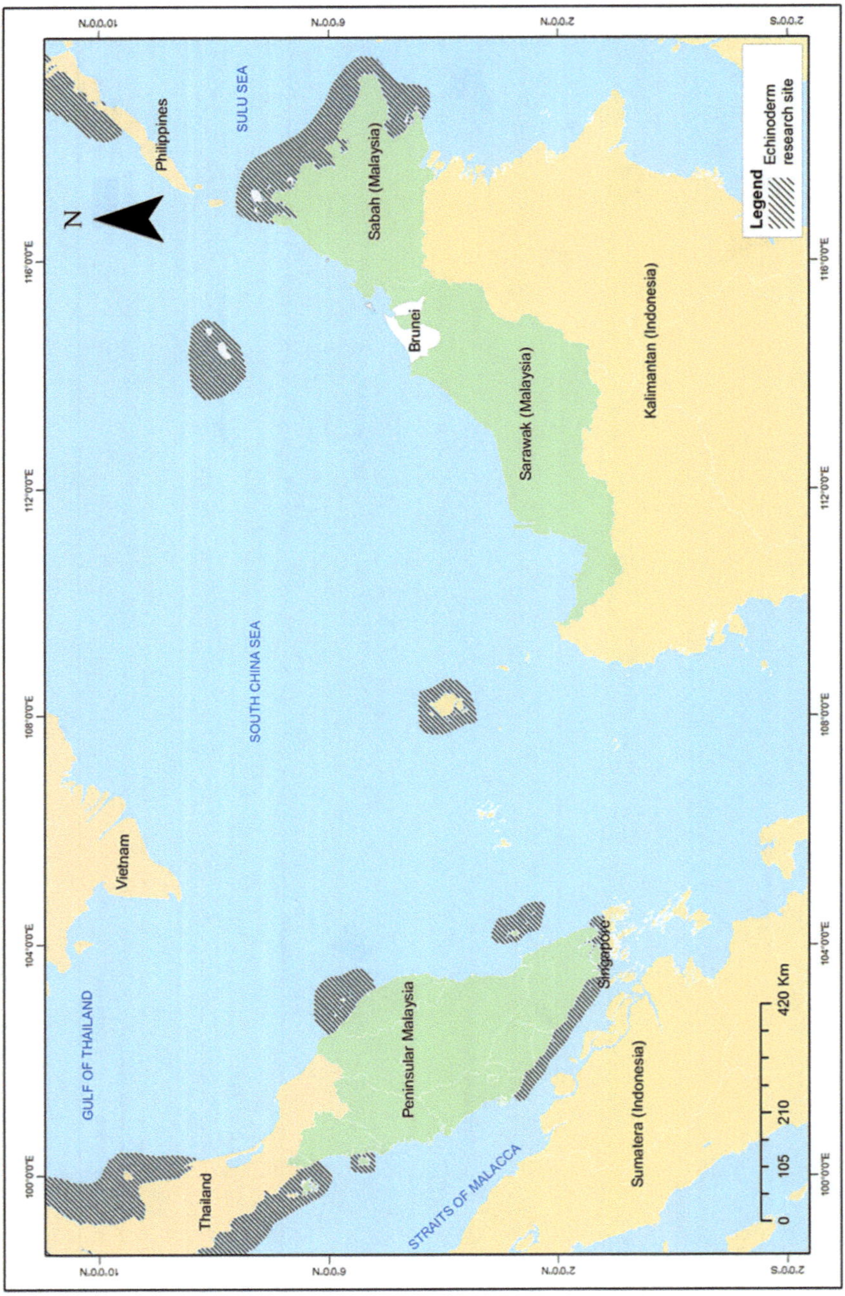

Fig. 9.8 Research sites of echinoderms in Malaysia

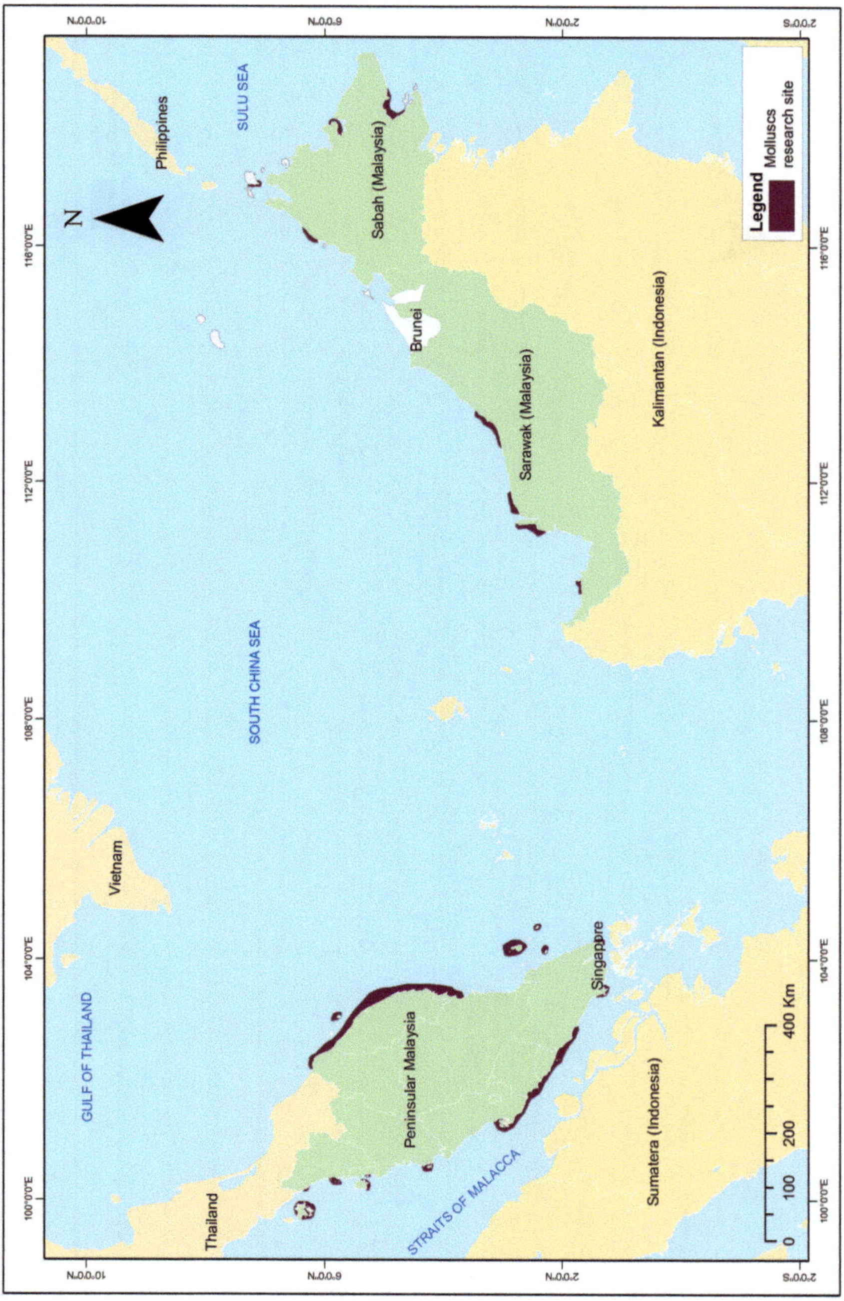

Fig. 9.9 Research sites of mollusks in Malaysia

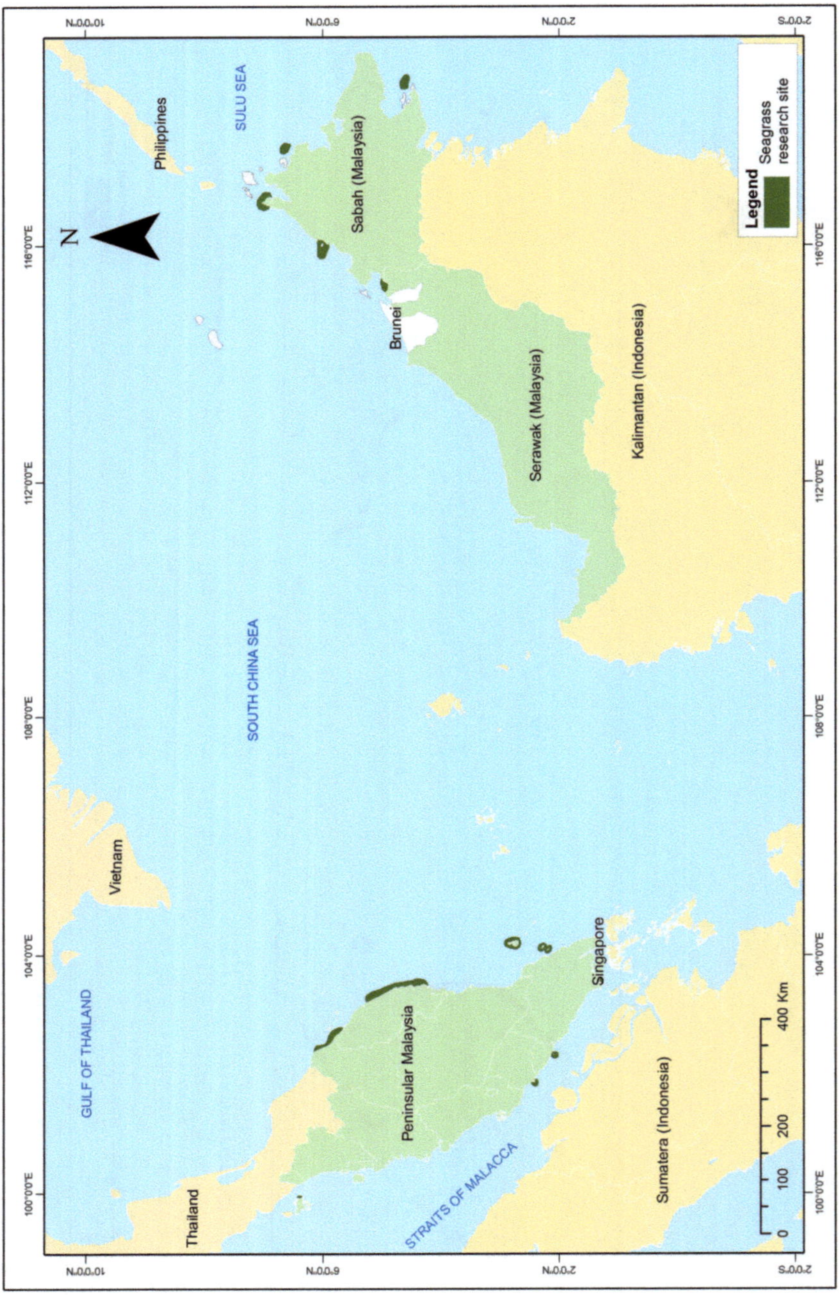

Fig. 9.10 Research sites of seagrass in Malaysia

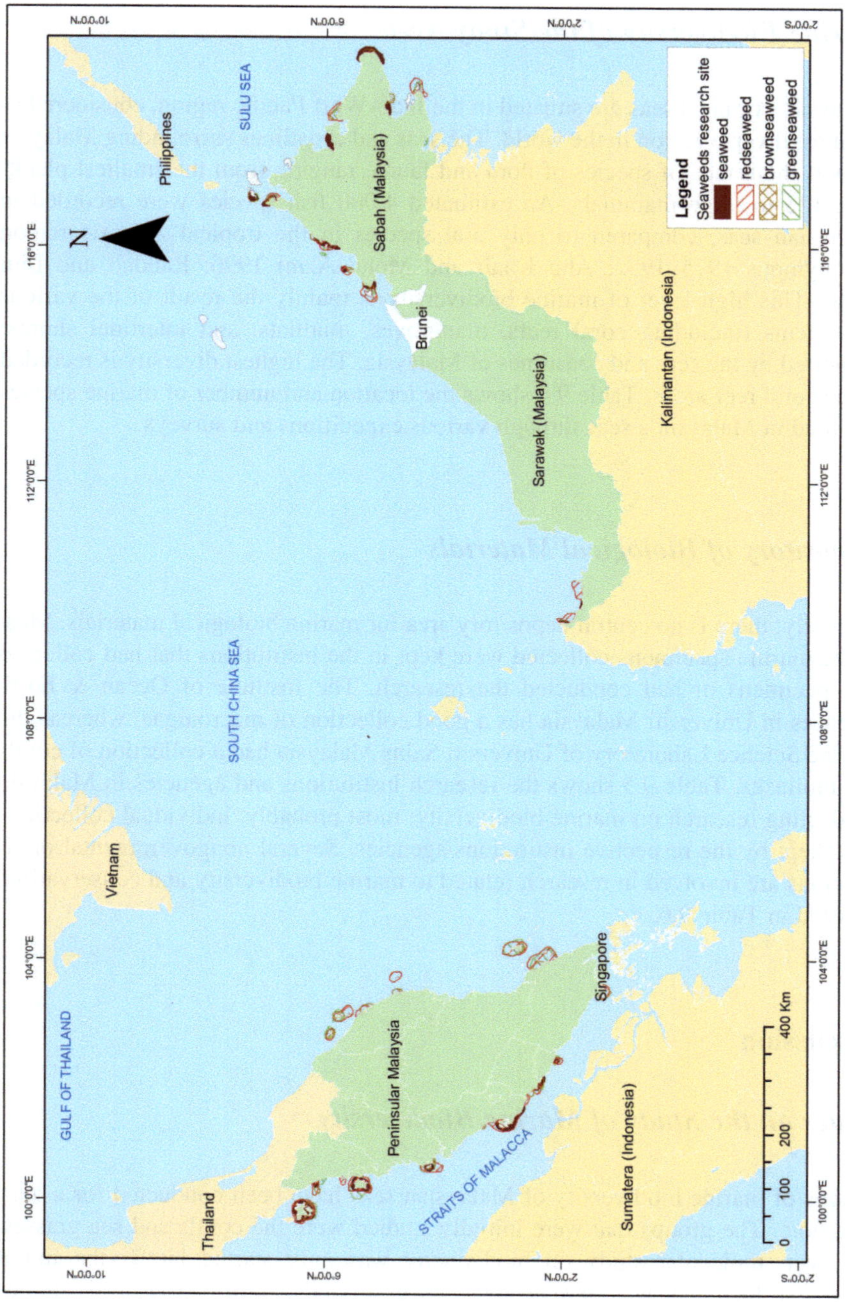

Fig. 9.11 Research sites of seaweeds in Malaysia

Results

Marine Biodiversity of the Study Area

Malaysia's tropical seas are situated in the Indo-West Pacific region, considered as the most diverse region in the world. The seas and coastlines surrounding Malaysia support thousands of species of flora and fauna, ranging from the smallest planktonic life to large mammals. An estimated 4,000 fish species were recorded in Malaysian seas, compared to only 500 species in the tropical Atlantic region (Anonymous 1975–1992; Abu Khair and Mohd Azmi 1996; Randall and Lim 2000). This high level of marine biodiversity is mainly the result of the various ecosystems (including coral reefs, mangroves, mudflats, and intertidal shores) supported by the seas and coastlines of Malaysia. The highest diversity is recorded in the coral reef areas. Table 9.4 shows the location and number of marine species recorded in Malaysia's seas through various expeditions and surveys.

Depository of Biological Materials

Currently, there is no central depository area for marine biological materials. Most of the marine specimens collected were kept in the institutions that had collected the specimens or had conducted the research. The Institute of Ocean & Earth Sciences in Universiti Malaysia has a good collection of macroalgae, whereas the Marine Science Laboratory of Universiti Sains Malaysia had a collection of corals and mollusks. Table 9.5 shows the research institutions and agencies in Malaysia conducting research on marine biodiversity; most probably, individual collections were kept by the respective institutions/agencies. Several nongovernmental organizations are involved in research related to marine biodiversity and conservation, as listed in Table 9.6.

Discussion

Issues on the Study of Marine Biodiversity

Studies of marine biodiversity of Malaysian seas have been conducted for a very long time. The groups that were initially studied were the corals and sea grasses. However, molecular study on biodiversity was only started lately; the initial published document was in year 2001, on gastropods. Information of all studies conducted on the marine biodiversity in Malaysian seas, based on published articles, is listed in the Appendix.

Table 9.4 Location and number of marine species recorded in Malaysia's seas

Classes	Location	Number of species
Bivalves	West Malaysia and Singapore	186
	East Coast Peninsular Malaysia	91
	Darvel Bay, Sabah	43
	South China Sea	12
	Malaysia	100
Cephalopods	Malaysia	10–18
Corals	Straits of Malacca	186
	West of South China Sea	657
	Sulu Sea	252
Crustaceans	East Coast of Peninsular Malaysia	7
	Southwestern South China Sea	14
	Sabah, Sarawak, and Brunei	22
Echinoderms	Gulf of Thailand	199
	Straits of Malacca	23
	Philippine Islands and adjacent seas	55
	Celebes Sea	94
	West of South China Sea	68
	South China Sea	982–1,010
	Central of South China Sea	41
	Andaman Sea	288
Fishes	South China Sea	795–3,365
	Malaysia	317–1,500
Gastropods	West Malaysia and Singapore	330
	East Coast Peninsular Malaysia	217
	Darvel Bay, Sabah	2–125
	South China Sea	218
	West Coast Malaysia Borneo	31
	Celebes Sea	2
Macroalgae	West of South China Sea	120
	Straits of Malacca	352
	East of South China Sea	51
	Malaysia	35–867
	Anambas and Natuna	77
Microalgae	West coast Sabah	4
	Johor	No data
	Malacca	2
	Singapore	2
	Sarawak	2
Microzooplankton	Gulf of Thailand and East Coast of Peninsular Malaysia	16
Phytoplankton	Gulf of Thailand and East Coast of Peninsular Malaysia	242
	Sabah, Sarawak, and Brunei waters	83–291
Protozoa	Gulf of Thailand and East Coast of Peninsular Malaysia	8
	Sabah, Sarawak, and Brunei waters	8
Seagrass	Malaysia	14
Scaphopods	Malaysia	20
Zooplankton	Sabah, Sarawak, and Brunei waters	39

Information obtained from various reports of expeditions and cruises

Table 9.5 Academic and research institutions in Malaysia conducting research on marine biodiversity

Name of institution	Marine Science Units/Departments
Universiti Kebangsaan Malaysia	• Marine Ecosystem Research Centre (EKOMAR) • School of Natural Resources and Environmental Sciences • Faculty of Science and Technology
Universiti Malaya	• Institute of Biological Sciences • Institute of Ocean and Earth Sciences (IOES)
Universiti Malaysia Terengganu	• Department of Marine Science • Institute of Oceanography (INOS) • Institute of Tropical Aquaculture
Universiti Putra Malaysia	• Laboratory of Marine Science and Aquaculture • Institute of Bioscience • Department of Biology • Bintulu Campus (Sarawak)
Universiti Sains Malaysia	• Marine Science Laboratory • School of Biological Sciences • Centre For Marine and Coastal Studies
International Islamic University	• Department of Biotechnology • Institute of Oceanography and Maritime Research
Universiti Teknologi MARA	• Marine Technology Program
Universiti Industri Selangor	
Department of Marine Parks Malaysia	
Johor Parks	
National Oceanographic Directorate	
Sabah Parks	

Table 9.6 Nongovernmental organizations involved in biodiversity-related research in Malaysia

Name of nongovernmental Organizations	Geographic coverage of activities
Malaysian Fisheries Society	National coverage
Malaysia Marine Science Society	National coverage
Malaysian Nature Society	National coverage
Maritime Institute of Malaysia	National coverage
World Fish Centre	International coverage
World Wildlife Fund Malaysia	National and international coverage

There were variances in the number of species recorded for the marine flora and fauna from different research or expeditions/cruises. These disagreements in the number of species do not show the true situation of the diversity but rather the collection efforts as well as the field of interest of the scientists.

Although Malaysia's seas are located at the heart of the world's most diverse region, the number of studies on the marine biodiversity is considered relatively limited. Most marine science research in Malaysia is applied, that is, related to areas such as fisheries and pollution monitoring. There is relatively very little basic research such as taxonomy studies. One of the major setbacks that Malaysian is facing in relationship to marine biodiversity research is the lack of classical taxonomy experts. Malaysia is facing problems in finding and training new and competent young marine taxonomists. Even among the younger generation of marine biologists, taxonomic research has practically no attraction as a career compared to careers in fields such as marine biotechnology or aquaculture.

In addition, most research funding is focusing on management and applied research rather than systematic research.

Geographic Gaps

One of the main issues of concern is the existence of gaps in the spatial and temporal coverage of marine biodiversity inventories. Most of the research conducted was in coastal and shallow-water areas. No deep-sea work has been done, especially in the depths of the southeast portion of the South China Sea and the continental slopes. Given the vast difference between the number of species known now and those waiting to be discovered, increased spatial coverage is critical. Following are some obvious areas where gaps exist:

1. The continental shelves north of Sarawak.
2. The deep seas north of Sabah.
3. The marine corridors of east Sabah.
4. The politically contentious areas of the South China Sea.
5. Areas beyond the air-diving depths of the Malaysian reef areas.
6. The turbid reefs of the Langkawi group and the Pangkor group of islands in the Straits of Malacca.

Closing the gaps in both temporal and spatial coverage depends on availability of ship time and increased budgetary provision for census-type work. Monitoring work that is crucial to determine the status of habitats and ecosystems is no longer funded by the governmental grant-awarding agencies. In this context, it is crucial to increase awareness among governmental funding agencies, international organizations, and donor agencies of the need to support biodiversity research.

Knowledge Gaps

Most research on marine biodiversity was on macrofauna and macroflora, with very limited studies being conducted on smaller groups such as Cyanobacteria, eukaryotic algae, Protozoa, Platyhelminthes, nematodes, copepods, isopods, and tunicates.

It is known that there is a large volume of unpublished information relating to marine and coastal biodiversity held in government agencies in Malaysia, sometimes compiled as statistics and reports, and also in baseline studies prepared by consultants for environmental impact assessments. Considerable time and effort, together with visits to the individual islands, is required to access this information. There are several incentives for researchers working in academia to publish their results in established peer-reviewed journals. In contrast, there are fewer rewards for those researchers working in government and private consultancies.

Lack of human, institutional, and technical capabilities in Malaysia is the main cause for the heterogeneous level of growth in our understanding of the biodiversity of our seas. The gaps in knowledge extend to several smaller taxa and to large parts of the shelf and deep-sea ecosystems beyond air-diving depths.

There is a need for the development of expertise on marine taxonomy in both the fields of classical taxonomy and its related molecular studies. Human resource development in the field of taxonomy lags behind other fields such as aquaculture, natural resource management, fisheries science, and product development in the responsible governmental department and ministries. One way to overcome this is the pooling of taxonomic resources in the region. Another is to conduct taxonomic workshops to initiate and instill interest in the field of taxonomy in the younger generation of scientists. Such an initiative has been successfully carried out by the Malaysian collaboration with the Japanese Society for the Promotion of Science (JSPS) and the National Geography of In-Shore Areas (NaGISA) project.

Studies involving complete life cycles are also glaringly few in number. Habitat and environmental requirements for life stages of marine species have only been investigated for a few species of commercial fishes such as the barramundi, tiger prawn, the seaweed *Glacilaria*, the blood cockle, edible oysters, and, for conservation—the giant clams.

There is also a significant absence of long-term studies. Most studies are 5 years in duration or shorter, coinciding with the way research grants are awarded in Malaysia. Here, research proposals are solicited continuously but end with the closure of each 5-year national "Malaysian Plan." New grants are again solicited at the beginning of each plan.

Two notable long-term studies that have been carried out are those on the fringing coral reefs of north Langkawi (Juliana et al. 2010), spanning more than 20 years from 1989 to the present, and the studies on the seagrass communities of Pulau Gazumbo, a small island in the Straits of Penang (Zulfigar et al. 2010). Although studies on both sites are still ongoing, these long-term studies so far have been supported from numerous grants. It is difficult to see the source of continued support for these studies. Their future continuity is therefore doubtful.

There is also an urgent need to promote regional cooperation on research to coordinate biodiversity studies in international seas and contentious areas.

Future Directions

One of the future directions in marine biodiversity research is to translate the biological resources into biological health, which involves production and conservation. This phase would combine basic and applied approaches, which include better understanding of the relationships between ecosystem processes and biodiversity as well as the role of biodiversity in production ecosystems.

Other future research will be in monitoring of biodiversity through establishing the utility of biodiversity surrogates and deriving ways of translating them into measures of biodiversity that are accurate, practical, and affordable. Future research will also be directed to issues pertaining to climate change.

Acknowledgments The present study was partly supported by Environment Research and Technology Development Fund #S-9 of Ministry of Environment Japan.

Appendix: Past Studies on the Marine Biodiversity of Malaysia's Seas

Year	Taxon	No. of species	Location	References
Corals				
1857	Corals	3	Strait of Malacca	Milne and Haime (1857)
1880	Corals	49	Southern of Strait of Malacca	Studer (1880)
1878	Three new species were reported	55	Southern of Strait of Malacca	Bruggemann (1878)
1970	18 Scleractinian and three other hard corals	21	Pulau Aur, Johor	Dunn (1970)
1973	Corals	Unknown	Pulau Gaya and Sapi	Lulofs (1973)
1974	*Gorgonian*, *Acropora*, *Pocillopora*, *Porites*, *Acropora*, *Pavona* sp., *Pocillopora*, Faviidae, *Montipora*, *Tubastrea*, *Dendrophyllia*, *Porites tenuis*, *Platygyra lamellina*, *Porites* sp., *Acropora*, *Pocillopora*, *Porites*	38	Pulau Perak, Pulau Jarak, Pulau Sembilan, Pulau Lalang, Pulau Saga, Pulau Rumbia, Straits of Malacca	Pillai and Scheer (1974)
1976–1978	-*Pocillopora*, *Acropora*, *Pavona*, *Porites*, Faviidae, *Echinopora*, *Pachyseris*, *Goniopora*, *Montipora*, *Galaxea*, *Millepora*, *Heliopora*, *Echinopora*, *Physogra*, *Seriatopora*, *Merulina*, hydroid, *Dendronephyta*, *Sarcophyton*	Unknown	Pulau Tiga, Pulau Dinawan, Pulau Sulug, Pulau Manukan, Pulau Mamutik, Sabah	Wood (1979)
	-*Acropora*, *Montipora*, *Porites*, *Faviidae*, *Millepora*, *Goniopora*, *Hydnophora*, *Pocillopora*, *Echinopora*, *Euphyllia*, *Astreopora*, *Galaxea*		K. Kinabalu: Pulau Gaya, Pulau Mengalum, Edgell Patches, Sapangar Bay Island and Pulau Sapi	
	-Faviidae, *Montipora*, *Pocillopora*, *Acropora*, *Heliopora*, *Turbinaria*, *Podabacia*, *Mycedium*, *Lytocarpus*		Pulau Malawali, Pulau Balak-Balak, Pulau Bandanan, Turtle Island, Pulau Gulisaan, Pulau Silingaan, Pulau Bakkungaan	
	Acropora, *Galaxea*, *Montipora*, *Turbinaria*, *Hydroids*, *Sarcophyton*, *Porites*, *Lytocarpus*, *Diploastrea*, *Mycedium*, *Pomacentrus*, *Pavona*, *Tubastrea micrantha*, *Goniastrea*, *Goniopora*, *Pachyseris*, *Pectinia*, *Parahalomitra*, *Alveopora*, *Millepora*		Bodgaya Island, Ligitan Island, Pulau Mabul, Pulau Sipadan	

1978	Thamnasteriidae, Pocilloporidae, Acroporidae, Agariciidae, Fungiidae, Astrocoeniidae, Poritidae, Faviidae, Merulinidae, Oculinidae, Mussidae, Pectiniidae, Caryophyliidae, Dendrophylliidae, Helioporidae, Milleporidae, Stylasteridae	170	Pulau Tioman (111), Pulau Tulai (65), Pulau Tinggi,/Pulau Mentinggi (91), Pulau Sembilang/Pulau Seri Buat(55), Pulau Babi Besar(14), Pulau Babi Hujung(53), Pulau Babi Tengah(15), Pulau Rawa(61), Pulau Gual, Pulau Harimau(42), Pulau Mensirip(27)	De Silva et al. (1980)
1979	52 genera of hard coral	Unknown	Pulau Redang and associated islands	Green et al. (1979)
1981	Scleractinian hermatypes, milliporinid, alcyonaceans, Fungia	42	Cape Rachado, Malaya (Tanjung Tuan,Melaka)	Hong and Sasekumar (1979)
1982	*Acropora*, *Montipora*, *Porites*, *Platygyra*, *Goniopora*, *Diploastrea*, *Plerogyra*,*Galaxea*, *Pocillopora*, *Pavona*, *Hydnophora*, *Favia*, *Fungia*, *Herpolitha* (35 genera)	Unknown	Pulau Payar Island group	De Silva and Ridzwan Abdul (1982)
				Aikanathan and Wong (1994)
1996	Gorgonians: Anthothelidae, Subergorgiidae, Melithaeidae, Acanthogorgiidae, Plexauridae, Ellisellidae	31	Singapore	Goh and Chou (1996)
2000	-*Montipora millepora*, *Acropora elseyi*, and *A. valenciennesi*, *Psammocora haimeana*, *Pavona frondifera*, *Echinophyllia orpheensis*, *Platygyra* ‘sp. 1’, *Catalaphyllia jardini*, *Turbinaria frondens*, *Leptoseris incrustans*, *Favites pentagona*	149	Pulau Redang	Harborne et al. (2000)
	Pocillopora eydouxi and *P. meandrina*, *Stylophora pistillata*, *Montipora confusa* and ‘sp. 2’, *Astreopora gracilis*, *Psammocora nierstraszi*, *Coscinaraea hahazimaensis*, *Pavona clavus*, *Leptoseris papyracaea*, *Diaseris distorta*, *Heliofungia actiniformis*, *Oxypora crassispinosa*, *Pectinia alcicornis*, *Favia maxima*, ‘sp. 1’, *F. veroni and* ‘sp. 2’,	183	Pulau Tioman	

(continued)

(continued)

Year	Taxon	No. of species	Location	References
	Goniastrea retiformis, *Montastrea* 'sp. 1', *Trachyphyllia geoffroyi*, *Euphyllia yaeyamaensis*, *Rhizopsammia verrilli*, *Millepora tenera* and possibly *Goniopora planulata*, *Leptoseris scabra*, *Acropora austere*			
	Montipora, *Acropora*, *Psammocora*, *Acanthastrea*, *Lobophyllia*, *Favites*	155	Pulau Tinggi	
2001	*Acropora*, *Montipora*, *Fungia*, *Porites*, *Pavona*, *Leptoseris*, *Turbinaria*, *Astreopora*, *Lobophyllia*	252	Pulau Banggi—9 reef sites	Fenner (2001)
2004	*Acropora*, soft coral	38	South China Sea	Zulfigar and Tan (2010)
2006	Hard coral (32 spp.), soft coral (5 spp.)	37	Bintulu, Sarawak	Idris et al. (2006)
	Heliophora	Unknown	Pulau Weh, Pulau Banyak, Pulau Rondo	IUCN (2006)
2007	Coral (18 genera)	Unknown	Pulau Tioman, Pulau Payar, Pulau Perhentian, Pulau Pangkor, Pulau Sibu	Toda et al. (2007)
2008	Astrocoeniidae, Pocilloporidae, Acroporidae, Poritidae, Siderastridae, Agariciidae, Fungiidae, Oculinidae, Pectinidae, Mussidae, Merulinidae, Faviidae, Trachyphyllidae, Caryophylliidae, Dendrophyllidae, Tubiporidae, Helioporidae, Milleporidae	156	Pulau Galang–Batam(RRI)	Mannuputi (2008)
	Acroporidae, Agariciidae, Dendrophyllidae, Pocilloporidae, Euphyllidae, Oculinidae, Faviidae, Mussidae, Fungiidae, Merulinidae, Pectiniidae, Poritidae, Siderastreidae	75	Pulau Bintan	Siringoringo (2008)
	Acroporidae, Agariciidae, Pocilloporidae, Euphyllidae, Oculinidae, Faviidae, Mussidae, Fungiidae, Merulinidae, Pectinidae, Poritidae,	154	Pulau Belitung	Siringoringo and Budiyanto (2008)

	Siderastreidae, Trachyphyllidae, Dendrophylliidae, Tubiporidae, Helioporid			
2009	Acroporidae, Agariciidae, Astrocoeniidae, Dendrophylliidae, Euphyllidae, Faviidae, Fungiidae, Merulinidae, Mussidae, Oculinidae, Pectiniidae, Pocilloporidae, Poritidae, Siderastreidae, Trachyphylliidae	161	Singapore	Huang et al (2009)
2009	Corals	1	Padang	Hoeksema (2009)
2010	Alcyoniidae, Helioporidae, Paraclcyoniidae, Xeniidae, Briareidae	25	Singapore	Benayahu and Chou (2010)
Echinoderms				
1971	Echinoidea	9	Gulf of Thailand	Clark and Rowe (1971)
		5	Straits of Malacca	
	Asteroidea	55	Philippine Islands and adjacent Seas	
		10	Celebes and Molucca Sea	
		7	Straits of Malacca	
1987	Crinoidea (13 spp.)	84	Bodgaya Islands and Sipadan Group Islands	George and George (1987)
	Asteroidea (19 spp.)			
	Ophiuroidea (20 spp.)			
	Echinoidea (10 spp.)			
	Holothuroidea (22 spp.)			
1983	Starfishes	1	Pulau Tioman	Chou (1983)
1991	Stichopodidae, Holothuroidea, Echinodermata	Unknown	Indo-Malayan Archipelago	Massin and Lane (1991)
1999	Holothuroidea	25	Johore Group Islands (South China Sea)	Zulfigar and Tan (1999)
2000	Echinoderms	982	South China Sea	Ng and Tan (2000)
	Holothuroidea	18	Johore Group Islands (South China Sea)	Zulfigar et al. (2000)
	Crinoidea (113 spp.)	982	South China Sea	Lane et al. (2000)
	Asteroidea (227 spp.)			

(continued)

(continued)

Year	Taxon	No. of species	Location	References
	Ophiuroidea (272 spp.)			
	Echinoidea (167 spp.)			
	Holothuroidea (203 spp.)			
	Holothuroidea	11	Pulau Payar (Straits of Malacca)	Zaidnuddin and Forbes (2000)
2002	Asteroidea	28	Anambas (South China Sea)	Purwati and Lane (2004)
2004	Crinoidea (35 spp.)	288	Andaman Sea	Putchakarn and Soncheng (2004)
	Asteroidea (52 spp.)			
	Ophiuroidea (87 spp.)			
	Echinoidea (50 spp.)			
	Holothuroidea (64 spp.)			
	Crinoidea (6 spp.)	190	Gulf of Thailand	
	Asteroidea (32 spp.)			
	Ophiuroidea (55 spp.)			
	Echinoidea (42 spp.)			
	Holothuroidea (55 spp.)			
2008	Echinoderm	25	Johore Group Islands (South China Sea)	Fujita et al. (2008)
2009	Asteroidea	20	Central South China Sea	Sim et al. (2009)
	Holothuroidea	21	Central South China Sea	Sim et al. (2009)
Macroalgae				
1978	Macroalgae	57	Pulau Redang	Green (1978)
1981	macroalgae	4	Cape Rachado, Malacca	Goh and Sasekumar (1981)
1991	Macroalgae	212	Malaysia	Phang and Wee (1991) and Phang (1994)

1994	Chlorophytes (19) Rhodophytes (21) Phaeophytes (13)	53	P. Tioman	Ahmad and Rusea (1994)
	Macroalgae	10	Port Klang, Selangor, Straits of Malacca	Aikanathan and Sasekumar (1994)
1999	Rhodophyceae	6	Penang	Masuda et al. (1999)
2000	Rhodophyceae	7	Langkawi	Masuda et al. (2000b)
	Rhodophyceae	6	Sandakan, Sabah	Terada et al. (2000)
	Rhodophyceae	5	Sandakan, Sabah	Masuda et al. (2000a)
2001	Rhodophyceae	5	Terengganu, Johore	Masuda et al. (2001b)
	Rhodophyceae	5	Terengganu	Masuda et al. (2001a)
2002	Rhodophyceae	5	Langkawi, Sandakan, Kota Kinabalu, Johore	Kawaguchi et al. (2002)
	Rhodophyceae	3	Sabah, Sarawak	Masuda et al. (2002)
2003	Rhodophyceae	4	Penang	Tani et al. (2003)
	Rhodophyceae	4	Penang, Melaka,Pulau Labuan, Sipadan	Yamagishi et al. (2003)
2004	Red algae (23 spp.) Green algae (29 spp.) Brown algae (22 spp.) Seagrass (3 spp.)	77	Anambas and Natuna	Liao et al. (2004)
2005	Chlorophytes (78 spp.) Rhodophytes (69 spp.) Phaeophytes (49 spp.) Cynophytes (13 spp.)	209	Malaysia	Mazlan et al. (2005)
	Chlorophytes (102 spp.)	319	Langkawi	Phang et al. (2005)

(continued)

(continued)

Year	Taxon	No. of species	Location	References
	Rhodophytes (186 spp.)			
	Phaeophytes (14 spp.)			
	Cynophytes (17 spp.)			
2006	Chlorophytes (10 spp.)	35	Golden Beach, Bintulu, Sarawak	Muta et al. (2006)
	Rhodophytes (19 spp.)			
	Phaeophytes (6 spp.)			
	Chlorophytes (12 spp.)	35	Malaysia	Muta et al. (2007)
	Rhodophytes (15 spp.)			
	Phaeophytes (8 spp.)			
2007	Chlorophytes (92 spp.)	260	Malaysia	Phang (2006)
	Rhodophytes (94 spp.)			
	Phaeophytes (57 spp.)			
	Cynophytes (17 spp.)			
2008	Macroalgae (805 spp.)	867, but only 805 can use for GIS	Peninsular Malaysia, Sabah, Sarawak	Du et al. (2008)
Harmful algae (HA)				
1977	*Pyrodinium bahamense* var. *compressum*	1	Brunei Bay, West Coast Sabah	Roy (1977)
1983	*Gymnodium* sp., *Rhizosolenia alta* forma *gracilima* (2 spp.)	5	South China Sea	Yu et al. (2007)
1984	*Rhizosolenia alta* forma *gracilima* (1 spp.)			
1985	*Rhizosolenia alta* forma *gracilima* (1 spp.)			
1987	*Trichodesmium hildebrandtii* (1 spp.)			
1989	*Pyrodinium bahamense* var. *compressum*	1	Brunei Bay, Kimanis Bay, Kota Kinabalu, Sabah	Usup and Asmat Ahmad (1989)
1991	*Rhizosolenia alta* forma *gracilima*, *Pseudo-nitzschia multiseries*, *Pseudo-nitzschia pungens*, *Leptocylindras danicus*,	18	South China Sea	Yu et al. (2007)

	Skeletonema costatum, *Pseudo-nitzschia delicatissima* (6)			
1992	*Noctiluca scintillans* (1 spp.)			
1996	*Mesodinium rubrum* (1 spp.)			
1998	*Mesodinium rubrum* (1 spp.)			
1999	*Peridinium quinquecorne*, *Pseudo-nitzschia sp.* (2 spp.)			
2000	*Rhizosolenia alta* forma *gracilima*, *Peridinium quinquecorne*, *Pseudo-nitzschia* sp., *Rhizosolenia fragillissima*, *Scrippsiella trochoidea*, *Leptocylindrus danicus*, *Pseudo-nitzschia pungens* (7 spp.)			
	Harmful algae	Unknown	Kimanis Bay, Sabah	Anton et al. (2000)
2001	*Fibrocapsa* sp., *Chattonella marina*, *Leptocylinras danicus*, *Rhizosolenia fragillissima*, *Pseudo-nitzschia* sp.	5	South China Sea	Yu et al. (2007)
	Harmful algae	133	South China Sea	Yan et al. (2001)
2002	Harmful algae	Unknown	Selat Tebrau, Johor	Usup et al. (2004)
	Harmful algae	Unknown	Malaysia	Usup et al. (2002b)
	Alexandrium minutum and *Alexandrium Tamiyavanichii*	2	Sebatu Melaka, Tumpat-SM	Usup et al. (2002a)
	Pyrodinium bahamense var. *compressum*, *Gymnodinium catenatum*	2	Singapore	Holmes and Teo (2002)
2003	*Scrippsiella trochoidea*, *Chattonella marina*	2	South China Sea	Yu et al. (2007)
	Alexandrium halim	1	Malaysia	Po-Teen et al. (2003)
2004	*Phaeocystis* cf. *globosa*	1	South China Sea	Tang et al. (2004)
	Trichodesmium erythraeum	1	South China Sea	Yu et al. (2007)
	Cochlodinium polykrikoides	1	South China Sea	Azanza (2005)
2005	*Cochlodinium polykrikoides*	1	Sulu Sea	
	Alexandrium taylori and *Alexandrium peruvianum*	2	Sarawak River and Kuching Bay	Lim et al. (2004)

(continued)

(continued)

Year	Taxon	No. of species	Location	References
2007	*Cochlodinium polykrikoides*	1	Sepanggar Bay, Kota Kinabalu, Sabah	Anton et al. (2007)
2008	*Pyrodinium bahamense* var. *compressum*	1	North West Sabah	Wang et al. (2008)
Seagrass				
1904	*Halophila beccarii, H. decipiens*	Unknown	Sarawak	Beccari (1904)
1924	*Enhauls acoroides*	Unknown	Peninsular Malaysia	Ridley (1924)
2006	Hydrocharitaceae	7	West and East Malaysia	Japar et al. (2006)
	Cymodoceaceae	6		
	Ruppiaceae	1		
Molluscs				
1960	Gastropods	23	Kalimantan	Abbott (1960)
1981	Gastropods	330	West Malaysia and Singapore	Purchon and Purchon (1981)
	Bivalves	186		
1984	Gastropods	217	East coast Peninsular Malaysia	Angot (1984)
	Bivalves	91		
1986	Mollusc	Unknown	Indo-Pacific	Reid (1986)
1990	Gastropods: Muricidae	1	Singapore and Malaysia	Tan and Sigurdsson (1990)
1991	Gastropod(Nudibranch)	1	Singapore	Sigurdssan(1991)
1993	Mollusca	Unknown	Sabah	Shabdin et al. (1993)
1996	Bivawes	75	Singapore	Sachidhanandam and Chou (1996)
1997	Mollusca	Unknown	Sarawak and Sabah	Hill et al. (1997)
1998	Gastropods	125	Darvel Bay, Sabah	Shabdin et al. (1998)
	Bivalves	43		
2000	Gastropods (Suicide)	130	South China Sea	Tan (2000)
	Gastropods	64		Zulfigar and Tan (2010)
	Bivalves	12		
	Gastropods (Nudibranchs)	24		

2000	Gastropods: Muricidae (130 spp.)	323	South China Sea	Ng and Tan (2000)
	Nudibranchs (193 spp.)			
2002	Gastropods (56 spp.)	83	South China Sea	Mudjiono (2009)
	Bivalves (27 spp.)			
	Gastropods	31	West coast Malaysia Borneo	Raven (2002)
		2	Celebes Sea	
2005	Gastropods	300	Malaysia	Mazlan et al. (2005)
	Bivalves	100		
	Scaphopods	20		
	Cephalopods	10		
2008	Bivalves	18	West Sumatra	Jabang et al. (2008)
		5	North Sumatera (Straits of Malacca)	
2009	Bivalves	10	Langkawi	Lam and Morton (2009)
2010	Gastropods (155 spp.)	258	Pulau Semakau (Singapore Straits)	Tan and Yeo (2010)
	Bivalves (97 spp.)			
	Polyplachopora (1 sp.)			
	Cephalopoda (5 spp.)			
Fishes				
1928	Fishes	Unknown	Straits of Malacea	Smedley (1928)
1932	*Mola lanceolata*	1	Noembing Island, near Bintan Island	Smedley (1932)
1974	Elasmobranchii	1500	Malaysia	Anon (1975–1992)
				Abu Khair and Mohd Azmi (1996)
1982	Marine fishes	Unknown	South China Sea and Andaman Sea	Tan et al. (1982)
1998	Phallostethid	1	Sulu Sulawesi, South China Sea, Pacific Ocean.	Parenti and Louie (1998)
2000	Marine fishes	3365	South China Sea	Randall and Lim (2000)

(continued)

(continued)

Year	Taxon	No. of species	Location	References
	Myxini (7 spp.)	3365	South China Sea	Randall and Lim (2000)
	Elasmobranchii (237 spp.)			
	Holocephali (4 spp.)			
	Actinopterygii (Osteichthyes) (3117 spp.)			
2004	Marine fisheries	430	Anambas and Natuna island	Adrim et al. (2004)
Molecular study				
2001	Gastropods	1	Kuantan, Malaysia	Tan and Liu (2001)
2002	Seagrass	1	P. Perhentian	Waycott et al. (2002)
2004	Marine snake	3	Sabah, Sarawak, P. Tioman	Alfaro et al. (2004)
2005	Seahorse	1	Johor	Teske et al. (2005)
2006	Gastropods	5	Bako (Sarawak), Redang Island (Terengganu), Santubong (Sarawak), Bak Bak (Kudat, Sabah), Pulau Manukan (Sabah)	Reid et al. (2006)
2006	Corals	1	Singapore	Kevin Lam et al. (2006)
2007	Bivalves	3	Penang	Taylor et al. (2007)
2008	Gastropods	13	Malaysia	Frey and Vermeij (2008)
2010	Sea cucumber	17	P. Tioman, Terengganu, Port Dickson	Kamarudin et al. (2010)
	Gastropods (Nudibranchs)	5	P. Tioman, P. Renggis, P. Chimbe, P. Labus	Pola and Gosliner (2010)
	Fishes	18	Straits of Malacca	Jamaluddin et al. (2010)
	Reef fishes	10	Sulu Sea	Manjaji-Matsumoto and Rodrigues (2010)

References

Abbott RT (1960) The genus *Strombus* in the Indo-Pacific. Indo-Pacific Mollusca 1(2):33–174

Abu Khair MM, Mohd Azmi A (1996) Marine fishes and fisheries of Malaysia and neighboring countries. Universiti Pertanian Malaysia Press, Kuala Lumpur, Malaysia, p 744

Adrim M, Chen IS, Chen ZP, Lim KKP, Tan HH, Yusof Y, Jaafar Z (2004) Marine fishes recorded from the Anambas and Natuna Islands, South China Sea. Raffles Bull Zool 11:117–130

Ahmad I, Rusea G (1994) Distribution and diversity of seaweeds in Pulau Tioman. Alga biotechnology in the Asia-Pacific region. Universiti of Malaya, Kuala Lumpur, Malaysia

Aikanathan S, Sasekumar A (1994) The community structure of macroalgae in a low shore mangrove forest in Selangor, Malaysia. Hydrobiologia 285:131–137

Aikanathan S, Wong E (1994) Marine Park island management conceptual plan for Peninsular Malaysia. Report produced under WWF Project MYS 256/93. Department of Fisheries Malaysia and the World Wide Fund for Nature (WWF) Malaysia, Kuala Lumpur

Alfaro ME, Karns DR, Voris HK, Abernathy E, Sellins SL (2004) Phylogeny of *Cerberus* (Serpentes: Homalopsinae) and phylogeography of *Cerberus rynchops*: diversification of a coastal marine snake in Southeast Asia. J Biogeogr 31:1277–1292

Angot M (1984) Seashells of the east coast of Peninsular Malaysia. In: Chua TE, Charles JK (eds) Coastal resources of east coast peninsular Malaysia: an assessment in relation to potential oil spills. Universiti Sains Malaysia, Pulau Pinang, pp 138–150

Anonymous (1975–1992) Annual fisheries statistic. Department of Fisheries, Ministry of Agriculture, Kuala Lumpur, Malaysia

Anton A, Alexander J, Estim A (2000) Harmful algal blooms in Malaysia: revisiting Kimanis Bay. In: 9th International Conference on Toxic Phytoplankton, Tasmania

Anton A, Teoh PL, Mohd-Shaleh SR, Mohammad-Noor N (2007) First occurrence of *Cochlodinium* blooms in Sabah. Harmful Algae 7(3):331–336

Azanza R (2005) Harmful algal bloom networks as 'safeguards' against seafood toxicity and massive fish kills (case of the Harmful Algal Bloom network of Southeast Asia or HABSEA). UNITAR Hiroshima Office for Asia and the Pacific, Series on Sea and Human Security Training Workshop on Food Security, Hiroshima, Japan

Beccari O (1904) Wanderings in the great forests of Borneo. Oxford University Press, Oxford/New York

Benayahu Y, Chou LM (2010) On some Octocorallia (Cnidaria: Anthozoa: Alcyonacea) from Singapore, with a description of a new *Cladiella* species. Raffles Bull Zool 58(1):1–13

Bruggemann F (1878) Uber einige Steinkorallen von Singapore. Abh Naturw Ver Bremen 5:539–549

Chou LM (1983) A note on the starfish *Acanthaster planci* on Pulau Tioman reefs. Singapore Inst Biol 8(4):28–30

Clark AM, Rowe FWE (1971) Monograph of shallow-water Indo-West Pacific echinoderms. Trustees of the British Museum (Natural History), London

De Silva MWRN, Betterton C, Smith RA (1980) Coral reef resources of the east coast of Peninsular Malaysia. In: Chua TE, Charles JK (eds) Coastal resources of east coast peninsular Malaysia. Universiti Sains Malaysia, Penang

De Silva MWRN, Ridzwan Abdul R (1982) Coral reef survey of Pulau Paya/Segantang group of islands, Kedah, Malaysia: expedition report and recommendations for management. Report produced under WWF Project Mal. 41/82. Universiti Pertanian Malaysia, Serdang

Du HL, Jilian Ooi LS, Rosmadi F, Phang SM (2008) Spatial patterns of seaweed distribution in Malaysia using GIS. Proc SPIE 7145:71452H

Dunn DF (1970) Some observation on marine life at Pulau Aur, Johore. Malay Nat J 23:58–67

Fenner D (2001) Reef corals of Banggi area. Greenforce, Kudat, Malaysia

Frey MA, Vermeij GJ (2008) Molecular phylogenies and historical biogeography of a circumtropical group of gastropods (genus: *Nerita*): implications for regional diversity patterns in the marine tropics. Mol Phylogenet Evol 48:1067–1086

Fujita T, Zulfigar Y, Tan SH (2008) Preliminary list of shallow-water ophiuroids (Echinodermata) collected from the Tioman Islands in the South China Sea, Malaysia. Mem Nat Mus Nat Sci 45:1–11

George JD, George J (1987) The coral reefs of the Bodgaya Islands (Sabah: Malaysia) and Pulau Sipadan: macroinvertebrates. Malay Nat J 40:225–260

Goh AH, Sasekumar A (1981) The community structure of the fringing coral reef, Cape Rachado, Malaya. Atoll Res Bull 92:1–244

Goh NKC, Chou LM (1996) An annotated checklist of the gorgonians (Anthozoa: Octocorallia) of Singapore, with a discussion of gorgonian diversity in the Indo-West Pacific. Raffles Bull Zool 44(2):435–459

Green JP (1978) A survey and proposal for the establishment of the Pulau Redang Archipelago National Park. Reports submitted to the World Wildlife Fund. Kuala Lumpur, Malaysia

Green JP, Harris S, Robertson G, Santavy D (1979) Some corals from Pulau Redang Archipelago. Malay Nat J 32:281–325

Harborne A, Fenner D, Barnes A, Beger M, Harding S, Roxburgh T (2000) Status report on the coral reef of the east coast of peninsular Malaysia. Coral Cay Conservation, Malaysia, p 89

Herre AWCT (1934) A new goby from Singapore Island. Bull Raffles Mus 9:85–86

Hill DS, Shabdin ML, Pilcher N (1997) Provisional checklist of the Mollusca of Sarawak and Sabah. Faculty of Resource Science & Technology and Institute of Biodiversity and Environmental Conservation, University ofMalaysia, Sarawak, Malaysia

Hoeksema BW (2009) Attached mushroom corals (Scleractinia: Fungiidae) in sediment-stressed reef conditions at Singapore, including a new species and a new record. Raffles Bull Zool 22:81–90

Holmes MJ, Teo SLM (2002) Toxic marine dinoflagellates in Singapore waters that cause food poisonings. Clin Exp Pharmacol Physiol 29:829–836

Hong GA, Sasekumar A (1979) The community structure of the fringing reef, Cape Rachado, Malaya. Atoll Res Bull 244:1–11

Huang DW, Karenne PP, Tun Chou LM, Todd PA (2009) An inventory of zooxanthellate scleractinian corals in Singapore, including 33 new records. Raffles Bull Zool 22:69–80

Idris MH, Muta HZ, Arshad A (2006) Status of coral reefs species at Patricia Shoals, Bintulu, Sarawak, Malaysia. J Appl Sci Res 2:816–820

IUCN (2006) Tsunami and earthquake damage to coral reefs of Aceh, Indonesia. IUCN, Gland

Jabang N, Jatna S, Mufti PP, Budiman A (2008) Biodiversity of intertidal bivalves in west Sumatra. Proceedings of the 11th National Science and Technology, Lampung University, Malaysia, pp 17–18

Japar Sidak B, Muta Harah Z, Arshad A (2006) Distribution and significance of seagrass ecosystems in Malaysia. Aquat Ecosyst Health Manag 9:203–214

Jamaluddin JAF, Ahmad AT, Basir S, Rahim MA, Nor SAM (2010) *Rastrelliger* systematics inferred from mitochondrial cytochrome *b* sequences. Afr J Biotechnnol 9:3063–3067

Juliana M, Zulfigar Y, Anisah LA, Norhanis MR, Teh CP, Zulfikar, Woo SP, Hasmah SI, Rusdi M, Aileen Tan SH (2010) Changes of the coral cover over 10 years in marine protected area: comparing two different sites at Pulau Payar Marine Park in the Northern Straits of Malacca. Paper presented in ESABII and NaGISA Joint Conferene in Commemoration of the CBD COP10, Nagoya and NaGISA Western Pacific Marine Biodiversity Conference. Nagoya University, Japan

Kamarudin KR, Hashim R, Usup G (2010) Phylogeny of sea cucumber (Echinodermata: Holothurioidea) as inferred from 16s mitochondrial rRNA gene sequences. Sains Malaysiana 39:209–218

Kawaguchi S, Kato A, Masuda M, Kogame K, Phang SM (2002) Taxonomic notes on marine algae from Malaysia. VII. Five species of Rhodophyceae, with the description of *Lomentaria gracillima* sp. nov. Bot Mar 45:536–547

Kevin Lam KY, Adrian Loo HB, Peter A, Todd, Chew FT, Chou LM (2006) Existence of intra-colonial paralogues of the ribosomal internal transcribed spacer (ITS) impedes studies of intra-colonial genetic variation in the scleractinian coral *Diploastrea heliopora* (Lamarck 1816). Raffles Bull Zool 54:485–489

Lam K, Morton B (2009) Oysters (Bivalvia: Ostreidae and Gryphaeidae) recorded from Malaysia and Singapore. Raffles Bull Zool 57:481–494

Lane D, Marsh LM, Vandenspiegel D, Rowe FWE (2000) Echinoderm fauna of the South China Sea: an inventory and analysis of distribution patterns. Raffles Bull Zool (Suppl) 8:459–493

Larson HK, Jaafar Z, Kelvin LKP (2006) An annotated checklist of the gobioid fishes of Singapore. Raffles Bull Zool 56(1):135–155

Liao LM, Uy FA, Hwyrosa NA (2004) Macrobenthic marine algae and sea grasses of the Anambas Expedition 2002. Raffles Bull Biodivers Suppl 11:19–23

Lim PT, Usup G, Leaw CP, Ogata T (2004) First report of *Alexandrium taylori* and *Alexandrium peruvianum* (Dinophyceae) in Malaysia waters. Harmful Algae 4:391–400

Lulofs RB (1973) A reef survey of Pulau Gaya and associated islands, Sabah. June 8–12th 1973. Unpublished report to the Sabah National Park Department. Kota Kinabalu, Malaysia, p 16

Manjaji-Matsumoto BM, Rodrigues KF (2010) Preliminary results of DNA barcoding of reef fishes from Sulu Sea, Sabah, Northern Borneo. Paper presented at the 2nd National Seminar on Prime Marine Scientific Expedition, Kuala Terengganu, Malaysia

Mannuputi A (2008) Profile of coral reef at Pulau Galang-Batam. LIPI Press, Bunga Rampai, Indonesia, pp 11–25

Massin C, Lane DJW (1991) Description of a new species of sea cucumber (Stichopodidae, Holothurioidea, Echinodermata) from the eastern Indo-Malayan Archipelago: *Thelonata rubralineata*. Micronesica 2:57–64

Masuda M, Abe T, Kawaguchi S, Phang SM (1999) Taxonomic notes on marine algae from Malaysia. I. Six species of Rhodophyceae. Bot Mar 42:449–458

Masuda M, Kogame K, Kawaguchi S, Phang SM (2000a) Taxonomic notes on marine algae from Malaysia. IV. Six species of *Ceramiales* (Rhodophyceae). Bot Mar 43:569–579

Masuda M, Kato A, Shimada S, Kawaguchi S, Phang SM (2000b) Taxonomic notes on marine algae from Malaysia. II. Seven species of Rhodophyceae. Bot Mar 43:181–190

Masuda M, Abe T, Kawaguchi S, Phang SM (2001a) Taxonomic notes on marine algae from Malaysia. VI. Five species of Ceramiales (Rhodophyceae). Bot Mar 44:467–477

Masuda M, Kogame K, Kawaguchi K, Phang SM (2001b) Taxonomic notes on marine algae from Malaysia. V. Five species of Rhodophyceae. Bot Mar 44:81–88

Masuda M, Abe T, Kogame K, Kawaguchi S, Phang SM, Daitoh M, Sakai T, Takahashi Y, Suzuki M (2002) Taxonomic notes on marine algae from Malaysia. VIII. Three species of *Laurencia* (Rhodophyceae). Bot Mar 45:571–579

Mazlan AG, Zaidi CC, Wan-Lotfi WM, Othman BHR (2005) On the current status marine biodiversity in Malaysia. Ind J Mar Sci 34:76–87

Milne EH, Haime J (1857) Histoire Naturelle des Coraliaires 2:1–663

Mudjiono (2009) Telaak Komunitas moluska di rafaan terumbu (reef flat) peraira kepulauan Natuna Besar, Kabupaten Natuna. Oseanologi den Limnologi di Indonesia 35(2):147–160

Muta HZ, Japar SB, Raesah A, Suzalina AA, Hisao O (2006) Marine macrophytes: macroalgae species and life forms from Golden Beach, Similajau National Park, Bintulu, Sarawak, Malaysia. Coast Mar Sci 30:243–246

Muta HZ, Wong SC, Japar SB, Arshad A, Hisao O (2007) Macroalgae diversity and life forms of intertidal rocky shores. Mar Res Indonesia 32:163–168

Ng PKL, Tan KS (2000) The state of marine biodiversity in the South China Sea. Raffles Bull Biodiver 8:3–7

Parenti LR, Louie KD (1998) *Neostethus djajaorum*, new species from Sulawesi, Indonesia, the first phallostethid fish (Teleostei: Atherinomorpha) known from east of Wallace's line. Raffles Bull Zool 46:139–150

Phang SM (1994) New records of Malaysia marine algae. In: Sasekumar A, Mashall N, Macintosh DJ (eds) Ecology and conservation of Southeast Asian marine and seawater environments including wetland hydrobiology. Kluwer, Ghent

Phang SM (2006) Seaweed resources in Malaysia: current status and future prospects. Aquat Ecosyst Health Manag 9:185–202

Phang SM, Wee YC (1991) Benthic marine algae. In: Kiew R (ed) State of nature conservation in Malaysia. Malayan Nature Society, Kuala Lumpur, Malaysia, pp 123–129

Phang SM, Wong CL, Lim PE, Yeong HY, Chan CX (2005) Seaweed diversity of the Langkawi Islands with emphasis on the northeastern region. Malay J Sci 24:77–94

Pillai CSG, Scheer G (1974) On a collection of Scleractinia from the Straits of Malacca. Proceedings of the Second International Coral Reef Symposium 1. Great Barrier Reef Comittee, Brisbane, Australia, p 20

Pola M, Gosliner TM (2010) The first molecular phylogeny of cladobranchian opisthobranchs (Mollusca, Gastropoda, Nudibranchia). Mol Phylogenet Evol 56:931–941

Po-Teen L, Chui-Pin L, Usup G (2003) Identification of *Alexandrium halim* (Dinophyceae) using epifluorescence microscopy. Ann Microscopy 3:102–107

Purchon RD, Purchon DEA (1981) The marine shelled Mollusca of West Malaysia and Singapore. Part I. General introduction and account of the collecting stations. J Mollusc Stud 47:290–312

Purwati P, Lane DJW (2004) Asteroidea of the Anambas Expedition 2002. Raffles Bull Zool 11:89–102

Putchakarn S, Soncheng P (2004) Echinoderm fauna of Thailand: history and inventory reviews. Sci Asia 30:417–428

Randall JE, Lim KKP (2000) A checklist of the fishes of the South China Sea. Raffles Bull Zool (Suppl) 8:569–667

Raven H (2002) Notes on molluscs from NW Borneo. 1. Stromboidea (Gastropoda, Strombidae, Rostellariidae, Seraphidae). Vita Malacologica 1:3–32

Reid DG (1986) The littorinid molluscs of mangrove forests in the Indo-Pacific region: the genus *Littoraria* (Mollusca: Gastropoda: Littorinidae). British Museum (Natural History), London

Reid DG, Lal K, Mackenzie-Dodds J, Kaligis F, Littlewood DTJ, Williams ST (2006) Comparative phylogeography and species boundaries in *Echinolittorina* snails in the central Indo-West Pacific. J Biogeogr 33:990–1006

Ridley HN (1924) The flora of the Malay Peninsula, vol IV. Monocotyledons. Asher, Amsterdam

Roy RN (1977) Red tide and outbreak of paralytic shellfish poisoning in Sabah. Med J Malaysia 31:247–251

Sachidhanandam U, Chou LM (1996) A generic account of marine soft-bottom benthic bivalves of Singapore. Raffles Bull Biodiver 44(2):509–530

Shabdin ML, Ismail G, Zaidi I, Sahrol H (1998) A checklist of marine gastropods and bivalves from Darvel Bay, Sabah, Malaysia. Malayan Nat J 52:193–203

Shabdin ML, Ramli R, Idris BAG (1993) Mollusca in Sabah: diversity and distribution (in Malay). In: Ipor IB, Assim ZB, Fasihuddin BA, Fariduddin MYM, Din LB (eds) Malaysian natural resources: sustainability, development & strategies. Faculty of Resource Science and Technology, Universiti Malaysia Sarawak, Sarawak, Malaysia, pp 521–528

Sigurdsson JB (1991) A nudibranch, *Murphydoris singaporensis*, new genus and species, from Singapore mangroves (Mollusca: Opisthobranchia: Goniodorididae). Raffles Bull Zool 29:259–263

Sim YK, Tan SH, Zulfigar Y (2009) The diversity and abundance of the sea cucumber (Echinodermata: Holothuroidea) from the Archipelagos of Beting Patinggi Ali, Beting Raja Jarom and Pulau Layang—layang, South China Sea. In: the 6th IMT-GT UNINET Conference 2008: Sustaining Natural Resources Towards Enhancing the Quality of Life within the IMT-GT Zone. The Gurney Resort Hotel, Penang, pp 357–360/25–30

Siringoringo RM (2008) Distribution condition of coral reef (Scleractinia) at coastal Pulau Bintan, Riau Province. LIPI Press, Bunga Rampai, Indonesia, pp 63–78

Siringoringo RM, Budiyanto A (2008) Coral potential at Pulau Belitung coastal. LIPI Press, Bunga Rampai

Smedly N (1927) Papers on Malayan aquatic biology. (ii) Notes on shore fauna of Morib, West Coast, Malay Peninsula. J Fed Malay Stat Mus 13:230–237

Smedley N (1928) Spolia Mentawiensia: notes on some fishes. J Malay Soc 6:64–65
Smedley N (1932) An ocean sunfish, *Mola lanceolata* (Lienard), in Malaysian waters. Bull Raffles Mus (7):17–21
Studer T (1880) Beitrag zur fauna der steinkorallen von Singapore. Mitth Bernisch Naturf Ges Bern 15–53
Tang DL, Kawamura H, Nhu HD, Takahashi W (2004) Remote sensing oceanography of a harmful algal bloom off the coast of Southern Vietnam. J Geophys Res 109:1–7
Tan KS (2000) Species checklist of Muricidae (Mollusca: Gastropoda) in the South China Sea. Raffles Bull Zool 8:495–512
Tan KS, Liu LL (2001) Description of a new species of *Thais* (Mollusca: Neogastropoda: Muricidae) from Taiwan, based on morphological and allozyme analyses. Zool Sci 18:1275–1289
Tan KS, Sigurdsson JB (1990) A new species of *Thais* (Gastropoda: Muricidae) from Singapore and Peninsular Malaysia. Raffles Bull Zool 38:205–211
Tan KS, Yeo RKH (2010) The intertidal molluscs of Pulau Semakau: preliminary results of 'Project Semakau'. Nat Singapore 3:287–296
Tan SM, Lim PY, Senta T, Hooi KK (1982) A color guide to the fishes of the South China Sea and the Andaman Sea. Primary Production Department/Marine Fisheries Research Department, SEAFDEC
Tani M, Yamagishi Y, Masuda M, Kogame K, Kawaguchi S, Phang SM (2003) Taxonomic notes on marine algae from Malaysia. IX. Four species of Rhodophyceae, with the description of *Chondria decidua* sp. nov. Bot Mar 46:24–35
Taylor JD, Williams ST, Glover E, Dyal P (2007) A molecular phylogeny of heterodont bivalves (Mollusca: Bivalvia: Heterodonta): new analyses of 18S and 28S rRNA genes. Zool Scripta 36:587–606
Terada R, Kawaguchi S, Masuda M, Phang SM (2000) Taxonomic notes on marine algae from Malaysia. III. Seven species of Rhodophyceae. Bot Mar 43:347–357
Teske PR, Hamilton H, Palsbøll PJ, Choo CK, Gabr H, Lourie SA (2005) Molecular evidence for long-distance colonization in an Indo-Pacific seahorse lineage. Mar Ecol Prog Ser 286:249–260
Toda T, Okashita T, Maekawa T, Kee Alfian AA, Kushairi M, Rajuddin M, Nakajima R, Chen W, Takahashi KT, Othman BHR, Terazaki M (2007) Community structures of coral reefs around Peninsular Malaysia. J Oceanogr 63:113–123
Usup G, Asmat Ahmad NI (1989) *Pyrodinium bahamense* var. *compressum* red tide studies in Sabah, Malaysia. In: Hallegraeff GM, Maclean JL (eds) Biology, epidemiology and management of *Pyrodinium* red tides. ICLARM, Manila, pp 97–110
Usup G, Cheah MY, Rozirwan, Ng BK, Leaw CP, Othman M, Faazaz AL (2004) Identification of the species responsible for the harmful algal bloom event in Selat Tebrau in 2002. Malay Appl Biol 35:59–62
Usup G, Leaw CP, Ahmad A, Lim PT (2002a) *Alexandrium* (Dinophyceae) species in Malaysian waters. Harmful Algae 1:265–275
Usup G, Pin LC, Teen LP, Ahmad A (2002b) Probable toxin producer responsible for the first occurrence of paralytic shellfish poisoning on the east coast of Peninsular Malaysia. Malay Appl Biol 31:29–35
Wang SF, Tang DL, He FL, Fukuyo Y, Azanza RV (2008) Occurrences of harmful algal blooms (HABs) associated with ocean environments in the South China Sea. Hydrobiologia 596:79–93
Waycott M, Freshwater DW, York RA, Calladine A, Kenworthy WJ (2002) Evolutionary trends in the seagrass genus *Halophila* (Thouars): insights from molecular phylogeny. Bull Mar Sci 71:1299–1308
Wood EW (1979) Ecological study of coral reefs in Sabah. World Wildlife Fund Project. Malaysia Report, vol 15, p 166
Yamagishi Y, Masuda M, Abe T, Uwai S, Kogame K, Kawaguchi S, Phang SM (2003) Taxonomic notes on marine algae from Malaysia. XI. Four species of Rhodophyceae. Bot Mar 46:534–547

Yan T, Zhou MJ, Zou JZ (2001) A national report on harmful algal blooms in China, Institute of Oceanology Chinese Academy of Sciences, Lingoau, People's Republic of China

Yu J, Tang DL, Oh IS, Yao LJ (2007) Response of harmful algal blooms to environmental changes in Daya Bay, China. Terr Atmos Ocean Sci 18:1011–1027

Zaidnuddin I, Forbes R (2000) Sea cucumbers of Pulau Payar, Kedah Marine Park Islands. In: Proceedings of National Symposium on Pulau Payar Marine Park, pp 174–177

Zulfigar Y, Tan SH (1999) Biodiversity of sea cucumber in the lower part of the South China Sea. In: The Joint Seminar on Marine and Fisheries Sciences. Melaka, Malaysia, 1–3 December 1999, pp 73–82

Zulfigar Y, Tan SH (2010) Research on the seas and island expedition 2004. Scientific Report, vol 1

Zulfigar Y, Tan SH, Fujita T, Terazaki M (2000) The distribution of sea cucumber associated with seagrass beds found in The South China Sea. 11th JSPS Joint Seminar on Marine Science, pp 362–368

Zulfigar Y, Abdullah AL, Razalli NM, Peng TC, Pinn WS, Mohamed J, Rusdi M, Tan SH (2010) Seagrass communities changes between 2003 and 2008 at Pulau Gazumbo in the Straits of Malacca. Poster presented at World Seagrass Conference. Phuket, Thailand. 21–25 November 2010

Part IV
Advanced Concepts and Methods of Integrated Biodiversity Monitoring

Chapter 10
Development of a National Land-Use/Cover Dataset to Estimate Biodiversity and Ecosystem Services

Munemitsu Akasaka, Akio Takenaka, Fumiko Ishihama, Taku Kadoya, Mifuyu Ogawa, Takeshi Osawa, Takehisa Yamakita, Shuichiro Tagane, Reiichiro Ishii, Shin Nagai, Hisatomo Taki, Takumi Akasaka, Hiroyuki Oguma, Toru Suzuki, and Hiroya Yamano

M. Akasaka (✉)
Graduate School of Agricultural and Life Sciences, The University of Tokyo, 1-1-1 Yayoi, Bukyo-ku, Tokyo 113-8657, Japan

Faculty of Agriculture, Tokyo University of Agriculture and Technology, 3-5-8 Saiwai-cho, Fuchu, Tokyo 183-8509, Japan
e-mail: muuak@cc.tuat.ac.jp

A. Takenaka • F. Ishihama • T. Kadoya • M. Ogawa • H. Oguma • H. Yamano
National Institute for Environmental Studies, 16-2 Onogawa, Tsukuba, Ibaraki 305-8506, Japan

T. Osawa
National Institute for Agro-Environmental Sciences, 3-1-3 Kannondai, Tsukuba, Ibaraki 305-8604, Japan

T. Yamakita
Graduate School of Agricultural and Life Sciences, The University of Tokyo, 1-1-1 Yayoi, Bukyo-ku, Tokyo 113-8657, Japan

Institute of Biogeosciences, Japan Agency for Marine-Earth Science and Technology (JAMSTEC), 2-15 Natsushima, Yokosuka, Kanagawa 237-0061, Japan

S. Tagane
Department of Biology, Kyushu University, 6-10-1 Hakozaki, Higashi-ku, Fukuoka 812-8581, Japan

R. Ishii • S. Nagai
Research Institute for Global Change, Japan Agency for Marine-Earth Science and Technology (JAMSTEC), 3173-25 Showa, Kanazawa-ku, Yokohama, Kanagawa 236-0001, Japan

H. Taki
Forestry and Forest Products Research Institute, 1 Matsunosato, Tsukuba, Ibaraki 305-8687, Japan

T. Akasaka
Department of Forest Science, Graduate School of Agriculture, Hokkaido University, N9W9, Kita-ku, Sapporo, Hokkaido 060-8589, Japan

T. Suzuki
Rakuno Gakuen University, 582 Midorimachi, Bunkyodai, Ebetsu, Hookaido 069-8501, Japan

S.-i. Nakano et al. (eds.), *Integrative Observations and Assessments*, Ecological Research Monographs, DOI 10.1007/978-4-431-54783-9_10, © Springer Japan 2014

Abstract Land-use/cover data are essential to estimate biodiversity and ecosystem services. Although many data and maps have been produced in Japan, there is significant variation in their characteristics, such as publication year, spatial resolution, and land-use/cover classification categories because the purposes of the data and maps differ (e.g., resource inventory). Unfortunately, little attention has so far been paid to biodiversity and ecosystem services, which require preparation of new land-use/cover data. First, we searched existing public national data in Japan published by government sectors and researchers and evaluated their characteristics. Second, we briefly reviewed user needs (e.g., temporal coverage, spatial resolution, and classification categories) to estimate biodiversity and ecosystem services. We then integrated the information, and we present herein a procedure to prepare a new national land-use/cover dataset from existing datasets to bridge the gaps between data producers and users by determining the temporal coverage needed to reveal historic change, determining the spatial resolution suitable for estimation, and setting classification categories needed for estimation. The resultant land-use/cover data could contribute significantly to research and management related to conservation of biodiversity and provision of ecosystem services.

Keywords Land use • Land cover • Vegetation map • Biodiversity • Ecosystem service • Japan

Introduction

Recent advances in species and environmental databases contribute toward the estimation of biodiversity and ecosystem services on a large spatial scale (Gaston 2000; Tittensor et al. 2010). However, it is nearly impossible to sample species composition, biodiversity, and ecosystem function and services continuously across a region. The practical and effective way is to use surrogates (Sarkar and Margules 2002). One of the most widely applied methods is the use of vegetation or habitat maps to stratify units of diversity (Dalleau et al. 2010).

Land-use/cover information is essential to estimating biodiversity and ecosystem services (Oliver et al. 2004; Foley et al. 2005). Japan could contribute significantly to this issue because both land-use/cover data and species distribution data have been collected intensively by government sectors and citizens. Several studies have shown a close link between land-use/cover and biodiversity at both regional and national scales (Kadoya et al. 2009; Yamaura et al. 2009), thus encouraging the use of land-use/cover data to estimate biodiversity and ecosystem services.

Although many land-use/cover data and maps have been produced in Japan, there is significant variation in their characteristics, such as publication year, spatial resolution, and land-use/cover classification categories because the purposes of the data and maps differ (e.g., resource inventory). Although some could be used to estimate species distribution (Kadoya et al. 2009), the data and maps have paid little attention so far to estimating biodiversity and ecosystem services, which will

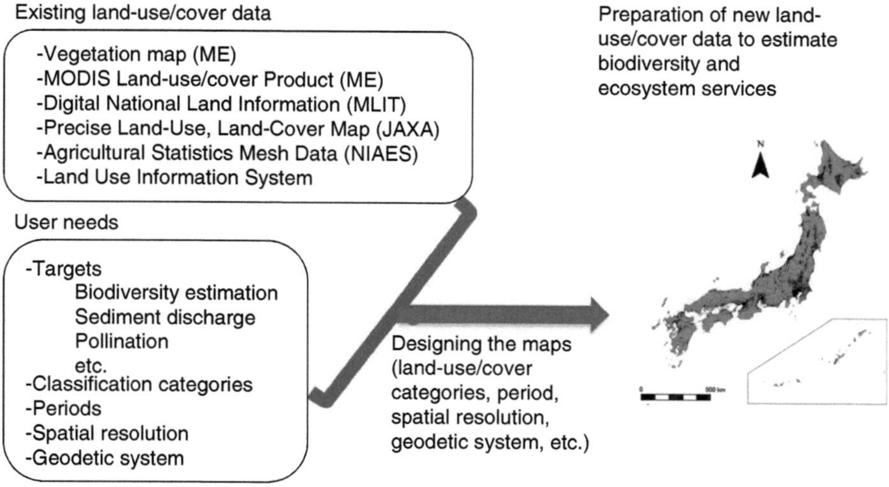

Fig. 10.1 Procedure to prepare a national land-use/cover dataset to estimate biodiversity and ecosystem services in the present study. *ME* Ministry of the Environment, *MLIT* Ministry of Land, Infrastructure, Transport and Tourism, *JAXA* Japan Aerospace Exploration Agency, *NIAES* National Institute for Agro-Environmental Sciences

require preparation of new land-use/cover data. This chapter presents a procedure to bridge the gap between data producers and users by compiling available information on Japan's national land-use/cover and users' needs for information on biodiversity and ecosystem services (Fig. 10.1). The procedure may be applied to other regions, and the resultant datasets could contribute significantly to research and management for conservation of biodiversity and ecosystem services.

National Land-Use/Cover Data in Japan

We collected information on national land-use/cover datasets published by governmental sectors and researchers. We found significant variation in the characteristics of this information, such as land-use/cover classification categories, publication year, and spatial resolution (Table 10.1). Probably the most fundamental issues are the spatial unit and the geodetic system of the dataset. Therefore, before describing each dataset, characteristics of the spatial unit and the geodetic system are described.

Japan's Standard Mesh System and Geodetic Systems

Although satellite-derived products normally have square pixels that have the same dimensions, Japan's map-based products adopted the Standard Mesh System, which is based on latitude and longitude (Matsuda 1980; Yaguchi 1990). The Standard Mesh

Table 10.1 Hierarchical classification categories of land-use/cover and characteristics of existing land-use/cover data (Ogawa et al., 2013). ○ and × denote existence and absence of the category, respectively.

First tier	Second tier	Third tier	Data provider	Ministry of the Environment						Ministry of Land, Infrastructure, Transport and Tourism		
			Data name	Actual Vegetation Map				MODIS Land Use/Cover Product***		Digital National Land Information		
			Reference	http://www.biodic.go.jp/trialSystem/top.html				Kakuta et al. (2009)		http://nlftp.mlit.go.jp/ksj-e/index.html		
			Preparation year	1973	1979-1986	1989-1998	1999-**	2007	2009	1976/1977	1987/1988	991/1992, 1997/1998, 2006/2007, 2009/2010
			Data format	Paper (1/200,000)	Polygon	Polygon	Polygon	250 m mesh	250 m mesh	~100 m (3" x 4.5") mesh	~100 m (3" x 4.5") mesh	~100 m (3" x 4.5") mesh ~1 km (30" x 45") mesh
			Characteristics summary*\Geodetic system (TD: Tokyo Datum; WGS: World Geodetic System)	TD	WGS	WGS	WGS	WGS	WGS	TD	TD	TD and WGS (after 2006/2007)
Grassland				○	○	○	○	○	○	×	×	×
	Natural grassland			○	○	○	○	×	×	2	2	2
		Alpine grassland	Alpine heathland and wind-exposed grassland and alpine scrub communities, with VN of 9 and Vegetation Class	○	○	○	○	×	×	2	2	2
		Snow patch community	Snow patch community with VN of 10	○	○	○	○	×	×	2	2	2
		Sand dune vegetation	Coastal vegetation according to the description of plant community, with VN of 10 Sand dune vegetation	○	○	○	○	×	×	2	2	2
		Coastal cliff herb vegetation	Coastal cliff herb vegetation	○	○	○	○	×	×	2	2	2
		Other natural grassland	Grassland with VNs of 9 and 10 (except for marsh and coastal vegetation)	○	○	○	○	×	×	2	2	2
	Secondary grassland			○	○	○	○	×	×	2	2	2
		Secondary grassland (tall)	Glassland with VN of 5 Tall vegetation with VN of 4	○	○	○	○	×	×	2	2	2
		Secondary grassland (short)	Grassland with VN of 4 (except for tall vegetation) Grass-dominated communty with VN of 7	○	○	○	○	×	×	2	2	2
	Artificial grassland			○	○	○	○	×	×	2	2	2
		Cultivated meadow (without golf course)	Cultivated meadow according to the description of plant community, with VN of 2 Vegetation on land slide area and road side slope with VN	○	○	○	○	×	×	1	1	1
		Cultivated meadow (including golf course)	Cultivated meadow according to the description of plant community, with VN of 2	○	○	○	○	○	○	2	2	○
	Bamboo grassland		Sasa grassland and bamboo community with VN of 5 and	○	○	○	○	×	×	2	2	2
Forest				○	○	○	○	○	○	○	○	○
	Natural forest			○	○	○	○	○	○	×	×	×
		Evergreen conifer forest (natural)	Evergreen conifer trees with VN of 9	○	○	○	○	×	×	×	×	×
		Deciduous conifer forest (natural)	*Larix kaempferi* community	○	○	○	○	×	×	×	×	×
		Evergreen broad-leaved forest (natural)	Evergreen broad-leaved trees with VN of 9	○	○	○	○	×	×	×	×	×
		Deciduous broad-leaved forest (natural)	Deciduous broad-leaved trees with Vegetation Naturalness (VN) of 9	○	○	○	○	×	×	×	×	×
		Evergreen conifer + Evergreen boroad-leaved forest (natural)	Evergreen conifer and evergreen boroad-leaved trees with VN of 9	○	○	○	○	×	×	×	×	×
		Evergreen conifer + Deciduous boroad-leaved forest (natural)	Evergreen conifer and deciduous boroad-leaved trees with VN of 9	○	○	○	○	×	×	×	×	×
	Secondary forest			○	○	○	○	○	○	×	×	×
		Evergreen conifer forest (secondary)	Evergreen conifer trees with VN of 7	○	○	○	○	×	×	×	×	×
		Deciduous conifer forest (secondary)	No plant community in the actual vegetation map is suitable for this category	×	×	×	×	×	×	×	×	×
		Evergreen broad-leaved forest	Evergreen broad-leaved trees with VN of 7 and 8	×	×	×	×	×	×	×	×	×
		Deciduous broad-leaved forest (secondary)	Deciduous broad-leaved trees with VN of 7 and 8 Deciduous broad-leaved forest according to the description of plant community	○	○	○	○	×	×	×	×	×
		Evergreen conifer + Evergreen boroad-leaved forest (secondary)	No plant community in the actual vegetation map is suitable for this category	×	×	×	×	×	×	×	×	×
		Evergreen conifer + Deciduous boroad-leaved forest (secondary)	Evergreen conifer and deciduous boroad-leaved trees with VN of 7	○	○	○	○	×	×	×	×	×
	Planted forest			○	○	○	○	○	○	×	×	×
		Evergreen conifer forest (planted)	Evergreen conifer trees with VN of 3 and 6	○	○	○	○	×	×	×	×	×
		Deciduous conifer forest (planted)	Planted deciduous conifer trees	○	○	○	○	×	×	×	×	×
		Evergreen broad-leaved forest (planted)	Evergreen broad-leaved trees with VN of 3 and 6	○	○	○	○	×	×	×	×	×
		Deciduous broad-leaved forest (planted)	Deciduous broad-leaved trees with VN of 6	○	○	○	○	×	×	×	×	×
	Other forests			○	○	○	○	×	×	×	×	×
		Natural scrub	Scrub with VN of 9 (except for coastal and apline	○	○	○	○	×	×	×	×	×

	Coastal scrub forest	Scrub according to the description of plant community, with VN of 10	○	○	○	○	×	×	×	×	×
	Bamboo forest	Bamboo forest according to the description of plant community, with VN of 7	○	○	○	○	×	×	×	×	×
Wetland			○	○	○	○	×	×	2	2	2
	Salt marsh vegetation	Salt marsh vegetation with VN of 10 and VC of 8	○	○	○	○	×	×	×	×	×
	Freshwater marsh vegetation	Freshwater marsh vegetation with VN of 10 and VC of 8 *Phragmitetea* community with VN of 5	○	○	○	○	×	×	×	×	×
Shore			○	○	○	○	○	○	○	○	○
	Water grass	*Lemnetea* and *Potamogetonetea* community with VN of 10	○	○	○	○	×	×	×	×	×
	Seagrass	*Zosteretea marinae* community	○	○	○	○	×	×	×	×	×
	Mangrove	Mangrove	○	○	○	○	×	×	×	×	×
Unique substrate			○	○	○	○	○	○	2	2	2
	Natural bare land	Natural bare land	○	○	○	○	×	×	×	×	×
	Vegegation on limestone	Vegegation on limestone	○	○	○	○	×	×	×	×	×
	Vegetation in volcanic desert, Vegetation in solfatara formation	Vegetation in volcanic desert and vegetation in solfatara formation	○	○	○	○	×	×	2	2	2
	Cliff and rocky vegetation	Cliff and rocky vegetation	○	○	○	○	×	×	2	2	2
	Coral reef vegetation	Coral reef vegetation	○	○	○	○	×	×	×	×	×
Agricultural land			○	○	○	○	○	○	○	○	○
Paddy field			○	○	○	○	○	○	○	○	○
	Paddy field	Paddy field community with VN of 2	○	○	○	○	○	○	○	○	○
	Uncultivated paddy field	Weed communities in uncultivated paddy field with VN of	○	○	○	○	×	×	×	×	×
Field			○	○	○	○	○	○	1	1	1
	Field	Weed communities in field with VN of 2 and 5	○	○	○	○	×	×	×	×	×
	Uncultivated field	Weed communities in uncultivated field with VN of 4	○	○	○	○	×	×	×	×	×
Other fields			○	○	○	○	×	×	×	×	×
	Roadside weed	Roadside weed communities with VN of 2 and 4	○	○	○	○	×	×	×	×	×
	Tea	*Thea sinensis* garden with VN of 3	○	○	○	○	×	×	○	○	○
	Orchard and mulberry and others	Plant community with VN of 3 (except for *Thea sinensis*	○	○	○	○	×	×	○	○	○
Urban and residential district			○	○	○	○	○	○	○	○	○
	Urban and residential district with many	Urban and residential district with many trees with VN of 2	○	○	○	○	×	×	×	×	×
	Urban and residential district	Urban and residential district with VN of 1	○	○	○	○	×	×	×	×	×
	Artificial bare land	Bare land with VN of 1 (except for residential district)	○	○	○	○	×	×	×	×	×
Open water		Open water	○	○	○	○	×	×	×	×	×
Snow			×	×	×	×	×	×	2	2	2
Others			×	×	×	×	○	○	○	○	○
Unknown			○	○	○	○	×	×	×	×	×

Notes
* Based on the actual vegetation map. See Ogawa et al. (2013) for details
** ~50% coverage of total national land (as of March 2013)
*** Overall accuracy = 62 %
**** Overall accuracy = 55 %
◎ Fourth tier categories are provided (wheat, cereals, potato, bean, industrial crop, vegetable, flowers and ornamental plants, seeds and saplings, and forage crop)
1 Code 2 (Table 4)
2 Code A (Table 4)

(continued)

Table 10.1 (continued)

First tier	Second tier	Third tier	Characteristics summary*\Geodetic system (TD: Tokyo Datum; WGS: World Geodetic System)	Data provider: Japan Aerospace Exploration Agency (JAXA); Data name: Precise Land-Use, Land-Cover Map****; Reference: http://www.eorc.jaxa.jp/ALOS/lulc/lulc_jindex.htm; Preparation year: 2011; Data format: 50 m mesh	Data name: Land Use Information System; Reference: Nishikawa (2002); Preparation year: 1850, 1900, 1950, 1985; Data format: ~2 km (1' x 1.5') mesh	Data provider: National Institute for Agro-Environmental Sciences; Data name: Agricultural Statistics Mesh Data; Reference: http://agrimesh.dc.affrc.go.jp/; Preparation year: 1970, 1975, 1980, 1985, 1990, 1995; Data format: ~1 km (30" x 45") mesh
				WGS	TD	TD
Grassland				○	○	×
	Natural grassland			×	×	×
		Alpine grassland	Alpine heathland and wind-exposed grassland and alpine scrub communities, with VN of 9 and Vegetation Class	×	×	×
		Snow patch community	Snow patch community with VN of 10	×	×	×
		Sand dune vegetation	Coastal vegetation according to the description of plant community, with VN of 10 Sand dune vegetation	×	×	×
		Coastal cliff herb vegetation	Coastal cliff herb vegetation	×	×	×
		Other natural grassland	Grassland with VNs of 9 and 10 (except for marsh and coastal vegetation)	×	×	×
	Secondary grassland			×	×	
		Secondary grassland (tall)	Glassland with VN of 5 Tall vegetation with VN of 4	×	×	×
		Secondary grassland (short)	Grassland with VN of 4 (except for tall vegetation) Grass-dominated commmunty with VN of 7	×	×	×
	Artificial grassland			×	×	×
		Cultivated meadow (without golf course)	Cultivated meadow according to the description of plant community, with VN of 2 Vegetation on land slide area and road side slope with VN	×	×	×
		Cultivated meadow (including golf course)	Cultivated meadow according to the description of plant community, with VN of 2	×	○	×
	Bamboo grassland		Sasa grassland and bamboo community with VN of 5 and	×	×	×
Forest				○	○	×
	Natural forest			×	×	×
		Evergreen conifer forest (natural)	Evergreen conifer trees with VN of 9	×	×	×
		Deciduous conifer forest (natural)	*Larix kaempferi* community	×	×	×
		Evergreen broad-leaved forest (natural)	Evergreen broad-leaved trees with VN of 9	×	×	×
		Deciduous broad-leaved forest (natural)	Deciduous broad-leaved trees with Vegetation Naturalness (VN) of 9	×	×	×
		Evergreen conifer + Evergreen boroad-leaved forest (natural)	Evergreen conifer and evergreen boroad-leaved trees with VN of 9	×	×	×
		Evergreen conifer + Deciduous boroad-leaved forest (natural)	Evergreen conifer and deciduous boroad-leaved trees with VN of 9	×	×	×
	Secondary forest			×	×	×
		Evergreen conifer forest (secondary)	Evergreen conifer trees with VN of 7	×	×	×
		Deciduous conifer forest (secondary)	No plant community in the actual vegetation map is suitable for this category	×	×	×
		Evergreen broad-leaved forest	Evergreen broad-leaved trees with VN of 7 and 8	×	×	×
		Deciduous broad-leaved forest (secondary)	Deciduous broad-leaved trees with VN of 7 and 8 Deciduous broad-leaved forest according to the description of plant community	×	×	×
		Evergreen conifer + Evergreen boroad-leaved forest (secondary)	No plant community in the actual vegetation map is suitable for this category	×	×	×
		Evergreen conifer + Deciduous boroad-leaved forest (secondary)	Evergreen conifer and deciduous boroad-leaved trees with VN of 7	×	×	×
	Planted forest			×	×	×
		Evergreen conifer forest (planted)	Evergreen conifer trees with VN of 3 and 6	×	×	×
		Deciduous conifer forest (planted)	Planted deciduous conifer trees	×	×	×
		Evergreen broad-leaved forest (planted)	Evergreen broad-leaved trees with VN of 3 and 6	×	×	×
		Deciduous broad-leaved forest (planted)	Deciduous broad-leaved trees with VN of 6	×	×	×
	Other forests			×	○	×
		Natural scrub	Scrub with VN of 9 (except for coastal and apline	×	×	×

(continued)

	Coastal scrub forest	Scrub according to the description of plant community, with VN of 10	×	×	×
	Bamboo forest	Bamboo forest according to the description of plant community, with VN of 7	×	○	×
Wetland			×	○	×
	Salt marsh vegetation	Salt marsh vegetation with VN of 10 and VC of 8	×	×	×
	Freshwater marsh vegetation	Freshwater marsh vegetation with VN of 10 and VC of 8 *Phragmitetea* community with VN of 5	×	×	×
Shore			○	○	×
	Water grass	*Lemnetea* and *Potamogetonetea* community with VN of 10	×	×	×
	Seagrass	*Zosteretea marinae* community	×	×	×
	Mangrove	Mangrove	×	×	×
Unique substrate			○	○	×
	Natural bare land	Natural bare land	×	×	×
	Vegegation on limestone	Vegegation on limestone	×	×	×
	Vegetation in volcanic desert, Vegetation in solfatara formation	Vegetation in volcanic desert and vegetation in solfatara formation	×	×	×
	Cliff and rocky vegetation	Cliff and rocky vegetation	×	×	×
	Coral reef vegetation	Coral reef vegetation	×	×	×
Agricultural land			○	○	○
Paddy field			○	○	○
	Paddy field	Paddy field community with VN of 2	○	○	○
	Uncultivated paddy field	Weed communities in uncultivated paddy field with VN of	×	×	×
Field			○	○	○
	Field	Weed communities in field with VN of 2 and 5	×	×	◎
	Uncultivated field	Weed communities in uncultivated field with VN of 4	×	×	×
Other fields			×	×	○
	Roadside weed	Roadside weed communities with VN of 2 and 4	×	×	×
	Tea	*Thea sinensis* garden with VN of 3	×	○	○
	Orchard and mulberry and others	Plant community with VN of 3 (except for *Thea sinensis*	×	○	○
Urban and residential district			○	×	×
	Urban and residential district with many	Urban and residential district with many trees with VN of 2	×	×	×
	Urban and residential district	Urban and residential district with VN of 1	×	×	×
	Artificial bare land	Bare land with VN of 1 (except for residential district)	×	×	×
Open water		Open water	×	×	×
Snow			○	×	×
Others			○	○	○
Unknown			×	×	×

Notes
* Based on the actual vegetation map. See Ogawa et al. (2013) for details
** ~50% coverage of total national land (as of March 2013)
*** Overall accuracy = 62 %
**** Overall accuracy = 55 %
◎ Fourth tier categories are provided (wheat, cereals, potato, bean, industrial crop, vegetable, flowers and ornamental plants, seeds and saplings , and forage crop)
1 Code 2 (Table 4)
2 Code A (Table 4)

(continued)

System is arranged from the largest mesh to the smallest. The largest mesh is called the First-order Mesh System, and the unit areas are divided into grids of 0°40′0″ (2/3°) (latitude) × 1°0′0″ (longitude), which correspond to the 1:200,000-scale district maps. The origin of the mesh is 0°N, 100°E. The mesh located at longitude X° and latitude Y° is numbered according to four data points, that is, *xxyy*, where *xx* and *yy* are calculated as $X - 100$ and $Y/(2/3)$, respectively (Fig. 10.2). The Second-order Mesh System has grids of 0°5′0″ × 0°7′30″ (i.e., ~10 × ~10 km), corresponding to a 1:25,000-scale topographic map. The Third-order Mesh System has grids of 0°0′30″ × 0°0′45″ (i.e., ~1 × ~1 km), obtained by dividing the Second-order Mesh into 10 × 10 smaller cells. This Third-order Mesh System is sometimes called the base mesh system in Japan. These meshes are also numbered in a consistent manner (Fig. 10.2). In addition to the three stratified mesh systems, the Fivefold Mesh System that merges 5 × 5 units of the Third-order Mesh (i.e., ~5 × ~5 km) and 1/10 of the Third-order Mesh (~100 × ~100 m) is also used. Many national spatial datasets and censuses are based on this system, and the system is still in use today.

Another issue is the geodetic system of the datasets. Japan developed its own geodetic system called Tokyo Datum about 100 years ago. The earth was represented by the Bessel ellipsoid of revolution, and at the origin in Tokyo, the astronomical latitude, longitude, and azimuth were determined based on the neighboring first-order triangulation point. Using emerging space geodetic techniques, this geodetic system was revised and adapted to the World Geodetic System in 2002. This new geodetic system is called the Japanese Geodetic Datum 2000 (JGD2000) (Matsumura et al. 2004). Caution is thus required when integrating Japan's spatial data prepared before 2002 with those prepared after 2002. An offset of about 450 m is found between the two geodetic systems.

Attributes of the Land-Use/Cover Data

Actual Vegetation Map (Ministry of the Environment)

Land-use/cover is related to vegetation. Actual vegetation maps (Fig. 10.3a) were prepared under the National Survey on the Natural Environment organized by the Ministry of the Environment of Japan, which was designed to reveal the status of land, surface water, and coastal areas across the country (Tsuzawa 1990). The National Survey on the Natural Environment has been conducted about every 5 years since 1973 under the authority of Article 4 of the Nature Conservation Law of Japan, which declares that "Once every 5 years, the country makes a basic survey of Japan's topography, geology, vegetation, wild animals, and any other information needed to develop policies for protecting the natural environment." The National Survey on the Natural Environment includes the (1) vegetation survey, (2) species diversity survey (medium- and large-sized mammal survey), and (3) ecosystem diversity survey (neritic ecosystem survey). The actual vegetation maps include both vegetation types, classified based on phytosociology, and

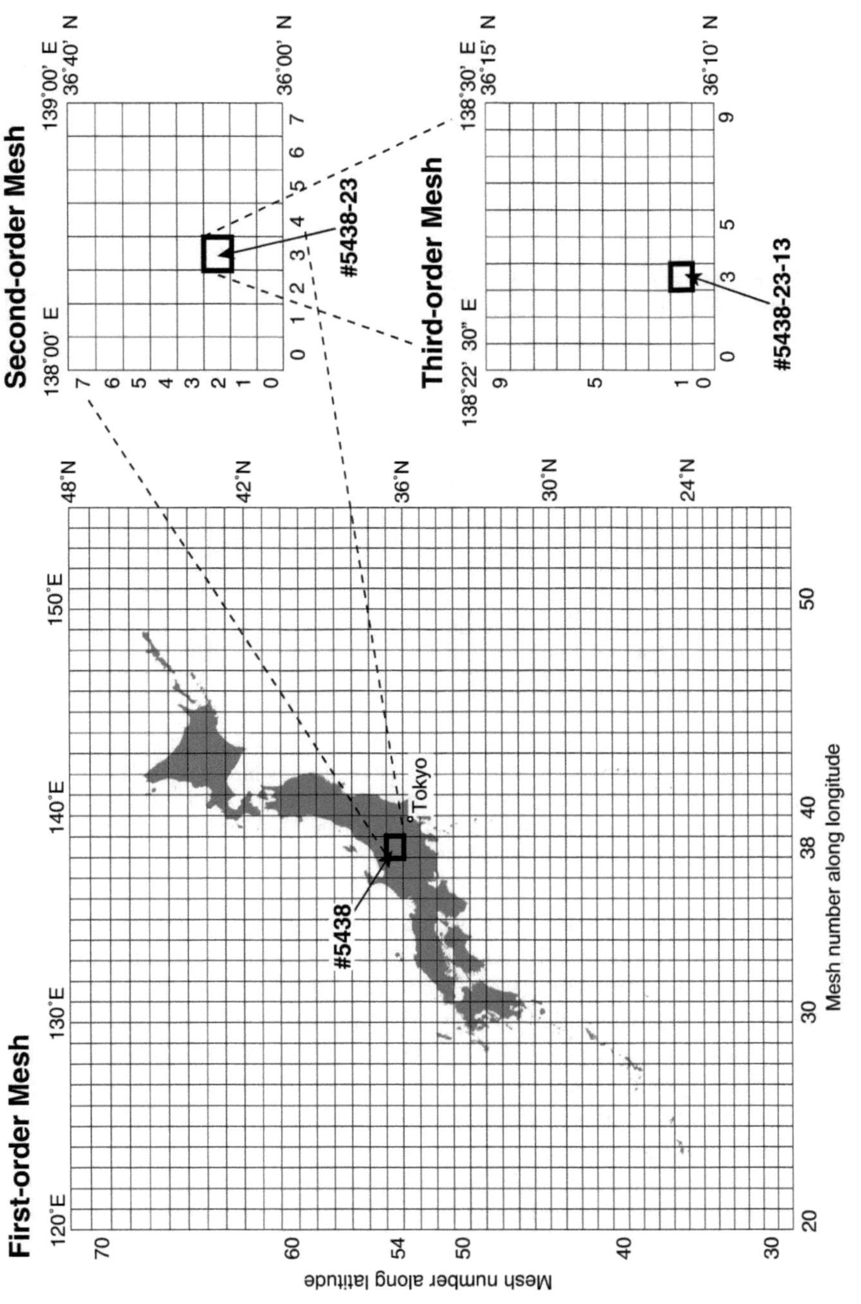

Fig. 10.2 Japan's Standard Mesh System and examples of setting numbers for the first-order, second-order, and third-order meshes. (Adapted from Yaguchi 1990)

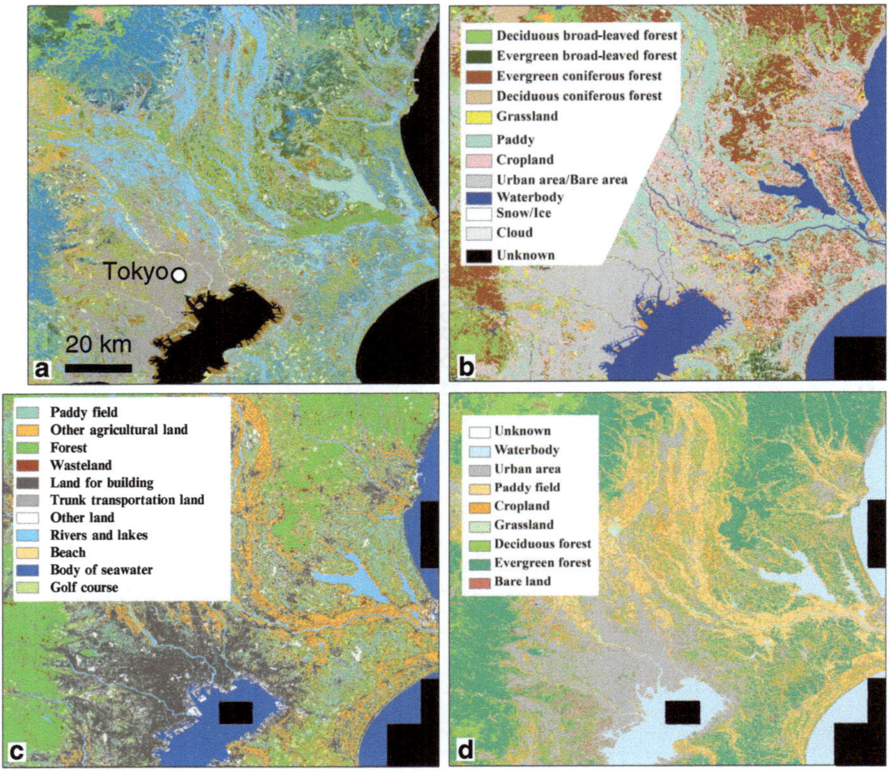

Fig. 10.3 Examples of land-use/cover data in Japan. The maps are enlarged to show land-use/cover around Japan's capital, Tokyo. **a** Actual vegetation map (Ministry of the Environment). **b** MODIS Land-use/cover Product (Ministry of the Environment). **c** Digital National Land Information (Ministry of Land, Infrastructure, Transport and Tourism). **d** Precise Land-Use, Land-Cover Map (Japan Aerospace Exploration Agency)

vegetation naturalness (Table 10.2), which is related to the degree of human disturbance of the vegetation. To date, three national actual vegetation maps are in existence, based on the first (1973), second and third (1979–1986), and fourth and fifth (1989–1998) vegetation surveys. The last two maps show consistency in their classification categories, scale (1:50,000), and use of GIS polygons in the World Geodetic System. These maps may be downloaded from the J-IBIS (Japan Integrated Biodiversity Information System) website (http://www.biodic.go.jp/english/J-IBIS.html).

The first survey, made in 1973, set 362 plant communities, along with ten categories of vegetation naturalness (Table 10.2) based on the degree of land development. Although the results were published in 1:200,000-scale actual vegetation maps, no GIS data were available. The second survey was made in 1979. In addition to vegetation naturalness, a three-level hierarchical classification scheme for plant communities was adopted, which identified 707 plant communities and grouped these into 299 categories, which were further grouped into ten

Table 10.2 Vegetation naturalness used in the actual vegetation maps (Ministry of the Environment

Class	Description
10	Natural vegetation of grassland and moorland
9	Natural vegetation of forest
8	Substitutional vegetation close to natural vegetation of forest
7	Substitutional vegetation of secondary forest
6	Planted forest
5	Substitutional vegetation of high-profile grassland
4	Substitutional vegetation of low-profile grassland
3	Fruit orchards, mulberry plantations, tea gardens, and other horticultural areas
2	Paddies, fields, and other arable land, residential area with abundant trees
1	Urban land, developed tracts, and other zones where plant life is virtually nonexistent

Larger numbers indicate high vegetation naturalness (adapted from Tsuzawa 1990)

Table 10.3 Vegetation class used in the actual vegetation maps (Ministry of the Environment)

Class	Description
1	Natural vegetation in alpine zone
2	Natural vegetation in *Vaccinio–Piceetea* region
3	Substitutional communities in *Vaccinio–Piceetea* region
4	Natural vegetation in *Fagetea crenatae* region
5	Substitutional communities in *Fagetea crenatae* region
6	Natural vegetation in *Camellietea japonicae* region
7	Substitutional communities in *Camellietea japonicae* region
8	Riverside, moor, salt marsh, and dune
9	Plantation and cultural land
10	Others (urban district, bare land, etc.)

This listing serves as the first tier of the hierarchical classification of plant communities (adapted from Tsuzawa 1990)

vegetation classes (Table 10.3). This survey covered about half the country, and the rest was covered by the third survey in 1983–1986 using the same classification categories. The results were published as 1:50,000-scale actual vegetation maps, and the polygons may be downloaded from the J-IBIS website. The fourth and fifth surveys were conducted in 1989–1993 and 1994–1998, respectively. Satellite images taken by Landsat MSS and Landsat TM were used to detect areas modified since the second and third surveys, and intensive surveys based on ground-truthing and interpretation of aerial photographs were conducted in these areas. As with the second and third surveys, the vegetation naturalness and the three-level hierarchical classification scheme were adopted. The results were published in 1:50,000-scale actual vegetation maps, and the polygons may also be downloaded from the J-IBIS website. The subsequent sixth and seventh surveys have been conducted since 1999, and the data currently cover about half the country. In the sixth and seventh surveys the divisions of the plant communities were revised, although the basic structure was retained.

MODIS Land-Use/Cover Product (Ministry of the Environment)

As a part of the seventh National Survey on the Natural Environment, the feasibility of satellite remote sensing data was examined. Land-use/cover data were prepared for 2007 and 2009 based on the MODIS (MODerate resolution Imaging Spectroradiometer) product at a spatial resolution of 250 m in the World Geodetic System (Fig. 10.3b). The data included 11 categories (Kakuta et al. 2009) (Table 10.1).

MODIS is a sensor onboard the Terra and Aqua satellites, which were launched by the US National Aeronautics and Space Administration (NASA) in 1999 and 2002, respectively, and are still in operation. Several products were developed using the data. MOD13Q is a product for NDVI (normalized difference vegetation index, calculated as [Red − NIR]/[Red + NIR], where Red and NIR indicates signal for red and near-infrared wavelength bands of the sensor) that can detect vegetation activities. A time-series MOD13Q product was used to identify land that has distinct vegetation phenology, such as evergreen and deciduous forests and paddy fields. Accuracy assessment of the classification by using ground-survey data during the sixth and seventh National Survey on the Natural Environment revealed approximately 60 % overall accuracy (Kakuta et al. 2009).

Digital National Land Information (Ministry of Land, Infrastructure, Transport and Tourism)

The Digital National Land Information (Fig. 10.3c), which is also called the National Land Numeric Information (http://nlftp.mlit.go.jp/ksj-e/index.html), was generated from topographic maps and aerial photographs to support national land-use and development planning (Matsuda 1980; Yaguchi 1990). Land-use/cover data were prepared in 1976–1977, 1987–1988, 1991–1992, 1997–1998, 2006–2007, and 2009–2010 at two spatial resolutions, the Third-order Mesh (~1 × ~1 km) and 1/10 of the Third-order Mesh (~100 × ~100 m). Although the data were based on the Tokyo Datum up to the 1997–1998 data, the 2006–2007 and 2009–2010 data were based on the JGD2000, according to the revision of the Japan's geodetic system in 2002 (Matsumura et al. 2004). The data comprise 16 categories, but these categories differ among the preparation years (Table 10.4). Whereas the dominant category was assigned to the ~100-m mesh, areas of each category were indexed to the Third-order Mesh System. The data may be downloaded from the website (http://nlftp.mlit.go.jp/ksj-e/index.html) developed by the Ministry of Land, Infrastructure, Transport and Tourism.

The Standard Mesh System (Fig. 10.2) was developed through the preparation of Digital National Land Information. First, 1:25,000-scale topographic maps issued by the Geographical Survey Institute of Japan were scanned, and the land-use/cover was digitized. When necessary, the results were revised by interpretation of aerial photographs. The mesh representing 1/10 of the Third-order Mesh System was overlain on the digitized data, and the dominant land-use/cover category for each

Table 10.4 Land-use/cover categories used in the Digital National Land Information (Ministry of Land, Infrastructure, Transport and Tourism)

Code/year	Land use/cover category 1976/1977	1987/1988	1991/1992, 1997/1998, 2006/2007, 2009/2010	Remarks
1	Paddy field	Paddy field	Paddy field	
2	Field	Field	Other agricultural land	
3	Orchard	Orchard	N/A	
4	Other tree plantation	Other tree plantation	N/A	Mulberry plantation, tea plantation
5	Forest	Forest	Forest	
6	Wasteland	Wasteland	Wasteland	Bamboo bush (height < 2 m), bare land, wetland, cliff, rock substrates, perpetual snow, diggings
7	Land for building A[1]	Land for building	Land for building	[1]Large (> 50 m), dense buildings
8	Land for building B[2]	N/A	N/A	[2]Small (<50 m), individual buildings
9	Trunk transportation land	Trunk transportation land	Trunk transportation land	railway, station, road
A	Other land	Other land	Other land	Vacant land, bareland
B	Lake	Internal water area	Rivers and lakes	
C	Rivers A[3]	N/A	N/A	[3]Dry riverbed
D	Rivers B[4]	N/A	N/A	[4]Artificial areas in dry riverbeds
E	Beach	Beach	Beach	
F	Body of seawater	Body of seawater	Body of seawater	
G	N/A	N/A	Golf course[5]	[5]New code that was originally included in the code A

Reference for the categories	
1976/1977	http://nlftp.mlit.go.jp/ksj/jpgis/codelist/LandUseCd-77.html
1987/1988	http://nlftp.mlit.go.jp/ksj/jpgis/codelist/LandUseCd-88.html
1991/1992, 1997/1998, 2006/2007, 2009/2010	http://nlftp.mlit.go.jp/ksj/jpgis/codelist/LandUseCd-YY.html

After 1987–1988, codes 7 and 8 were integrated into code 7, and codes B to D were integrated into code B mesh was assigned. These data were then integrated into the Third-order Mesh System (Matsuda 1980; Yaguchi 1990). After the 1991–1992 data, satellite (Landsat TM) images were also used to complement the topographic maps.

Precise Land-Use, Land-Cover Map (Japan Aerospace Exploration Agency)

The Earth Observation Research Center (EORC) of the Japan Aerospace Exploration Agency (JAXA), in collaboration with Tsukuba University, generated national land-use/cover data in 2012 by using data from the Advanced Land Observing Satellite (ALOS) collected using Advanced Visible and Near Infrared Radiometer type-2 (AVNIR2) at a spatial resolution of 50 m in the World Geodetic System (JAXA EORC 2012) (Fig. 10.3d). The data comprise 12 categories (Table 10.1), which can be downloaded at the ALOS website (http://www.eorc.jaxa.jp/ALOS/lulc/lulc_jindex.htm).

ALOS was launched by JAXA in 2007, and it was in operation until 2011. AVNIR2 is an optical sensor that has observation bands in visible and near-infrared regions. The spatial resolution of the image is 10 m. As with the MODIS land-use/cover product, vegetation phenology was used for classification. ALOS AVNIR2 data were subject to atmospheric correction and orthorectification based on digital elevation models published by the Geographical Survey Institute of Japan. Normalized Difference Vegetation Index (NDVI) images derived from the data were then composited for each month to cover the whole country. As with the MODIS Land-use/Cover Product, land-use/cover was discriminated based on the seasonal NDVI trajectories. The overall accuracy showed ~55 % (JAXA EORC, 2012).

Land Use Information System

The Institute of Geography, Hokkaido University of Education, constructed a land-use/cover dataset for Japan. The land-use/cover datasets for ca. 1900, ca. 1950, and ca. 1985 were constructed by 1992, and the entire data were integrated into a system called the Land Use Information System (LUIS). Topographic maps (1:50,000 scale) issued by the Geographical Survey Institute of Japan were scanned and digitized. They adopted 35 land-use/cover categories described on the maps (Table 10.5), and the information was integrated at a spatial scale of 1° (latitude) × 1.5° (longitude). As the production years suggest, the geodetic system was Tokyo Datum. The major results of the project have been presented in an atlas of environmental change in modern Japan (Nishikawa 2006).

Table 10.5 Land-use/cover categories used in the Land-use Information System (Nishikawa 2006)

Dry paddy field
Paddy field
Marshy paddy field
Field and vacant land
Mulberry plantation
Tea plantation
Orchard
Other tree plantation
Grassland
Broad-leaved forest
Coniferous forest
Mixed forest
Bamboo forest
Wasteland
Dwarf pine
Gravely sand
Wetland
Individual tree
Dead tree
Oriental paper bush
Palm tree
Nursery tree
Saltpan
Village
City
River
Lake
Sea
Road
Railway
School
Public office
Temple
Shrine
Unclassified

Agricultural Statistics Mesh Data (National Institute for Agro-Environmental Sciences)

For agricultural areas, the GHG Agro-stat MeteoCrop Soil-information Data Base (gamsDB) was prepared by the National Institute for Agro-Environmental Sciences under the Ministry of Agriculture, Forestry and Fisheries (Osawa et al. 2012). The database includes meteorological data, soil data, and greenhouse-gas-flux data of agricultural lands, as well as agricultural statistics data. The statistics data, called Agricultural Statistics Mesh Data (ASMD), comprise a spatial dataset that includes crop field areas (13 categories; Table 10.1) and livestock numbers (4 categories) at a spatial resolution of the Third-order Mesh (~ 1 km) based on the Tokyo Datum. The data are available for every 5-year period from 1975 to 1995, and currently the year 2000 data are in preparation. These data may be found at the gamsDB website (http://agrimesh.dc.affrc.go.jp/) (Osawa et al. 2012).

ASMD was originally prepared to estimate nitrogen load from agricultural lands. Two available data categories were used for generating ASMD. One was the land-use data, which were prepared as ~100-m mesh data (1/10 of the Third-order Mesh) in the Digital National Land Information developed by the Ministry of Land, Infrastructure, Transport and Tourism, and the other was the agricultural census data of rural communities and their attached maps. These data were integrated into the Third-order Mesh. A detailed preparation procedure is described by Kohyama et al. (2003).

User Needs for Potential Application to Biodiversity and Ecosystem Service Research

Surrogacy of the maps for biodiversity and ecosystem services was shown to increase when proper classification categories were presented (Dalleau et al. 2010). Categories of the land-use/cover data required should be specified to fill the gap between producers and users. We reviewed potential application of the datasets to estimation of biodiversity and ecosystem services to specify land-use/cover categories.

One of the main topics in biodiversity is estimation of species distribution at a wider spatial extent, which is particularly important when the target is a rare or keystone species. Human interference could have significant influence, both negative and positive (e.g., the Satoyama Index; Kadoya and Washitani 2011), on species distribution and biodiversity. Thus, it is essential to discriminate natural and artificial lands. For forest areas, it is necessary to discriminate among natural, secondary, and planted forests, as well as between evergreen and deciduous forests. Grassland areas generally include natural and artificial areas (cultivated meadows, golf courses), which should also be discriminated.

Although all agricultural areas are a product of human interference, the trajectory of the interference history may affect biodiversity. Specifically, detection of abandoned paddy fields may be needed to estimate biodiversity in the Satoyama areas. In addition to biodiversity, agricultural areas are closely related to material discharge and ecosystem services. Sediment, nutrient, and pesticide discharge from agricultural lands may affect river and coastal ecosystems (Fabricius 2005), and making the land–sea connection is strongly encouraged to establish integrated coastal reserve planning (Stoms et al. 2005); this requires detailed classification of land-use/cover, especially for crop types, as each crop has its own crop calendar that affects sediment, nutrient, and pesticide dynamics. With respect to ecosystem service perspectives, one outstanding example is pollination. To estimate the value of pollination, detailed classification on crop types may also be required.

We should also pay attention to spatial resolution, as it affects both computation time and accuracy of estimation (Luoto et al. 2007). Coarser spatial resolution (>10 km) may be feasible for national-scale analysis that uses more computer resources, whereas finer spatial resolution (<1 km) may be required for regional study based on watersheds or regional sectors. Additionally, the spatial unit should correspond to the unit of census surveys. For example, rare species of vascular plants were surveyed based on the Second-order Mesh (~10 × ~10 km), whereas wild animals were surveyed based on the Fivefold Mesh System (~5 × ~5 km) during the National Survey on the Natural Environment.

Another important item to be considered is the preparation year of the data. Japan's economic growth (high economic growth from 1954 to 1973 and stable economic growth until 1991) significantly modified the national landscape, potentially resulting in biodiversity loss. On the other hand, growing awareness of biodiversity and ecosystem services in the 2000s could have positive effects on conservation activities. Therefore, data for three periods (during economic growth, after economic growth, and at present) may be required.

Toward Generation of National Land-Use/Cover Data for Biodiversity and Ecosystem Services

Based on the compilation of available land-use/cover data and a brief review of the potential applications, we have determined the periods, spatial resolution, and classification categories for generalized national land-use/cover datasets to estimate biodiversity and ecosystem services. Table 10.6 summarizes the results for each dataset based on the publication year, spatial resolution, data format, geodetic system, and land-use/cover categories required.

The periods should be the 1970s, the 1990s, and the 2000s to achieve temporal coverage for the periods during economic growth, after economic growth, and the present, respectively. Unfortunately, no national land-use/cover map was available before the 1970s except for the Land Use Information System (Table 10.1). Spatial resolutions should be the Second-order Mesh (~10 × ~10 km), the Fivefold Mesh (~5 × ~5 km), and the Third-order Mesh (~1 × ~1 km). The Second-order Mesh may be used for national-scale analysis, whereas the Third-order Mesh may

Table 10.6 Attributes needed for the new land-use/cover data

Period	1970s, 1990s, 2000s
Spatial resolution and data format	Second-order Mesh (~10 × ~10 km), Fivefold Mesh (~5 × ~5 km), Third-order Mesh (~1 × ~1 km)
Geodetic system	World Geodetic System
Land-use/cover categories	See Table 10.1

be used for regional-scale analysis. The Fivefold Mesh would be needed to analyze the distribution of wild animals because the census survey of wild animals is based on the Fivefold Mesh. These procedures would reduce errors in the locations of the meshes or polygons because data generated from topographic maps and satellite images may contain errors as great as several tens of meters. Data before 2002 using the Tokyo Datum should be reprojected to the World Geodetic System. It should be noted, however, the original information in the meshes based on the Tokyo Datum would not be retained were reprojection to the World Geodetic System used. Therefore, meshes with Tokyo Datum may be used when comparing other data with the census data based on the Tokyo Datum.

The classification categories differ among the datasets, and thus rearrangements and/or integration would be needed according to user needs, as described in the previous section. The actual vegetation maps prepared by the Ministry of the Environment have the most detailed categories for plant communities and vegetation naturalness. It was thus possible to integrate the vegetation categories and the vegetation naturalness into land-use/cover, as demonstrated by Ogawa et al. (2013) and shown in Table 10.1. As with the scheme for the actual vegetation maps, a systematic, hierarchical classification scheme was adopted. Although the vegetation map data satisfy most of the requirements for land-use/cover, they lack detailed categories for agricultural areas. The only dataset that includes detailed categories for crops is the Agricultural Statistics Mesh Data (ASMD) (Table 10.1). Thus, integration of the two datasets would provide the most powerful and useful land-use/cover data.

Other national land-use/cover datasets do not satisfy the requirement that natural and artificial areas should be classified (Table 10.1). Both the Digital National Land Information and the Land Use Information System did not classify natural and planted forests, nor did it consider evergreen and deciduous forests, probably because they were generated from the topographic maps that had no categories for these distinctions. However, they would be useful in detecting overall changes in national land-use/cover at a very fine (~100 × 100 m) spatial resolution (Digital National Land Information) or with a long (~100 years) time frame (Land Use Information System), although the classification categories are not satisfactory for biodiversity and ecosystem service research.

Satellite-derived products (Precise Land-Use, Land-Cover Map, and the MODIS land-use/cover product) also lack the essential categories. Additionally, these products have inevitable misclassifications that arose from digital image processing. Nonetheless, the use of satellite images, especially those obtained by multispectral satellite sensors, would be the most cost-effective way to derive synoptic mapping of land-use/cover and habitats (Yamano 2013). Note that the vegetation maps by the Ministry of the Environment have been updated by satellite (Landsat MSS and Landsat TM) images during the fourth and fifth National Surveys on the Natural Environment (Tsuzawa 1990), and Digital National Land Information after 1991–1992 used satellite (Landsat TM) images to update the data.

Considering the publication year, spatial resolution, and land-use/cover categories, suitable data would be available from the 1970s and 1990s. However, the critical issue is the lack of a suitable dataset for the 2000s, as the current available land-use/cover products (Precise Land-Use, Land-Cover Map for 2012, the MODIS land-use/cover products for 2007 and 2009, and the Digital National Land Information for 2007) do not satisfy the requirements. As the subsequent sixth and seventh National Survey on the Natural Environment covers about half the country, filling the remaining half by combining the available products and additional interpretation of aerial photographs and/or high spatial resolution satellite images may be worthwhile until the full preparation of the vegetation maps can be completed following the completion of the National Survey on the Natural Environment.

Conclusions

Compiling information on available national land-use/cover data and user needs revealed an optimized preparation procedure for the national land-use/cover data to estimate biodiversity and ecosystem services in Japan (Fig. 10.1, Table 10.1). It could be possible to reveal changes in biodiversity and ecosystem services during Japan's period of peak economic development by using datasets for the 1970s and 1990s. Issues for preparation of the recent data, however, remain.

Global-scale or Asian-scale assessment of biodiversity and ecosystem services is an urgent task. Although several global-scale land-use/cover datasets have been prepared (Sato and Tateishi 2001), biodiversity and ecosystem service-oriented data are still scarce. Another important issue is quality assessment of the data. Although ground-based sampling was possible for national surveys in Japan, the use of satellite data may be encouraged for generating global-scale datasets. As misclassification is inevitable in the analysis of satellite data, evaluation of the products based on validation efforts would be an additional essential step in preparing the data. Iwao et al. (2006) validated several global-scale land-use/cover datasets by referring to the Degree Confluence Project data (http://www.confluence.org/) that were collected at the latitude and longitude integer degree intersections (the "confluences") and documented the state of the surroundings. Their procedure may be applied to other global-scale land-use/cover data. Our procedures to bridge the gap between data producers and users, combined with quality evaluation of the land-use/cover data, could be further applied to other data at a larger spatial extent.

Acknowledgments This study is a contribution to the S-9 and D-1008 projects supported by the Environment Research and Technology Development Fund, Ministry of the Environment, Japan, and the Biodiversity Priority Program, National Institute for Environmental Studies (NIES). Data prepared according to the present procedure may be downloaded from the website of the Center for Environmental Biology and Ecosystem Studies, NIES (http://www.nies.go.jp/biology/kiban/index.html).

References

Dalleau M, Andréfouët S, Wabnitz CC, Payri C, Wantiez L, Pichon M, Friedman K, Vigliola L, Benzoni F (2010) Use of habitats as surrogates of biodiversity for efficient coral reef conservation planning in Pacific Ocean islands. Conserv Biol 24:541–552

Fabricius KE (2005) Effect of terrestrial runoff on the ecology of corals and coral reefs: review and synthesis. Mar Pollut Bull 50:125–146

Foley JA, DeFries R, Asner GP, Barford C, Bonan G, Carpenter SR, Chapin FS, Coe MT, Daily GC, Gibbs HK, Helkowski JH, Holloway T, Howard EA, Kucharik CJ, Monfreda C, Patz JA, Prentice IC, Ramankutty N, Snyer PK (2005) Global consequences of land use. Science 309:570–574

Gaston KJ (2000) Global patterns in biodiversity. Nature (Lond) 405:220–227

Iwao K, Nishida K, Kinoshita T, Yamagata Y (2006) Validating land cover maps with degree confluence project information. Geophys Res Lett 33:L23404

JAXA EORC (2014) Overview of the precise land-use, land-cover map (Japan, ver. 14.02) derived from ALOS data. http://www.eorc.jaxa.jp/ALOS/lulc/LC_v14.02readme.pdf. Accessed 23 Mar 2014

Kadoya T, Washitani I (2011) The Satoyama index: a biodiversity indicator for agricultural landscapes. Agric Ecosyst Environ 140:20–26

Kadoya T, Ishii HS, Kikuchi R, Suda S, Washitani I (2009) Using monitoring data gathered by volunteers to predict the potential distribution of the invasive alien bumblebee *Bombus terrestris*. Biol Conserv 142:1011–1017

Kakuta S, Honma R, Someya T, Tsukamoto Y, Yanagawa T, Shimizu O (2009) Nationwide general vegetation survey by using multi-temporal MODIS data. Ann Conf Remote Sens Soc Jpn, Nagoya, pp. 167–168

Kohyama K, Hojito M, Sasaki H, Miyaji H (2003) Generation of agricultural statistics mesh data using Digital National Land Information. Jpn J Soil Sci Plant Nutr 74:415–424

Luoto M, Virkkala R, Heikkinen RK (2007) The role of land cover in bioclimatic models depends on spatial resolution. Glob Ecol Biogeogr 16:34–42

Matsuda H (1980) Digital National Land Information. GeoJournal 4:313–318

Matsumura S, Murakami M, Imakiire T (2004) Concept of the new Japanese geodetic system. Bull Geogr Surv Inst 51:1–9

Nishikawa O (ed) (2006) Atlas: environmental change in modern Japan. Asakura, Tokyo

Ogawa M, Takenaka A, Kadoya T, Ishihama F, Yamano H, Akasaka M (2013) Land-use classification and mapping at a whole scale of Japan based on a national vegetation map. Jpn J Conserv Ecol 18:69–76

Oliver I, Holmes A, Dangerfield JM, Gilling M, Pik AJ, Britton DR, Holley M, Montgomery ME, Raison M, Logan V, Pressey RL, Beattie AJ (2004) Land systems as surrogates for biodiversity in conservation planning. Ecol Appl 14:485–503

Osawa T, Kohyama K, Kuwagata T, Sudo S (2012) Web interface for integration of divers agro-environmental databases. Agric Inform Res 21:1–10

Sarkar S, Margules CR (2002) Operationalizing biodiversity for conservation planning. J BioSci 27:299–308

Sato HP, Tateishi R (2001) Global land use, land cover and vegetation classification systems: a review. J Geogr Surv Inst 96:1–33

Stoms DM, Davis FW, Andelman SJ, Carr MH, Gaines SD, Halpern BS, Hoenicke R, Leibowitz SG, Leydecker A, Madin EMP, Tallis H, Warner RR (2005) Integrated coastal reserve planning: making the land-sea connection. Front Ecol Environ 3:429–436

Tittensor DP, Mora C, Jetz W, Lotze HK, Ricard D, Berghe EV, Worm B (2010) Global patterns and predictors of marine biodiversity across taxa. Nature (Lond) 466:1098–1101

Tsuzawa M (1990) National surveys on the natural environment ("Green Census"). J Geogr 99:64–71

Yaguchi A (1990) Digital National Land Information. J Geogr 99:3–11

Yamano H (2013) Multispectral application. In: Goodman JA, Purkis SJ, Phinn SR (eds) Coral reef remote sensing: a guide for multi-level sensing, mapping and assessment. Springer, Dordrecht, pp 51–78

Yamaura Y, Amano T, Koizumi T, Mitsuda Y, Taki H, Okabe K (2009) Does land-use change affect biodiversity dynamics at a macroecological scale? A case study of birds over the past 20 years in Japan. Anim Conserv 12:110–119

Chapter 11
Land Cover Classification Using Multi-Temporal Satellite Images in a Subtropical Region

Mitsunori Ishihara, Hitoshi Hasegawa, Seiji Hayashi, and Hiroya Yamano

Abstract This study was performed to develop and apply a land cover classification method in subtropical regions by using multi-temporal satellite images acquired from various satellites in the Todoroki River watershed, Ishigaki Island, Okinawa, Japan (24º22.6′N, 124º14.5′E). Six satellite images, including the winter and summer periods from 2006 to 2008, were used. This data set included data from several satellites: ALOS/AVNIR2, SPOT/HRV, HRG, and FORMOSAT-2. The land cover classification was performed using the decision tree method, considering the crop calendar and the time-series Normalized Difference Vegetation Index (NDVI) calculated from the satellite data. Our method showed better classification accuracy than another traditional method (ISODATA). In addition, the classification result includes dynamics of crop growth patterns. The land cover classification method proposed in the present study can be applied to other tropical and subtropical regions that have similar crops.

Keywords Crop growth • Ishigaki Island • Japan • Land cover • Satellite data

---

M. Ishihara
National Institute for Environmental Studies, 16-2 Onogawa, Tsukuba, Ibaraki 305-8506, Japan

National Institute for Agro-Environmental Sciences, 3-1-3 Kannondai, Tsukuba, Ibaraki 305-8604, Japan

H. Hasegawa
Kokushikan University, 4-28-1 Setagaya, Setagaya-ku, Tokyo 154-8515, Japan

S. Hayashi • H. Yamano (✉)
National Institute for Environmental Studies, 16-2 Onogawa, Tsukuba, Ibaraki 305-8506, Japan
e-mail: hyamano@nies.go.jp

S.-i. Nakano et al. (eds.), *Integrative Observations and Assessments*, Ecological Research Monographs, DOI 10.1007/978-4-431-54783-9_11, © Springer Japan 2014

Introduction

Land cover provides essential data for estimating resource inventory, biodiversity, and ecosystem services (Akasaka et al. 2014) and setting up land management plans. It is also used to estimate sediment discharge, which significantly affects river and coastal marine ecosystems, by identifying bare lands. Fine sediment called "red soil" covers half the area of Okinawa Islands, Japan (Mkadam et al. 2006). In subtropical regions, red soil runoff from bare land causes water pollution of rivers and environmental degradation in coastal areas (e.g., Omija et al. 1998). To counteract these issues, it is important to understand both land cover (e.g., crop type) and the timing of bare soil exposure after harvesting, which is the source of red soil, and to estimate the amounts of red soil outflow quantitatively (Blanco et al. 2010).

Although land cover classifications have been performed widely based on satellite images, classification using a single image, however, may not work, because growth of crops shows significant temporal variation. In subtropical regions, sugarcane, which is the main crop in Okinawa, has three planting patterns including summer, winter and ratoon planting. It is thus required to use multi-temporal images to capture these patterns to achieve high classification accuracy for land cover with the dynamics of crop growth patterns.

To date, many multispectral satellite sensors have been deployed, and thus classification using multi-temporal satellite images that consider the crop calendar is possible (e.g., Murakami et al. 2001; Inoue et al. 2007). If the crop calendar could be known in the study area, the method does not need training data for the classification, which is particularly effective in remote areas with limited access. In this study, we developed and applied a land cover classification method in consideration of the crop calendar using multi-temporal images acquired from various satellites for the Todoroki River watershed, Ishigaki Island, Okinawa, Japan.

Materials and Methods

Study Area

The study area was the Todoroki River watershed (approximately 4 × 5 km; centered 24º22.6′N, 124º14.5′E), Ishigaki Island, Okinawa, located in the south-western part of Japan. Annual mean air temperature was 24.0 °C and annual mean rainfall was 1,899 mm, measured at the nearby study area of the Japan Meteorological Agency Automated Meteorological Data Acquisition System set at 24°20.6′N, 124°11.2′E. The major land cover includes forest (evergreen), grassland, sugarcane, and paddy fields, with a small portion of pineapple plantation.

Table 11.1 Characteristics of the satellite images used in this study

Satellite sensor	Date	Season	Spatial resolution (m)	Cloud cover (%)
SPOT HRV	2006.3.4	Winter	20	0
ALOS AVNIR2	2006.7.23	Summer	10	3
FORMOSAT MS	2007.1.31	Winter	8	1
ALOS AVNIR2	2007.9.24	Summer	10	50
SPOT HRG	2008.3.3	Winter	10	7
ALOS AVNIR2	2008.9.26	Summer	10	34

Satellite Data and Processing

Six satellite images, including data from winter and summer for the periods from 2006 to 2008, were used (Table 11.1). This data set included data from several satellites: ALOS/AVNIR2 (spatial resolution: 10 m), SPOT/HRV and HRG (20 and 10 m), and FORMOSAT-2 (8 m). These images contained partial cloud cover because there were no cloud-free images. Therefore, we created cloud masks by visual assessment and removed these cloud-covered areas from the analysis. All the images were georeferenced with RMS (root mean square) errors smaller than 1.0 pixel and resampled to a spatial resolution of 8 m according to FORMOSAT MS, using the nearest-neighbor method. After these geometric corrections, simplified atmospheric correction was performed using the darkest pixel method (MacFarlane and Robinson 1984). This method removes the atmospheric path radiance by subtracting the DN (digital number) value of the darkest pixel (over deep water) in the scene for all the other pixels. We converted these DN values obtained after geometric and atmospheric correction to radiance values using the conversion equation of each sensor. Then, we calculated the Normalized Difference Vegetation Index (NDVI) by the following equation using radiance values of all images:

$$\text{NDVI} = (R_{\text{NIR}} - R_{\text{Red}})/(R_{\text{NIR}} + R_{\text{Red}}) \tag{11.1}$$

where R_{Red} and R_{NIR} are the red and near-infrared radiance values, respectively. According to these procedures, NDVI was calculated in all images, and these images were used to classify the detailed land cover type.

Ground-Based Survey

The survey was conducted (1) to gain an understanding of the crop calendar to determine thresholds of NDVI for classification using multi-temporal satellite images and (2) to validate the classification results.

We collected information regarding the crop cultivation pattern (the crop calendar) through interviews with farmers in the study area. In addition, land cover was surveyed using Geographic Information System (GIS) software and a Global Positioning System (GPS) over the entire study area in May 2004, August 2004, December 2004, and February 2005 (Hasegawa 2006).

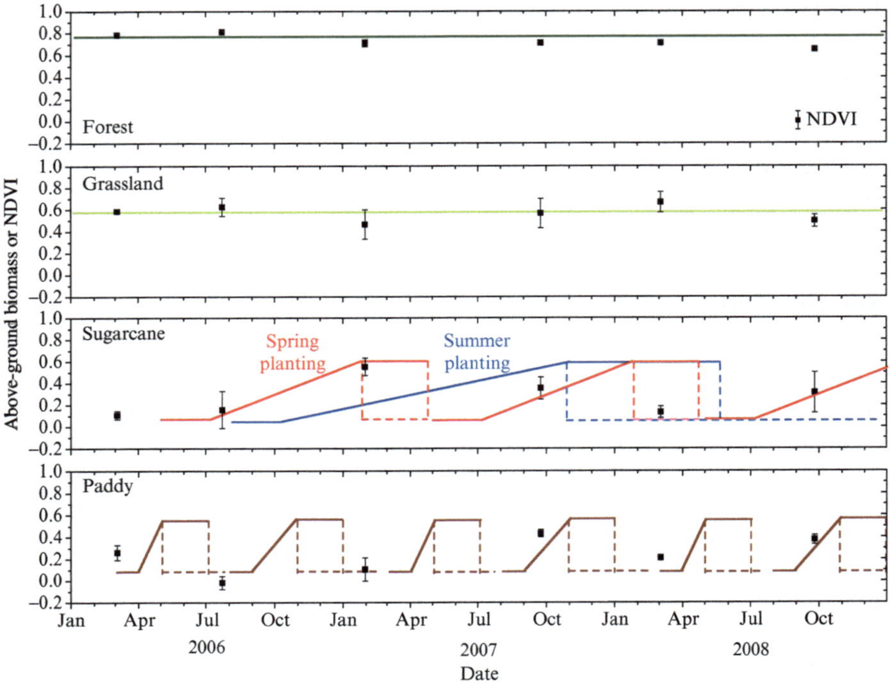

Fig. 11.1 Crop calendar in the Todoroki River watershed and typical average Normalized Difference Vegetation Index (NDVI) values of each land cover category from 2006 to 2008. *Solid lines* are typical NDVI values, *hatched lines* are timing of harvesting, and *points* are average NDVI values extracted from the satellite images in each land cover category

Land cover maps made by the latter survey (Hasegawa 2006) were used to validate our classification. These land cover maps had 40 classes and more than 3,000 polygons. These classes contained information on growing conditions in the agricultural areas. Therefore, we reclassified 40 classes into 6 classes (forest, grassland, sugarcane, pineapple, paddy, and other), taking into consideration growing conditions in the four land cover maps.

Classification and Accuracy Assessment

Classification categories were forest (evergreen), grassland, sugarcane, pineapple and paddy. Land cover classification was performed by decision tree classification (e.g., Pal and Mather 2003). To compare the results, we performed unsupervised ISODATA classification using one scene only (FORMOSAT-2 on January 31, 2007). The structure of the decision tree was designed according to the crop calendar of these classification categories made from the ground survey information of the land cover maps (Hasegawa 2006) and interview (Fig. 11.1). Unfortunately,

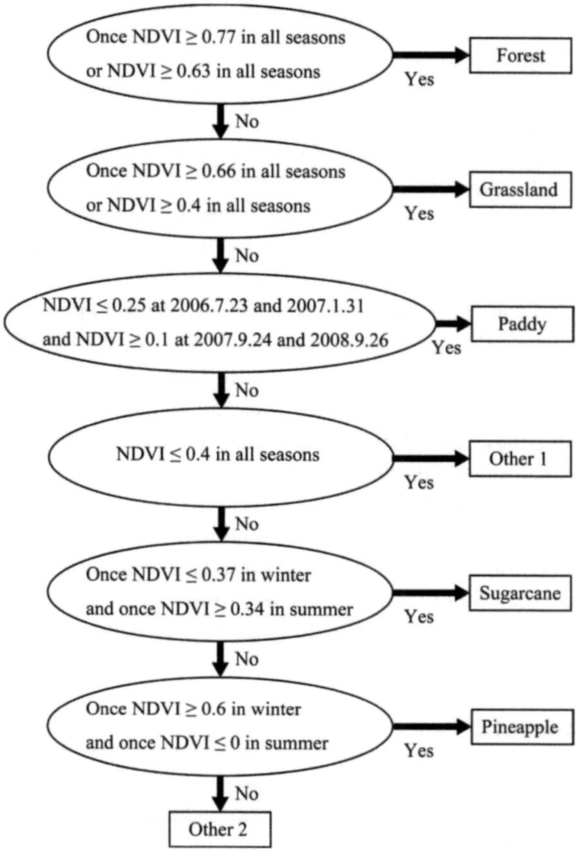

Fig. 11.2 The thresholds of the decision tree

it was difficult to show the crop calendar of pineapple plantations in the figure because the cultivation period for pineapple is 3 years or more and the crop is harvested two or more times. Therefore, we did not show the crop calendar of pineapple in Fig. 11.1. Moreover, we determined the thresholds of the decision tree by considering the features of the temporal change patterns of NDVI values in each category calculated from the land cover map from the ground survey and the satellite data (Fig. 11.2). The land cover classification was performed after determination of the structure of the decision tree.

The accuracy of the land cover classification results for the multi-temporal images was assessed from the four merged land cover maps (Hasegawa 2006) made by the ground survey. To examine the effectiveness of the classification, the unsupervised ISODATA classification was performed based on the FORMOSAT-2 image obtained on January 31, 2007, and the accuracy assessment was carried out based on the land cover map made by the ground survey in February 2005. The data sets for verification differ among the two methods, decision tree classification and unsupervised ISODATA classification. Because the unsupervised method was classified using one scene only, the verification data closest to the season when the image was acquired were used.

Results and Discussion

Figure 11.1 shows temporal patterns of typical NDVI values of each land cover category from 2006 to 2008. All NDVIs showed patterns of change similar to the crop calendar obtained from the ground survey. NDVI values of forest and grassland were constant from 0.6 to 0.8 in all seasons. In contrast, NDVI values of the other agricultural areas, such as sugarcane and paddy fields, decreased significantly according to the timing of harvesting. These temporal patterns of NDVI values are useful for decision tree classification.

Table 11.2 shows error matrices of the land cover classification results. In the case of the decision tree method, the overall accuracy and kappa coefficient were 64 % and 0.43, respectively. Both the producer's accuracy and user's accuracy exceeded 60 % in forest, grassland, and sugarcane. However, the classification accuracy was markedly low for paddy fields and pineapple plantations. The classification accuracies of these categories were thought to be low because the parcels of land associated with paddy fields and pineapple plantations were small in the study area. The maximum spatial resolution of the satellite images was 20 m in this study, and it was difficult to classify agricultural areas smaller than this resolution. On the other hand, the accuracy of the unsupervised method was poorer than that of the decision tree method. Thus, the decision tree method proposed in the present

Table 11.2 Error matrixes of the land cover classification results

	Reference						User's
	Forest	Grassland	Sugarcane	Pineapple	Paddy	Total	accuracy (%)
(a) Decision tree classification[a]							
Forest	270	139	11	0	1	421	64.1
Grassland	124	369	84	0	2	579	63.7
Sugarcane	28	76	251	5	8	368	68.2
Pineapple	0	0	5	1	0	6	16.7
Paddy	1	4	4	2	6	17	35.3
Total	423	588	355	8	17	1,391	
Producer's accuracy (%)	63.8	62.8	70.7	12.5	35.3		
(b) Unsupervised classification[b]							
Forest	275	49	9	0	2	335	82.1
Grassland	190	677	342	29	7	1,245	54.4
Sugarcane	117	230	233	25	13	618	37.7
Pineapple	0	0	0	0	0	0	0
Paddy	14	21	52	4	69	160	43.1
Total	596	977	636	58	91	2,358	
Producer's accuracy (%)	46.1	69.3	36.6	0	75.8		

[a]Overall accuracy, 64 %; kappa coefficient, 0.43
[b]Overall accuracy, 53 %; kappa coefficient, 0.30

study was more accurate than the general unsupervised method. Moreover, each classified pixel has its trajectory on the dynamics of crop growth patterns, including the timing of the occurrence of bare soil.

In conclusion, the decision tree method using multi-temporal images acquired from various satellites was found to be an effective method for precise land cover classification that includes temporal and annual changes in land cover. This method can be used for classification even when no ground survey information is available and can be used to calculate the spatial changes in agricultural areas, which are the source of red soil. Further, our classification method could be applied widely to other subtropical and tropical regions that have similar crops.

Acknowledgments This work was supported by a Grant-in-Aid for Scientific Research on Innovative Areas "Coral reef science for symbiosis and coexistence of human and ecosystem under combined stresses" (No. 20121004) of the Ministry of Education, Culture, Sports, Science and Technology (MEXT), Japan. This is a contribution to the S-9 Project supported by the Environment Research and Technology Development Fund, Ministry of the Environment, Japan.

References

Akasaka M, Takenaka A, Ishihama F, Kadoya T, Ogawa M, Osawa T, Yamakita T, Tagane S, Ishii R, Nagai S, Taki H, Akasaka T, Oguma H, Suzuki T, Yamano H (2014) Development of a national land use/cover dataset to estimate biodiversity and ecosystem services. In: Nakano S, Yahara T, Nakashizuka T (eds) The biodiversity observation network in the Asia-Pacific Region: integrative observations and assessments of Asian biodiversity. Springer, Tokyo

Blanco AC, Nadaoka K, Yamamoto T, Kinjo K (2010) Dynamic evolution of nutrient discharge under stormflow and baseflow conditions in a coastal agricultural watershed in Ishigaki Island, Okinawa. Jpn Hydrol Process 24:2601–2616

Hasegawa H (2006) The decline of coral reef conditions caused by extensive land modification: a case study of the Shiraho area on Ishigaki Island. PhD Thesis, Tokyo Metropolitan University, Okinawa

Inoue Y, Qi J, Olioso A, Kiyono Y, Horie T, Asai H, Saito K, Ochiai Y, Shiraiwa T, Douangsavanh L (2007) Traceability of slash-and-burn land-use history using optical satellite sensor imagery: a basis for chronosequential assessment of ecosystem carbon stock in Laos. Int J Remote Sens 28:5641–5647

MacFarlane N, Robinson IS (1984) Atmospheric correction of LANDSAT MSS data for a multidate suspended sediment algorithm. Int J Remote Sens 5:561–576

Mkadam MK, Yonaha T, Ali SV, Tokuyama A (2006) Dissolved aluminum and silica release on the interaction of Okinawan subtropical red soil and seawater at different salinities: experimental and field observations. Geochem J 40:333–343

Murakami T, Ogawa S, Ishitsuka N, Kumagai K, Saito G (2001) Crop discrimination with multitemporal SPOT/HRV data in the Saga Plains. Jpn Int J Remote Sens 22:1335–1348

Omija T, Nakasone K, Kobayashi R (1998) Water pollution caused by soil runoff at a coral reef in Shiraho, Ishigaki Island. Ann Rep Okinawa Pref Inst Health Environ 32:113–117

Pal M, Mather PM (2003) An assessment of the effectiveness of decision tree methods for land cover classification. Remote Sens Environ 86:554–565

Chapter 12
Impervious Surface Area as an Indicator for Evaluating Drainage Basins

Bunkei Matsushita, Fan Yang, and Takehiko Fukushima

Abstract Construction of an impervious surface in a watershed will impact biodiversity and the ecosystem function of the watershed. In this chapter, we first introduce the relationships between impervious surface area (ISA) and biodiversity of watersheds. Then, we review the existing methods for estimating ISA from remote sensing data. Finally, we give two case studies in Lake Kasumigaura Basin to show a long-term change of ISA and the altered hydrological characteristics of the basin.

Keywords Biodiversity • Drainage basin • Ecological characteristics • Hydrological characteristics • Indicator • Remote sensing • Thermal characteristics

Introduction

On October 31, 2011, the population of the world had reached 7 billion, and most of these people are concentrated in urban areas. From 2009 on, the urban population began to exceed the rural population in the world, and until 2020, such shift will happen even in the less-developed regions (UN 2009). The rapid growth of world population and urbanization causes great demands for housing, buildings, roads, parking lots, and other manmade surfaces, all of which are classified as impervious surfaces. Therefore, it can be considered that the Earth is being paved with

B. Matsushita (✉) • T. Fukushima
Faculty of Life and Environmental Sciences, University of Tsukuba, Tennoudai 1-1-1, Tsukuba, Ibaraki 305-8572, Japan
e-mail: matsushita.bunkei.gn@u.tsukuba.ac.jp

F. Yang
Graduate School of Life and Environmental Sciences, University of Tsukuba, Tennoudai 1-1-1, Tsukuba, Ibaraki 305-8572, Japan

S.-i. Nakano et al. (eds.), *Integrative Observations and Assessments*, Ecological Research Monographs, DOI 10.1007/978-4-431-54783-9_12, © Springer Japan 2014

impervious surfaces by human beings (Sutton et al. 2009). Elvidge et al. (2007) reported that 0.43 % of the land surface in the world had been covered by an impervious surface at the end of the twentieth century. In contrast, the percentage of impervious surface may reach a larger value in inhabited regions. It has been known that both the hydrological and ecological characteristics of a drainage basin can be disturbed by a constructed impervious surface (Elvidge et al. 2007), thus resulting in biodiversity and ecosystem function loss (Utz et al. 2011). A widely accepted scale for the impacts of existing impervious surface area (ISA) on a drainage basin is that the drainage basin is stressed if it contains 1–10 % ISA, impacted if it contains 10–25 % ISA, and degraded if it contains more than 25 % ISA (Arnold and Gibbons 1996; Schueler 1994).

The Link Between ISA and Biodiversity

The changed ISA can impact biodiversity through two ways: direct and indirect impacts (Fig. 12.1). The increased ISA caused by urbanization and urban sprawl may directly impact biodiversity in a watershed by replacing natural surface types such as vegetation and soil with impervious materials. Because the ISA has different surface characteristics from the natural land cover, its increase causes a great influence on the hydrological and thermal characteristics of watersheds, thus indirectly impacting biodiversity. The altered hydrological characteristics caused by increasing ISA can be further separated into physical and chemical alterations.

Alteration of Ecological Characteristics Caused by ISA (Direct Impacts)

Cutting vegetation leads to loss not only of plant production but also of natural habitats. For example, increased ISA from urbanization causes forest loss, and the remaining forest patches are usually isolated and too small to maintain viable populations of forest-dependent wildlife species (Er et al. 2005). Mason et al. (2007) also reported that diversity and abundance of neotropical migrants, insectivores, and forest-interior specialist birds have been decreased as a result of suburban development. On the other hand, the sealing of soils by impervious materials not only restricts or hampers energy exchange as well as water and gas movements in sealed areas but also increases pressure in adjacent nonsealed areas, thus degrading their ecological functions (Scalenghe and Marsan 2009).

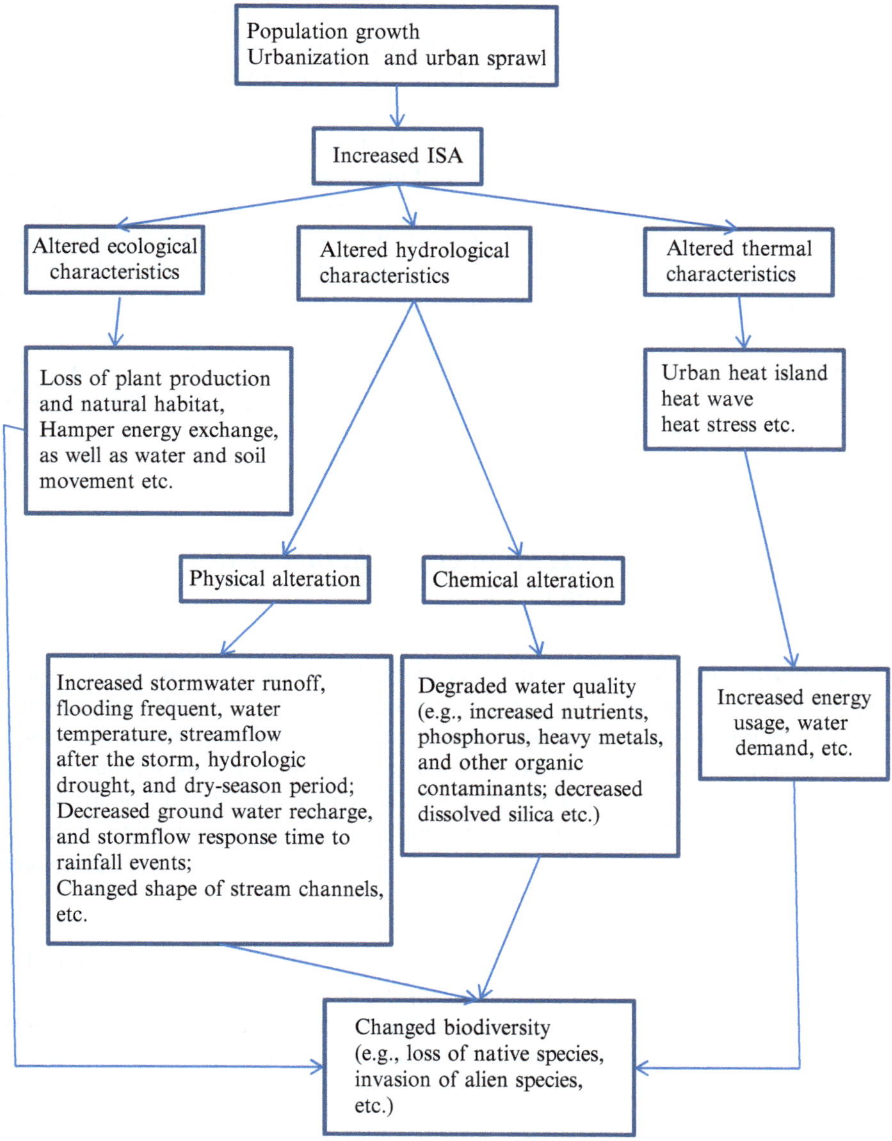

Fig. 12.1 Link between impervious surface area (*ISA*) and biodiversity

Physical Alteration of Hydrological Characteristics Caused by ISA (Indirect Impacts)

The physical alterations of hydrology include increased frequency and magnitude of stormwater runoff and flooding, changed shape of stream channels, increased water temperatures, faster receding of streamflow after the storm, longer dry periods between rainfall events, and reduced groundwater recharge. These altered streamflow regimes may further affect biodiversity of stream ecosystems. For example, Nelson et al. (2008) pointed that fish from urban stream systems are confronted with extreme flows during storms and runoff events because of stream and watershed hydrology altered by urbanization. Barrett et al. (2010) confirmed, through an artificial stream experiment, that high water velocities in urban streams result in decreased larval retention in streams. Helms et al. (2009) reported that fish species diversity, richness, and biotic integrity were lower in streams that received a high frequency of spate flows according to data from 18 watersheds in the Lower Piedmont of Western Georgia. Sung et al. (2011) found that the hydrological drought caused by watershed urbanization facilitates the invasion of alien species in riparian forests in a hot and semiarid region, thus causing negative effects on the biodiversity of native species. The riparian vegetation can generally provide shade and enhance habitat in streams (Collier et al. 2009).

Chemical Alteration of Hydrological Characteristics Caused by ISA (Indirect Impacts)

Hydrological consequences of ISA also include deterioration in surface water quality because urban pollutants are swept into aquatic environments, consequently causing chemical alteration of the hydrology in a watershed. The stormwater runoff from an impervious surface may contain excess nutrients, phosphorus, heavy metals, and other organic contaminants, which are potential stressors or stimulants for algae, invertebrates, and fish within the aquatic community (Johnson et al. 2011). Urban rivers daily receive tons of phosphorus and other pollutants from stormwater generated by impervious surfaces. Loucaides et al. (2007) compared dissolved silica loading from two adjacent coastal watersheds with different percentages of ISA during four rain and five non-rain events. They found that dissolved silica loading from the more impervious watershed is lower than that from a less impervious watershed because the discharge contained a significantly larger proportion of rainwater poor in dissolved silica than groundwater rich in dissolved silica in the more impervious watershed (Loucaides et al. 2007). Generally, dissolved silica availability is a factor that affects the composition of algal populations in aquatic ecosystems. Lewis and Grimm (2007) investigated the relationship between nutrient export and catchment features and found that loads per catchment were greater from more impervious catchments, probably because impervious catchments held more mobilizable nutrients and conveyed more rainfall water as overland flow. The export of nitrogen from urban catchments has been a global problem.

Alteration of Thermal Characteristics Caused by ISA (Indirect Impacts)

The ISA may also alter sensible and latent heat fluxes of watersheds, causing changed thermal characteristics (e.g., urban heat islands). The changed thermal characteristics have a great impact not only on both micro- and macroclimate but also on human health and the urban ecosystem because of increased energy usage and smog formation from increased urban temperature (Zoran et al. 2011; Haselbach et al. 2011). In addition, Guhathakurta and Gober (2010) reported that the increased ISA resulted in urban heat islands, which then reduced diurnal temperature difference, and thus increased water demand by humans and the ecosystem (e.g., mesic vegetation). As a result, all these changed thermal characteristics caused by the increased ISA may finally increase the loss of biodiversity.

How to Estimate ISA

A wide range of studies such as urban hydrology, urban climate, land-use planning, watershed resource management, global environmental change, and human–-environment interactions require accurate and up-to-date geospatial data of ISA (Jiang et al. 2009; Weng 2012). Remote sensing has permitted estimates of the ISA distribution as a proxy measure of the human ecological footprint (Sutton et al. 2009).

Figure 12.2 summarizes a general method for estimating ISA from remote sensing data. Remote sensing data can be separated into three categories according to spatial resolution: high (less than 10 m, e.g., aerial photograph, IKONOS, QuickBird, OrbView), medium (10–100 m, e.g., Landsat/TM or ETM+, SPOT/HRV, ALOS/AVNIR2), and coarse (larger than 100 m, e.g., NOAA/AVHRR, MODIS) spatial resolution data. Remote sensing data with a high spatial resolution are usually not a good choice for ISA estimation because of the scale of the study areas (cover area too small), facility of available data (very low temporal resolution), and cost (too expensive), even though it can provide reliable ISA data by visual analysis of images. In contrast, remote sensing data with medium spatial resolution are most appropriate for estimating ISA at a city or basin scale (Wu and Murray 2003; Yang et al. 2010). However, for ISA estimation at a national, continental, or global scale, coarse-resolution data will be a better choice when considering practical problems such as ready availability of data, time constraints, and high costs. Reliable ISA data that derive from high-resolution imagery are helpful for validating ISA estimation at the medium-resolution level, and then validated ISA from the medium-resolution imagery can be further used to validate ISA estimation at the coarse-resolution level (Lu et al. 2008; Weng 2012).

Image data pre-processing, including geometric correction, relative radiometric normalization, conversion of digital numbers to top-of-the-atmosphere reflectance,

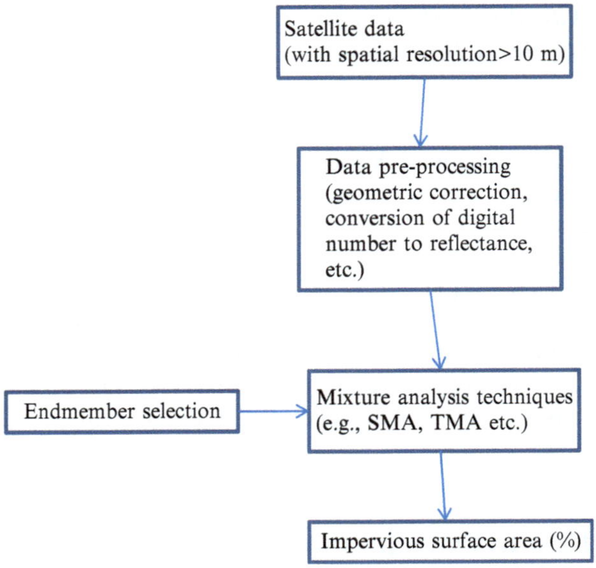

Fig. 12.2 Flowchart for estimating impervious surface area (ISA) from satellite data. *SMA*, spectral mixture analysis; *TMA*, temporal mixture analysis

and removal of water pixels, is required before ISA estimation (Yang et al. 2010). Atmospheric correction is not so important in this step because relative radiometric normalizations have been done between the different satellite images.

Because mixed pixels will become dominant in the medium- and coarse-resolution remote sensing images, pixel-based approaches usually fail to estimate ISA accurately. Therefore, over the past decades, a number of image analysis techniques accommodating mixing problems have been proposed, of which linear spectral mixture analysis (LSMA) is a well-established and effective technique to address this mixture problem (Somers et al. 2011). For a mixed pixel, the reflectance for each band can be written as follows (Adams et al. 1986):

$$R_b = \sum\nolimits_{j=1}^{N} f_j R_{j,b} + e_b \tag{12.1}$$

$$\sum\nolimits_{j=1}^{N} f_j = 1,\ f_j \geq 0 \tag{12.2}$$

where N is the number of endmembers (i.e., spectrally 'pure' materials), f_j is the fraction of endmember j, $R_{j,b}$ is the reflectance of endmember j in band b, and e_b is the unmodeled residual indicating disagreement between the measured and modeled spectra in band b. The fraction of each endmember can be obtained using the least-squares method.

The key to successful SMA is appropriate endmember selection, which involves identifying both number and type of endmembers as well as their corresponding spectral signatures (Somers et al. 2011). Ridd (1995) suggested that there are three basic constituent components (i.e., endmembers) to composite the earth's surface: vegetation, ISA, and soil (i.e., V-I-S model). Wu and Murray (2003) extended the three endmembers to four endmembers by further separating ISA into high albedo (e.g., concrete) and low albedo (e.g., asphalt) because of their very different spectral signatures. Roberts et al. (1998) and Powell et al. (2007) proposed a multiple endmember spectral analysis, which allows the number and type of endmembers to vary from pixel to pixel. However, the maximum number of endmembers for each unmixing process is still not larger than four, and the basic types of endmembers are similar to those described by Ridd (1995). It can be considered that three or four endmembers are appropriate for ISA estimation by using medium- and coarse-resolution satellite images because most of these kinds of satellite sensors have only a limited number of bands (i.e., multispectral imagery).

To obtain more accurate ISA estimation, another important issue is addressing endmember variability (Somers et al. 2011). Variation still exists in endmember spectral signatures, caused by spatial and temporal variability in scene components and differential illumination conditions, even within each endmember class. Therefore, the use of fixed endmember spectra (commonly the mean of available spectra for each endmember class) will linearly decrease the accuracy of subpixel fraction estimates (Settle 2006). Many techniques, such as multiple endmember spectral mixture analysis (MESMA; Roberts et al. 1998), normalized spectral mixture analysis (NSMA; Wu 2004), and pre-screened and normalized multiple endmember spectral mixture analysis (PNMESMA; Yang et al. 2010), have been developed to reduce the errors in LSMA related to endmember variability.

Somers et al. (2011) gave a comprehensive overview on addressing endmember variability, and Weng (2012) gave an extensive review on various remote sensing approaches to extract and estimate ISA. The reader is advised to read their works for more detailed information.

In addition to SMA, another unmixing technique called temporal mixture analysis (TMA) can also be used to estimate ISA, especially for using coarse spatial resolution satellite data because that they all have higher temporal resolution (about 1 day) (Knight and Voth 2011; Yang et al. 2012). The TMA is algebraically identical to the SMA (i.e., Eq. 1) but uses temporal spectra instead of electromagnetic spectra (Piwowar et al. 1998). Yang et al. (2012) suggested use of rearranged NDVI (normalized difference vegetation index) time-series data at a stable temporal zone (i.e., the last six maximum NDVI) to estimate ISA in Japan. The greatest advantage of the TMA-based method is that it can provide an alternative to further reduce endmember variability. Figure 12.3, a comparison of averaged relative standard deviation for each endmember among the SMA, NSMA, and TMA, shows that the endmembers in the TMA gave the smallest variation, followed by the endmember in the NSMA, with the endmember in the SMA showing the largest variation. This observation indicates that the TMA has particular potential to reduce endmember variability and thus can provide an improved ISA estimation.

Accounting for characteristic interaction among ground objects, some researchers suggest a nonlinear mixture model (Borel and Gerstl 1994; Somers et al. 2009). However, because the quantitative effects of nonlinear mixing on the accuracy of fraction estimation have not been well understood, as well as practically considering the computational time, most unmixing studies still assume linear mixing if the accuracy requirement for ISA estimation is not so high (Somers et al. 2011).

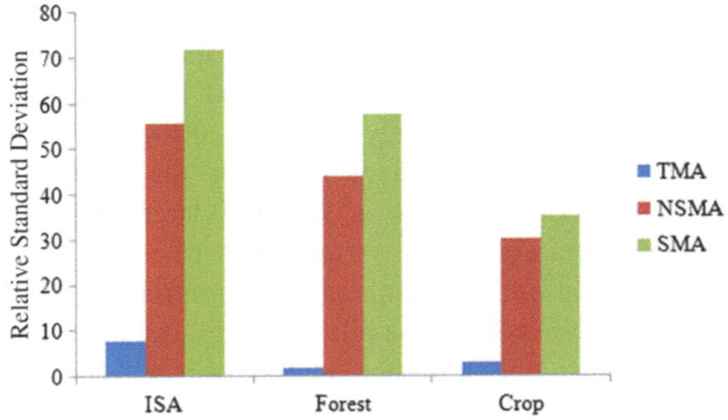

Fig. 12.3 Comparison of averaged relative standard deviation for each endmember among spectral mixture analysis (*SMA*), normalized spectral mixture analysis (*NSMA*), and temporal mixture analysis (*TMA*)

Case Studies in Lake Kasumigaura Basin

Lake Kasumigaura Basin

The Lake Kasumigaura Basin, located in the eastern part of the Kanto Plain of Japan, covers an area of about 2,157 km^2. The population of the Lake Kasumigaura Basin was approximately 1 million in 2000. With a surface area of 220 km^2 and an average depth of 4 m, the lake is the second largest in Japan. The climate of the area is similar to other regions on the Pacific side of Japan with an annual average air temperature of about 14 °C and an annual precipitation of 1,250 mm.

The dominant landscapes in the Lake Kasumigaura Basin are paddy fields (25.8 %), forests (22.7 %), plowed fields (21.2 %), and water (10.6 %) (Matsushita et al. 2006). The traditional industries in this area are agriculture, livestock management, and fishery production. However, many new industrial plants have been constructed, and industrial shipments from the area continue to rise every year. Because the study area is only 60 km from Tokyo, there has been a rapid rise in the number of residents, recreational facilities, and resort areas in recent years, which has caused rapid changes in the landscape structure in past decades. From 1979 to 1996, human-modified landscapes, such as artificial fields and golf courses, increased rapidly, while forests and croplands rapidly decreased (Fukushima et al. 2007). Increased patch (landscape element) numbers, Shannon's diversity index, and decreased mean patch area indicate that the Lake Kasumigaura Basin has become more fragmented and heterogeneous (Matsushita et al. 2006). With the completion of a new railroad linking Tsukuba and the Akihabara area of Tokyo in 2005, this trend is likely to persist as the population in the Lake Kasumigaura Basin continues to increase.

Twenty-Year Change of ISA in the Lake Kasumigaura Basin

Figure 12.4 shows ISA fractions in 22 subbasins, the residual basin, and the whole Lake Kasumigaura Basin for 1987, 2000, and 2007. From the results of the whole Lake Kasumigaura Basin, it can seen that the basin had already fallen into the impacted category (10–25 % ISA cover) in 1987, and its rank has been continuously sliding toward the degraded category in the subsequent two decades.

Results at the subbasin scale show that no subbasin has an ISA fraction less than 10 %, in which 9 subbasins (accounting for 25 % of the total basin) fell into the degraded category (ISA cover >25 %) and the remaining 13 subbasins and the residual basin into the impacted category in 2007. In contrast, there were 7 degraded subbasins (accounting for 16 % of the total basin) and 8 (accounting for 18 % of the total basin) for 1987 and 2000, respectively.

Overall, the Lake Kasumigaura Basin showed a high ISA fraction level early in 1987, and the ISA fraction has increased continuously since the 1980s, when Japan experienced very rapid economic development. Because the study area is only 60 km from Tokyo, the construction of the Joban highway that links Tokyo to the study area has led to a rapid rise in the number of residents, recreational facilities, and resort areas in the Lake Kasumigaura Basin during this period (Matsushita et al. 2006). Although the economic bubble burst in the early 1990s, the ISA fraction still continued to increase at an even faster pace from 2000 to 2007. These results suggest that if this increasing trend continues, Lake Kasumigaura Basin will reach the degraded category in the near future. Therefore, appropriate assessment techniques such as accurate estimation of the ISA fraction will provide critical information to policymakers for efficient development and management of the drainage basin as well as for sustainable growth in this region (Yang et al. 2010).

Relationships Between Hydrological Characteristics and ISA

Two rivers, Ono River and Koise River, were selected to analyze how ISA in the watersheds impacts the hydrological characteristics of rivers. These two rivers were chosen for two main reasons. The first reason is that the two river basins have different ISA fractions and change trend. The Koise River Basin has the lowest ISA fraction in the whole Lake Kasumigaura Basin, and the ISA fraction had almost no change during the period of 1987–2007 (11.7 % in 1987, 12.3 % in 2000, and 11.2 % in 2007; Fig. 12.4). In contrast, the Ono River Basin has a larger ISA fraction, which rapidly increased from 1987 to 2007 (25.7 % in 1987, 27.6 % in 2000, and 32.6 % in 2007; Fig. 12.4). The second reason is that their basin areas are similar: 222.3 km^2 for Koise River Basin and 176.4 km^2 for Ono River Basin.

Hourly runoff data and rainfall data for Koise River and Ono River were obtained from the Kasumigaura river office. The following rules were used to

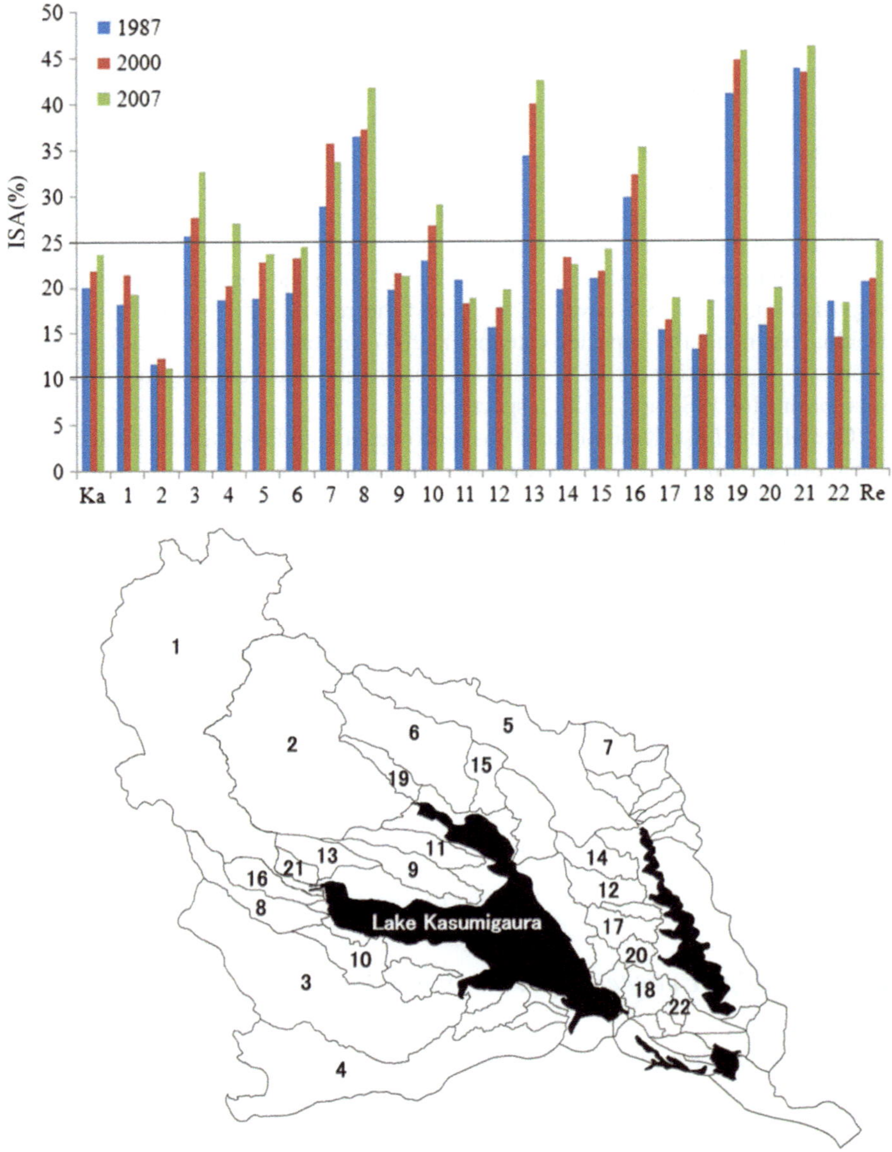

Fig. 12.4 *Top*: Percentages of impervious surface area (*ISA*) in Lake Kasumigaura Basin and its subbasins in 1987, 2000, and 2007. *Bottom*: Twenty-two subbasins included in Lake Kasumigaura Basin. The *numbers* in the *bottom figure* correspond to the subbasin codes in the *top figure*. *Ka* (Lake Kasumigaura Basin), *1* (Sakura River Basin), *2* (Koise River Basin), *3* (Ono River Basin), *4* (Shintone River Basin), *5* (Tomoe River Basin), *6* (Sonobe River Basin), *7* (Hokota River Basin), *8* (Hanamuro River Basin), *9* (Ichinose River Basin), *10* (Seimei River Basin), *11* (Hishiki River Basin), *12* (Yamada River Basin), *13* (Sakai River Basin), *14* (Takeda River Basin), *15* (Kajinashi River Basin), *16* (Bizen River Basin), *17* (Kura River Basin), *18* (Yokosu River Basin), *19* (Sanno River Basin), *20* (Gantsu River Basin), *21* (Shin River Basin), *22* (Mae River Basin), *Re* (Residual)

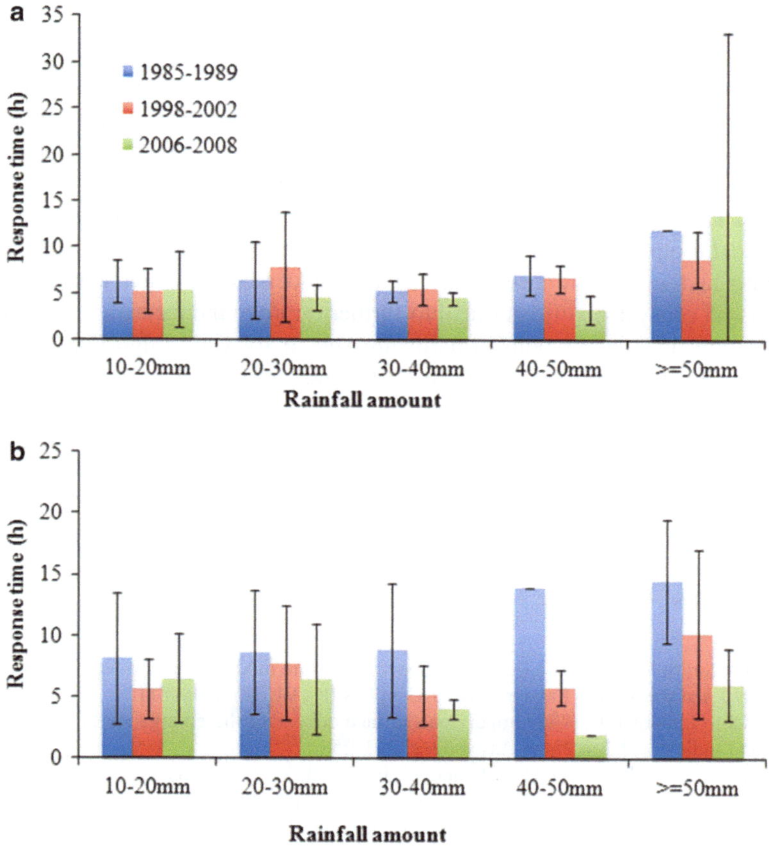

Fig. 12.5 Response times under various rainfall intensities for Koise River (**a**) and Ono River (**b**)

extract available rainfall events for the further analyses: (1) runoff peak occurs after rainfall peak; (2) rainfall ends before runoff peak; (3) rainfall within 8 h after end of the previous rainfall is considered as the same rainfall event as the previous rainfall; (4) there is no rainfall within 24 h before and after the selected rainfall events; (5) runoff and rainfall data observed during the periods of 1985–1989, 1998–2002, and 2006–2008 correspond with ISA data in 1987, 2000, and 2007, respectively. Using the extracted rainfall event data, response time, which is determined as the time difference between peak runoff and peak rainfall, was calculated to demonstrate runoff characteristics of each subbasin. Figure 12.5 shows the response times under various rainfall intensities in Koise River and Ono River. The response times for Ono River diminished from approximately 12 h to 1.7 h at different rainfall intensities, whereas the same tendency was not observed in the Koise River for all rainfall intensities. The results indicate that the response time of river discharge was reduced by the increased ISA in the watershed.

Conclusions

With the rapid growth of world population as well as urbanization and urban sprawl, the Earth's surface is being paved continually by ISA. The increased ISA has altered the ecological, hydrological, and thermal characteristics in many watersheds, and thus resulted in loss of native species biodiversity and invasion of alien species. Therefore, the ISA is expected to be an effective indicator of the human ecological footprint. Estimating the ISA distribution from satellite data can provide accurate and up-to-date geospatial data to understand human activity and its impacts on ecosystems. It is critical to further improve the measurement ability of remote sensing systems both in hardware and in software (algorithms) to allow these systems to be applied more efficiently and more widely.

Acknowledgments The present study was supported by the Environment Research and Technology Development Fund (S9-4-1) of the Ministry of the Environment, Japan.

References

Adams JB, Smith MO, Johnson PE (1986) Spectral mixture modeling: a new analysis of rock and soil types at the Viking Lander 1 site. J Geophys Res 91(B8):8098–8112

Arnold CL, Gibbons CJ (1996) Impervious surface coverage: the emergence of a key environmental indicator. J Am Plann Assoc 62(2):243–258

Barrett K, Helms BS, Guyer C, Schoonover JE (2010) Linking process to pattern: causes of stream-breeding amphibian decline in urbanized watersheds. Biol Conserv 143(9):1998–2005

Borel CC, Gerstl SA (1994) Nonlinear spectral mixing models for vegetative and soil surfaces. Remote Sens Environ 47:403–416

Collier KJ, Aldridge BMTA, Hicks BJ, Kelly J, Macdonald A, Smith BJ, Tonkin J (2009) Ecological values of Hamilton urban streams (North Island, New Zealand): constraints and opportunities for restoration. N Z J Ecol 33(2):177–189

Er KBH, Innes JL, Martin K, Klinkenberg B (2005) Forest loss with urbanization predicts bird extirpations in Vancouver. Biol Conserv 126:410–419

Elvidge CD, Tuttle BT, Sutton PC, Baugh KE, Howard AT, Milesi C, Bhaduri B, Nemani R (2007) Global distribution and density of constructed impervious surfaces. Sensors 7(9):1962–1979

Fukushima T, Takahashi M, Matsushita B, Okanishi Y (2007) Land use/cover change and its drivers: a case in the watershed of Lake Kasumigaura, Japan. Landsc Ecol Eng 3(1):21–31

Guhathakurta S, Gober P (2010) Residential land use, the urban heat island, and water use in Phoenix: a path analysis. J Plan Educ Res 30(1):40–51

Haselbach L, Boyer M, Kevern TJ, Schaefer VR (2011) Cyclic heat island impacts on traditional versus pervious concrete pavement systems. Transp Res Rec 2240:107–115

Helms BS, Schoonover JE, Feminella JW (2009) Assessing influences of hydrology, physicochemistry, and habitat on stream fish assemblage across a changing landscape. J Am Water Resour Assoc 45(1):157–169

Jiang LM, Liao MS, Lin H, Yang LM (2009) Synergistic use of optical and InSAR data for urban impervious surface mapping: a case study in Hong Kong. Int J Remote Sens 30(11):2781–2796

Johnson KA, Steinman AD, Keiper WD, Ruetz CR (2011) Biotic responses to low-concentration urban road runoff. J N Am Benthol Soc 30(3):710–727

Knight J, Voth M (2011) Mapping impervious cover using multi-temporal MODIS NDVI data. IEEE J Sel Top Appl Earth Obs Remote Sens 4(2):303–309
Lewis DB, Grimm NB (2007) Hierarchical regulation of nitrogen export from urban catchments: interactions of storms and landscapes. Ecol Appl 17(8):2347–2364
Loucaides S, Cahoon LB, Henry EJ (2007) Effects of watershed impervious cover on dissolved silica loading in storm flow. J Am Water Resour Assoc 43(4):841–849
Lu D, Tian H, Zhou G, Ge H (2008) Regional mapping of human settlements in southeastern China with multisensor remotely sensed data. Remote Sens Environ 112(9):3668–3679
Mason J, Moorman C, Hess G, Sinclair K (2007) Designing suburban greenways to provide habitat for forest-breeding birds. Landsc Urban Plan 80:153–164
Matsushita B, Xu M, Fukushima T (2006) Characterizing the changes in landscape structure in the Lake Kasumigaura Basin, Japan using a high-quality GIS dataset. Landsc Urban Plan 78(3):241–250
Nelson JA, Gotwalt PS, Simonetti CA, Snodgrass JW (2008) Environmental correlates, plasticity, and repeatability of differences in performance among blacknose dace (*Rhinichthys atratulus*) populations across a gradient of urbanization. Physiol Biochem Zool 81(1):25–42
Piwowar JM, Peddle DR, LeDrew EF (1998) Temporal mixture analysis of arctic sea ice imagery: a new approach for monitoring environmental change. Remote Sens Environ 63(3):195–207
Powell RL, Roberts DA, Dennison PE, Hess LL (2007) Sub-pixel mapping of urban land cover using multiple endmember spectral mixture analysis: Manaus, Brazil. Remote Sens Environ 106(2):253–267
Ridd MK (1995) Exploring a V-I-S (vegetation-impervious surface-soil) model for urban ecosystem analysis through remote sensing: comparative anatomy for cities. Int J Remote Sens 16(12):2165–2185
Roberts D, Gardner M, Church R, Ustin S, Scheer G, Green R (1998) Mapping chaparral in the Santa Monica Mountains using multiple endmember spectral mixture models. Remote Sens Environ 65(3):267–279
Scalenghe R, Marsan FA (2009) The anthropogenic sealing of soils in urban areas. Landsc Urban Plan 90(1-2):1–10
Schueler TR (1994) The importance of imperviousness. Watershed Protect Tech 1(3):100–111
Settle J (2006) On the effect of variable end member spectra in the linear mixture model. IEEE Trans Geosci Remote Sens 44:389–396
Somers B, Cools K, Delalieux S, Stuckens J, Van Der Zande D, Verstraeten WW et al (2009) Nonlinear hyperspectral mixture analysis for tree cover estimates in orchards. Remote Sens Environ 113:1183–1193
Somers B, Asner GP, Tits L, Coppin P (2011) Endmember variability in spectral mixture analysis: a review. Remote Sens Environ 115(7):1603–1616
Sung CY, Li MH, Rogers GO, Volder A, Wang ZF (2011) Investigating alien plant invasion in urban riparian forests in a hot and semi-arid region. Landsc Urban Plan 100(3):278–286
Sutton PC, Anderson SJ, Elvidge CD, Tuttle BT, Ghosh T (2009) Paving the planet: impervious surface as proxy measure of the human ecological footprint. Prog Phys Geogr 33(4):510–527
UN (2009) World Urbanization Prospects The 2009 revision highlights. Prospects. New York
Utz RM, Eshleman KN, Hilderbrand RH (2011) Variation in physicochemical responses to urbanization in streams between two Mid-Atlantic physiographic regions. Ecol Appl 21(2):402–415
Weng Q (2012) Remote sensing of impervious surfaces in the urban areas: requirements, methods, and trends. Remote Sens Environ 117:34–49
Wu C (2004) Normalized spectral mixture analysis for monitoring urban composition using ETM+ imagery. Remote Sens Environ 93(4):480–492
Wu C, Murray AT (2003) Estimating impervious surface distribution by spectral mixture analysis. Remote Sens Environ 84(4):493–505

Yang F, Matsushita B, Fukushima T (2010) A pre-screened and normalized multiple endmember spectral mixture analysis for mapping impervious surface area in Lake Kasumigaura Basin, Japan. ISPRS J Photogramm Remote Sens 65(5):479–490
Yang F, Matsushita B, Fukushima T, Yang W (2012) Temporal mixture analysis for estimating impervious surface area from multi-temporal MODIS NDVI data in Japan. ISPRS J Photogramm Remote Sens 72:90–98
Zoran M, Savastru D, Miclos S, Tautan MN, Baschir L (2011) Multisensor satellite remote sensing data for heat waves assessment in metropolitan region. J Optoelectron Adv Mater 13(9-10):1159–1166

Chapter 13
Biodiversity of Kargil Cold Desert in the Ladakh Himalaya

M.D. Behera, S. Matin, and P.S. Roy

Abstract Biological diversity has long been a source of wonderment and scientific curiosity, but it is now increasingly a source of concern. We attempted to classify, map, and characterize vegetation and its biodiversity content using satellite imagery and field enumeration in the cold desert of the Kargil-Ladakh region located in the Western Himalaya, India. Fair-weather, multi-temporal IRS-P6 LISS III data were subjected to visual interpretation, which estimated rocky barren land as the dominant land cover, followed by vegetation (alpine pasture, alpine scrub, agroforest, and subalpine forest) and snow cover. Asteraceae was reported as the most speciose family, numbering more species than genera. *Artemisia* emerged as the most speciose genus. Species diversity decreased with increasing altitude, but the maximum was found at transition zones of agroforest and alpine scrub. The highest species similarity was observed between alpine scrub and alpine pastures, indicating pronounced intermixing and interaction. The detailed updated map (at a scale of 1:50,000) with extensive field observation would overcome a long-pending gap in ecological study in the region and may also be used as baseline data for conservation and planning, and the region may serve as a natural ecological laboratory for climate change studies.

Keywords Bioprospecting • Landscape dynamics • Phytosociology • Remote sensing • TIV • Visual interpretation

M.D. Behera • S. Matin (✉)
CORAL, Indian Institute of Technology, Kharagpur 721302, India
e-mail: shafiquematin@gmail.com

P.S. Roy
University of Hyderabad, Hyderabad 500046, India

S.-i. Nakano et al. (eds.), *Integrative Observations and Assessments*, Ecological Research Monographs, DOI 10.1007/978-4-431-54783-9_13, © Springer Japan 2014

Introduction

Biodiversity, the variety of life, is distributed heterogeneously across the Earth. The most striking feature of the Earth is the existence of life; and the most striking feature of life is its diversity. This biological diversity has long been a source of wonderment and scientific curiosity, but it is now increasingly a source of concern. Some areas teem with biological variation (e.g., moist tropical forests and coral reefs), whereas others are virtually devoid of life (e.g., deserts and polar region), and others are somewhere in between. Determining why these differences occur has long been a core objective for ecologists and biogeographers. The reasons for this interest are twofold. First, it reflects increased opportunity provided by improvements in available data and analytical tools, the former resulting mostly from extensive collation of existing specimen and species occurrence records, the establishment of dedicated distribution mapping schemes, and the use of remote sensing technology (to measure vegetation and other environmental variables). Second, it reflects concern over the future of biodiversity and the resultant need to determine its current status, to predict its likely response to global environmental change, and to identify the most effective schemes for in situ conservation and sustainable use. Biodiversity changes across environmental gradients such as latitude, altitude, and moisture, and hence inventorying and monitoring of biodiversity, is also done at different organizational levels from genes to ecological systems (landscapes), and at different spatial scales from a few square meters to continents.

Here, we have attempted to characterize the biodiversity of the Kargil district through in situ enumeration and utilization of geoinformatics techniques, in part of the Ladakh cold desert in Jammu and Kashmir state in India. Kargil is a mountainous region situated in the Himalayas along the northernmost extreme of India. The district covers an area of 17,303 km^2 with a population of 1.25 lakhs. It is situated between 30° and 35° N latitude and 75° and 70°E longitude. The district is referred to as a "cold desert" because it combines features of arctic and desert climates, including wide diurnal and seasonal fluctuations in temperature, from −40 °C in winter to +35 °C in summer, and extremely low precipitation, with an annual 10 to 30 cm, primarily from snow (LAHDC 2005). The high altitude and low humidity mean the radiation level is among the highest in the world (up to 6–7 kWh/mm). Soil is thin, sandy, and porous. Because of these peculiar physical and climatic conditions, the Kargil region shows a unique floral assemblage. The whole region is composed of high rocky mountains (Fig. 13.1) and arid desert, snowbound and devoid of natural vegetation. It occupies a unique position because of its high altitude, ranging from 2,500 to 7,000 m above sea level (Fig. 13.2). The region is an almost treeless expanse. The scarcity of precipitation causes plants to be generally found growing along moist river margins or in moist rock crevices. The aspect characteristics of the landscape have a vital role in vegetation growth. Many slopes on the eastern side show a higher degree of vegetation cover than on western slopes (Fig. 13.3). These forests also support the lives of various ecologically important wild animal species such as *Aconitum heterophyllum*, *Uncia uncia*, *Ovis vignei vignei*, *Cuon alpinus laniger*, *Bos grunniens*, and *Otocolobus manul* (Table 13.1).

Biodiversity Assessment in Himalayan Cold Deserts at Different Physiognomic Gradients

Species diversity and its distribution along the altitudinal gradient has been a subject of ecosystem analysis. Community composition, structure, diversity, and dynamics are related to gradients in obvious ecological factors (e.g., elevation,

Fig. 13.1 Image chips and corresponding field photographs

Fig. 13.1 (continued)

slope, aspect). Direct gradient analyses have provided many insights into the nature of ecological communities, the factors influencing the distribution of individual species, the adaptive significance of particular growth forms and their contribution to competitive success in different environments, and potential controls on productivity, biomass, and biological diversity (Whittaker 1956, 1960; Peet 1978; Gentry 1982, 1988, 1992). Gradient analyses in alpine and subalpine forest can be particularly challenging, given the large number of species involved, the frequent absence of local taxonomic treatments, and the difficulty of access to the diagnostic

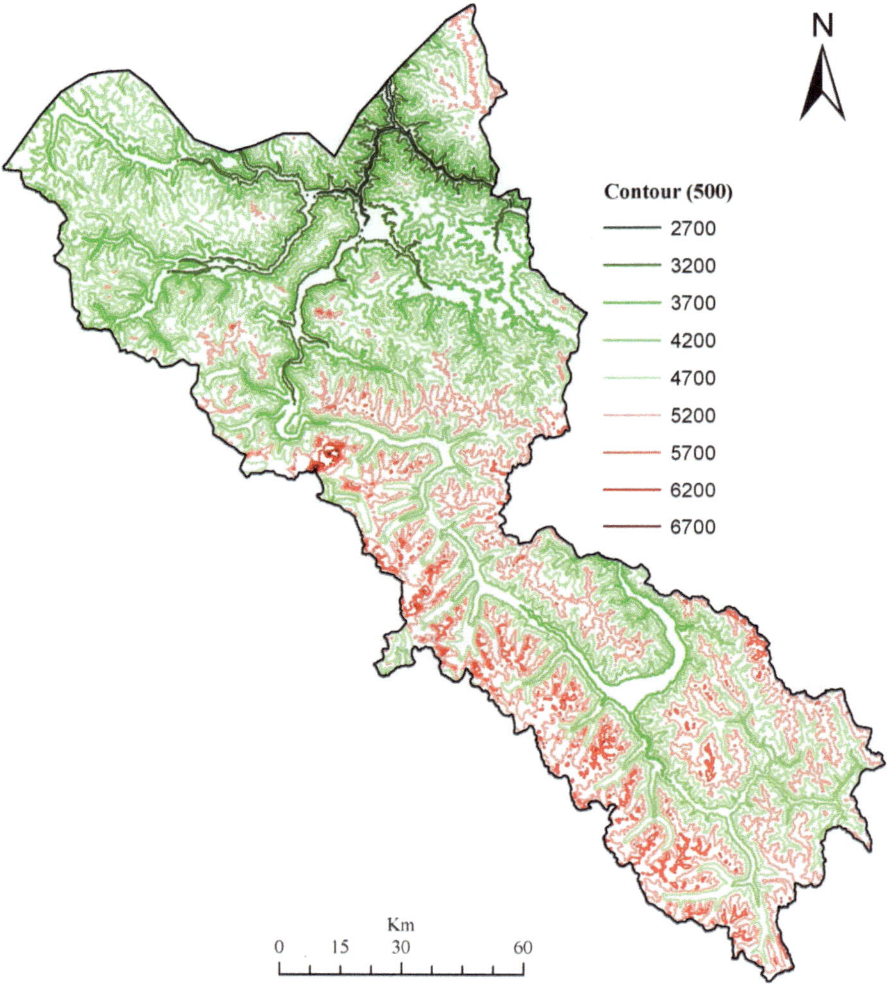

Fig. 13.2 Contours at 500-m interval prepared from SRTM DEM

parts of many canopy trees and epiphytes. Rahbek (1995) commented that approximately half the studies detected a mid-altitude peak in species richness in a critical literature review on species richness patterns in relationship to altitude. Vetaas and Grytnes (2002) have also reviewed these aspects in the Nepalese Himalaya. Although the plant community of a region is a function of time, nevertheless altitude, latitude, slope, aspect, rainfall, and humidity have roles in the formation of community composition. Very few researchers have explored the region of western Himalaya through biodiversity studies, and most of them were reliant on field surveys. Klimeš (2003) studied life forms and clonality of vascular plants along an altitudinal gradient in the eastern cold deserts of Ladakh and explored special habitats and threatened plants of Ladakh and Kashmir Himalaya, respectively.

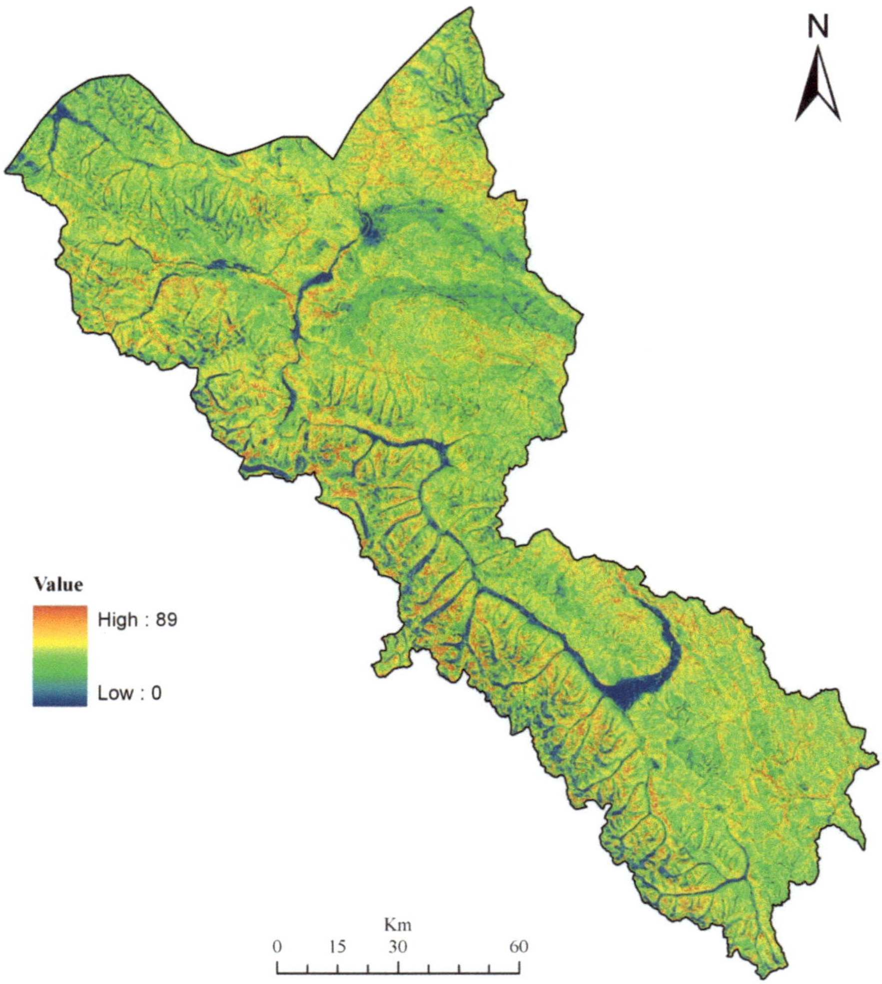

Fig. 13.3 Slope map shows absence of tablelands except in valleys

Geneletti and Dawa (2009) carried out environmental impact assessment of mountain tourism in Ladakh; Fox (1983) studied wildlife conservation and land-use changes in the trans-Himalayan region of Ladakh; and Humbert-Droz and Dawa (2004) studied the biodiversity of Ladakh and suggested a strategic action plan. There have been a few remote sensing-based biodiversity characterization studies in the western Himalaya. Joshi et al. (2006) studied biodiversity characterization in Ladakh, with a focus on the Nubra Valley. Porwal et al. (2003) carried out stratification and mapping of *Ephedra gerardiana* Wall. in the cold deserts of Himachal Pradesh. Matin et al. (2012) demonstrated a method to integrate the fauna component with existing flora data for complete diversity analysis, utilizing the GPS-based

Table 13.1 List of threatened species as per the International Union for Conservation of Nature (IUCN) (3.1), Jammu and Kashmir Wildlife (Protection) Act, 1978, and Indian Wildlife (Protection) Act, 1972

Sl.	Plant species	Status	Animal species	Status
1	*Aconitum heterophyllum*	CR EN	*Alticola roylei*	LR
2	*Aconitum violaceum*	CR EN	*Anser indicus*	LC
3	*Anaphalis contorta*	R	*Ardea Cinerea*	LC
4	*Arnebia euchroma*	R-EN	*Bos grunniens*	VU, SII
5	*Astragalus strictus*	R	*Canis lupus chanku*	LR
6	*Convolvulus arvensis*	R	*Capra aegagrus hircus*	VU
7	*Dactylorhiza hatagirea*	CR EN	*Capra ibex sibirica*	LC
8	*Ephedra gerardiana*	R	*Cuon alpinus laniger*	VU
9	*Ferula jaeschkeana*	VU	*Equus kiang kiang*	LC
10	*Heracleum pinnatum*	EN	*Gypaetus barbatus*	LC
11	*Hippophae rhamnoides*	LR, NT	*Gyps himalayensis*	LC
12	*Hyoscyamus niger*	LR, NT, VU	*Lutra lutra*	LR
13	*Lonicera heterophylla*	R	*Marmota caudata*	LR
14	*Lonicera spinosa*	R	*Mustela altaica*	NT
15	*Meconopsis aculeata*	CR	*Mustela erminea*	LR
16	*Physochlaina praealta*	VU	*Ochotona roylei*	LR
17	*Podophyllum hexandrum*	EN	*Otocolobus manul*	EN
18	*Populus ciliata*	R	*Ovis vignei vignei*	EN/ VU
19	*Populus deltoides*	EX	*Procapra picticaudata*	NT
20	*Populus euphratica*	R	*Pseudois nayaur nayaur*	LC
21	*Populus nigra*	EX	*Streptopelia orientalis*	LC
22	*Rheum webbianum*	VU	*Uncia uncia*	EN
23	*Rumex patientia*	VU	*Ursus arctos*	LR, S-I
24	*Taraxacum officinale*	CR EN	*Vulpes ferrilata*	LR

CR, Critical; CR EN, critically endangered; EN, endangered; R, rare; VU, vulnerable; LR, low risk; EX, exotic; NT, near threatened; LC, least concern; S, schedule
Source: Matin et al. (2012)

location information on *Medicago sativa* and *Plantago annua* to simulate their potential distribution in the year 2020 (SRES A1B scenario, IPCC) using the Maxent model in part of Ladakh Himalaya.

Mountain Alpine Ecosystems and Kargil Ladakh

Alpine ecosystems occur at a range of air density, water availability, and seasonality in the treeless high terrain of mountains worldwide. Mountainous regions have a diversity of landforms, topography, climate, soils, slope exposure, geology, and other abiotic factors. Each combination of these factors affect site temperature and moisture status. Because plant distributions are influenced primarily by temperature and moisture (as controlled by their ecological amplitude), any significant change in abiotic factors causes a change in plant composition. Composition varies along geographic scales: boreal dwarf-shrub heaths, temperate sedge heaths, subtropical dwarf shrubs and tussock grasslands, and tropical giant forest lands. Along local topographic gradients plant cover changes from windswept dwarf-shrub heath, to dense grass-sedge heath, to snow bank vegetation. These cold and relatively little exploited alpine ecosystems, nonetheless, are among those where climate warming impacts are forecast to be pronounced and detectable early on. The variability of alpine environments is predominantly dependent upon altitude (air density), water availability, and seasonality, which are specific to each mountain region. These factors determine, besides the available flora and fauna, the altitudinal zonation and the structure and functioning of the ecosystems. In Kargil, three vegetation belts are observed. The first belt begins at the valley foothills and climbs a few meters in which introduced agroforest is associated with planted tree species of *Populas* and *Salix*. The second belt of scrubland ranging between lower foothills and the medium slopes is mostly associated with the medicinally important species of *Hippophae rhamnoides*, *Hordeum vulgare*, *Juglans regia*, and *Brassica napus*. The third belt appears after 3,500 m, where many alpine grasses grow. The climatic conditions of the upper part of the Kargil mountains resemble the cold-arid climate prevailing in the Southern Hemisphere. For this reason, most of the species belongs to the grass family (Poaceae) with species such as *Poa annua*, *Saussurea ceratocarpa*, and *Melica persica*.

Ecological Gradient

The terrain being hilly and sloped, the available land for agriculture is meager. Many valleys are dotted with formations of alluvial fans, marsh meadows, boulder-strewn riverbeds, and small sandy meanders. Remaining areas, which are not inhibited by humans, are characterized by seasonal to perennial vegetation. These areas are either in the narrow valleys or at higher altitudes, but altogether represent very scanty vegetation. The altitude increases toward the south. The alpine and subalpine flora of the high-altitude desert region of Ladakh, Jammu, and Kashmir has the greatest affinity toward the Central Asiatic region (Dhar and Kachroo 1983). More than 20 threatened species have been identified in the region (Table 13.1), of great medicinal and ecological importance (Anonymous 2011). The vegetation of these zones is characterized by scattered low bushes, with extensive coverage of tussock grasslands, herbaceous formations, sedge meadows, and stony deserts with cushion-like growing forms. Three major vegetation types have been reported in the region. (I) Subalpine forests have been found in very few areas in Kargil region. The altitude range for these areas is 3,000 to 4,000 m, and domination is mainly *Pinus roxburghii*. (II) Alpine scrub occurs mostly in drier areas and can also be found in areas near river streams. The major vegetation association is *Hippophae* and *Myricaria* scrub. (III) Alpine pasture is mainly moist alpine grasslands or pastures occurring at higher altitudes and moister areas near the snow line. *Poa annua*, *Medicago sativa*, and *Plantago lanceolata* are the dominant species, forming the major land cover of the region. (IV) Agroforest comprises clusters of *Populas* and *Salix* plantations in the areas near rivers and streams. These species are mostly planted adjacent to the agriculture fields by local people and the forest department. The presence of the three natural vegetation types, that is, moist alpine forest, scrubland, and pastureland, almost corresponds to the dominance of the three life forms, trees, shrubs, and herbs, in altitudinal gradients that offer prospects for various ecological analyses. Because the landscape has scanty anthropogenic influence, which is restricted to valleys, the majority of the landscape is suitable for basic spatial analysis of patch dynamics studies. The complex physiognomy leading to varied microclimates provides a basis to analyze the vegetation distribution and other structural and functional behavior. The influence of snow is the major dominant factor triggering other ecological processes that need to be understood as a potential linkage to biodiversity characterization. In the changing climatic scenario, while there is much hue and cry about glacier melting, increase in tropical areas, vegetation shifts, species tolerance and adaptation, etc., the region offers a perfect laboratory for ecological studies and global change research.

Vegetation Versus Elevation and Seasonality

Exposure is one of the key factors responsible for biological richness in mountainous areas. There is a considerable difference in the distribution of the vegetation communities between the north-facing and south-facing slopes of the mountains, which is related to both solar radiation and precipitation. As a general rule, the south-facing slopes of the mountains receive more sun radiation than do the northern slopes, and that is why the density (number of species) is greater in the southern aspect. North-facing slopes of the mountains receive a lesser amount of sun time, and hence snow cover time increases accordingly, resulting in less ground coverage of vegetation. On the southern slopes of the Kargil, sun radiation intensity is higher on the slopes facing south than the other slopes. For that reason, high occurrence of dry scrub is common on the south-facing slopes. For example, spiny species of *Hippophae* are found mostly in south-facing plots.

With the increase of altitude, temperature, relative humidity, and air pressure decrease continuously, but solar radiation intensifies. Depending on these circumstances, the temperature of southern slopes is a few degrees higher than the northern slopes. In Kargil, the vegetation and land cover show drastic changes with increasing altitude. At lower altitudes, mainly along the valleys, human-induced vegetation and land-use classes such as agroforestry and settlements show considerable spread in the region. As altitude increases, however, the distribution of agroforests decreases, taken over by moist alpine pastures and moist alpine scrub. The landscape is dominated by barren land and snow. The distribution of snow starts at around 4,500 m and continues to the highest altitudinal point in the region. Mountains with an alpine zone occur at all latitudes, from the wet tropics to the polar regions. Apart from a steady decrease of temperature with increasing elevation, at an average rate of 0.60 °C/100 m, air pressure also decreases (Nagy and Grabherr 2009). The latter becomes particularly relevant in mountains such as the Himalayas, where the highest peaks reach more than 8,000 m. Low CO_2 might be one of the causes for the absence of diversity in families from the high ground or their generally low diversity compared to the lowlands (Nagy and Grabherr 2009). In contrast, low carbon dioxide pressure seems to have no limiting effect on plants: other factors such as low temperature set the limits (Korne 2003). The macroclimate of the life zone to which a mountain region belongs determines the climatic conditions in its alpine zone. Species that are sensitive to frost require permanent snow protection and as a result have developed remarkable snow tolerance. For example, various species of *Ranunculus* in alpine areas are known to be able to survive up to 33 months permanently under snow (Moser et al. 1977). This freezing tolerance may be related to adaptation to low temperatures (Sømme 1997).

Landform Characteristics Versus Flora

The vegetation cover in mountain areas is of immense importance as it affects local and regional climate and reduces erosion. Understanding the distribution and patterns of vegetation distribution along with the local influencing factors is important, and these factors have been studied by many researchers (Turner and Gardner 1992; Henebry 1993). Topography is the principal controlling factor in vegetation growth; climatological factors play a secondary role. Elevation, aspect, and slope are the three main topographic factors that control the distribution and patterns of vegetation in mountain areas. Among these three factors, elevation is the most important (Day and Monk 1974). Elevation (Fig. 13.2), along with slope (Fig. 13.3) and aspect (Fig. 13.4), in many respects determines the microclimate and thus the large-scale spatial distribution and patterns of vegetation (Day and Monk 1974; Marks and Harcombe 1981). Several observations can be mentioned regarding the effects of elevation and aspect on vegetation growth in mountainous areas. Vegetation distribution in the region exhibits a vertical gradient from the climatic changes with elevation. Second, vegetation growth in such regions can be significantly affected by aspect, where most occurs in the northern aspect (Fig. 13.4), but the highest diversity and equitability are observed in the southwestern aspect. The impact of aspect on the vegetation growth is most significant in the vertical zone of 3,500 to 6,000 m. The best vegetation in this zone is distributed between eastern and northeastern aspects. In other words, the best vegetation growth is on the shady side of the mountain, where much less evapotranspiration (ET) is expected. The reduced ET on the shaded side is important for vegetation growth in the Kargil region because it is located in a cold desert region. It is also observed that better vegetation growth occurs over a larger elevation range on the side facing southwest and south. The much wider vertical zone with better vegetation growth on the shady side may significantly affect the local water cycle and climate.

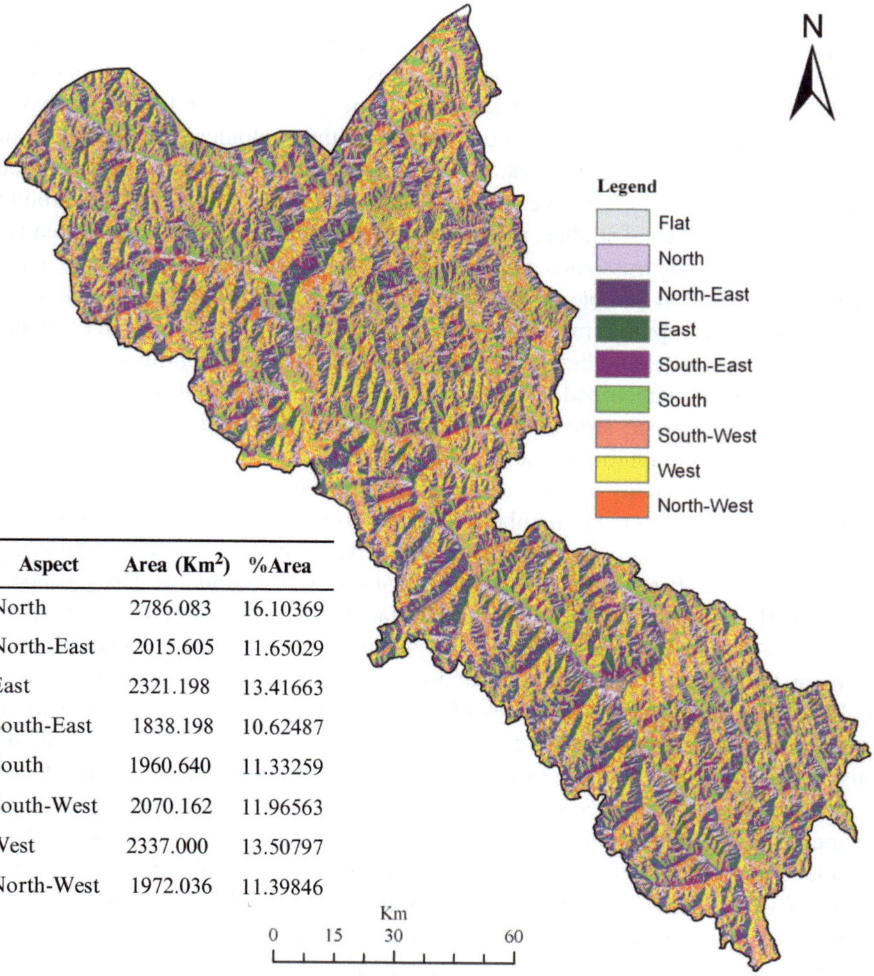

Aspect	Area (Km^2)	%Area
North	2786.083	16.10369
North-East	2015.605	11.65029
East	2321.198	13.41663
South-East	1838.198	10.62487
South	1960.640	11.33259
South-West	2070.162	11.96563
West	2337.000	13.50797
North-West	1972.036	11.39846

Fig. 13.4 Aspect map revealed that North and South-East aspects occupy maximum and minimum areas respectively

Community Analysis and Alpine Ecosystem

The plant community data imply knowledge of structure and composition of the component species. The biophysical components of a plant's environment interact to form a temperature and moisture regime based primarily on gradients of elevation, slope, and aspect. These components can either balance or offset each other: an example is the situation wherein a species switches aspect with changes in elevation. A plant species whose distribution switches from one aspect to another may seem unusual, but it really is not, because both situations may provide similar temperatures and moisture. Some ecologists have referred to this concept as an effective environment because it demonstrates that one set of site factors can be ecologically equivalent to another. Differing combinations of site factors can have a similar influence on an ecosystem consequent to the ecological principle of compensating effects (Allen and Hoekstra 1992). Community analysis was carried out to understand the floristic and vegetation pattern in the region. To carry out ground-level phytosociological studies, a sampling intensity of 0.002 % of forest area and 0.001 % for scrub/grasslands was envisaged following the species–area curve principle. Thus, 154 sample plots proportionally covered 30, 44, and 80 plots for agroforest, moist alpine scrub, and moist alpine pastures, respectively. In each plot, names of trees (GBH and height measurement), herbs, shrubs, climbers, seedlings, and saplings were recorded for all species. A phytosociological database was created, and the basic structural parameters were computed, including relative frequency, relative dominance, and relative density.

Spatial heterogeneity of the environment increases the number of different habitats, permitting a greater number of different resource use strategies, preventing competition equilibrium and exclusion, and hence may result in higher diversity. The scale of this heterogeneity is different for different organisms. Disturbance is also widely regarded as one of the major factors for species diversity (Connell 1978; Noss 1996). The disturbance may result from forest fragmentation, fires, grazing, trampling, or windblow. From different disturbance factors, larger areas of habitat may transform into a series of smaller ones, each of which will support fewer selected species. Therefore, the natural heterogeneity coupled with spatially distinct disturbance regimes brings complexity in habitats, which leads to the formation of different species assemblages. Biodiversity assessment at the landscape level requires a phytosociological database on different aspects such as species type, species richness, diversity, abundance, and basal cover. The assessment of variability of such a wide range of parameters and fixing sample size for large heterogeneous habitats is difficult. A certain sample size could be fixed and accuracy checked against samples size after the sampling exercise. This method provides an open-ended design whereby sample size could be enhanced, and the exercise can be further continued to determine accuracies within the framework of time and cost limits. In distribution of sample size in studies conducted throughout the world, the plot sizes of 0.04 ha (20 × 20 m size within a vegetation type depends on area of each sample unit), 0.1 ha (32 × 32 m), and 1 ha as belt transect or square plots could be considered. To study biodiversity at stratification based on remotely sensed data, 0.01 ha as plot size for diverse evergreen forest types, 0.04 ha for deciduous forests, belt transects for fragmented habitats, and a modified Whittaker's plot for shrubs/scrub/grasslands/sparse trees are preferred. The random distribution of small size plots (0.04/0.01 ha) across various habitats will be helpful in statistically analyzing the diversity patterns across given vegetation types. Alternatively, a selected number of large-size plots of 1 ha across major microhabitats of given vegetation types will be helpful in understanding the role of microlevel topographic and biotic factors in species diversity and richness.

According to the discrete community theory, plant communities have distinct boundaries within which species composition is homogeneous, and the transition between them is relatively rapid. As such, the number of species would be underestimated when extrapolated from a small-scale species–area curve to a larger scale because of the occurrence of new species across community boundaries (Wilson and Chiarucci 2000). Alpha diversity can be adequately described by the coefficient c, which itself can be accurately predicted from the total number of species; and the coefficient z can be used to predict the number of species per individual and vice versa. Thus, these two coefficients are strong and complementary measures of species diversity. The higher the z, greater would be the impact of further harvest of individuals; in other words, the areas with high z would be more susceptible to selective or random felling of trees even if there is no contraction of area.

The sample plots of 0.04 ha/0.1 ha were randomly distributed across each stratum. The sample plot was reached on the ground based on GPS locations and also important ground bearings available from the survey of India (SOI) map. As the location was reached, the sample plot of desired dimension was laid. For sampling shrub species, two plots of 5 × 5 m in two opposite corners were laid. For herbaceous plants, five plots of 1 × 1 m (four in corners and one in center) were laid. A nested sampling design was followed within each sample point to record shrubs, herbs, climbers, epiphytes, and lianas. At each sample point, the circumference at breast height (CBH) of all tree species was recorded and marked with chalk to avoid duplication. To ascertain that measurements were taken at proper height, a 1.37-m stick was used. For shrubs, the CBH was measured about 30 cm above ground. The total number of seedling of various species was counted, and average girth of each species was recorded. For shrubs, total number of tillers for each species was counted and for each species an average circumference at ground height level was estimated. For herbaceous layer or ground flora, the nested quadrat method with 1 × 1 m plot size was taken in two (north and south) corners. Only the number of individuals of each life form was recorded. Based on spatial variability, a belt transect or a square plot of large size was laid on a case-by-case basis. The transect or square plot was gridded at 20 × 20 m, and all the information on trees, shrubs, herbs, climbers, and lianas was recorded independently for each subgrid. This method facilitated understanding the impact of microlevel spatial gradients, structure, and diversity patterns.

Biological diversity can be quantified in many ways. The two main factors taken into account when measuring diversity are richness, the number of species, and evenness, the relative abundance of individuals among the species. Species richness is described as the number of species present in a sample or habitat per unit area. The simplest species richness index is based on the total number of species and the total number of individuals in a given sample or habitat; it indicates that higher the value the greater is the species richness. Maximum diversity is observed in agroforest, based on the Shannon diversity index (1948), as 1.82, followed by moist alpine scrubs at 1.51 and moist alpine pastures with the value 0.62 (Fig. 13.5a). Equitability and dominance have been observed at different aspects and correlated with Shannon diversity. Although Pielou equitability (1969) depends upon the Shannon diversity

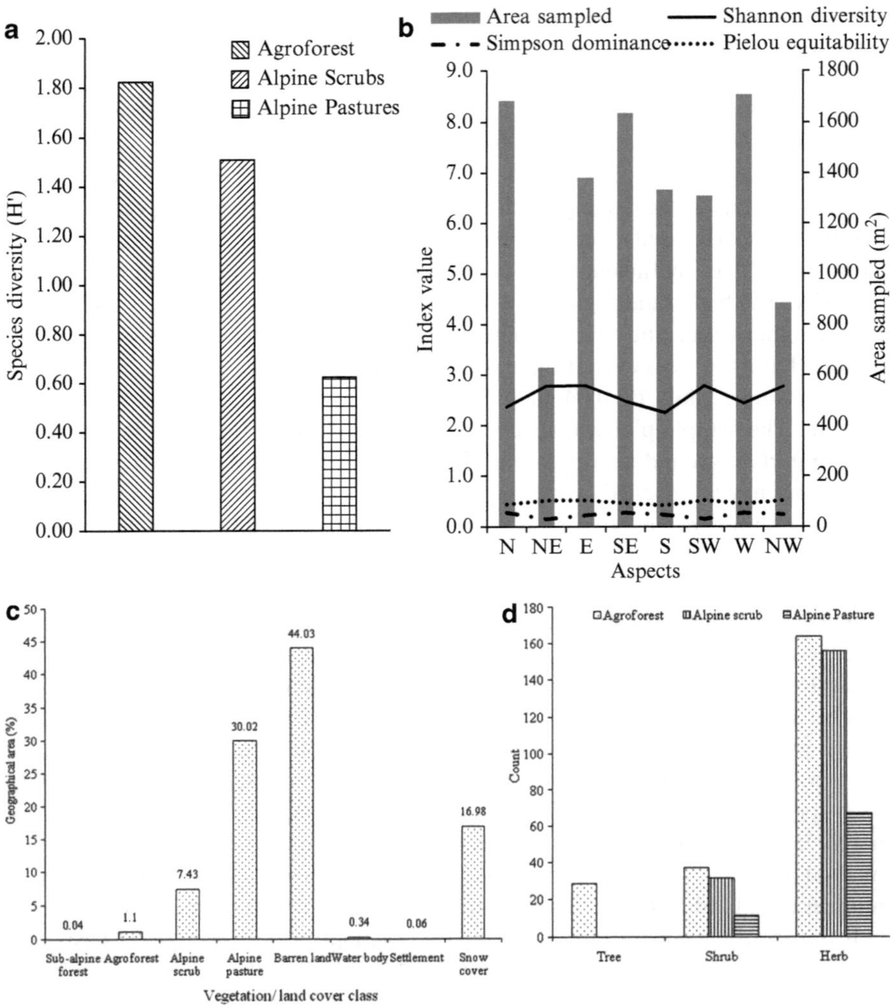

Fig. 13.5 Shannon diversity index for different vegetation classes (**a**) and Shannon diversity, Simpson dominance, and Pielou equitability at different aspects (**b**). Maximum diversity and equitability were observed at the eastern aspect and maximum dominance at the northern aspect. (**c**) Area estimates of forest vegetation and land cover map of Kargil. (**d**) Life form distribution pattern in three vegetation cover types as per field sampling

index (J'), the J' found was high in South-West aspect (SW), followed by Eastern (E) and North-Eastern aspect (NE) (Fig. 13.5b). The Simpson index (1948) is the most reliable and easy way to understand dominance in species occurrence. Its maximum value goes up to one, observed in case of single species. Each individual represents a small fraction of the total value. For Kargil, the maximum dominance was observed in *Lamium album* for agroforest (0.080), *Crepis multicaulis* (0.09) in scrubland, and *Saussurea ceratocarpa* (0.4) in pastureland.

Vegetation Type Mapping and Geoinformatics Technology

Classification is a relevant procedure in studying the vegetation mosaic and the interaction between them and the landscape. The information contained in vegetation types can be studied using community analysis and other ecological measurements. Development of simple, user-friendly vegetation classification based on ecological principles is a prerequisite for vegetation type mapping. Vegetation in general and (forest in particular) is an extremely complex system (Anonymous 2008), and the orderly arrangement of vegetation comes under classification, which is nothing but a process of sorting out similar classes into a few logical classes. Here the scheme primarily considers physiognomy, structure (vertical stratification, height, and spacing), phenology (green and brown wavelength), function or life forms, flora, climate, topography, geomorphology, soil, and biotic factors for classifying vegetation entities. Mapping vegetation through remotely sensed images involves various considerations, processes, and techniques. The increasing availability of remotely sensed images from the rapid advances in remote sensing technology expands the horizon of our choices of imagery sources. Various sources of imagery are known for their differences in spectral, spatial, radioactive, and temporal characteristics and thus are suitable for different purposes of vegetation mapping. Generally, it is necessary to first develop a vegetation classification for classifying and mapping vegetation cover from remotely sensed images at either a community level or species level. Then, correlations of the vegetation types (communities or species) within this classification system with discernible spectral characteristics of remotely sensed imagery are established. These spectral classes of the imagery are finally translated into the vegetation types in the process of image interpretation.

Remote sensing is the science of acquiring information about the Earth's surface without actually being in direct contact with it (Lillesand et al. 2007). It began with aerial photography, but technological development has enabled the acquisition of information at other wavelengths including near-infrared (NIR), thermal infrared (TIR), and microwave regions of the spectra, known as spectral bands. Vegetation has a unique spectral signature that enables it to be distinguished readily from other types of land cover in an optical/near-infrared image. Ideally, reflectance is low in visible spectra (blue and green regions) as a result of absorption by chlorophyll, and a sudden increase is called the red edge. The red edge refers to the region of rapid change in reflectance of chlorophyll in the near-infrared range, which strongly reflects at wavelengths greater than 700 nm. The change can be from 5 % to 50 % reflectance between 680 to 730 nm. In the near-infrared (NIR) region, the reflectance is much higher than that in the visible band because of cellular structure in leaves. Hence, vegetation can be identified by high NIR but generally low visible light reflectance. Of four vegetation classes in Kargil (Table 13.2), only agroforest and Chir/Pine forest consists of tree species that have compact intercellular structures and lesser amounts of leaf chlorophyll content; hence, the reflectance is medium in the NIR band and less in green and red bands. Shortwave infrared (SWIR), which is a water absorption band, shows higher reflectance only for scrubs

Table 13.2 Vegetation types as compared with Champion and Seth (1968)

Vegetation class	Spectral response on image	Altitudinal range (m)	Dominant species composition	Champion and Seth (1968)
Subalpine forest	Bright red tone with coarse texture	2,700–3,800	*Juniperus semiglobosa*, *Pinus wallichiana*	(14) Subalpine forest
Agroforest	Bright red color, located along water streams	2,500–3,600	*Populus* spp., *Salix* spp.	(NA) *Populus*/*Salix* forest
Moist alpine scrub	Light brown to deep red color, located along valleys and also along hill slopes	2,800–5,100	*Hippophae rhamnoides*, *Myricaria germanica*	(16/C1) Alpine scrub
Moist alpine pasture	Light pink to light red color, located along highest altitudinal range of vegetation classes	2,800–5,800	*Ephedra gerardiana*, *Poa alpina*, *Rheum webbianum*	(15/C3) Alpine pastures

and much less for agroforest. Scrub, which mostly consists of species such as *Bromus japonicas*, *Hyppophae rhamnoides*, and *Myricari germanica*, exhibits maximum reflection in the SWIR band because of the lesser water content. Species such as *Poa annua*, *Phleum alpinum*, and *Artemisia brevifolia* have a huge amount of fresh chlorophyll and water in their leaves and hence show higher reflectance in the red region and less in the NIR region because of the compact intercellular arrangements of leaves.

The vegetation and land-cover map clearly depicts the distribution pattern of the region (Fig. 13.6). It is always difficult to find good cloud-free data for interpretation. However, we found satellite images of October 2005, which was absolutely cloud free, and hence these images were used in this study. According to mapping, the largest area is covered by barren land (44.03 %), followed by vegetation, that is, 38.59 % (Fig. 13.5c). The vegetation cover represents distinct strata, with moist alpine pastures covering an area of 30.02 % (5,194.5 km^2), followed by moist alpine scrub covering 7.43 % (1,285.54 km^2), agroforests 1.1 % (190.55 km^2), and subalpine forests 0.04 % (6.24 km^2). Subalpine forest showed a few patches along the boundary of the Baramullah district distributed between 2,700 and 3,800 m altitude with *Pinus wallichiana* and *Juniperus semiglobosa* as the dominant species. Champion and Seth (1968) put this class under subalpine forest class under section 14 (Table 13.2). Agroforests are found distributed along river streams and settlements between 2,500 and 3,600 m altitude. They showed flourishing growth of *Populus* and *Salix* species, along with *Poa annua*, *Medicago sativa*, *Phleum alpinium*, *Bromus japonicus*, *Digitaria stewertiana*, etc. The areas were well differentiated on the satellite data because of a peculiar bright red tone. Moist alpine scrub was found in the altitudinal range of 2,800 to 5,100 m. At some places, *Hippophae* and *Myricaria* formations were observed. *Aconitum heterophyllum*, *Aconitum violaceum*, *Anaphalis brusus*, *Agrostis gigantea*, *Alisma graminium*, etc., were dominantly distributed. On satellite imagery, moist alpine scrub reflected a light to bright red tone with smooth texture (Table 13.2).

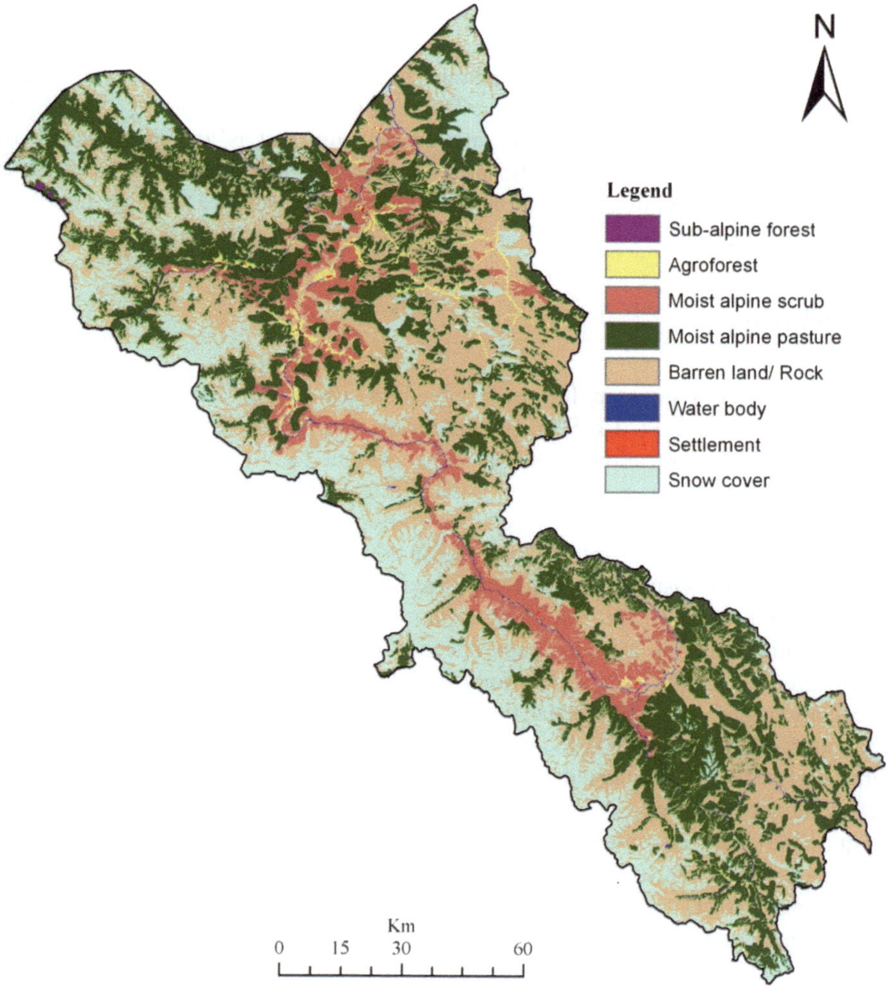

Fig. 13.6 Vegetation/land-cover map derived from IRS-P6, LISS-III satellite data

Champion and Seth (1968) put this class under moist alpine scrub under section 16/C1. Moist alpine pastures were found distributed over 5,194.5 km^2 occurring between 2,800 and 5800 m altitude, reaching close to the snow cover at places. Similarly, moist alpine scrub reflected light red to light brown color on the satellite imagery, often causing considerable ambiguity to delineate between scrub and pastureland. Scrublands were dominated by *Poa annua*, *Medicago sativa*, *Artemisia brevifolia*, *Plantago lanceolata*, and *Poa pratensis*. Champion and Seth (1968) put this class under moist alpine pasture category under section 15/C3. The present study with medium-resolution satellite data has characterized vegetation/land-cover classes in the Kargil region with reasonably good accuracy. Being a cold desert region, the vegetation and land-cover map showed dominance of barren land and snow cover with lesser areas under vegetation cover. The overall classification accuracy for vegetation was observed to be 93 %, which is acceptable for further landscape analysis including extrapolation of biodiversity attributes. The reason for the 15 % inaccuracy could be delineation between similar classes at the transition that is apparent because of fuzziness. A large area of the landscape is occupied by barren land (44 %), which is found to be dominant over vegetation (39 %) and snow cover (17 %) (Fig. 13.5c).

Conclusion

The vegetation type map has been prepared using four individual scenes of georeferenced IRS P6 LISS III satellite data and validated with ground visits using exact GPS locations. Various enhancement techniques such as standard deviational and contrast stretching, available in ERDAS Imagine software, were used to make data more interpretable. A hybrid approach considering NDVI classified images and false color composites (FCC) of satellite data was implemented for visual interpretation. Visual interpretation involved classification of the satellite data under various vegetation and land-cover categories with the help of interpretation keys to prepare a final vegetation and land-use map of the study area. Field data collected during two field trips were used to calculate primary vegetation analyses to understand phytosociological status in the region such as Shannon diversity (1948), Pielou equitability (1969), and Simpson dominance index (1949). The dispersed landscape of Kargil includes only one vegetation class under tree species, although a few small patches of subalpine forest (Chir/Pine) were also mapped (0.04 %) in the southwestern part. About 30 % of the geographic area is mapped as moist alpine pasture followed by moist alpine scrub, agroforest, and other land-cover classes. The ecosystem diversity ranges from subalpine forest to moist alpine pastures. The agroforest covers only 1.1 % of the total land area of the district, found mostly along the rivers in the foothills of valleys; at lower slopes moist species of scrub are prominent. At top altitudinal range, alpine meadow (as moist alpine pasture) was mapped. With the help of fusion of conventional as well as remote sensing techniques, vegetation characterization in hilly areas can be done easily, accurately, and effectively. Satellite data of the optimal vegetative period are very important because of the seasonal growing nature of many pasture and scrub species and also the lesser period of ground coverage by snow. NDVI-, VI-, and NDWI-based classification integrated with visual interpretation helps to identify and differentiate vegetation from nonvegetation classes in difficult terrain, where vegetation is either dispersed or sensor reflectance is much less.

Three hundred and twenty four species were encountered during two field visits in Kargil. Of these, only 16 are tree species. About half (157) of the total species found are economically important (Annex: Economically important species list, Anonymous 2011) in many ways, and 81 species known to be of medicinal importance

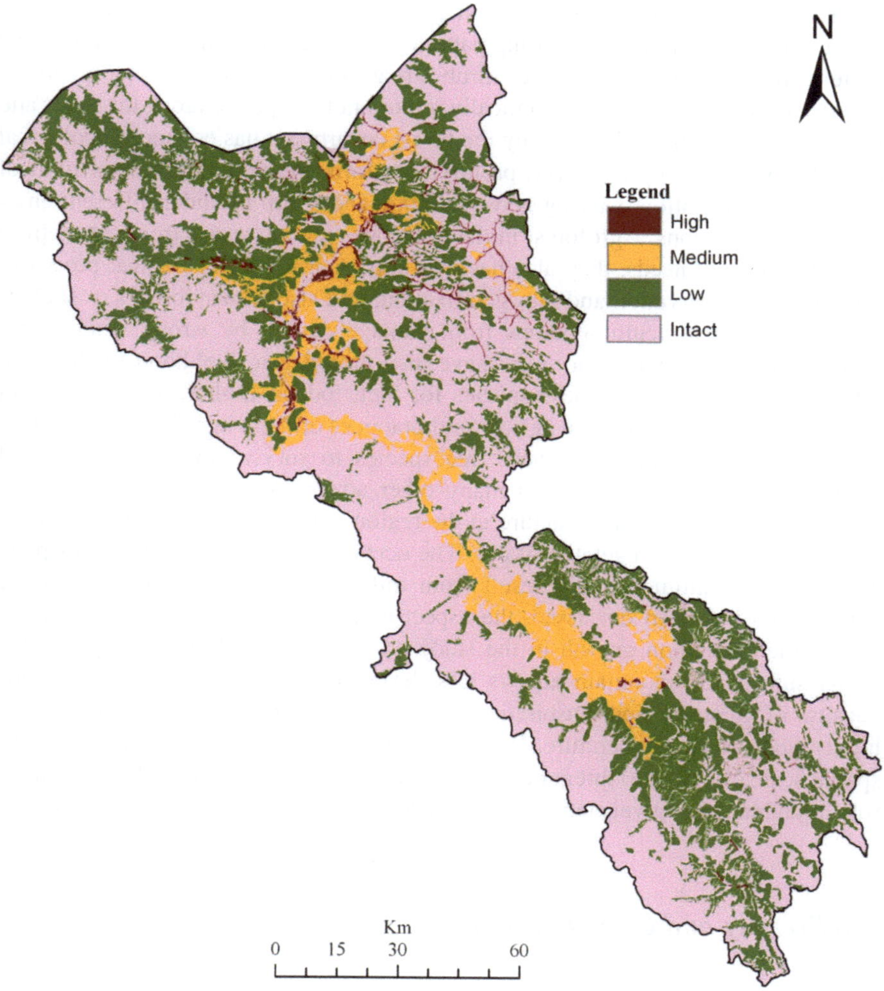

Fig. 13.7 Biodiversity value image

(Annex: List of medicinal plants, Anonymous 2011). A total of 226, 189, and 79 species were found in agroforest, moist alpine scrub, and moist alpine pasture, respectively. Family diversity was maximum in agroforest, with 46 species, followed by 36 in moist alpine scrub and 20 in moist alpine pasture. Study of genera of all species has revealed that 137 genera have been recorded in agroforest, 108 in moist alpine scrub, and only 46 in moist alpine pasture (Fig. 13.5d). Various species of *Hippophae* (*H. rhamnoides*), locally called sea buckthorn, has a higher medicinal importance with highest total importance value. The diversity status in the region can be understand more broadly and precisely if more plotting can be done at vegetative areas of high altitude and a majority of steeper slopes (Fig. 13.7).

Physiognomy plays a vital role in shaping the vegetation pattern in the district. Species diversity showed decreasing pattern with increasing altitude, similar to the eastern Himalaya. Trees and larger shrubs are generally rare in this region. Plantation (such as agroforest) of economically important tree species along the roads and rivers and within the arable land by the forest department has enhanced ecological stability in the region. However, pasturelands form the upper altitudinal limit of vegetation distribution, varying between 3,000 and 5,000 m under different environmental conditions. Agroforest in the mountains are a treasure trove; apart from fulfilling primary needs, they also have an important role in ecological processes such as soil conservation and nitrogen fixation, and also as windbreaks or shelterbelts to protect agricultural crops from damage caused by strong winds. Tree species such as *Populas* and *Salix* have also provided a livelihood to the local villagers. Dependence on these species for fuel, fodder, timber, and many other passive uses has increased the agricultural practices of these agroforests in the past few decades. The ability of these tree species to survive in a great range of temperature (−45 to 46 °C) and many other adverse conditions helps these to acquire native community structure in the region with the observed highest diversity value among all vegetation types. The occurrence of a large number of plant species at low altitude might result from mild climatic conditions and diverse habitats. As has been discussed, the aspect of the mountain slopes significantly affects the vegetation growth in the study area. The vegetation coverage on the sunny side in the mountain area is less developed than that on the shady side because of greater evapotranspiration and differences in their solar radiation and higher land surface temperature. The vegetation type land-cover map generated at a spatial scale of 1:50,000 integrated with ground sampling points can have a crucial role in understanding vegetation patterns and ecology at varying levels and scales.

Biodiversity and Climate Change

Climate change is a major cause of biodiversity loss (CBD 2011). Past changes in the global climate resulted in major shifts in species ranges and marked reorganization of biological communities, landscapes, and biomes. The present global biota was affected by fluctuating concentrations in the Pleistocene (the past 1.8 million years) of atmospheric carbon dioxide, temperature, and precipitation (CBD 2011), and coped through evolutionary changes, species plasticity, range movements, or the ability to survive in small patches of favorable habitat (refugia). Biodiversity and climate change are interconnected, not only through climate change effects on biodiversity, but also through changes in biodiversity that can affect climate change. Observed changes in climate have already adversely affected biodiversity at the species and ecosystem levels, and further deteriorations in biodiversity are inevitable with further changes in climate (Malhi et al. 2010). Such changes, which have resulted in major shifts in species ranges and marked reorganization of biological communities, landscapes, and biomes, occurred in a landscape that was not as fragmented as it is today and with little or no pressure from human activities in places such as Kargil. Changes in climate during the past few decades of the twentieth century have already affected biodiversity. The observed changes in the climate system (e.g., increased atmospheric concentrations of carbon dioxide, increased land and ocean temperatures, changes in precipitation, and sea level rise), particularly the warmer regional temperatures, have affected the timing of reproduction of animals and plants and migration of animals, the length of the growing season, species distributions and population sizes, and the frequency of pest and disease outbreaks in all kinds of living organisms.

References

Allen TFH, Hoekstra TW (1992) Towards a unified ecology. Columbia University Press, New York

Anonymous (2008) Biodiversity characterisation at landscape level using remote sensing & geographic information system. National Remote Sensing Centre (NRSC), DOS/DBT, Government of India, Balanagar, Hyderabad

Anonymous (2011) Biodiversity characterisation at landscape level for Kargil district using remote sensing & GIS techniques. IIT-CORAL/SAML/PR-1/(LLS-2011)

CBD (2011) Convention on biodiversity, Canada (http://www.cbd.int/)

Champion HG, Seth SK (1968) A revised survey of the forest types of India. Govt. of India Press Manager of Publications (Delhi)

Connell JH (1978) Diversity of tropical rainforest and coral reefs. Science 199:1304–1310

Daubenmire R, Daubenmire JB (1968) Forest vegetation of Eastern Washington and Northern Idaho. Washington State University Extension, Pullman (1968 as XT0060). Reprinted 2002. MISC0249

Day F, Monk CD (1974) Vegetation patterns on a southern Appalachian watershed. Ecology 55:1064–1074

Dhar U, Kachroo P (1983) The alpine flora of Kashmir. Scientific Publishers, Jodhpur

Fox JL (1983) Constraints on winter habitat selection by the mountain goat (*Oreamnos americanus*) in Alaska. Ph.D. Thesis, University of Washington, Seattle

Geneletti D, Dawa D (2009) Environmental impact assessment of mountain tourism in developing regions: a study in Ladakh, Indian Himalaya. Environ Impact Assess Rev 29:229–242

Gentry AH (1982) Patterns of neotropical plant species diversity. Evol Biol 15:1–84

Gentry AH (1988) Changes in plant community diversity and floristic composition on environmental and geographical gradients. Ann Mo Bot Gard 75:1–34

Gentry AH (1992) Tropical forest biodiversity: distributional patterns and their conservational significance. Oikos 63:19–28

Henebry GM (1993) Detecting change in grasslands using measures of spatial dependence with Landsat TM data. Remote Sens Environ 46:223–234

Heywood VH (1995) Global biodiversity assessment. Cambridge University Press, New York

Humbert-Droz B, Dawa S (2004) Biodiversity of Ladakh. Strategy and action plan. LEDEG, Sampark, Kolkata

Joshi PK, Rawat GS, Padilya H, Roy PS (2006) Biodiversity characterization in Nubra valley, Ladakh with special reference to plant resource conservation and bioprospecting biodiversity and conservation. Biodivers Conserv 15:4253–4270

Klimeš L (2003) Life-forms and clonality of vascular plants along an altitudinal gradient in Eastern Ladakh (NW Himalayas). Basic Appl Ecol 4(4):317–328

Korne C (2003) Carbon limitation in trees. J Ecol 91:4–17

LAHDC (2005) Ladakh autonomous hill council, Leh. http://leh.nic.in/default.htm
Lillesand TM, Kiefer RW, Chipman JW (2007) Remote sensing and image interpretation, 5th edn. Wiley, Singapore
Malhi Y, Silman M, Salinas N, Bush M, Meir P, Saatchi S (2010) Introduction: elevation gradients in the tropics: laboratories for ecosystem ecology and global change research. Global Change Biol 16:3171–3175
Marks PL, Harcombe PA (1981) Vegetation of the Big Thicket National Preserve, Texas. Ecol Monogr 51:287–305
Matin S, Chitale VS, Behera MD, Mishra B, Roy PS (2012) Fauna data integration and species distribution modelling as two major advantages of geoinformatics based phytobiodiversity study in today's fast changing climate. Biodivers Conserv 21:1229–1250
Moser W, Brzoska W, Zachhuber K, Larcher W (1977) Ergebnisse des IBP-Projekts Hoher Nebelkogel 3184 m. Sitzungsber Österr Akad Wiss Math-naturwiss Klasse Abt I 186:387–419
Myers N, Mittermeier RA, Mittermeier CG, Da-Fonseca GAB, Kent J (2000) Biodiversity hotspots for conservation priorities. Nature (Lond) 403:853–858
Nagy L, Grabherr G (2009) The biology of alpine habitats. Oxford University Press, Oxford
Noss RF (1996) Ecosystems as conservation targets. Trends Ecol Evol 11:351
Paczoski J (1928) Biologiczna struktura lasu (in Polish: the biological structure of forest). Sylwan 3:193–221
Peet RK (1978) Forest vegetation of the Colorado Front Range: patterns of species diversity. J Veg 37:65–78
Pielou EC (1969) An introduction to mathematical ecology. Wiley, New York
Porwal MC, Sharma L, Roy PS (2003) Stratification and mapping of Ephedra gerardiana Wall. in Poh (Lahul and Spiti) using remote sensing and GIS. Curr Sci 84:208–212
Rahbek C (1995) The elevational gradient in species richness: a uniform pattern? Ecography 18:200–205
Shannon CE (1948) A mathematical theory of communication. Bell Syst Tech J 27:379–423
Simpson EH (1949) Measurement of diversity. Nature (Lond) 163:688
Singh E, Singh MP (2010) Biodiversity and phytosociological analysis of plants around the municipal drains in Jaunpur. Int J Biol Life Sci 6(2):77–82
Sømme L (1997) Adaptations to the alpine environment in insects and other terrestrial arthropods. In: Wielgolaski FE (ed) Polar and alpine tundra. Ecosystems of the world, vol 3. Elsevier, Amsterdam, pp 11–25
Stone WC, Frayer JRA (1935) A botanical study of pasture mixtures. Sci Aric 15:777–805
Turner MG, Gardner RH (1992) Quantitative methods in landscape ecology: the analysis and interpretation of landscape heterogeneity. Springer, New York
UNEP-WCMC (2000) Global biodiversity: Earth's living resources in the 21st century. World Conservation Press, Cambridge
Vetaas OR, Grytnes JA (2002) Distribution of vascular plant species richness and endemic richness along the Himalayan elevation gradient in Nepal. Glob Ecol Biogeogr 1:291–301
Whittaker RH (1956) Vegetation of the Great Smoky Mountains. Ecol Monogr 26:1–80
Whittaker RH (1960) Vegetation of the Siskiyou Mountains, Oregon and California. Ecol Monogr 30:279–338
Wilson JB, Chiarucci A (2000) Do plant communities exist? Evidence from scaling-up local species–area relations to the regional level. J Veg Sci 11:773–775

Chapter 14
Water Circulation in a Fringing Reef and Implications for Coral Distribution and Resilience

Hiroya Yamano, Hiroshi Hata, Toshihiro Miyajima, Ken Nozaki, Ken Kato, Akira Negishi, Masayuki Tamura, and Hajime Kayanne

Abstract Shiraho Reef, Ishigaki Island, Southwest Japan is characterized by (1) a lack of prominent channels, which causes ponding at low tide, (2) its location in a monsoon area, which causes the dominant wind direction to change seasonally, and (3) its orientation facing the Pacific Ocean, where swells originate in distant stormy areas in addition to those produced by local winds. The water circulation observed during three separate periods and the estimated residence time in Shiraho Reef are very sensitive to all three factors. Advanced satellite technology contributes significantly to the long-term monitoring of ocean swell states, and the water circulation observed during the three periods can be classified using swell state (high and low swell) and wind conditions (northerly and southerly winds). During high swell states, the water residence time ranged from 3 to 8 h. The shorter time occurred when ponding did not occur, whereas the longer time occurred when ponding was observed.

H. Yamano (✉)
National Institute for Environmental Studies, 16-2 Onogawa, Tsukuba, Ibaraki 305-8506, Japan
e-mail: hyamano@nies.go.jp

H. Hata
Hazama Technical Research Institute, 515-1 Karima, Tsukuba, Ibaraki 305-0822, Japan

T. Miyajima
Atmosphere and Ocean Research Institute, The University of Tokyo, 5-1-5 Kashiwanoha, Kashiwa, Chiba 277-8564, Japan

K. Nozaki • K. Kato • A. Negishi
Energy Electronics Institute, National Institute of Advanced Industrial Science and Technology, AIST Central 2, 1-1-1 Umezono, Tsukuba, Ibaraki 305-8568, Japan

M. Tamura
Department of Urban and Environment Engineering, Kyoto University, Kyotodaigakukatsura, Nishikyo-ku, Kyoto 615-8540, Japan

H. Kayanne
Department of Earth & Planetary Science, The University of Tokyo, 7-3-1 Hongo, Bunkyo-ku, Tokyo 113-0033, Japan

S.-i. Nakano et al. (eds.), *Integrative Observations and Assessments*, Ecological Research Monographs, DOI 10.1007/978-4-431-54783-9_14, © Springer Japan 2014

The water circulation and residence time likely affect the coral distribution and resilience to bleaching.

Keywords Bleaching • Coral reef • Distribution • Ishigaki Island • Japan • Resilience • Water circulation

Introduction

Fringing reefs are shallow, contain abundant organisms, and have a high rate of organic production. They are threatened by the anthropogenic effects of nutrient and sediment loading, as well as warming seas (e.g., Smith and Buddemeier 1992). To understand the role and fate of fringing reefs, it is necessary to understand their water circulation, because this is a major factor in determining the community structure and the exchange or the dispersal of materials.

Lagoon circulation is driven mainly by tide, wind, and radiation stress induced by ocean swells breaking on the reef crest (Andrews and Pickard 1990). In fringing reefs, physical environment of lagoons is significantly affected by offshore environment (Nadaoka et al. 2001a,b; Tamura et al. 2007), and breaking of ocean swells on the reef crest is considered the major factor affecting the water circulation on reef flats (Roberts et al. 1975; Prager 1991; Symonds et al. 1992; Wolanski et al. 1993; Lugo-Fernandez et al. 1998a,b; Hearn 1999). Other factors are also important locally. The tide is considered the main driving force in fringing reefs in the Great Barrier Reef of Australia because of the large tidal range (Parnell 1988). To understand the water circulation throughout the year in reefs located in monsoon areas, such as the Ryukyu Islands of Japan and the Maldives, where the wind direction changes seasonally, wind direction must be considered (Yamano et al. 1998; Kench et al. 2009).

Although all reefs possess a general wave-resistant feature, the reef crest, local reef topography varies from site to site and affects the water circulation and exchange in a reef. Parnell (1988) showed that bay morphology significantly affected the circulation pattern in addition to tides in the Great Barrier Reef of Australia. Shallow reefs sometimes isolate the lagoon from the ocean during low tides because of exposure of the reef crest. This phenomenon has been examined in detail in a platform reef (Ludington 1979; Frith 1981; Frith and Mason 1986). Because fringing reefs are shallow, the circulation is sensitive to both bay morphology and reef topography. Thus, it is necessary to understand the effect of topography on water circulation in a fringing reef in addition to the effects of ocean swells, tides, and winds.

In this study, we observed water circulation in a shallow fringing reef (Shiraho Reef), where the backreef waters are isolated from the ocean during low tides. We classified the water circulation pattern according to ocean swells and winds. Then, we discuss the effect of reef-flat topography on circulation in Shiraho Reef.

Finally, we present implications of the water circulation and residence time for the coral distribution and resilience to bleaching.

Study Site

Shiraho Reef is located on the east coast of Ishigaki Island, southwest Japan, and faces the Pacific Ocean (Fig. 14.1). Ishigaki Island is situated in a monsoon area, and the dominant winds change from southerly in summer to northerly in winter (Hasegawa and Yamano 2004). Over the entire year, northerly winds are dominant (Fig. 14.1), and the speed of a northerly wind is generally greater than that of a southerly wind. Figure 14.2a is an aerial photograph of Shiraho Reef, illustrating the well-developed fringing reef with distinct topographic zonation from land to ocean: backreef moat and reef crest. Shiraho Reef extends about 3 km along the shore, and it is from 700 to 850 m from the shoreline to the reef crest. The southern boundary of the reef is a shallow area called "Watanji," which is subaerially exposed during low tides and separates the backreef moat from the ocean. Thus, the southern end of the reef lacks prominent channels (deep passage) through which water flows to the ocean (see Kabira Reef situated in the same island; Fig 14.2b), while there are small, shallow (1–2 m deep) channels at the center of the reef and at its southern boundary (Fig. 14.2a). The northern end of Shiraho Reef has a prominent channel that connects the backreef moat to the open ocean (Nadaoka et al. 2001a; Tamura et al. 2007).

Nakamori et al. (1992) estimated the depth distribution of the reef from aerial photographs: this was updated by Kayanne et al. (2000) using actual field measurements (Fig. 14.2a). The depth from mean sea level is around 2 m in the moat, and 0.2–0.6 m over the reef crest, and "Watanji." Although the depth in the moat is similar to that of Kabira Reef, the depth at the reef crest is shallower than at Kabira Reef (0.6–1.0 m for the crest), which is situated on the northwestern coast of Ishigaki Island (Yamano et al. 2000). The areas with depths less than 1.0 m are subaerially exposed during low spring tides. Overall, Shiraho Reef is considered a semi-enclosed fringing reef, and a specific feature of this reef is the ponding of backreef water during low tides (slackwater period) (Nakamori et al. 1992; Kayanne et al. 1995). The water volume in Shiraho Reef as estimated from the depth distribution is about 547,000 m^3.

As with other reef systems, a distinct ecological zonation is found in Shiraho Reef (Nakamori et al. 1992; Iryu et al. 1995; Kayanne et al. 2002). From shore to ocean, five zones are observed: (1) seagrass beds with small patches of *Porites* corals close to the beach, (2) sand bottom, (3) branching *Acropora* and *Montipora* corals at the back of the reef crest, (4) larger benthic Foraminifera on the reef crest, which is subaerially exposed during low tides, and (5) robust branching *Acropora* corals at the reef edge facing the ocean. Regarding the third zone, the blue coral *Heliopora coerulea*, instead of *Acropora* and *Montipora* corals, is found in the southern part around station (Sta.) 5 (Fig. 14.2a). As large *H. coerulea* populations

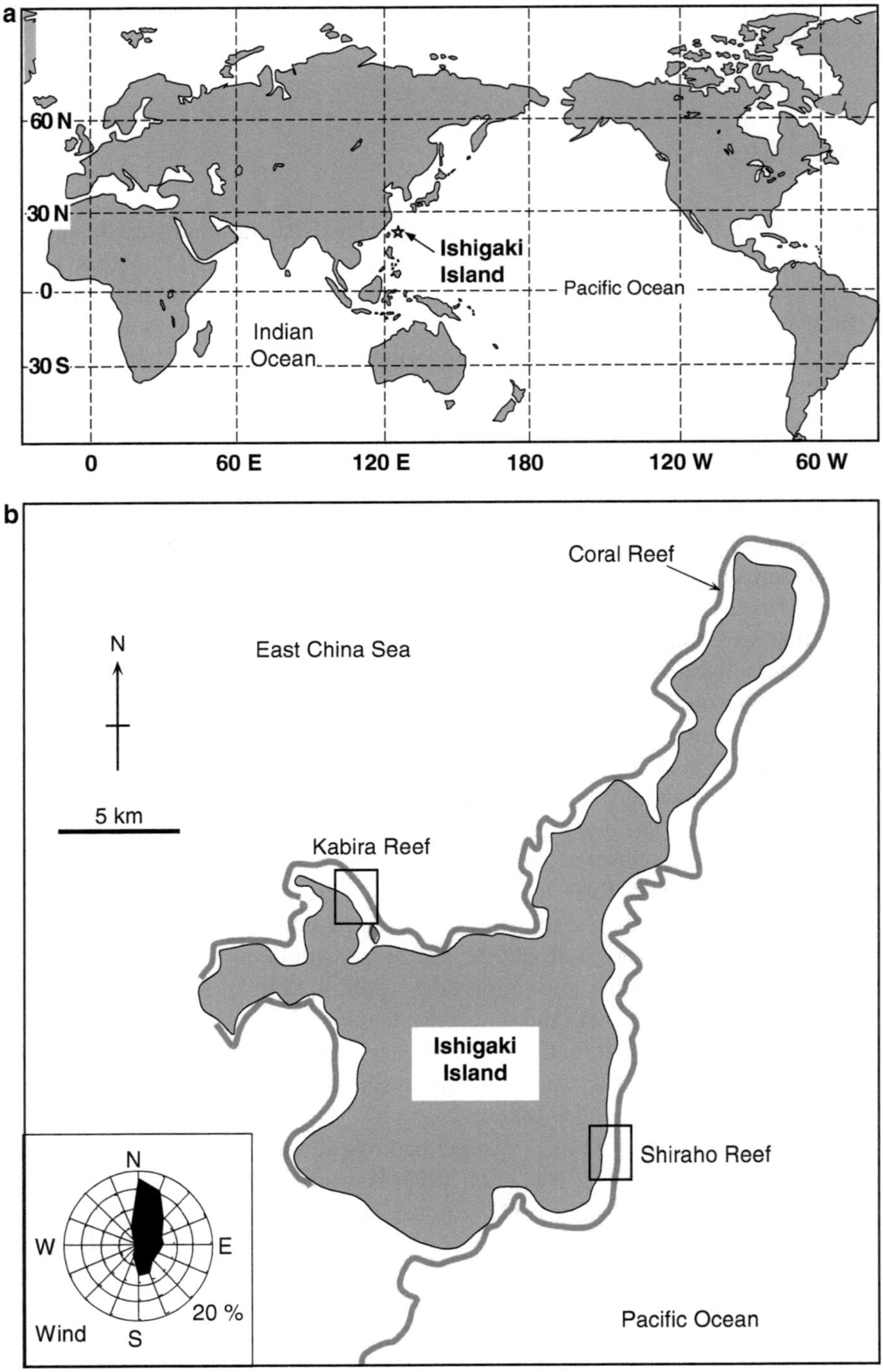

Fig. 14.1 Location of Ishigaki Island (**a**) and map of Ishigaki Island (**b**), showing the location of Shiraho Reef. The location of Kabira Reef on the northwestern coast is also indicated

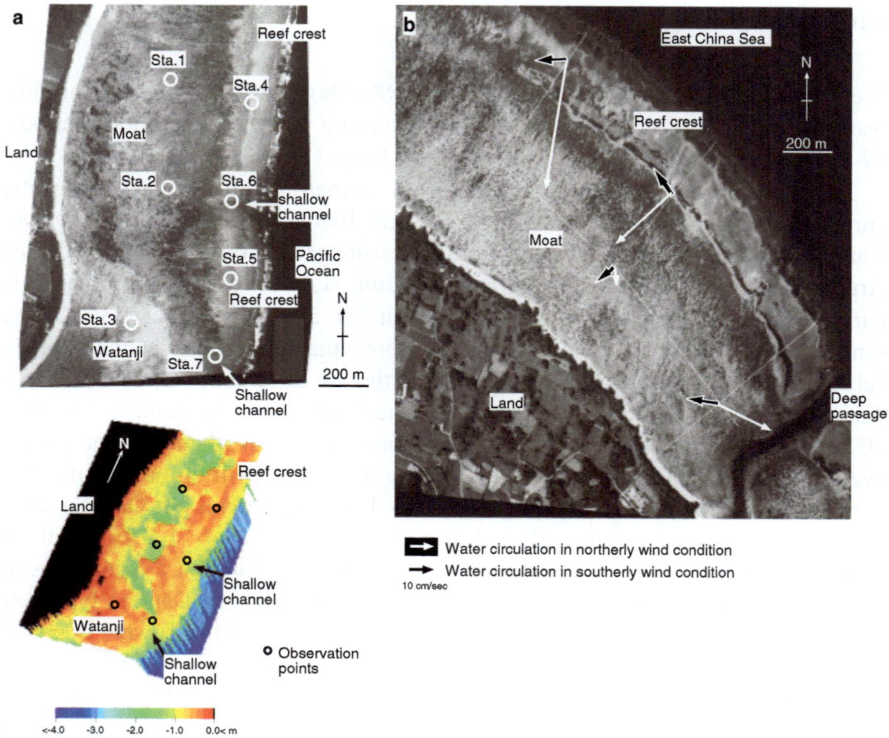

Fig. 14.2 Aerial photograph of Shiraho (**a**) and Kabira (**b**) reefs taken in 1994, showing the location of observation points. *Black patches* close to the beach and the reef crest indicates seagrass and corals, respectively. The *lower figure* shows oblique view of the relief of Shiraho Reef based on Kayanne et al. (2000). Depth is based on mean sea level. The area less than 1 m deep (*yellow* to *red*) is subaerially exposed in low tides. Vectors at Kabira Reef represent the dominant patterns of water circulation throughout the year. (Adapted from Yamano et al. 1998)

are rare elsewhere (Takino et al. 2010), this *H. coerulea* population is a specific feature in Shiraho Reef, which encourages conservation activities of the corals (Suzuki et al. 2012).

Severe bleaching events occurred in 1998 and 2007 because of anomalously high sea surface temperatures (Kayanne et al. 2002; Dadhich et al. 2012). Recovery from the bleaching event depends on the population or colony size; larger populations and colonies recovered more quickly than smaller populations or colonies (Kayanne et al. 2002). Although overall recovery of coral populations was observed after the 1998 bleaching (Kayanne et al. 2002), significant decline of *Acropora* and *Montipora* corals was observed following the 2007 bleaching (Suzuki et al. 2012).

Methods

We established seven observation stations in Shiraho Reef (Fig. 14.2a). Currents, water depths, and winds were observed during three observation periods: 17–23 June 1997, 16–20 March 1999, and 28 June–2 July 1999.

In 1997, current directions and velocities were measured using electromagnetic current meters (ACM-8M and ACM-16M; Alec Electronics) set at stations (Sta.) 1 and 3–7; water level was measured with a pressure gauge (WLR-7, AANDERAA; Argonaut-XR, Sontek) at Sta. 2. In March and June–July 1999, one electromagnetic current meter and one pressure gauge (Argonaut-XR; Sontek) were set at Sta. 2. The sensors of the current meters were set 50 cm above the bottom. In March and June–July 1999, we tracked drifting buoys and bottles by differential GPS and a laser distance meter set at Sta. 2. The drogues of the drifting buoys extended from 0 to 50 cm below the sea surface. As the water depth is less than 50 cm at some points, even in backreef areas, we used bottles during low tides. Drifting buoy and bottle experiments were conducted during high, ebb, low, and flood tides. Data on wind direction and velocity were collected at a point approximately 1.5 km south of Shiraho Reef (Fig. 14.2a) at 5-min intervals in June 1997 and at 1-h intervals in March and June–July 1999, using Kadec-Kaze (Kona System). Data on typhoons that can affect circulation in a fringing reef (Nadaoka et al. 2001a) were obtained from Shimazaki et al. (2006).

Significant wave height (SWH) off Shiraho Reef was obtained from altimeter data from the TOPEX/POSEIDON satellite, which routinely measures the SWH (Fu et al. 1994). The SWH data obtained by satellite are reliable and can be used to calculate the water flow over a reef crest (Tartinville and Rancher 2000) and to estimate water renewal time for atoll lagoons (Andréfouët et al. 2001; Pagés and Andréfouët 2001). The data provided by the Physical Oceanography Distributed Active Archive Center of the NASA Jet Propulsion Laboratory (ftp://podaac.jpl.nasa.gov/pub/sea_surface_height) were used. For each observation period (June 1997 and March and June–July 1999), we extracted the SWH for the area bounded by 23.5–24.5ºN and 125–126ºE in the Pacific Ocean off Shiraho Reef. For comparison, the same procedure was used for the East China Sea off Kabira Reef (24.5–25.5ºN and 123.5–124.5ºE).

The water residence time for the southern section of the reef was estimated from the 1997 observation data because the observation stations surround this reef section. The residence time was calculated by subtracting the amount of water flowing into the reef during 12.5 h (half a tidal cycle) from the water volume. The volume transport of each section was calculated by multiplying the current velocity by the cross-sectional area through Shiraho Reef. The cross-sectional area was determined by multiplying water depth by width. At Sta. 1, water depths were obtained from the depth data observed at Sta. 2, as both points were located in the backreef moat. The water depths for Sta. 3–7 were obtained from the predicted tide levels using the tide table for Port Ishigaki published by the Japan Meteorological Agency.

Results and Discussion

Winds

Figure 14.3a shows the current vectors at each observation point, and the wind vectors and water depths at Sta. 2 during the observation period in June 1997. The distribution of wind directions and velocities during the period 17–23 June can be subdivided into two short periods. Southerly winds with a maximum speed of 4 m/s dominated from 17 June to 03:00 on 20 June. Northerly winds with a maximum speed of 4 m/s occurred from 03:00 on 20 June to 23 June. From 16:00 on 19 June until 08:00 on 20 June, a significant current was measured at every observation point that could have been caused by an approaching typhoon (Typhoon OPAL).

Figure 14.3b shows the current vectors at Sta. 2 and the wind vectors and water depths at Sta. 2 during the observation period in March 1999. The distribution of wind directions and velocities during the period from 15 to 20 March can be subdivided into two short periods. Northerly winds with a maximum speed of 3 m/s dominated until 00:00 on 17 March and after 07:00 on 20 March. Southerly winds with a maximum speed of 6 m/s occurred from 00:00 on 17 March until 07:00 on 20 March.

Figure 14.3c shows the current vectors at Sta. 2 and the wind vectors and water depths at Sta. 2 during the observation period in June to July 1999. Throughout the observation period, southerly winds dominated.

Ocean Swells

Figure 14.4 shows the long-term data and the monthly average SWH off Shiraho (Pacific Ocean) and Kabira (East China Sea) Reefs observed by TOPEX/POSEIDON. The SWH off Shiraho Reef was 1.1–4.7 m, 1.6–2.3 m, and 0.9–1.0 m in June 1997, March 1999, and June–July 1999, respectively (Fig. 14.4a). As two typhoons (Typhoons OPAL and PETER) approached Ishigaki Island in June 1997, a large SWH is expected. The monsoon climate produces strong northerly winds in winter and weak southerly winds in summer. The SWH in the East China Sea was correlated with wind direction and speed, whereas relatively stable SWH was observed in the Pacific Ocean (Fig. 14.4b).

Water Circulation

Ocean swells and winds contribute significantly to water circulation on reef flats (Roberts et al. 1975; Yamano et al. 1998). The SWH off Shiraho Reef was not correlated with local winds around Shiraho Reef (Fig. 14.4), although Yamano

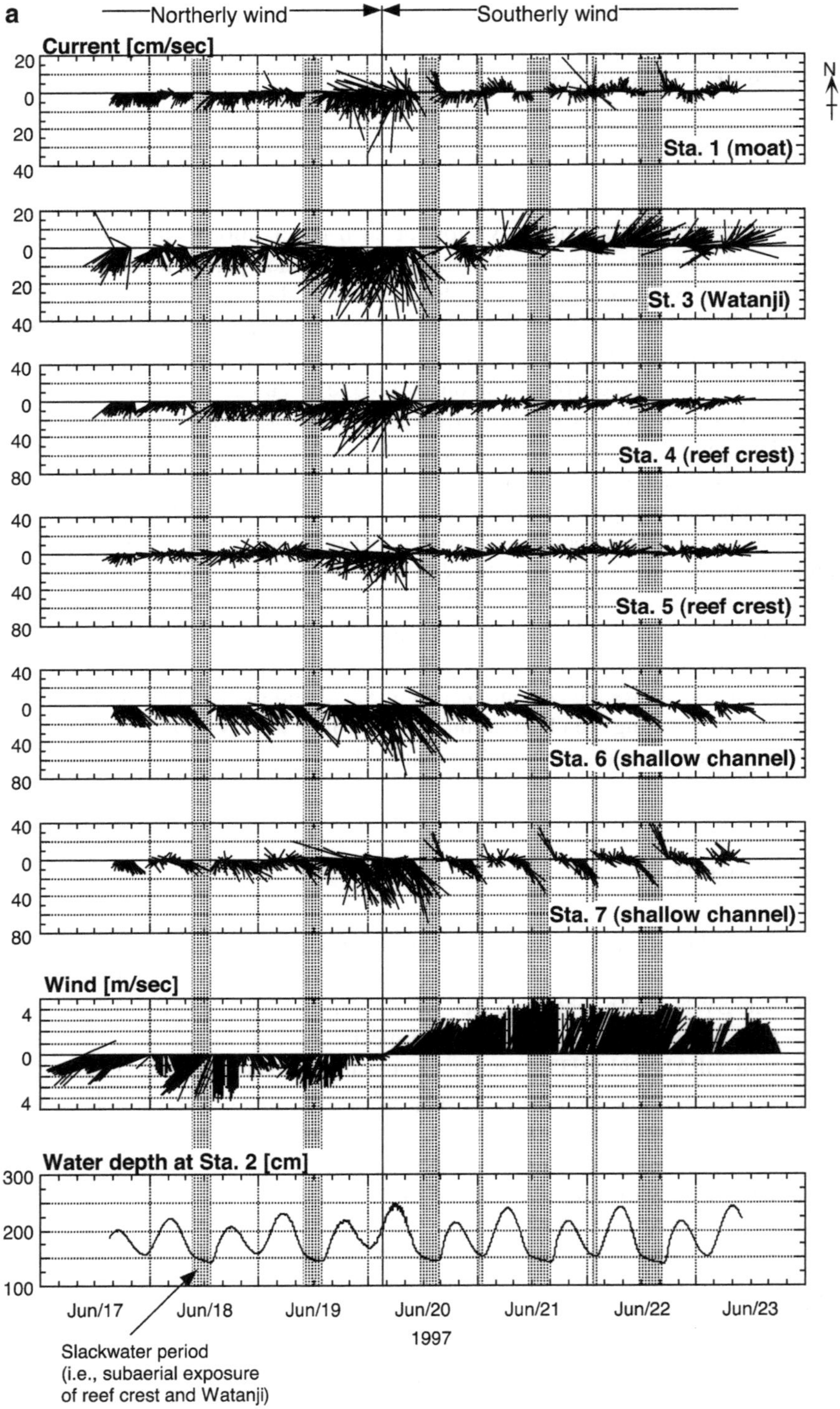

Fig. 14.3 Current and wind vectors and water level during June 1997 (**a**), March 1999 (**b**), and June–July 1999 (**c**)

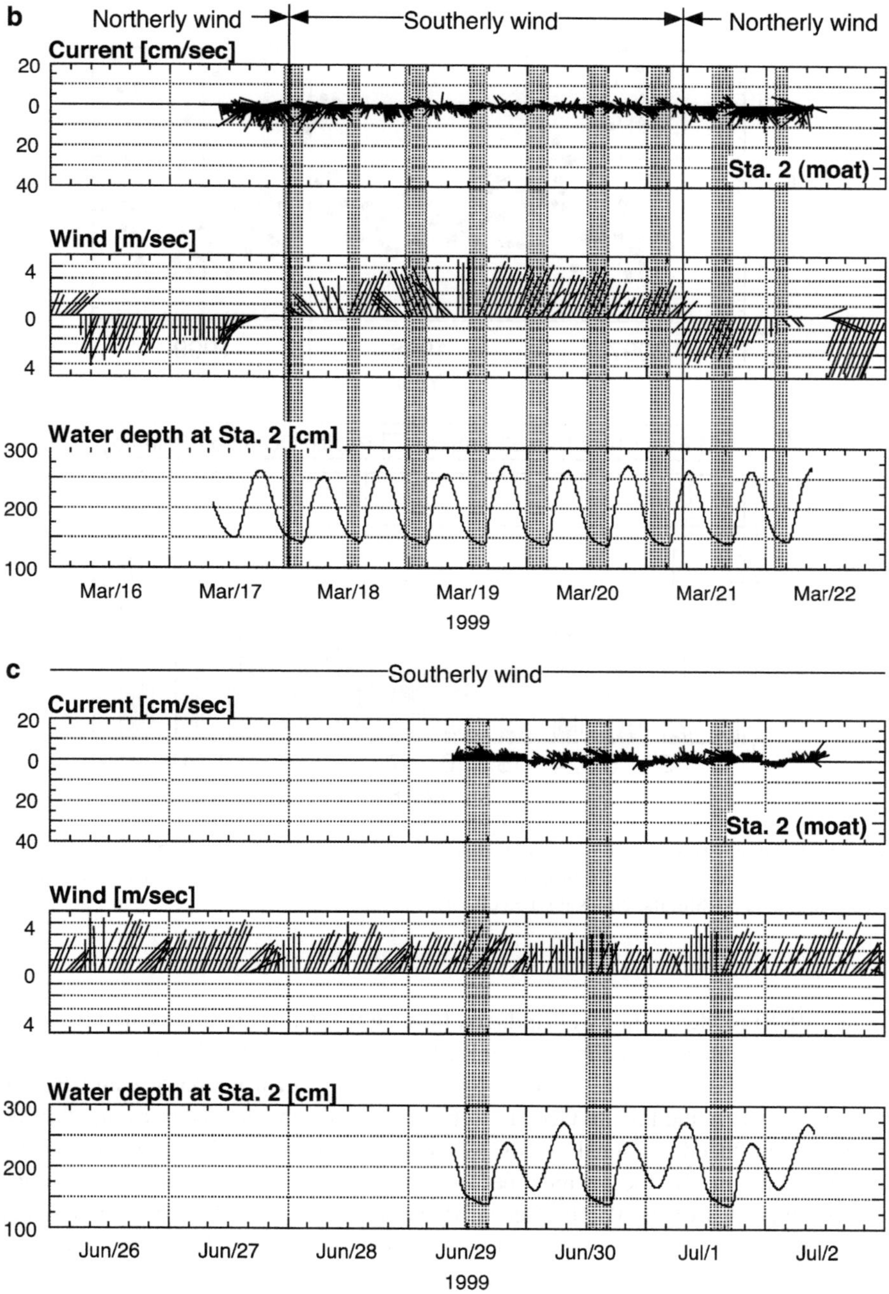

Fig. 14.3 (continued)

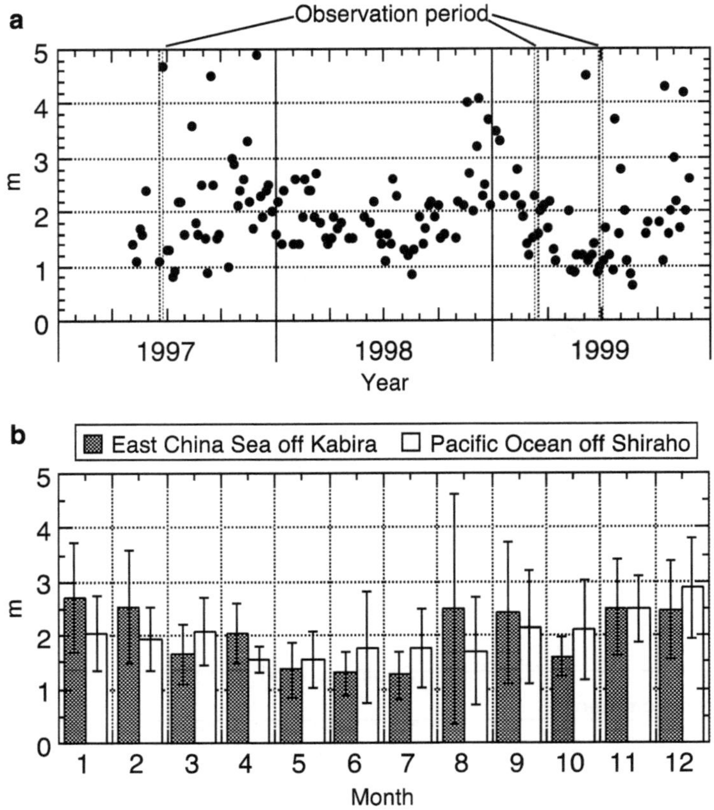

Fig. 14.4 Time series of satellite-observed significant wave heights off Shiraho Reef (Pacific Ocean) (**a**) and monthly mean significant wave heights (mean ± SD) for the Pacific Ocean and East China Sea off Shiraho and Kabira Reefs (**b**)

et al. (1998) modeled the swell-induced flow on the Kabira reef crest and backreef water circulation as a function of wind speed observed in situ. We believe that this results from the difference in the fetch areas of the East China Sea and Pacific Ocean. Swells in the Pacific Ocean originate from distant stormy areas, in addition to those produced by local winds, whereas swells in the East China Sea are generated mainly by local winds, because the area of the East China Sea is much smaller than that of the Pacific Ocean. Because the data on SWH off Kabira Reef (Fig. 14.4b) showed a marked wind dependence, the swell-induced flow could be modeled using wind speed alone, and the water circulation pattern was classified using wind only. In Shiraho Reef, however, the circulation pattern must be classified using both swell state and wind direction and speed. Therefore, we considered four conditions: high swell/northerly wind, high swell/southerly wind, low swell/northerly wind, and low swell/southerly wind. As Figs. 14.3 and 14.4 suggest,

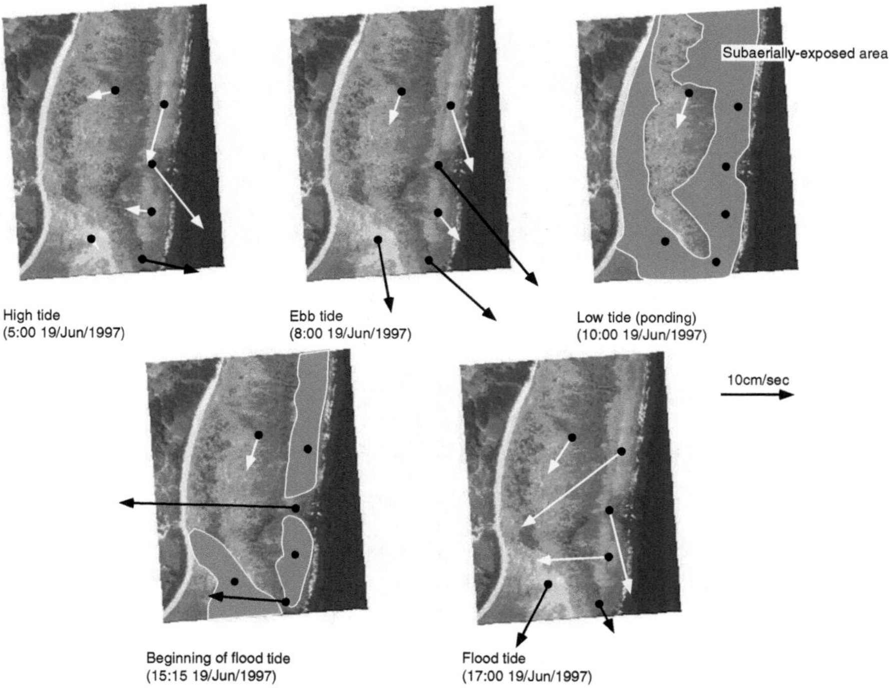

Fig. 14.5 Water circulation on Shiraho Reef in June 1999 during high, ebb, low, beginning of flood, and flood tides. The condition is "High swell/northerly wind"

low swell/northerly wind was not observed, and therefore only the other three conditions were considered. However, northerly winds are generally strong and produce large SWH; the low swell/northerly wind condition is rare in Ishigaki Island.

The threshold value for high and low swell states was based on Kraines et al. (1998), who successfully modeled the wave-driven flow over a reef crest and set the ocean SWH to 0.65–0.9 m for relatively calm observation periods and 1.2–1.4 m for relatively severe observation periods. We set the threshold value separating low and high swell states at 1.1 m. To study the circulation pattern, we calculated the mean velocity value every 15 min. Under each wind condition, we describe the circulation patterns for five tide conditions: high, ebb, low, beginning of flood, and flood tides (Figs. 14.5, 14.7). The tide was semi-diurnal, and ponding occurred when the water level was less than 160 cm. We also show the trajectory of drifting buoys and bottles, which indicate the circulation in Shiraho Reef (Figs. 14.6, 14.8, 14.9) under five tide conditions.

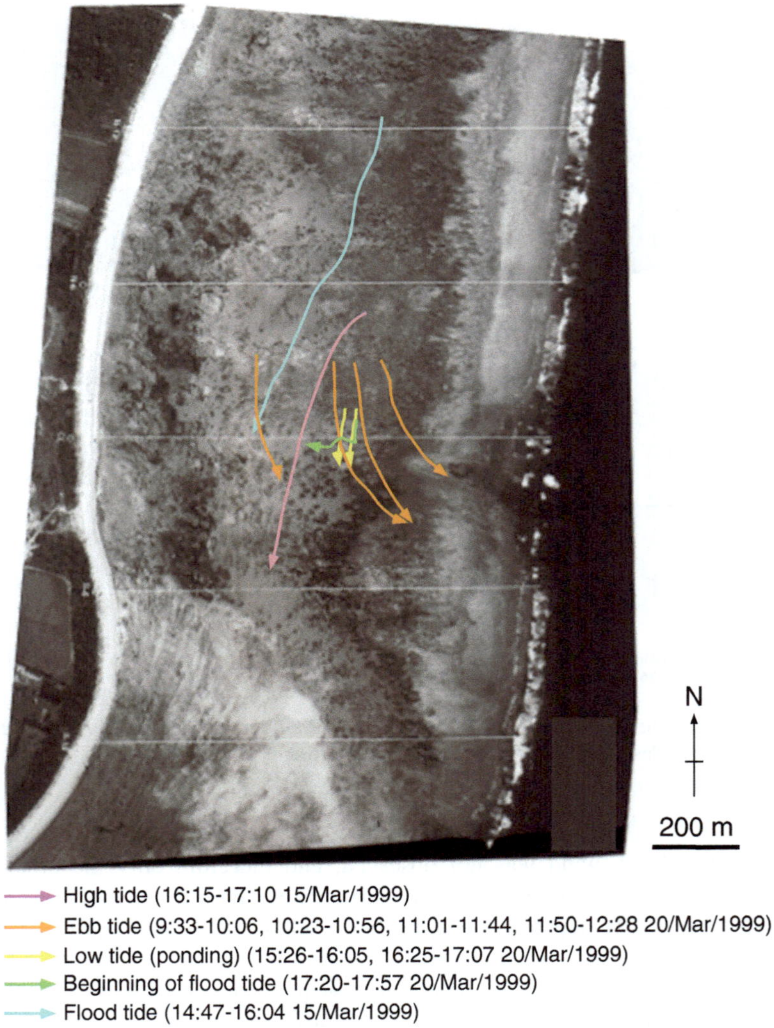

Fig. 14.6 Trajectory of drifting buoys and bottles in March 1999. Date and release and recovery times are shown. The condition is "High swell/northerly wind"

High Swell/Northerly Wind (June 1997 and March 1999)

During low tide, the reef crest and "Watanji" were exposed, there was no inflow, and ponding occurred (Figs. 14.5, 14.6). The current flowed slowly (<10 cm/s) southward in the moat. At the beginning of flood tide (after ponding occurred), ocean waters entered the moat through channels (Sta. 6 and 7) and then over the reef crest (Sta. 4 and 5). However, the current in the channels reversed and flowed into the ocean within 20 min. From flood to high tide, the current moved southward, and

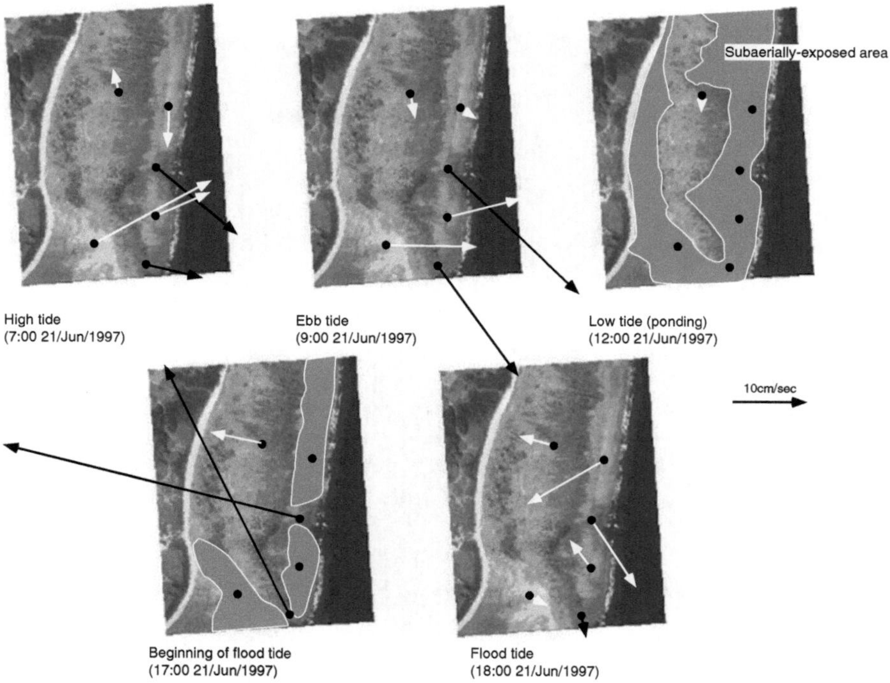

Fig. 14.7 Water circulation on Shiraho Reef in June 1999 during high, ebb, low, beginning of flood, and flood tides. The condition is "High swell/southerly wind"

this was likely generated by the northerly winds. The channels were the major passages through which water flowed into the ocean. During ebb tide, water flowed out through the channels and over the reef crest.

The water residence time calculated for this condition in June 1997 ranged from 3 to 8 h, with a mean value of 5 h. The shorter time occurred when ponding did not occur, and the longer time occurred when ponding was observed.

High Swell/Southerly Wind (June 1997 and March 1999)

Except for "Watanji" (Sta. 3), the circulation pattern was similar to that during northerly winds (Figs. 14.7 and 14.8). The current on "Watanji" was generally northward, and was likely generated by the southerly winds; the current was southward during northerly winds. This change in the southern part of the moat appeared to affect the current around Sta. 2. During low tide, the water at Sta. 2 was almost stagnant (Fig. 14.8). During ebb tide, water from both the northern and southern parts of the reef flowed out from the channel at Sta. 6.

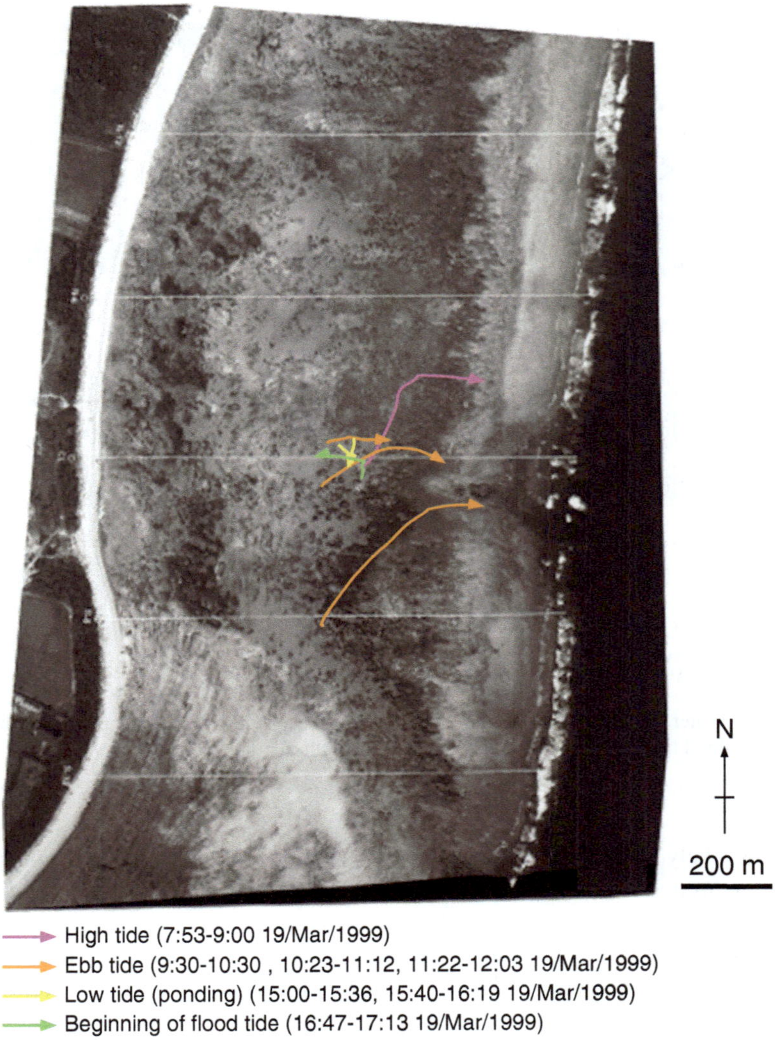

Fig. 14.8 Trajectory of drifting buoys and bottles in March 1999. Date and release and recovery times are shown. The condition is "High swell/southerly wind"

The water residence time calculated for this condition in June 1997 also ranged from 3 to 8 h, with a mean value of 5 h. The shorter time occurred when ponding did not occur; the longer time occurred when ponding was observed.

Low Swell/Southerly Wind (June–July 1999)

Current meter data were obtained only at Sta. 2 (Fig. 14.3c), so only the circulation around Sta. 2 is described. During low tide, ponding occurred. The current flowed slowly (<10 cm/s) northward in the moat (Fig. 14.9). At the beginning of flood tide

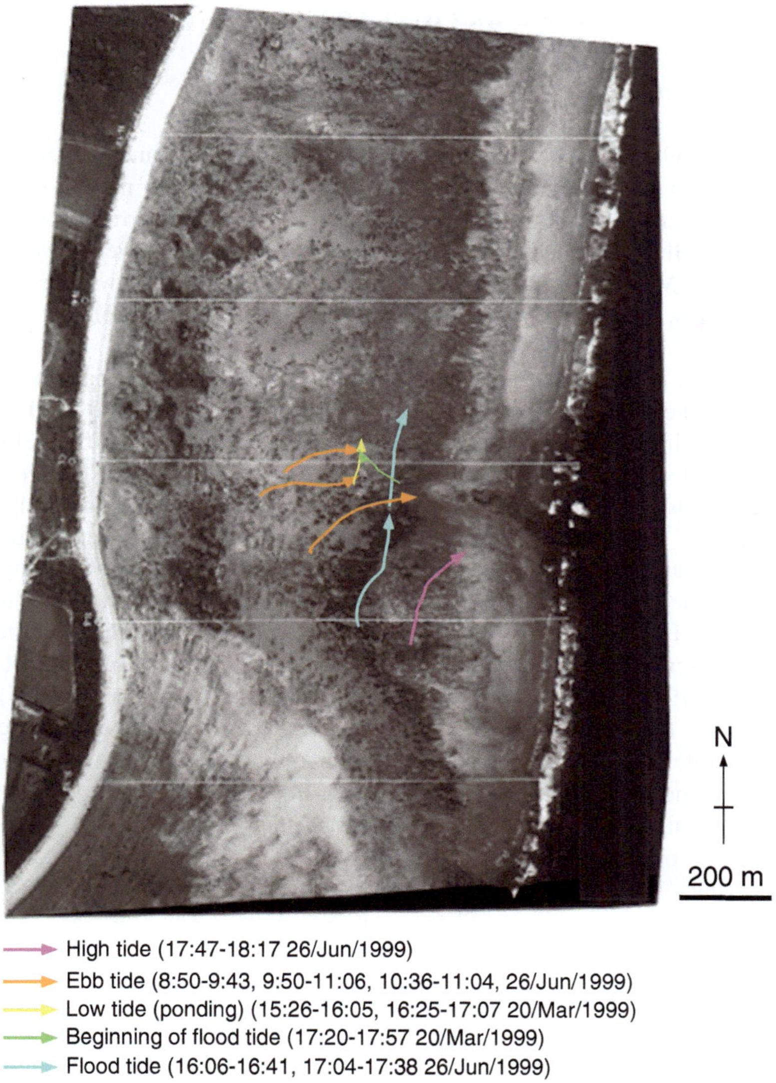

Fig. 14.9 Trajectory of drifting buoys and bottles in March and June 1999. Date and release and recovery times are shown. The condition is "Low swell/southerly wind"

(after ponding), ocean waters entered the moat through the channel (Sta. 6). From flood to high tide, the current moved northward, and was likely generated by the southerly winds. During ebb tide, water flowed out through the channel at Sta. 6. The residence time could not be estimated for June–July 1999.

Effects of Ocean Swells and Reef Topography

Water Circulation

During the high swell/southerly wind condition, although the wind blew northward, the current in the slackwater period was almost stagnant or moved southward (Fig. 14.8). During the low swell/southerly wind condition, the current moved northward, paralleling the wind direction (Fig. 14.9). Thus, high swells affected the circulation in March 1999, which could be the result of strong inflow current caused by breaking waves on the northern part of the reef crest as modeled by Tamura et al. (2007).

The effect of the semi-enclosed topography of Shiraho Reef can be seen by comparing the circulation in Shiraho and Kabira Reefs (Yamano et al. 1998), which have similar climatological settings. The circulation in Kabira Reef shows marked wind influence (Fig. 14.2b). Because the swells off Kabira Reef are highly correlated with the wind speed (Fig. 14.4b), the northerly and southerly wind conditions at Kabira Reef correspond with the high swell/northerly wind and low swell/southerly wind conditions at Shiraho Reef, respectively. During each condition, the circulation pattern at Shiraho Reef shows significant tide dependence (Figs. 14.5, 14.6, 14.7, 14.8, 14.9). Topography controls tidal modification and the water flushing in from the ocean just after flooding of the reef crest (Ludington 1979; Frith 1981; Frith and Mason 1986; Kayanne et al. 1995). The current over the reef crest also appears to be modified by this semi-enclosed environment. Currents over the reef crest are mainly caused by wave-overtopping and setup, and generally show unidirectional flow from the ocean to the shore (Roberts et al. 1975; Symonds et al. 1992; Lugo-Fernandez et al. 1998a, b). However, the opposite flow, from shore to ocean, was observed at Sta. 4 and 5 during ebb tide (Figs. 14.5, 14.7).

Water Residence Time

The residence time in Shiraho Reef varied significantly because of ponding. The residence time was 3 h with no ponding and reached 8 h with ponding. Thus, it is necessary to observe currents during various tide, wind, and ocean swell conditions to obtain better estimates of the circulation and water residence time. Nakamori et al. (1992) calculated a water residence of 12 h from observations of currents at two points (Fig. 14.2a) in the backreef area and wind data. This longer residence time was likely an artifact of observations conducted mainly in the slackwater period. The water residence time calculated for Kabira Reef using the same procedure was 3 h for high swells (northern wind) (estimated from Yamano et al. 1998). The residence time at Shiraho Reef was similar to that at Kabira Reef when no ponding occurred, indicating the significant effect of ponding.

Implications for Coral Distribution

This hydrodynamic environment may affect coral distribution. The slow current velocity at low tide and resulting longer residence time is thought to result in the occurrence of large, dense *Heliopora coerulea* communities in the backreef area, which is a specific feature of Shiraho Reef, probably because calm waters during the slackwater period foster locally successful settlement of *H. coerulea* larvae (Harii and Kayanne 2002, 2003).

In addition to ecological factors (e.g., population/colony size; Kayanne et al. 2002), variation in physical environments would affect coral susceptibility to thermal stress. Although Shiraho Reef lacks prominent channels in the southern part, the two shallow channels (Fig. 14.2a) contribute to water exchange between the moat and ocean (Figs. 14.3a, 14.5, 14.7). Spatial distribution of water temperatures estimated by a two-dimensional thermal model simulation for Shiraho Reef showed lower thermal stress around the shallow channels (Dadhich et al. 2012). Further, high water flow may facilitate resilience of corals to thermal stress (Nakamura and van Woesik 2001; Nakamura et al. 2003). *Montipora* corals (surveyed by Kayanne et al. 2002) distributed near the shallow channels appeared to have been affected less than those in backreef areas in Shiraho Reef by high-temperature-induced mass coral bleaching in 2007 (Dadhich et al. 2012), suggesting the importance of understanding water circulation in a reef.

In conclusion, water circulation and residence time in fringing reefs, which affect coral distribution, are highly affected by reef topography and surrounding oceanographic and climatological environments. In addition to large-scale observation and analyses (e.g., Shimazaki et al. 2006; Yamano et al., this volume), reef-scale observation and analyses, including detailed mapping (e.g., Suzuki et al. 2012) and hydrodynamic modeling (e.g., Tamura et al. 2007; Dadhich et al. 2012), would be required to set up effective management and conservation plans at a particular reef.

Acknowledgments We are deeply indebted to the Fuyo Ocean Development and Engineering Company, the Shiraho Fishery Association, and E. Shimabukuro and M. Taira for logistics help and maintaining the instruments. Thanks are extended to Y. Tanaka, Y. Umezawa, A. Watanabe, A. Tanaka, Y. Tachibana, and M. Kohda for helping us during the field survey. A. Muramoto helped us to arrange the data. We are indebted to the Physical Oceanography Distributed Active Archive Center of NASA Jet Production Laboratory for providing the TOPEX/POSEIDON altimeter data. Some current meters were provided by M. Yamamuro. This study was supported by the Core Research for Evolutional Science and Technology (CREST), the Japan Science and Technology Corporation. This is a contribution to the S-9 Project supported by the Environment Research and Technology Development Fund, Ministry of the Environment, Japan.

References

Andréfouët S, Pagés J, Tartinville B (2001) Water renewal time for classification of atoll lagoons in the Tuamotu Archipelago (French Polynesia). Coral Reefs 20:399–408

Andrews JC, Pickard GL (1990) The physical oceanography of coral-reef systems. In: Dubinsky Z (ed) Ecosystems of the world, vol 25. Coral reefs. Elsevier, Amsterdam, pp 11–48

Dadhich AP, Nadaoka K, Yamamoto T, Kayanne H (2012) Detecting coral bleaching using high-resolution satellite data analysis and 2-dimensional thermal model simulation in the Ishigaki fringing reef, Japan. Coral Reefs 31:425–439

Frith CA (1981) Circulation in a platform reef lagoon, One Tree Reef, southern Great Barrier Reef. Proc 4th Int Coral Reef Symp 1:347–354

Frith CA, Mason LB (1986) Modelling wind driven circulation One Tree Reef, Southern Great Barrier Reef. Coral Reefs 4:201–211

Fu L-L, Christensen EJ, Yamarone CA Jr, Lefebvre M, Menard Y, Dorrer M, Escudier P (1994) TOPEX/POSEIDON mission overview. J Geophys Res 99:24369–24381

Harii S, Kayanne H (2002) Larval settlement of corals in flowing water using a racetrack flume. Mar Tech Soc J 36:76–79

Harii S, Kayanne H (2003) Larval dispersal, recruitment, and adult distribution of the brooding stony octocoral *Heliopora coerulea* on Ishigaki Island, southwest Japan. Coral Reefs 22:188–196

Hasegawa H, Yamano H (2004) Ishigaki Island. In: Ministry of the Environment and Japanese Coral Reef Society (eds) Coral reefs of Japan. Ministry of the Environment, Tokyo, pp 212–218

Hearn CJ (1999) Wave-breaking hydrodynamics within coral reef systems and the effect of changing relative sea level. J Geophys Res 104:30007–30020

Iryu Y, Nakamori T, Matsuda S, Abe O (1995) Distribution of marine benthic organisms and its geological significance in the modern reef complex of the Ryukyu Islands. Sediment Geol 99:243–258

Kayanne H, Suzuki A, Saito H (1995) Diurnal changes in the partial pressure of carbon dioxide in coral reef water. Science 269:214–216

Kayanne H, Tanaka Y, Harii S, Yamano H, Haltuch MA, Yamamuro M, Ide Y, Miyamoto N, Akimoto F, Ito N, Nakane T (2000) Distribution and biomass of primary producers in Shiraho Reef, Japan integrated in GIS. 9th Int Coral Reef Symp Abstract 235

Kayanne H, Harii S, Ide Y, Akimoto F (2002) Recovery of coral reef populations after the 1998 bleaching on Shiraho Reef in the southern Ryukyus, NW Pacific. Mar Ecol Prog Ser 239:93–103

Kench PS, Parnell KE, Brander RW (2009) Monsoonally influenced circulation around coral reef islands and seasonal dynamics of reef island shorelines. Mar Geol 266:91–108

Kraines SB, Yanagi T, Isobe M, Komiyama H (1998) Wind-wave driven circulation on the coral reef at Bora Bay, Miyako Island. Coral Reefs 17:133–143

Ludington CA (1979) Tidal modifications and associated circulation in a platform reef lagoon. Aust J Mar Freshw Res 30:425–430

Lugo-Fernandez A, Roberts HH, Suhayda JN (1998a) Wave transformations across a Caribbean fringing-barrier coral reef. Cont Shelf Res 18:1099–1124

Lugo-Fernandez A, Roberts HH, Wiseman WJJ (1998b) Tide effects on wave attenuation and wave set-up on a Caribbean coral reef. Estuar Coast Shelf Sci 47:385–393

Nadaoka K, Nihei Y, Kumano R, Yokobori T, Omija T, Wakaki K (2001a) A field observation on hydrodynamic and thermal environments of a fringing reef at Ishigaki Island under typhoon and normal atmospheric conditions. Coral Reefs 20:387–398

Nadaoka K, Nihei Y, Wakaki K, Kumano R, Kakuma S, Moromizato S, Omija T, Iwao K, Shimoike K, Taniguchi H, Nakano Y, Ikema T (2001b) Regional variation of water temperature around Okinawa coasts and its relationship to offshore thermal environments and coral bleaching. Coral Reefs 22:373–384

Nakamori T, Suzuki A, Iryu Y (1992) Water circulation and carbon flux on Shiraho coral reef of the Ryukyu Islands, Japan. Cont Shelf Res 12:951–970

Nakamura T, van Woesik R (2001) Water-flow rates and passive diffusion partially explain differential survival of corals during the 1998 bleaching event. Mar Ecol Prog Ser 212:301–304

Nakamura T, Yamasaki H, van Woesik R (2003) Water flow facilitates recovery from bleaching in the coral *Stylophora pistillata*. Mar Ecol Prog Ser 256:287–291

Pagés J, Andréfouët S (2001) A reconnaissance approach for hydrology of atoll lagoons. Coral Reefs 20:409–414

Parnell KE (1988) The hydrodynamics of fringing reef bays in the Great Barrier Reef Marine Park. Proc 6th Int Coral Reef Symp 2:503–508

Prager EJ (1991) Numerical simulation of circulation in a Caribbean-type backreef lagoon: a preliminary study. Coral Reefs 10:177–182

Roberts HH, Murray SP, Suhayda JN (1975) Physical processes in a fringing reef system. J Mar Res 33:233–260

Shimazaki H, Yamano H, Yokoki H, Yamaguchi T, Chikamori M, Tamura M, Kayanne H (2006) Global mapping of factors controlling reef-island formation and maintenance. Proc 10th Int Coral Reef Symp, Okinawa, pp 1577–1584

Smith SV, Buddemeier RW (1992) Global change and coral reef ecosystems. Annu Rev Ecol Syst 23:89–118

Suzuki R, Hasegawa H, Maekawa S, Shibata T, Sagawa T, Goto Y, Ichikawa K (2012) Making maps for the conservation of shallow lagoon in Shiraho Coral Reef of Ryukyu Islands, Japan. MAP. J Jpn Cartogr Assoc 50:1–16

Symonds G, Black KP, Young IR (1992) Wave-driven flow over shallow reefs. J Geophys Res 100:2639–2648

Takino T, Watanabe A, Motooka S, Nadaoka K, Yasuda N, Taira M (2010) Discovery of a large population of *Heliopora coerulea* at Akaishi Reef, Ishigaki Island, southwest Japan. Galaxea J Coral Reef Stud 12:85–86

Tamura H, Nadaoka K, Paringit EC (2007) Hydrodynamic characteristics of a fringing coral reef on the east coast of Ishigaki Island, southwest Japan. Coral Reefs 26:17–34

Tartinville B, Rancher J (2000) Wave-induced flow over Mururoa Atoll reef. J Coast Res 16:776–781

Wolanski E, Delesalle B, Dufour V, Aubanel A (1993) Modeling the fate of pollutants in the Tiahura Lagoon, Moorea, French Polynesia. 11th Aust Conf Coast Ocean Eng, pp 583–589.

Yamano H, Kayanne H, Yonekura N, Nakamura H, Kudo K (1998) Water circulation in a fringing reef located in a monsoon area: Kabira Reef, Ishigaki Island, Southwest Japan. Coral Reefs 17:89–99

Yamano H, Kayanne H, Yonekura N, Kudo K (2000) 21-year changes of backreef coral distribution: causes and significance. J Coast Res 16:99–110

Yamano H, Hongo C, Sugihara K, Yara Y, Nakao Y, Fujii M (this volume) Current status of the national coral database in Japan: dataset development, applications, and future directions. In: Nakano, S, Yahara T, Nakashizuka T (eds) The biodiversity observation network in the Asia-Pacific Region: integrative observations and assessments of Asian biodiversity. Springer, Tokyo

Chapter 15
An Integrated Indicator of Biodiversity in Agricultural Ponds: Definition and Validation

Shinya Ishida, Taku Kadoya, and Noriko Takamura

Abstract One of the promising approaches to monitoring biodiversity is assessing the status of pressures driving the biodiversity state. To achieve this, we need to identify the principal pressures that cause simultaneous biodiversity loss across taxonomic groups and clarify how multiple pressures act synergistically or at least simultaneously to decrease biodiversity in the focal ecosystem. Here, we introduce a framework for an integrated biodiversity indicator that takes into consideration the estimated relative importance of multiple pressures. The indicator is defined as a function of the pressure(s) and is parameterized to explain a number of individual states of biodiversity. We showed that the framework can be successfully applied to a real ecosystem, a series of 64 agricultural ponds. We focused on macrophytes, Odonata, and benthic macroinvertebrates as the individual states of biodiversity of the ponds and on three types of pressure: eutrophication, habitat destruction, and invasive alien species. We then evaluated the relationships among pressures with direct effects and the individual states of biodiversity and used a hierarchical Bayesian approach to calculate the integrated biodiversity indicator. We found that the integrated indicator could explain the behaviors of several individual states of biodiversity. To demonstrate the applicability of our approach, we adapted the integrated indicator to another dataset of 35 different agricultural ponds in which the integrated indicator was calculated using the relative importance of multiple pressures estimated from the previous 64-pond study. We found that we could successfully extrapolate the integrated indicator to the 35 agricultural ponds. These results demonstrate the advantages of the framework in providing a more practical method for assessing biodiversity in freshwater lentic environments and in quantifying the relative importance of the major threats to biodiversity to prioritize strategies in conservation planning and policy making.

S. Ishida • T. Kadoya (✉) • N. Takamura
Center for Environmental Biology and Ecosystem Studies,
National Institute for Environmental Studies, Ibaraki, Japan
e-mail: kadoya@nies.go.jp

S.-i. Nakano et al. (eds.), *Integrative Observations and Assessments*, Ecological Research Monographs, DOI 10.1007/978-4-431-54783-9_15,

Keywords Agricultural ecosystem • Aquatic macrophyte • Benthic macroinvertebrate • Biodiversity surrogate • Multi-species • Odonata

Introduction

Monitoring temporal changes of biodiversity on a large spatial scale is essential for planning effective biodiversity conservation strategies. Biodiversity is, however, a complex, multiscale, multifaceted entity (Noss 1990) that includes the richness and diversity of numerous species across multiple taxa. Consequently, for a meaningful assessment, more comprehensive consideration is necessary, but the amount of labor, funds, and time needed for field surveys on a large spatial scale is often prohibitive. Thus, the development of an integrated indicator that could explain the state of biodiversity is needed for large-scale monitoring (Duelli and Obrist 2003; Scholes and Biggs 2005).

One promising approach for monitoring biodiversity is by assessing the status of pressures driving the biodiversity state. However, different taxonomic groups or guilds can behave differently in response to changes, even with the same driver or pressure (Perfecto et al. 2003; Schulze et al. 2004; Kadoya et al. 2009; Heino 2010). In addition, in a real system multiple pressures causing ecological change can interact synergistically to accelerate biodiversity loss (Didham et al. 2007; Brook et al. 2008; Darling and Cote 2008).

In shallow lakes and ponds, the condition of the ecosystem can change from a clear-water state to a phytoplankton-dominated one (Scheffer and Carpenter 2003). Studies have shown that the most influential factor in this type of regime shift is eutrophication from excessive nitrogen and phosphorus as a result of agricultural fertilizers, sewage effluent, and urban stormwater runoff (Scheffer and Carpenter 2003). After a regime shift, the abundance and richness of most aquatic taxonomic groups decrease simultaneously, mainly as a result of less light penetration through the water column (because of phytoplankton blooms), which inhibits the growth of submerged plants that provide a substrate for a rich and varied aquatic life (Scheffer 2004). For shallow lakes and ponds, a difference in taxonomic group response to the same pressure, therefore, may not be significant. In closed freshwater ecosystems, invasive alien species of carnivorous fishes, crayfish, and aquatic plants are a major threat to biodiversity (Revenga et al. 2005; Dudgeon et al. 2006). Carnivorous invasive species can affect biodiversity directly through predation and indirectly by reducing the amount of larger zooplankton and aquatic macrophytes. Because abundant large zooplankton and aquatic macrophytes tend to suppress algae through grazing and competition, respectively, a decrease in their populations resulting from invasive species may facilitate a regime shift from a macrophyte-dominated to an algae-dominated state synergistically with eutrophication. Therefore, to assess the biodiversity state of such closed freshwater ecosystems with respect to environmental pressures, quantifying the synergic, or at least simultaneous, effects of multiple pressures is essential.

In this chapter, we introduce a framework for an integrated biodiversity indicator that takes into consideration the estimated relative importance of multiple environmental pressures for shallow lakes and ponds (Kadoya et al. 2011). The framework is based primarily on an assessment of agricultural ponds, which support diverse populations of aquatic animal and plant species (Knutson et al. 2004; Oertli et al. 2005) but have recently suffered from severe biodiversity degradation from the aforementioned pressures.

First, we briefly summarize the definition of the integrated biodiversity indicator that was previously developed (Kadoya et al. 2011). Second, we then show how the framework can be applied to ecosystems, and finally we discuss the advantages of the framework and future directions for further applications.

Definition of the Integrated Indicator

The conceptual framework for the integrated indicator is illustrated in Fig. 15.1. Each individual biodiversity state was modeled as a simple linear function of the integrated biodiversity indicator as follows:

$$Link(Ind_{ij}) = a_{1i}Int_j + a_{2i}, \quad (15.1)$$

where Ind_{ij} is the observed ith individual state of biodiversity at site j, *Link* is a link function for the indicator, and a_{1i} and a_{2i} are the regression coefficient and

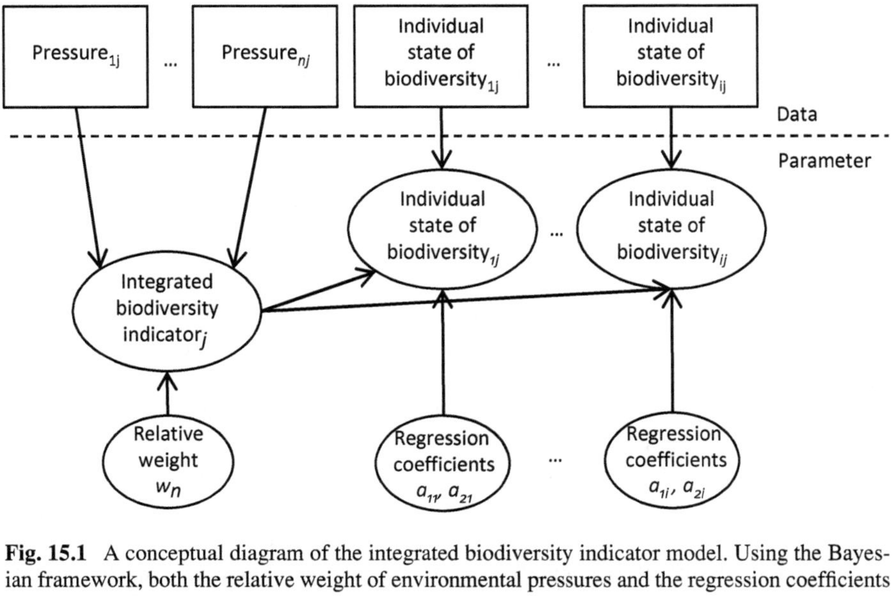

Fig. 15.1 A conceptual diagram of the integrated biodiversity indicator model. Using the Bayesian framework, both the relative weight of environmental pressures and the regression coefficients determining the dependency of individual states of biodiversity on the integrated indicator are simultaneously estimated based on the observed data. (Modified from Kadoya et al. 2011)

intercept, respectively, which determine the dependency of the *i*th individual biodiversity state on the integrated biodiversity indicator *Int* and are to be estimated. We assumed identical functions and log-link functions for continuous data (e.g., functional diversity of aquatic plants, larval odonates, and benthic macroinvertebrates) and count data (e.g., species richness or the number of endangered species of the focal taxa), respectively. The continuous and count data were assumed to follow normal and Poisson distributions, respectively.

The integrated biodiversity indicator *Int* was modeled as a weighted mean of environmental pressures:

$$Int_j = I_1 w_1 Press_{j1} + I_2 w_2 Press_{j2} + , \ldots , + I_n w_n Press_{jn}, \qquad (15.2)$$
$$\sum_n w_n = 1,$$

where I_n is the indicator function of the observed *n*th pressure $Press_n$, which determines the direction of effects of the pressure on the integrated indicator (i.e., $I_n = 1$ for pressures with a positive effect and $I_n = -1$ for pressures with a negative effect). The direction of the pressure effect was set a priori on the basis of existing knowledge. w_n is a weight for the *n*th pressure, which determines the relative contribution of the pressure to the integrated biodiversity indicator and is to be estimated. Thus, once we estimate w_n based on the data of individual biodiversity states (Ind_s) and pressures ($Press_n$), we can calculate a value of the integrated indicator (*Int*) for each pond from Eq. (15.2).

The model can be fitted to observed pressure factors and individual biodiversity states at each pond by using the Bayesian simulation-based method (Markov chain Monte Carlo, MCMC) in WinBUGS 1.4 (Lunn et al. 2000) in combination with R release 2.9.2 (package R2 WinBUGS). Readers who are interested in the detailed procedure of the parameter estimation can consult Kadoya et al. (2011).

Application and Validation of the Framework to Ecosystems

Kadoya et al. (2011) applied the afore-developed framework to a series of 64 agricultural ponds, selecting aquatic macrophytes, Odonata, and benthic macroinvertebrates as the focal taxa, all of which are considered to be representative taxa in freshwater ecosystems (Engelhardt and Ritchie 2001; Heino 2002; Heino et al. 2003; Declerck et al. 2005; Ilmonen and Paasivirta 2005; Kadoya et al. 2009; Simaika and Samways 2009). They calculated eight individual states of biodiversity based on comprehensive surveys of flora and fauna at all 64 ponds: total richness of (1) aquatic plants, (2) adult Odonata, and (3) benthic macroinvertebrates; number of endangered species of (4) aquatic plants and (5) Odonata; and functional diversity of (6) aquatic plants, (7) larval Odonata, and (8) benthic macroinvertebrates. They also focused on three environmental pressures:

Table 15.1 Estimated posterior median and 95 % confidence interval (CI) of parameters based on 64 ponds

Parameter		Median	95 % CI
Relative weight			
Chlorophyll *a*	w_1	0.440	(0.327, 0.562)
Proportion of bank protection	w_2	0.280	(0.183, 0.383)
Bluegill	w_3	0.279	(0.193, 0.362)
Regression coefficient			
Aquatic plant richness	a_{11}	0.701	(0.500, 0.928)
Adult odonate richness	a_{12}	0.373	(0.274, 0.477)
Benthic richness	a_{13}	0.281	(0.182, 0.384)
Endangered aquatic plant richness	a_{14}	1.550	(0.896, 2.364)
Endangered odonate richness	a_{15}	2.600	(1.340, 4.294)
Aquatic plant functional diversity	a_{16}	0.081	(0.031, 0.131)
Larval odonate functional diversity	a_{17}	0.122	(0.076, 0.168)
Benthic functional diversity	a_{18}	0.020	(–0.030, 0.072)
Intercept			
Aquatic plant richness	a_{21}	1.514	(1.389, 1.636)
Adult odonate richness	a_{22}	2.677	(2.609, 2.743)
Benthic richness	a_{23}	2.640	(2.573, 2.708)
Endangered aquatic plant richness	a_{24}	–0.566	(–1.054, –0.187)
Endangered odonate richness	a_{25}	–1.782	(–2.778, –1.037)
Aquatic plant functional diversity	a_{26}	0.221	(0.186, 0.256)
Larval odonate functional diversity	a_{27}	0.276	(0.244, 0.310)
Benthic functional diversity	a_{28}	0.550	(0.513, 0.588)

See text for a detailed description of each parameter
Source: Modified from Kadoya et al. (2011)

eutrophication, habitat destruction, and invasive alien species, each of which includes several candidate factors describing the pressure state. They then evaluated the relationships among direct pressures and the eight individual states of biodiversity and used a hierarchical Bayesian approach to calculate the integrated biodiversity indicator (Table 15.1). Among the pressure factors examined, chlorophyll *a* concentration, proportion of bank protection, and bluegill density were the most influential in each of three pressure categories. They also demonstrated that the integrated indicator explained the behaviors of the eight individual states of biodiversity (Fig. 15.2).

We describe here our study to test the versatility of the framework. Specifically, we validated the ability of the integrated indicator, which was calculated from previously derived parameters determined by Kadoya et al. (2011) (Table 15.1), to explain the behavior of individual biodiversity states collected from a different series of 35 ponds.

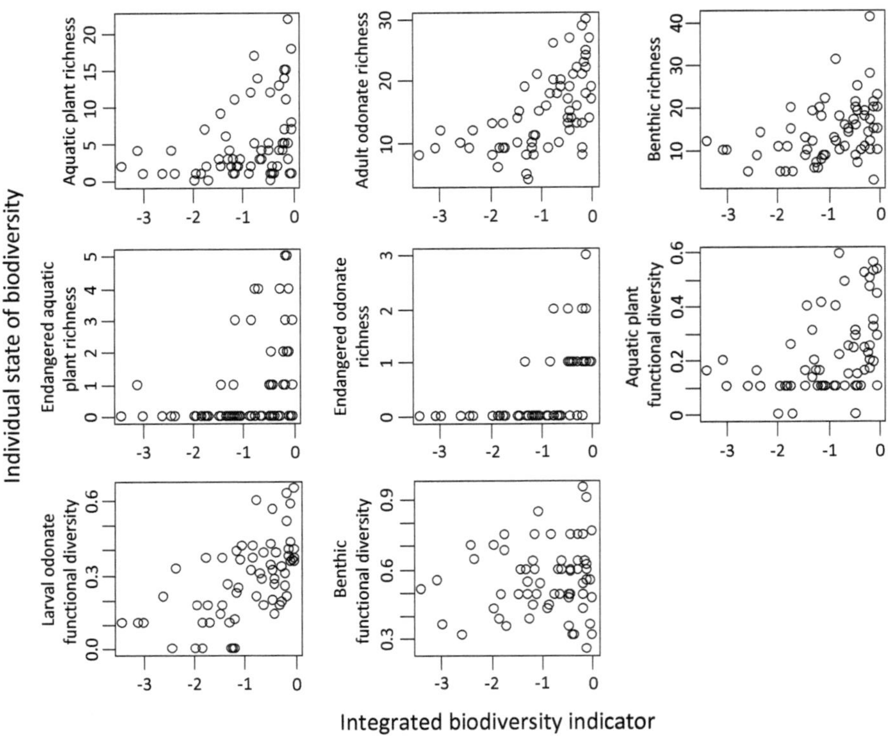

Fig. 15.2 Relationships between the estimated integrated biodiversity indicator and observed individual states of biodiversity of the 64 agricultural ponds surveyed in 2006–2008. See Table 15.1 for the regression coefficients. (Modified from Kadoya et al. 2011)

Dataset from 35 Ponds

Study Area

The survey was conducted over an area of approximately 700 km^2 in southwestern Hyogo Prefecture, Japan (34°46′N, 134°56′E), where the predominant land uses are paddy fields (36.7 %), broad-leaved forests (35.3 %), and urban areas (15.3 %). The study area has a warm temperate climate with a mean annual temperature of 14.4 °C (minimum, 3.5 °C in January; maximum, 26.4 °C in August) and mean annual precipitation of 1,198.3 mm (data provided by the Miki Climatological Observatory located within the study area at 145 m a.s.l.). Before 1910, many irrigation ponds had been created in Hyogo Prefecture (8,395 km^2) to provide water for paddy cultivation, and more than 55,000 agricultural ponds, corresponding to about 20 % of all ponds in Japan, were present by the 1950s. However, by 1997 more than 11,000 ponds had been lost, mainly as a result of urban or residential development (Takamura 2007). Even where ponds have not been destroyed, their biodiversity has drastically decreased during recent decades.

The 35 ponds were chosen on the basis of dominant vegetation (emergent, floating-leaved, and no apparent vegetative cover) and dominant surrounding land use (predominantly urban, rural, or forest), with two to six replicates for each combination. There was no duplication with the former 64 ponds. The 35 ponds had surface areas ranging from 1,018 to 29,204 m^2 (mean ± SD, 9,636 ± 6,789 m^2) and maximum depths ranging from 0.6 to 3.9 m (1.7 ± 0.8 m). The elevation of the ponds averaged 50 ± 24.8 m a.s.l.

Census of Aquatic Organisms

Aquatic Macrophytes

Native aquatic macrophytes growing throughout each pond were extensively surveyed at the approximate peak of macrophyte biomass (August–September) in 2001 by using an inflatable boat, wading, and using a rake. All native species of emergent, floating-leaved, free-floating, and submerged aquatic macrophytes encountered were recorded. Nomenclature and classification criteria followed those of Kadono (1994). *Chara* spp. and *Nitella* spp. were identified to the genus level, and each genus was treated as one species in the calculation of individual biodiversity state.

Adult and Larval Odonates

The number of adult odonate species was censused by walking the shoreline of each pond between 0900 and 1500 on a clear day with little cloud cover. Odonate species were caught and identified when these were not distinguishable by sight. When walking along the shoreline was not feasible, the route census was conducted by using an inflatable boat. The census was conducted six times: 13–20 May 2001, 17 June–5 July 2001, 21–29 July 2001, 8–24 September 2001, 6–27 October 2001, and 27 April–to 6 May 2002. We used the maximum number of odonate species from the six surveys so as not to count the same individual more than once.

The larvae survey was conducted twice, from 16 December 2001 to 13 April 2002 for collecting overwintering larvae and from 16 May to 16 June 2002 for collecting larvae that hatch in the spring. We set a quadrat (0.9 × 0.9 m, 0.45 m height) for each 60 m of shoreline length in the littoral zone and swept a D-frame dipnet (2-mm mesh) through the entire surficial sediments, including decayed aquatic plants, within the quadrat. We sorted larval odonates >5 mm in length and identified them to species level under a binocular microscope. We recorded all individuals from each survey at each pond.

Benthic Macroinvertebrates

The benthic macroinvertebrate surveys were conducted once for each pond during 15–21 May 2002. We collected three sediment samples in the center of each pond using an Ekman–Birge-type bottom sampler (mouth opening of 150 × 150 mm) and from three to nine samples near the shore, according to the pond's size, using a D-frame dipnet (0.2-mm mesh). The sediment from the bottom sampler was washed using a 0.2-mm-mesh sieve to eliminate the finer particles, and the samples were preserved in 10 % formalin and indentified to the genus or species level under a binocular microscope.

Calculation of Individual States of Biodiversity

Using the collected data, we calculated eight individual states: total species (taxonomic) richness of (1) aquatic macrophytes, (2) adult Odonata, and (3) benthic macroinvertebrates; number of endangered species of (4) aquatic macrophytes and (5) Odonata; and functional diversity of (6) aquatic macrophytes, (7) larval Odonata, and (8) benthic macroinvertebrates (Table 15.1).

Endangered status was based on both prefectural (Hyogo Prefecture 2003) and national (Ministry of the Environment of Japan 2007) listings for aquatic macrophytes and on the national Red List (Ministry of the Environment of Japan 2007) for odonates.

We used Petchey and Gaston (2006) functional diversity indicator. On the basis of published information (Kadono 1994; Ohtaki and Ishido 2007; Yashiro 2007), we used the following as functional traits for aquatic macrophytes: growth form (five types: emergent, floating leaved, free floating, submerged, and submerged-free floating), life cycle (two types: annual and perennial), and formation of tubers, formation of rhizomes, and evergreenness. For the larval odonates, we used body size as the functional trait [six metrics (based on Sugimura et al. 1999): body length, head width, head length, antenna length, wing bud length, and hind leg length] because body size is considered to be the principal trait determining the trophic level of the species (Wissinger 1992). For the benthic macroinvertebrates, we used feeding group as the functional trait. Based on the published information (chironomids, Monakov 2003; aquatic insects, Merritt et al. 2008; other aquatic macroinvertebrates, Thorp and Covich 2010), we considered the following six feeding groups as functional traits: algae feeders, zooplankton feeders, plant feeders, predators, scavengers, and detritivores. These categories were not mutually exclusive, and more than one category was assigned to a species where appropriate.

Assessment of Direct Pressures

We assessed chlorophyll *a* concentration, proportion of bank protection, and bluegill density as described next. These were determined to be the most influential pressures on the biodiversity of agricultural ponds (Kadoya et al. 2011).

Chlorophyll *a* Concentration

We measured the concentration of chlorophyll *a* during the summer (20–26 July 2001) at the approximate center of each pond. Phytoplankton for determining chlorophyll *a* concentrations was collected onto GF/F filters (pore size, 0.7 μm) and measured spectrophotometrically (U-2800A; Hitachi, Tokyo) after extraction in methanol (99 %; Marker et al. 1980).

Pond Bank Protection

The proportion of pond bank covered by concrete was obtained from georeferenced 1:2,500 geographic maps on ArcGIS 9.1 (ESRI, Redlands, CA, USA).

Bluegill Density

At each pond, we surveyed bluegill (*Lepomis macrochirus*), a predatory invasive alien species and one of the 100 worst invasive alien species in Japan (Ecological Society of Japan 2002).

A Y-shaped fixed net (6.0-m sleeve diameter and 0.7-m mouth opening; 4-mm mesh) and five box-net traps (0.4 × 0.25 × 0.4 m with 2-mm mesh) were set for one night in the littoral zone of each pond during 2–6 July 2001. We counted all of the bluegills caught in the net and traps.

Statistical Analysis

Using the data from the 35 ponds and the model parameters (i.e., relative weights of pressures) previously estimated from the dataset of the 64 ponds (Table 15.1), we calculated the integrated biodiversity indicator for the 35 ponds. Using generalized liner models (GLMs), we then tested whether the integrated indicator explained the behavior of the individual biodiversity states of the 35 ponds.

Table 15.2 Average, standard deviation (SD), minimum (Min), and maximum (Max) of individual states of biodiversity and the three principal pressures among 35 agricultural ponds surveyed in 2001–2002

Data	Mean	SD	Min	Max
Individual states of biodiversity				
Aquatic plant richness	3.23	3.05	0	12
Adult odonate richness	15.77	6.77	6	29
Benthic richness	8.71	7.68	0	31
Endangered aquatic plant richness	0.69	1.21	0	6
Endangered odonate richness	0.80	0.96	0	3
Aquatic plant functional diversity	0.14	0.15	0	0.49
Larval odonate functional diversity	0.24	0.24	0	0.62
Benthic functional diversity	0.48	0.28	0	1
Pressures				
Chlorophyll *a* (μg/l)	140.4	205.2	1.6	941.7
Proportion of bank protection	0.81	0.3	0	1
Bluegill (CPUE)	30.34	91.08	0	522

CPUE, catch per unit effort

The best models were selected based on Akaike's information criterion (AIC; Burnham and Anderson 2002); the lowest AIC model is the most relevant (i.e., has the higher explanatory power). A log-link function with Poisson error distribution was applied to fit the model to species richness, whereas the identical link function with the Gaussian error distribution was applied to fit the model to functional diversity.

Results of the Validation

The biodiversity states and the pressure status varied substantially among the 35 ponds (Table 15.2). The relationships between the integrated biodiversity indicator and the individual states of biodiversity were graphically examined before the GLM analyses and showed that the biodiversity state can be explained by the single integrated biodiversity indicator, except for the one pond that had the lowest value of the integrated indicator (Fig. 15.3). We excluded the single pond from subsequent analyses because the pond had apparently different conditions both for pressures and for biodiversity.

Model comparisons based on AICs showed that the models that included the integrated indicator as an explanatory variable were always superior to null models (Table 15.3). In addition, the integrated indicator was always significantly or marginally positively related to each of the eight individual biodiversity states (Table 15.4). These results support the applicability of the framework across different series of pond ecosystems.

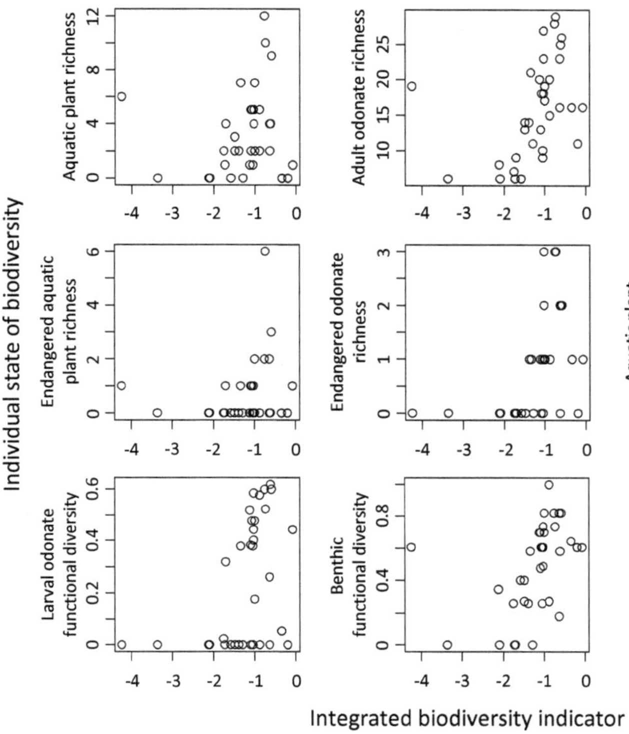

Fig. 15.3 Relationships between the estimated integrated biodiversity indicator and observed individual states of biodiversity of the 35 ponds surveyed in 2001–2002. See Table 15.4 for regression coefficients

Applicability of the Integrated Indicator

The integrated indicator calculated using the relative importance of multiple pressures estimated in the 64 ponds was successfully extrapolated to the different series of 35 ponds, suggesting that our framework is applicable for biodiversity assessment at larger spatial scales. The success of the integrated indicator can be partly ascribed to the commonality of principal pressures (i.e., chlorophyll *a* concentration, proportion of bank protection, and bluegill density) among the agricultural ponds.

Each environmental pressure in our study suppresses biodiversity in a different manner. Eutrophication, as indicated by the chlorophyll *a* concentration, is one of the main triggers inducing a regime shift from a macrophyte-dominated to a phytoplankton-dominated state (Scheffer and Carpenter 2003). Bank protection artificially stabilizes the margins of ponds, and as a result hydrologic ecotones and vegetation zones within and around the ponds are lost; this loss can have significant negative impacts on emergent and floating-leaved macrophytes and,

Table 15.3 Model comparisons between the model with the integrated indicator as the explanatory variable and the null model based on Akaike's information criterion (AIC)

Model	AIC	ΔAIC
Aquatic plant richness		
Integrated indicator model	175.78	0
Null model	183.91	8.13
Adult odonate richness		
Integrated indicator model	220.25	0
Null model	256.16	35.91
Benthic richness		
Integrated indicator model	283.20	0
Null model	344.96	61.76
Endangered aquatic plant richness		
Integrated indicator model	81.59	0
Null model	86.88	5.29
Endangered odonate richness		
Integrated indicator model	77.87	0
Null model	85.17	7.3
Aquatic plant functional diversity		
Integrated indicator model	–29.12	0
Null model	–27.74	1.38
Larval odonate functional diversity		
Integrated indicator model	–0.87	0
Null model	3.64	4.51
Benthic functional diversity		
Integrated indicator model	–2.07	0
Null model	13.49	15.56

The individual state of biodiversity for each of the 34 agricultural ponds was included in a generalized linear model as the response variable (a pond with the lowest value of the integrated indicator was eliminated)

incidentally, on larval odonates and benthic invertebrates that utilize macrophytes as shelter or oviposition material. In addition, the concrete materials used for bank protection may alkalinize the pond water, enhancing water blooms and adversely affecting the growth of some submerged macrophytes. Bluegill negatively affected macroinvertebrate fauna, including larval odonates and benthic macroinvertebrates, by direct predation (Ecological Society of Japan 2002). Moreover, the juvenile stage of the species predates large zooplankton; this predation could facilitate a macrophyte-dominated to algae-dominated regime shift by reducing the grazing pressure by large zooplankton on phytoplankton (Scheffer 2004).

Our analysis thus demonstrated that different types of environmental pressure in the agricultural ponds affect the individual states of biodiversity simultaneously (and synergistically, in part) in the same direction. Our framework, therefore, appears to be sufficiently versatile to assess the biodiversity state of freshwater

Table 15.4 Parameter estimates, standard error (SE), and *p* value of generalized linear models with the individual biodiversity state of each of the 34 agricultural ponds as the response variable and the integrated indicator as the explanatory variable

Response variable	Parameter	Estimate	SE	*p* value
Aquatic plant richness	Intercept	1.743	0.205	<0.001
	Integrated indicator	0.563	0.186	0.002
Adult odonate richness	Intercept	3.268	0.092	<0.001
	Integrated indicator	0.481	0.082	<0.001
Benthic richness	Intercept	3.070	0.120	<0.001
	Integrated indicator	0.888	0.118	<0.001
Endangered aquatic plant richness	Intercept	0.700	0.423	0.098
	Integrated indicator	1.109	0.438	0.011
Endangered odonate richness	Intercept	0.921	0.382	0.016
	Integrated indicator	1.139	0.398	0.004
Aquatic plant functional diversity	Intercept	0.227	0.054	<0.001
	Integrated indicator	0.076	0.042	0.077
Larval odonate functional diversity	Intercept	0.431	0.082	<0.001
	Integrated indicator	0.163	0.063	0.014
Benthic functional diversity	Intercept	0.810	0.081	<0.001
	Integrated indicator	0.287	0.062	<0.001

A pond with the lowest value of the integrated indicator was eliminated

lentic ecosystems, such as lakes and ponds, that have similar ecosystem properties as those in our case study.

The successful validation of our integrated framework may be dependent on the similarity among ponds of the same region in climate condition, land-use history, and management, which potentially affect the biodiversity of ponds. Examination of the framework using datasets from different regions, with different conditions, is needed for further validation.

We excluded 1 of the 35 ponds from the analysis as it had an extremely low value of the integrated indicator. This pond had moderate values of some individual biodiversity states, despite the worst environmental conditions among the 35 ponds; the value of chlorophyll *a* was the second highest (662.1 μg/l), the proportion of bank protection was the highest (100 %), and the population of bluegill was the highest (522 CPUE). One explanation for this anomaly is that other environmental factors that were not considered in the integrated indicator raised the level of biodiversity in the pond. The pond is fringed by a medium-size broadleaf forest (635 m^2, within 10 m from the pond edge). This buffer may have positive effects on adult odonates, as it provides resting sites, and on benthic invertebrates, as a source of litter that they feed on to the pond. In addition, a field survey showed that the bank protection of the pond was aging and had partly crumbled, which extended the habitat for some aquatic macrophytes. An alternative explanation is that the pond was in a transition period for biodiversity and we observed a time-delayed response of biodiversity to worsening environmental conditions. In support of this explanation, the pond had a number of adult odonate species but no larvae; when bluegill invade a pond, larval odonates likely decrease more rapidly than adults.

Our framework assumed that the relationship between individual biodiversity states and environmental pressures was at equilibrium; further studies are needed to test the robustness of the framework to the violation of this assumption.

Conclusions and Future Directions

In this study, we demonstrated that the framework to derive the integrated indicator of Kadoya et al. (2011) is sufficiently applicable to a pond ecosystem. One of the major advantages of the framework is that it enables us to use an integrated indicator instead of assessing a number of biodiversity states, each of which provides different information owing to the multifaceted nature of biodiversity. Another advantage is that, because the integrated indicator is defined as a function of environmental pressures, biodiversity states can be monitored by assessing the change of pressures. Observing the organisms directly is not necessary, which should make biodiversity monitoring at larger spatial scales more feasible than conventional organism-based indices (Mazaris et al. 2008; Menetrey et al. 2011) because some pressures with direct effects (e.g., chlorophyll *a*) are easier to assess using remote sensing techniques. A third advantage of our framework is that we can clarify how multiple pressures affect biodiversity by estimating the relative importance of each, which would be valuable for identifying the major threats to biodiversity and prioritizing strategies in conservation planning and policy making.

There is increasing concern that the multiple pressures of ecological change will interact synergistically to accelerate biodiversity loss (Brook et al. 2008; Laurance and Useche 2009). Thus, a fourth advantage of the framework is that synergistic effects on biodiversity can be easily incorporated, at least in part, by adding the interacting terms in Eq. (15.2) where appropriate.

Several approaches have been developed for evaluating biodiversity. One of the commonly used causal frameworks for describing the interactions between society and the environment is the Driver, Pressure, State Impact, and Response Model (European Environment Agency 2007) in which relationships between drivers and biodiversity are explicitly evaluated. The framework proposed in the present study could directly link the data to such structural indicator schemes with statistically rigorous criteria, making our framework a potentially powerful tool.

Considering the wide variety of pond ecosystem types, such as semi-natural and artificial ponds or permanent, semi-permanent, and temporary ponds, future research should test whether the framework is applicable to different pond types. In addition, large-scale monitoring of the biodiversity in agricultural ponds is necessary. For instance, we have begun to assess the status of environmental pressures at a larger spatial scale, using remote sensing techniques or existing national land-use information. Once the appropriateness of extrapolations to other areas has been ascertained, we could quantitatively and efficiently assess the states of biodiversity in agricultural ponds at regional and national scales, which would significantly contribute to decisions for effective conservation planning.

Acknowledgments We thank T. Aoki, M. Nakagawa [National Institute for Environmental Studies, Tsukuba (NIES)], and S. Higuchi (Kobe University) for field assistance and water analysis. We are grateful to H. Ito (NIES), K. Satake (NIES), and R. Ueno (NIES) for their advice on the analyses of aquatic invertebrates. We also thank M. Akasaka (Tokyo University of Agriculture and Technology), S.S. Matsuzaki (NIES), and J. Nishihiro (University of Tokyo) for their critical and constructive comments on an earlier version of the manuscript. This study was supported, in part, by a grant from the Japan Society for the Promotion of Science (to N. T.; no. 21310151) and by the Environment Research and Technology Development Fund (S9-4) of the Ministry of the Environment, Japan.

References

Brook BW, Sodhi NS, Bradshaw CJA (2008) Synergies among extinction drivers under global change. Trends Ecol Evol 23:453–460

Burnham KP, Anderson DR (2002) Model selection and multimodel inference: a practical information–theoretic approach, 2nd edn. Springer, New York

Darling ES, Cote IM (2008) Quantifying the evidence for ecological synergies. Ecol Lett 11:1278–1286

Declerck S, Vandekerkhove J, Johansson L, Muylaert K, Conde-Porcuna JM, Van der Gucht K, Perez-Martinez C, Lauridsen T, Schwenk K, Zwart G, Rommens W, Lopez-Ramos J, Jeppesen E, Vyverman W, Brendonck L, De Meester L (2005) Multi-group biodiversity in shallow lakes along gradients of phosphorus and water plant cover. Ecology 86:1905–1915

Didham RK, Tylianakis JM, Gemmell NJ, Rand TA, Ewers RM (2007) Interactive effects of habitat modification and species invasion on native species decline. Trends Ecol Evol 22:489–496

Dudgeon D, Arthington AH, Gessner MO, Kawabata ZI, Knowler DJ, Leveque C, Naiman RJ, Prieur-Richard AH, Soto D, Stiassny MLJ, Sullivan CA (2006) Freshwater biodiversity: importance, threats, status and conservation challenges. Biol Rev 81:163–182

Duelli P, Obrist MK (2003) Biodiversity indicators: the choice of values and measures. Agric Ecosyst Environ 98:87–98

Ecological Society of Japan (ed) (2002) Handbook of alien species in Japan. Chijinsyokan, Tokyo

Engelhardt KAM, Ritchie ME (2001) Effects of macrophyte species richness on wetland ecosystem functioning and services. Nature (Lond) 411:687–689

European Environment Agency (2007) Halting the loss of biodiversity by 2010: proposal for a first set of indicators to monitor progress in Europe. In: EEA technical report. European Environment Agency, Copenhagen

Heino J (2002) Concordance of species richness patterns among multiple freshwater taxa: a regional perspective. Biodivers Conserv 11:137–147

Heino J (2010) Are indicator groups and cross-taxon congruence useful for predicting biodiversity in aquatic ecosystems? Ecol Indic 10:112–117

Heino J, Muotka T, Mykra H, Paavola R, Hamalainen H, Koskenniemi E (2003) Defining macroinvertebrate assemblage types of headwater streams: implications for bioassessment and conservation. Ecol Appl 13:842–852

Hyogo Prefecture, Japan (2003) Precious nature of Hyogo (in Japanese). Hyogo Kankyo Souzou Kyoukai, Kobe

Ilmonen J, Paasivirta L (2005) Benthic macrocrustacean and insect assemblages in relation to spring habitat characteristics: patterns in abundance and diversity. Hydrobiologia 533:99–113

Kadono Y (1994) Aquatic plants of Japan (in Japanese). Bunichi-Sougoushppan, Tokyo

Kadoya T, Suda S, Washitani I (2009) Dragonfly crisis in Japan: a likely consequence of recent agricultural habitat degradation. Biol Conserv 142:1899–1905

Kadoya T, Akasaka M, Aoki T, Takamura N (2011) A proposal of framework to obtain an integrated biodiversity indicator for agricultural ponds incorporating the simultaneous effects of multiple pressures. Ecol Indic 11:1396–1402
Knutson MG, Richardson WB, Reineke DM, Gray BR, Parmelee JR, Weick SE (2004) Agricultural ponds support amphibian populations. Ecol Appl 14:669–684
Laurance WF, Useche DC (2009) Environmental synergisms and extinctions of tropical species. Conserv Biol 23:1427–1437
Lunn DJ, Thomas A, Best N, Spiegelhalter D (2000) WinBUGS: a Bayesian modelling framework: concepts, structure, and extensibility. Stat Comput 10:325–337
Marker AFH, Nusch EA, Rai H, Riemann B (1980) The measurement of photosynthetic pigments in freshwaters and standardization of methods: conclusions and recommendations. Arch Hydrobiol 14:91–106
Mazaris AD, Kallimanis AS, Sgardelis SP, Pantis JD (2008) Does higher taxon diversity reflect richness of conservation interest species? The case for birds, mammals, amphibians, and reptiles in Greek protected areas. Ecol Indic 8:664–671
Menetrey N, Oertli B, Lachavanne JB (2011) The CIEPT: a macroinvertebrate-based multimetric index for assessing the ecological quality of Swiss lowland ponds. Ecol Indic 11:590–600
Merritt RW, Cummins KW, Berg MB (eds) (2008) An introduction to the aquatic insects of North Amercia, 4th edn. Kendall Hunt, Dubuque
Ministry of the Environment of Japan (2007) Natinal Red List of Japan. http://www.biodic.go.jp/english/rdb/rdb_f.html
Monakov AV (2003) Feeding of freshwater invertebrates. Knobi, Belgium
Noss RF (1990) Indicators for monitoring biodiversity: a hierarchical approach. Conserv Biol 4:355–364
Oertli B, Biggs J, Cereghino R, Grillas P, Joly P, Lachavanne JB (2005) Conservation and monitoring of pond biodiversity: introduction. Aquat Conserv 15:535–540
Ohtaki S, Ishido T (2007) Illustrated Japanese water plants (in Japanese). Hokuryuukan, Tokyo
Perfecto I, Mas A, Dietsch T, Vandermeer J (2003) Conservation of biodiversity in coffee agroecosystems: a tri-taxa comparison in southern Mexico. Biodivers Conserv 12:1239–1252
Petchey OL, Gaston KJ (2006) Functional diversity: back to basics and looking forward. Ecol Lett 9:741–758
Revenga C, Campbell I, Abell R, de Villiers P, Bryer M (2005) Prospects for monitoring freshwater ecosystems towards the 2010 targets. Philos Trans R Soc Lond B Biol Sci 360:397–413
Scheffer M (2004) Ecology of shallow lakes. Kluwer, Dordrecht
Scheffer M, Carpenter SR (2003) Catastrophic regime shifts in ecosystems: linking theory to observation. Trends Ecol Evol 18:648–656
Scholes RJ, Biggs R (2005) A biodiversity intactness index. Nature (Lond) 434:45–49
Schulze CH, Waltert M, Kessler PJA, Pitopang R, Shahabuddin, Veddeler D, Muhlenberg M, Gradstein SR, Leuschner C, Steffan-Dewenter I, Tscharntke T (2004) Biodiversity indicator groups of tropical land-use systems: comparing plants, birds, and insects. Ecol Appl 14:1321–1333
Simaika JP, Samways MJ (2009) An easy-to-use index of ecological integrity for prioritizing freshwater sites and for assessing habitat quality. Biodivers Conserv 18:1171–1185
Sugimura M, Ishida S, Kojima K, Ishida K, Aoki H (1999) Dragonflies of the Japanese Archipelago in color (in Japanese). Hokkaido University Press, Sapporo
Takamura N (2007) Biodiversity monitoring at agricultural ponds. In: Washitani I, Kito S (eds) Biodiversity monitoring: collaboration to build capacity for ecosystem management (in Japanese). University of Tokyo Press, Tokyo
Thorp JH, Covich AP (eds) (2010) Ecology and classification of North American freshwater invertebrates, 3rd edn. Academic, Burlington
Wissinger SA (1992) Niche overlap and the potential for competition and intraguild predation between size-structured populations. Ecology 73:1431–1444
Yashiro K (2007) An illustrated guide of Cyperaceae (in Japanese). Zenkoku Nouson Kyouiku Kyoukai, Tokyo

Chapter 16
Ubiquitous Genotyping for Conservation of Endangered Plant Species

Yuji Isagi and Shingo Kaneko

Abstract With the increased human impact on nature, biodiversity is decreasing in a variety of ecosystems, and rapid and effective conservation measures are required. Genetic information, including genetic diversity, clonal structure, gene flow, inbreeding, relatedness, evolutionary significant unit (ESU), and phylogeny can provide information that is crucial for conserving species. The rapid advances in genetic sciences and methodology have enabled us to obtain large amounts of genetic information at a low cost, even for non-model organisms. By taking advantage of such remarkable advances in genetic sciences, it has become possible to genotype all individuals of an endangered species (ubiquitous genotyping) that grow in the wild and to construct novel schemes to conserve biodiversity by comprehensively monitoring the genetic traits of all individuals in the wild. Genetic characteristics such as genetic diversity, clonality, genetic differentiation, and relatedness were analyzed, and the results will be reflected in appropriate and cost-effective conservation activities. By utilizing the information from ubiquitous genotyping, we can evaluate the exact number of wild individuals and genetic diversity, manage endangered species by genetic tagging of individuals, understand the current status of crossing and appropriate crossing for biological conservation, prevent genetic pollution, evaluate local populations for conservation values, identify the origin of endangered species, evaluate its conservation values, and monitor genetic traits between generations. Such merits and examples of implementing biological conservation with information from ubiquitous genotyping for several critically endangered plant species are shown.

Y. Isagi (✉)
Graduate School of Agriculture, Kyoto University, Kyoto 606-8502, Japan
e-mail: isagiy@kais.kyoto-u.ac.jp

S. Kaneko
Faculty of Symbiotic Systems Science, Fukushima University, Fukushima 960-1296, Japan

S.-i. Nakano et al. (eds.), *Integrative Observations and Assessments*, Ecological Research Monographs, DOI 10.1007/978-4-431-54783-9_16, © Springer Japan 2014

Keywords Biological conservation • Clonality • Conservation values • Genetic differentiation • Genetic diversity • Genetic pollution • Genetic tagging

Introduction

The Present Situation of Endangered Species and the Importance of Genetic Diversity for Biological Conservation

With the increased human impact on nature, biodiversity is decreasing in a variety of ecosystems, and rapid and effective conservation measures are required. In the International Union for Conservation of Nature (IUCN) Red List (2008), 8,457 plant species were treated as threatened, accounting for about 3 % of the 298,506 described plant species in the world. However, we should carefully interpret this relatively small ratio of threatened plant species, because only 4 % of plants have been assessed for their conservation status (Vie et al. 2009). Of the vascular plants that are known in the Japanese Archipelago, where the conservational status of all described vascular plant species has been surveyed, 7,000 species have been identified and about 40 % of them are endemic to the archipelago. Additionally, 523 species are listed as critically endangered; 491, endangered; and 676, vulnerable. In total, about 24 % of the vascular plant species in Japan are threatened. Most of the endangered plant species in the Japanese archipelago are composed of fewer than several hundreds of wild individuals.

Because every individual organism has a finite lifespan, the regeneration processes of populations must be considered to conserve threatened species in the long term. In small populations of an endangered species, several genetic uncertainties such as founder effects, genetic drift, inbreeding, loss of genetic diversity, and accumulation of deleterious mutations increase the extinction risk (Frankham 2005). There have been many reports that recognize the significant correlation between genetic diversity within a species and the risk of extinction of the species (Spielman et al. 2004; Leimu et al. 2006; Reed and Frankham 2003).

Problems If Conservation Activities Do Not Utilize Genetic Information

Genetic information can provide information that is crucial for conserving species. This information includes genetic diversity, clonal structure, gene flow, inbreeding, relatedness, evolutionary significant unit (ESU), and phylogeny. However, genetic information seldom is utilized in most conservation activities for endangered species. The conservation activities do not make use of the genetic information because of the belief that obtaining the information requires additional, sophisticated work and more cost, and these activities sometimes do not even recognize the importance of genetic information. Further, most conservation activities prioritize on increasing the number of individuals of the targeted endangered species. As a consequence, many problematic activities have been undertaken for the purpose of biological conservation because of the lack or complete absence of genetic information. These problems, for example, include propagation and reintroductions that do not represent the full genetic diversity of the species, transplantation without considering intraspecific genetic differentiation, breeding without consideration of relatedness, and hybridization with related species that can cause genetic pollution. These activities, although most of them have been done with good intentions, may in fact cause fatal effects for the endangered plant species.

Impact of Rapid and Remarkable Advances in Genetic Science on Biological Conservation

As shown by the fulfillment of the Human Genome Project, rapid advances in genetic sciences and methodology enable us to obtain large amounts of genetic information at a low cost. For example, an entire human genome now can be sequenced at a cost of less than $1,000. Similar dramatic changes to acquire genetic information are underway not only for model species but also for wild non-model species. With these advances, it becomes easy to develop suitable genetic markers and obtain genetic information for wild endangered species. For example, developing appropriate microsatellite markers, which are useful for discrimination of individuals, parentage analysis, and evaluation of genetic diversity, required a relatively large amount of effort (Squirrell et al. 2003) even though several easier and sophisticated methods had been invented (e.g., Lian et al. 2006). However, the advent of next-generation sequencing changed the situation. Using even the smallest next-generation sequencer, a sufficient number of microsatellite loci can be developed within several days (Gardner et al. 2011). Not only marker development but also other kinds of analysis such as restriction site-associated DNA (RAD; Baird et al. 2008) and genotyping-by-sequencing with next-generation sequencing methods (Elshire et al. 2011) have promising potential to revolutionize obtaining of useful genetic information for biological conservation.

By taking advantage of such remarkable advances in genetic sciences, it has become possible to genotype all individuals of an endangered species that grow in the wild and to construct novel schemes to conserve biodiversity by comprehensively monitoring the genetic traits of all individuals in the wild. We initiated a new kind of approach to conserve endangered species: ubiquitous genotyping (Fig. 16.1). The word "ubiquitous" means "to be everywhere," and "ubiquitous genotyping" here means to genotype all individuals of the endangered species in the wild. In the habitat of a target endangered species, we first determined the location

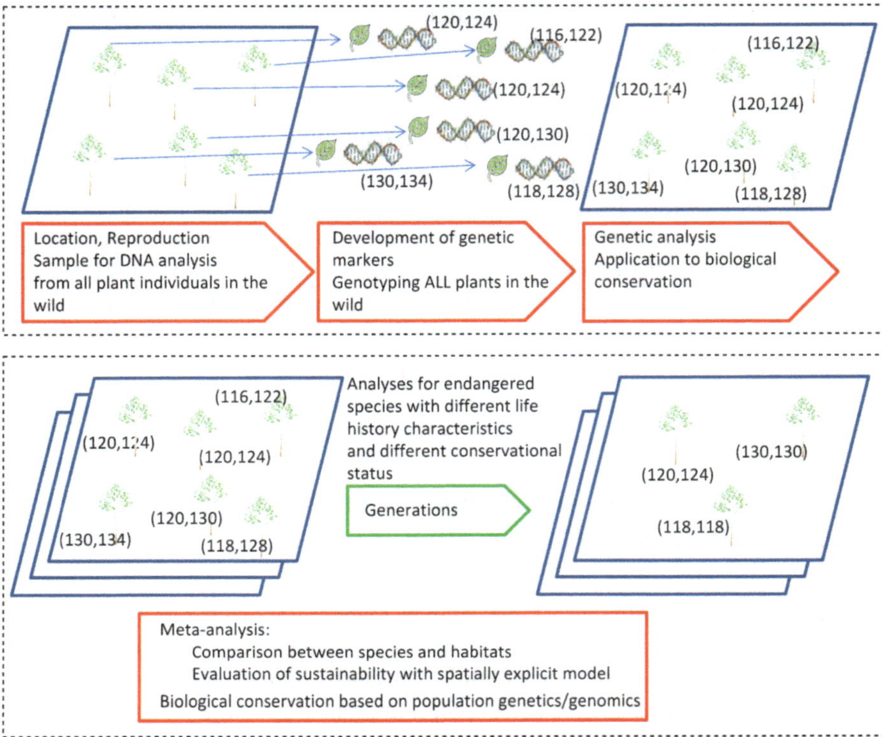

Fig. 16.1 Ubiquitous genotyping: a conservation scheme that genotypes all individuals of an endangered species

of all plants, and then we took a small amount of leaf sample for DNA extraction from all individuals. In the laboratory, highly polymorphic genetic markers, such as microsatellites, were developed, and all individuals were genotyped. Genetic characteristics such as genetic diversity, clonality, genetic differentiation, and relatedness were analyzed, and the results will be reflected in appropriate and cost-effective conservation activities. By gathering the data from ubiquitous genotyping for a variety of endangered species with different life history characteristics and different conservational status, the sustainability of a local population or species can be estimated through a spatially explicit model.

Merits of Ubiquitous Genotyping

So far, there have been only a limited number of studies that genotyped all remnant individuals for the conservation and evaluation of sustainability of endangered species (Lawrence et al. 2008; Drummond et al. 2000). By utilizing the information from ubiquitous genotyping, we can evaluate genetic diversity of populations, select appropriate plant individuals for artificial crossing, determine appropriate locations for transplantation based on genetic structure, detect genetic pollution, prevent illegal trading by genetic tagging, and evaluate the long-term viability of populations. The merits and examples of implementing biological conservation with information from ubiquitous genotyping for several critically endangered plant species are as follows.

Evaluation of the Exact Number of Wild Individuals and Genetic Diversity

Genetic tagging of all remnant individuals by ubiquitous genotyping clarifies the genetic identity of individuals and the number of clones in the wild. To evaluate endangerment, each threatened species was categorized as critically endangered, endangered, or vulnerable on the basis of the current number and fluctuations in the number of individuals (ramet). However, genetic information must be indispensable for the appropriate evaluation of endangerment of species because many plants can undergo vegetative reproduction, which multiplies the number of ramets much more than that of the genets. Using ubiquitous genotyping, the number of genets for an endangered species, an index that is simple but important for biological conservation, can be estimated quite accurately.

For ex situ conservation, which tries to protect endangered species by placing individuals outside their threatened natural habitats, the most appropriate individuals for the attempt can be selected on the basis of information from ubiquitous genotyping. In the case of a species that consists of only of a small number of genets, it is possible to conserve all genetic diversity of the species by determining the number and location of clones in the wild.

Example: *Veronicastrum villosulum*

Veronicastrum villosulum (Plantaginaceae) (Fig. 16.2a) is a critically endangered perennial species with only about 130 known wild plants in a last remaining wild population in Tokushima Prefecture, Japan. Microsatellite analysis revealed that the 130 wild plants were all genetically identical (Kato et al., unpublished data). The wild population stands along a riverside, which is prone to flooding (Fig. 16.2b). Fortunately or unfortunately, the population consists of only one clone, and all genetic diversity in the wild population can be conserved ex situ by transplanting 1 plant from the habitat. After an extensive genetic survey for planted individuals in botanical gardens and elementary schools, we found three other clones with unique alleles (Kato et al., unpublished data) and have kept them at the Hiroshima Botanical Garden, achieving complete ex situ conservation of the

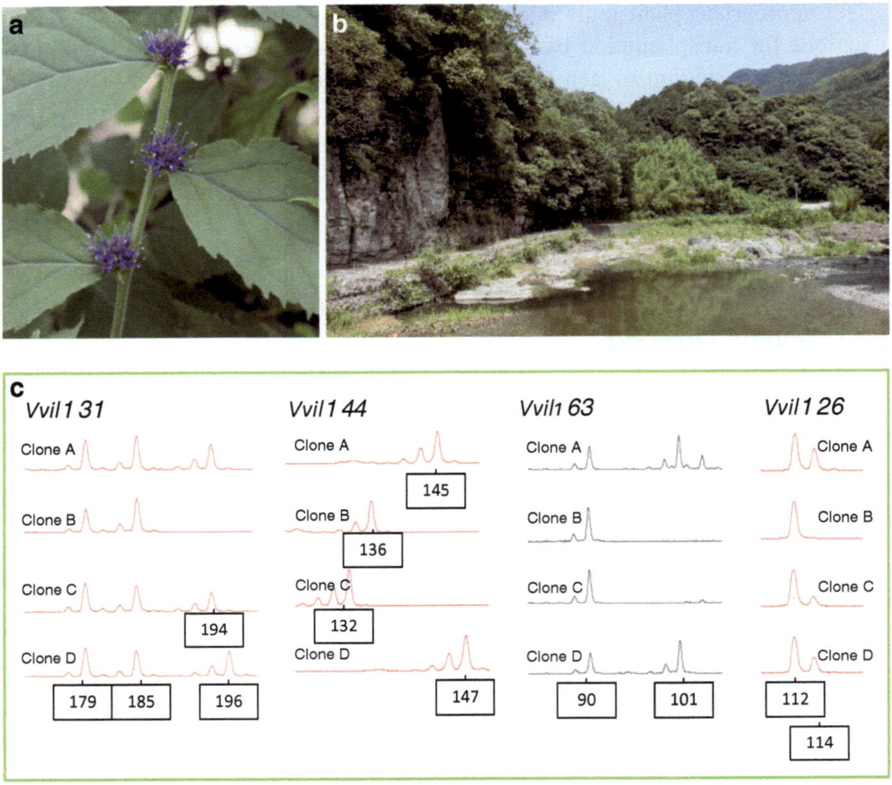

Fig. 16.2 (**a**) Flowering plants of *Veronicastrum villosulum*. (**b**) Sole wild population of *V. villosulum* in Tokushima Prefecture, Japan. (**c**) Example genotypes of the four extant clones of *V. villosulum* at four microsatellite loci

genetic diversity of *V. villosulum*. Because the four clones grown at the Hiroshima Botanical Garden are all genotyped (Fig. 16.2c), we can conduct not only conservation of all genetic diversity within the endangered plant species but also monitor genetic changes between generations and detect genetic pollution.

Management by Genetic Tagging of Individuals

Genetic tagging is a method that identifies individuals on the basis of their genotypes. Because genetic tagging is universally applicable, noninvasive, not lost over time, and provides unambiguous identification and efficient matching of tags, genetic tagging is more suitable to identify and track the movements of every individual than that by conventional tagging techniques, including man-made tags and natural markings (Palsboll 1999).

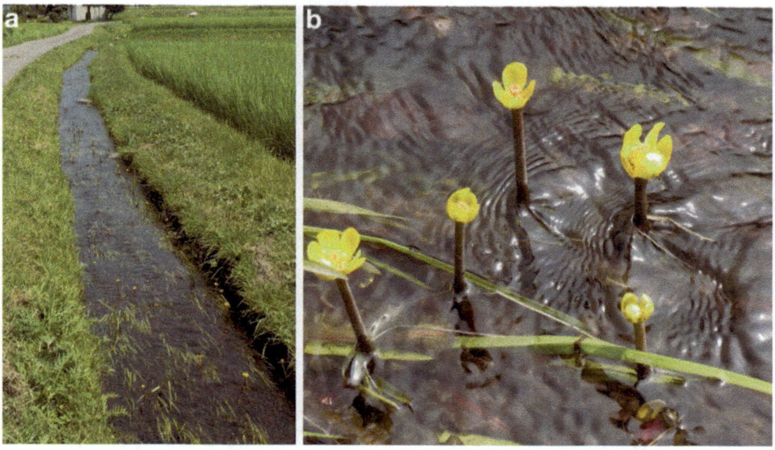

Fig. 16.3 (**a**) Largest population of *Nuphar submerusa* in the city of Nikko. (**b**) In contrast to congeners, all leaves of *N. submerusa* are submerged

Endangered species attract people, and rare plants are often picked from their habitats illegally and traded on the market. Identifying the origin of the organism on the market is one of the effective pressure tactics to prevent illegal digging.

Example: *Nuphar submerusa*

Nuphar submerusa (Nymphaeaceae) is a relatively new aquatic plant species described by Shiga et al. (2006) (Fig. 16.3). When the species was described, the habitats were limited to only three sites in Tochigi Prefecture, central Japan. Because the rhizome of this species elongates and ramets propagate vegetatively, it is difficult to count the number of clones exactly by visual examination. After extensive leaf sampling for DNA analysis and ubiquitous genotyping, it became clear the three wild populations included 31, 18, and 3 clones (Shiga et al., 2013). Although the distances between populations were less than about 40 km, we found that the genetic differentiation was large and that each population had to be conserved separately.

Nuphur submerusa is a recently discovered, beautiful, and rare species; hence, the species is subjected to illegal digging and trading on the Internet. Shiga et al. (2013) obtained and genotyped four individual plants of *N. submerusa* and *Nuphar* sp. from two distributors on the Internet. By comparing the genotypes with those of all extant wild plants, we inferred the original habitat of the plants from the Internet (Shiga et al., 2013). One plant from company A was genetically identical to the plants growing in the city of Nikko. Genotypes of the other three plants from company B were all identical to a clone in the city of Sano, but the clone was not

N. submerusa; the clone was a hybrid between *N. submerusa* and *Nuphar japonicum*. We speculate that company B might have taken the genet from Sano, divided the genet into three pots, and sold them on the Internet with three different names. We notified the society through symposia and media (newspapers and TV programs) that the information about the location and genotypes that was obtained by ubiquitous genotyping for all clones of *N. submerusa* in the wild enabled us to specify the origin of the plants in the market. As a result, the trading of these dug plants on the Internet has now stopped. Complete genetic tagging by ubiquitous genotyping is effective to prevent illegal digging of endangered plant species (Shiga et al., 2013).

Understanding the Current Status of Crossing and Appropriate Crossing for Biological Conservation

To conserve endangered species, we should try to prevent inbreeding that might cause inbreeding depression and reduced fitness. Ubiquitous genotyping clarifies the current status of crossing between individuals within a population and the origin of offspring (selfing, inbreeding, or outbreeding) can be inferred. In small populations or species with a small number of individuals, crossing between less related individuals is effective to prevent inbreeding depression. As an extreme example, in 1995, the introduction of Texas cougars (*Puma concolor stanleyana*) into the last small population of Florida panthers (*Puma concolor coryi*), which showed expression of recessive deleterious genes, significantly restored the fitness of the Florida panther population (Hedrick 2001). Of course, there must be positive and negative effects of introducing genetically diverse individuals. In the case of the Florida panthers, the expression of recessive deleterious genes was efficiently prevented by crossing between subspecies. (Subsequently, Culver et al. (2000) found that North American populations, including Florida panthers and Texas cougars, were genetically homogeneous.)

For endangered species with a small number of individuals, the most appropriate artificial crossing can be performed based on information from ubiquitous genotyping when it becomes difficult for the species to undergo normal crossing because of ecological and genetic reasons, such as collapse of pollinator fauna or no genetic diversity within a local population. Although family trees are generally not available for wild species, information from ubiquitous genotyping can be used to select appropriate individuals as pollen of seed parents by inferring relatedness between populations, and defective breeding plans that reproduce only particular lineages can be prevented.

Fig. 16.4 (**a**) Large *Morus boninensis* endemic to the Bonin Islands with a stem diameter up to 1 m. (**b**) Fowering *Adenophora palustris*. (**c**) Seeds of *A. palustris* that were collected and stored under refrigeration from 1981

Prevention and Detection of Genetic Pollution

Endangered species can become extinct not only by disappearance of individuals but also by hybridization and introgression (Levin et al. 1996; Rhymer and Simberloff 1996). In addition to the extinction that might happen by means of inconsonant genes from different species, endangered species might become extinct genetically by introgression. Hybridization with widely circulated horticultural species or crossing between ESUs within species could be accurately detected and prevented using information from ubiquitous genotyping.

Example: *Morus boninensis*

The Bonin Islands are typical oceanic islands that are located 1,000 km south of Tokyo. About 20 % of the vascular plant species are endemic, and many of them are threatened. *Morus boninensis* (Fig. 16.4a) is a critically endangered endemic plant species that is found on the Bonin Islands. Only about 170 wild trees of *M. boninensis* grow on the three islands of the Bonin Islands. In addition to the small number of remnant individuals, hybridization with the introduced species, *Morus acidosa*, has been confirmed by genetic analysis (Tani et al. 2003).

To conduct genetically proper propagation, appropriate selection of mother trees for seed orchards, which will not disturb the ESU, is essential. Tani et al. (2006) looked for all remnant trees in the wild and conducted ubiquitous genotyping. For the propagation of *M. boninensis*, seeds from a limited number of mother trees on one island (Otouto-jima Island) were mainly used because interspecific hybridization occurred on the other two islands (Chichi-jima and Haha-jima Islands). Tani et al. (2006) found that the propagation program caused low genetic diversity in man-made populations, and they proposed that the spatial genetic structure be considered for adequate propagation programs of the biological conservation of *M. boniensis*.

Evaluation of Local Populations for Conservation Values

Genetic diversity and differentiation within and between local populations can be evaluated. In most conservation activities, there are constraints and resource limitations, and not all local populations necessarily can be protected. We can utilize the information from ubiquitous genotyping to set a range of priority target local populations to be conserved. The necessity, allowable distance, and amount of transplantation between local populations can also be properly estimated.

Example: *Adenophora palustris*

Adenophora palustris (Campanulaceae) is a critically endangered plant species (Fig. 16.4b). About 1,000 plants are growing in five wild populations and one restored population. Extensive field research and ubiquitous genotyping with microsatellite markers (Masumoto et al. 2011) clarified that all Japanese populations of this species consisted of 247 genets (Isagi et al., unpublished data). The numbers of ramets and genets differed remarkably among local populations. Population size has diminished remarkably during the past 20–30 years, and analysis using STRUCTURE (Pritchard et al. 2000) indicated distinct characteristics of each local population: some populations maintained the original genetic traits while others seemed to suffer genetic drift. For example, the population in Fukutomi-cho, Hiroshima Prefecture, consisted of only a few clones; the bearing rates were low, but each clone maintained the ancestral characteristics (low F_{ST} value). In contrast, the population with a relatively large number of clones in Toyosaka-cho, Hiroshima Prefecture, showed increased genetic drift (high F_{ST} value). The largest population of this species in Sera-cho, Hiroshima Prefecture, had many clones, did not suffer severe effects from genetic drift, and was worthy of higher conservation priority than the other populations.

One interesting fact that was uncovered during the exhaustive genotyping was that the preserved seeds (collected in 1981) in the refrigerator of a private botanical garden (the Shigei Herb Botanical Garden) did not suffer any effects of genetic drift

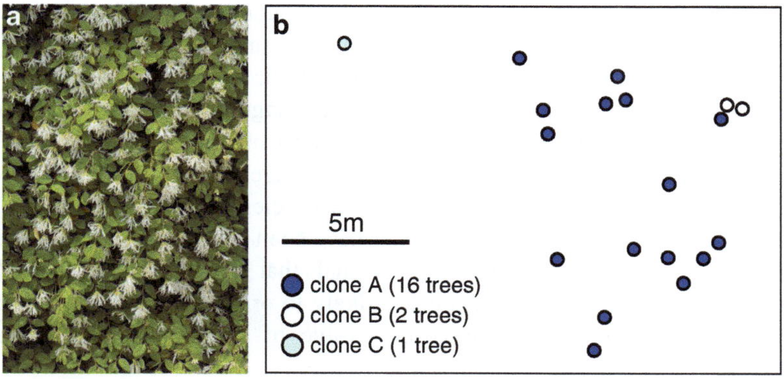

Fig. 16.5 (**a**) *Loropetalum chinense*: flowering of one of the three "wild" populations in Japan. (**b**) Spatial distribution of trees in a population of *L. chinense*. Scattered distribution of an identical clone of an *L. chinense* population, indicating a man-made origin

and maintained the high genetic diversity that might have been maintained originally until the wild population of this species started to shrink (Fig. 16.4c). These seeds are very useful for appropriate population restoration. The Shigei Herb Botanical Garden had already used the seeds to recreate a population of *A. palustris*, and this population did not show traces of genetic drift.

Identification of the Origin of Endangered Species and Evaluation of Its Conservation Values

Endangered plant species sometimes grow in nonnative habitats because they were transplanted by humans. We can detect the signature of transplantation as an unnatural spatial distribution of genetic diversity using thorough genetic analysis. Exact information of the origin of endangered species enables us to evaluate the conservation value of the populations in an appropriate manner.

Example: *Loropetalum chinense*

Loropetalum chinense is an evergreen tree species (Fig. 16.5a) that is found in India, China, and Japan. The species is common in China, but it is endangered in Japan, where only 104 individuals are growing in three wild populations. Local governments and nonprofit organizations (NPOs) have been conserving each of the three wild populations eagerly. Ubiquitous genotyping revealed that each population consisted of only a few clones, and eight clones existed in the three wild populations in Japan (Isagi et al., unpublished data). In each population, each of the adult trees was growing several meters away from each other (Fig. 16.5b).

This tree species does not conduct vegetative reproduction by rhizome elongation. Therefore, the "wild" populations in Japan must be propagated and established by human activities, such as cutting and transplantation.

Although *L. chinense* is ranked as critically endangered in the Red List from the Ministry of the Environment in Japan, the result of ubiquitous genotyping clearly indicated that all the "wild" populations in Japan were not native; rather, they were artificial populations. Many endangered plant species are growing in rural landscapes in Japan, and conservation activities have been conducted. However, endangered species with a small number of individuals that grow in managed woodlands or grasslands near human settlements are likely to originate from transplantation. We need to reexamine the conservation value of such species by ubiquitous genotyping.

Monitoring of Genetic Traits Between Generations

Changes between generations in allele frequencies, fitness, and breeding systems such as outcrossing or inbreeding can be monitored in detail. These items are essential to evaluate the sustainability of populations. By comparing genotypes between generations, we can also infer the existence of undiscovered individuals in the wild, as shown in the following example.

Example: *Stachyurus macrocarpus* var. *prunifolius*

Stachyurus macrocarpus var. *prunifolius* is a critically endangered species that is endemic to the Bonin Islands, Japan. The species was described in 1939, and only 1 tree was identified on Haha-jima Island. The tree flowered every year, but it did not fruit. Because of this, people who were concerned about the conservation of this tree were worried about the future of this species. Fortunately, 14 wild trees were discovered in 2007 (Abe and Hoshi 2008), and 2 of the trees were found to bear fruits. The seeds germinated well, and more than 100 seedlings were obtained. Hence, for this species, there are 15 adult trees in the wild and about 100 seedlings in the nursery. Further, several questions concerning the practical use of these seedlings for conservation arose, including "How about genetic traits of these seedlings?" and "How should we treat these seedlings?"

Ubiquitous genotyping using microsatellite markers (Kaneko et al. 2009) revealed that the seedlings in the nursery did not inherit some alleles from the wild adult population (Fig. 16.6). This finding is not surprising because the seedlings all came from seeds of two known wild adults, and these two adult trees did not represent the genetic diversity of the entire population of adult trees in the wild. We have to take care when we plan to plant the seedlings in the wild. Another interesting finding was that the seedlings had alleles that were not found in the adult trees, which means that there must be some undiscovered adults in the wild.

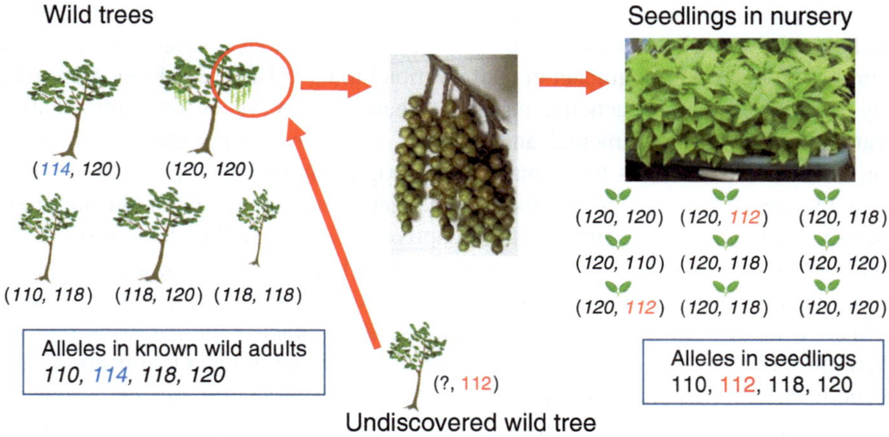

Fig. 16.6 Genotypes and microsatellite locus: *Smap058* of wild trees and seedlings of *Stachyurus macrocarpus* var. *prunifolius*. The population of seedlings in the nursery does not have all the alleles that are found in the known wild adult tree population. Allele *112*, which is unique to the seedling population, indicates the existence of an undiscovered adult (pollen donor) in the wild

The two known seed parents received pollen grains from undiscovered trees in the wild, and the seedlings inherited the alleles that were unique to the seedlings (Fig. 16.6). This finding enhanced motivation for people to find other trees in the wild, and new wild individuals were recently found (Abe, personal communication). Comprehensive genetic analysis by ubiquitous genotyping can indicate the existence of undiscovered individuals in the wild (Kaneko et al. 2013).

Similar approaches were conducted for an endangered seabird, the Chatham Island Taiko (*Pterodroma magentae*). By conducting comprehensive sampling and genotyping microsatellite and mitochondrial DNA, Lawrence et al. (2008) inferred that there were still undiscovered burrows of this bird species.

Future Perspectives

Using the information from ubiquitous genotyping, which determines the position and genotypes of all remnant individuals of an endangered species, the sustainability of populations and the impact of population disappearance on species survival can be simulated by spatially explicit models. Through meta-analysis for species with different life history characteristics and conservational status, we can obtain scientific predictions corresponding to differences between species or populations, and thereby provide appropriate conservation guidelines and allocate resources efficiently and reasonably to conservation activities.

The method of DNA sequencing has been improving in speed, accuracy, and cost-effectiveness. Especially, the advent of next-generation sequencing changed the situation dramatically. Sequencing or genotyping thousands of loci in many individuals will likely become routine, even for non-model organisms, and that surely will make a large impact in conservation biology. The breakthrough brought by next-generation sequencing enabled enhanced surveys of not only neutral variations but also detrimental and adaptive variations. Comprehensive surveys on neutral loci enable us to estimate the demographic histories, population sizes, and founder or bottleneck events of endangered species in detail. In addition, surveys on functional (detrimental and adaptive) loci combined with bioinformatics will help us to infer the response/tolerance to environmental changes and population sustainability. Information from many neutral and functional loci is also effective to distinguish ESUs, which are useful to practice proper conservation activities. Not only the estimation of genetic diversity, individual identification, and genetic differentiation between populations inferred with neutral loci, but also gene expression and adaptation researched with functional loci over the whole genome for every individual of an endangered species, will improve the quality and validity of biological conservation for critically endangered species.

Acknowledgments We thank T. Abe, Y. Hoshi, N. Inoue, K. Kato, I. Masumoto, M. Mizutani, T. Sera, and T. Siga for their assistance with this study. This work was supported by Grants in Aid for Scientific Research (Grant No., 20241056) from the Japan Society for the Promotion of Science and the Environment Research and Technology Development Fund (D-0903 and S-009) of the Ministry of the Environment, Japan.

References

Abe T, Hoshi Y (2008) Ecological characteristics of the endangered shrub *Stachyurus macrocarpus* var. *prunifolius* Tuyama revealed by a population discovery in Haha-jima Island (Ogasawara). Jpn J Conserv Ecol 13:219–223

Baird NA, Etter PD, Atwood TS, Currey MC, Shiver AL, Lewis ZA, Selker EU, Cresko WA, Jhonson EA (2008) Rapid SNP discovery and genetic mapping using sequenced RAD markers. PLoS ONE 3:e3376

Culver M, Johnson WE, Pecon-Slattery J, O'Brien SJ (2000) Genomic ancestry of the American puma (*Puma concolor*). J Hered 91:186–197

Drummond RSM, Keeling DJ, Richardson TE, Gardner RC, Wright D (2000) Genetic analysis and conservation of 31 surviving individuals of a rare New Zealand tree, *Metrosideros bartlettii* (Myrtaceae). Mol Ecol 9:1149–1157

Elshire RJ, Glaubitz JC, Sun Q, Poland JA, Kawamoto K, Buckler ES, Mitchell SE (2011) A robust, simple genotyping-by-sequencing (GBS) approach for high diversity species. PLoS ONE 6:e19379

Frankham R (2005) Genetics and extinction. Biol Conserv 126:131–140

Gardner MG, Fitch AJ, Bertozzi T, Lowe AJ (2011) Rise of the machines—recommendations for ecologists when using next generation sequencing for microsatellite development. Mol Ecol Resour 11:1093–1101

Hedrick PW (2001) Conservation genetics: where are we now? Trends Ecol Evol 16:629–636

Kaneko S, Abe T, Isagi Y (2009) Development of microsatellite markers for *Stachyurus macrocarpus* and *Stachyurus macrocarpus* var. *prunifolius* (Stachyuraceae), critically endangered shrub species endemic to the Bonin Islands. Conserv Genet 10:1863–1865

Kaneko S, Abe T, Isagi Y (2013) Complete genotyping in conservation genetics, a case study of a critically endangered shrub, *Stachyurus macrocarpus* var. *prunifolius* (Stachyuraceae) in the Ogasawara Islands, Japan. J Plant Res 126(5):635–642

Lawrence HA, Taylor GA, Crockett DE, Millar CD, Lambert DM (2008) New genetic approach to detecting individuals of rare and endangered species. Conserv Biol 22:1267–1276

Leimu R, Mutikainen P, Koricheva J, Fischer M (2006) How general are positive relationship between plant population size, fitness, and genetic variation? J Ecol 94:942–952

Levin DA, Francisco-Ortega J, Jansen R (1996) Hybridization and the extinction of rare plant species. Conserv Biol 10:10–16

Lian C, Wadud MA, Geng Q, Shimatani K, Hougetsu T (2006) An improved technique for isolating codominant compound microsatellite markers. J Plant Res 119:415–417

Masumoto I, Kaneko S, Ohtake K, Isagi Y (2011) Development of microsatellite markers for *Adenophora palustris* (Campanulaceae), a critically endangered wetland plant species in Japan. Conserv Genet Resour 3:163–165

Palsboll PJ (1999) Genetic tagging: contemporary molecular ecology. Biol J Linn Soc 68:3–22

Pritchard JK, Stephens M, Donnelly P (2000) Inference of population structure using multilocus genotype data. Genetics 155:945–959

Reed DH, Frankham R (2003) Correlation between fitness and genetic diversity. Conserv Biol 17:230–237

Rhymer JM, Simberloff D (1996) Extinction by hybridization and introgression. Annu Rev Ecol Syst 27:83–109

Shiga T, Ishii J, Isagi Y, Kadono Y (2006) *Nuphar submersa* (Nymphaeaceae), a new species from central Japan. Acta Phytotaxon Geobot 57:113–122

Shiga T, Yokogawa M, Kaneko S, Isagi Y (2013) Genetic identification for trading plants of endangered macrophytes, *Nuphar submersa* and *N. xfluminalis* (Nymphaeaceae), based on genotype data of all remnant individuals growing in the wild. Jpn J Conserv Ecol 18:33–44

Spielman D, Brook BW, Frankham R (2004) Most species are not driven to extinction before genetic factors impact them. Proc Natl Acad Sci USA 101:15261–15264

Squirrell J, Hollingsworth PM, Woodhead M, Sussell J, Lowe AJ, Gibby M, Powell W (2003) How much effort is required to isolate nuclear microsatellites from plants? Mol Ecol 12:1339–1348

Tani N, Kawahara T, Yoshimaru H, Hoshi Y (2003) Development of SCAR markers distinguishing pure seedlings of the endangered species *Morus boninensis* from *M. boniensis* × *M. acidosa* hybrids for conservation in Bonin (Ogasawara) Islands. Conserv Genet 4:605–612

Tani N, Yoshimaru H, Kawahara T, Hoshi Y, Nobushima F, Yasui T (2006) Determination of the genetic structure of remnant *Morus boninensis* Koidz. trees to establish a conservation program on the Bonin Islands. Jpn BMC Ecol 6:14

Vie J-C, Hilton-Taylor C, Stuart SN (eds) (2009) Wildlife in a changing world. An analysis of the 2008 IUCN Red List of Threatened Species. IUCN, Gland

Chapter 17
Eco-Evolutionary Genomic Observation for Local and Global Environmental Changes

Shin-Ichi Morinaga, Takaya Iwasaki, and Yoshihisa Suyama

Abstract Local and global environmental changes produce shifts in species distribution as well as adaptive evolution in wild species. Recent advances in genomic analysis tools, such as the next-generation DNA sequencing technique, have allowed genetic observation of ecological and evolutionary responses to the environmental changes. In this review, we propose three approaches for 'eco-evolutionary genomic observation' to assess and predict the effect of environmental changes on biodiversity. First, paleo-ecogenomics, the genomic and metagenomic analyses of ancient pollen grains or sediments, can be employed to identify long-term evolutionary changes, thereby facilitating the reconstruction of ancient biological communities and interactions. Second, chronological genomics, which uses museum or herbarium specimen samples to study population genomics, can be used to elucidate short-term evolutionary changes and detect the adaptive genes favored by natural or human-induced selection. Third, ecological genome niche modeling, based on the distribution data of adaptive variants detected by genome-wide scans and the geographic patterns of various environmental factors, will be useful for prediction of evolutionary responses to future environmental changes.

All authors equally contributed.

S.-I. Morinaga (✉)
Graduate School of Arts and Sciences, The University of Tokyo, 3-8-1, Komaba, Meguro-ku, Tokyo 153-8902, Japan

CREST, Japan Science and Technology Agency, K's Gobancho, 7, Gobancho, Chiyoda-ku, Tokyo 102-0076, Japan
e-mail: morinaga@bio.c.u-tokyo.ac.jp

T. Iwasaki
Graduate School of Arts and Sciences, The University of Tokyo, 3-8-1, Komaba, Meguro-ku, Tokyo 153-8902, Japan

Y. Suyama
Field Science Center, Graduate School of Agricultural Science, Tohoku University, 232-3 Yomogida, Naruko-onsen, Osaki, Miyagi 989-6711, Japan

S.-i. Nakano et al. (eds.), *Integrative Observations and Assessments*, Ecological Research Monographs, DOI 10.1007/978-4-431-54783-9_17, © Springer Japan 2014

These new approaches provide a point of view for observation of evolution caused by environmental changes in chronological order.

Keywords Adaptive evolution • Chronological genomics • Eco-evolutionary genomics • Ecological genome niche modeling • Evolutionary responses • Genetic differentiation and diversity • Local and global environmental changes • Paleo-ecogenomics • Species distribution

Introduction

Environmental changes should produce shifts in species distribution, and these changes can also cause adaptive evolution in wild species. For example, severe climatic oscillations during the Quaternary period produced great changes in species distribution (Bennett 1997; Hewitt 2000). Known to be one of the coldest periods in the Quaternary, the last glacial maximum (~25,000–17,000 years ago) resulted in a global cooling that was particularly severe for temperate species, which resulted in marked changes in their distribution ranges (Hewitt 2004). The response of temperate species to climatic oscillations has been considered a dynamic process consisting of repeated retreats into refugia during glacial periods and range expansions from these refugia during interglacial periods. In addition, wild species are affected by human-induced environmental change and disturbance (Kinnison and Hendry 2001; Hoffmann and Daborn 2007). In response to global warming, species are likely to establish their distribution ranges at higher altitudes or higher latitudes, moving to locations within their optimal climatic range (Jump and Penuelas 2005). As with other human-induced environmental changes, global urbanisation has produced great changes such as fragmentation and extinction of populations by direct displacement from habitats (Hahs et al. 2009). As a result, it has caused a decline in biodiversity in urban areas (McKinney 2008). These various environmental changes can also cause changes in the kind and strength of selective pressures on populations.

Population genetic analysis using neutral genetic markers has been applied to assess the effects of environmental changes on the distribution, adaptation, and extinction patterns of wild species (Avise 2000; Hoffmann and Willi 2008). During the past few decades, phylogeographic studies revealing the migration histories of species based on geographic patterns of genetic differentiation and diversity have been conducted on many species in various areas. The results have made clear that considerable dynamic changes in species distribution ranges occurred in the Quaternary, especially during the period from the last glacial maximum to the present (Avise 2000; Hewitt 2000, 2004). On the other hand, predictions of population genetic responses to the recent climate and environmental changes have focused on the levels of gene flow and genetic diversity (Jump and Penuelas 2005). Fragmented landscapes resulting from human activity lead to the fragmentation of the distribution ranges of many species and increase the genetic isolation of each population

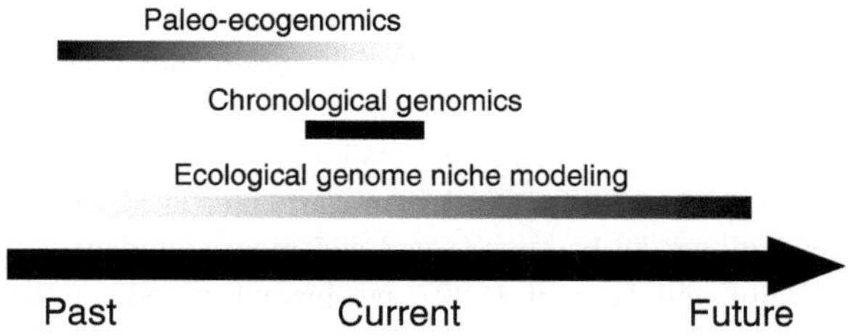

Fig. 17.1 Three eco-evolutionary genomic observation approaches for assessment and prediction of the effect of environmental changes on organisms

(Williams et al. 2003). Fragmentation has increased genetic drift and inbreeding and decreased gene flow among neighboring populations (Young et al. 1996). Furthermore, the inbreeding depression caused by population fragmentation brings about a decline in fitness, a reduction in population size, and a further increase in population fragmentation, and finally leads to the extinction of local populations (extinction vortex; Gilpin and Soulé 1986).

Recent advances in genomic analysis techniques, such as next-generation DNA sequencing, provide tools for determining the genetic basis of ecological and evolutionary responses to environmental changes. The technique uses highly parallel DNA sequencing, in which hundreds of thousands, or millions, of reads are produced in a single run (Mardis 2008). Evolutionary ecological studies based on genome and transcriptome analyses for model and non-model organisms have been conducted with this technique (Renn and Siemens 2010; Ekblom and Galindo 2011). In particular, several genes underlying habitat adaptation have been identified, providing a platform for addressing some of the ecological and evolutionary questions (Stapley et al. 2010). Moreover, metagenomics has revolutionized ecology by facilitating investigations of the structure and functions of microbial communities in complex ecosystems (Simon and Daniel 2011). In various wild species and ecosystems, the eco-evolutionary responses to climatic and environmental changes can be observed, assessed, and predicted using advanced genomic analysis tools.

In this review, we propose that 'eco-evolutionary genomic observation' will provide the tools for assessment and prediction of the effect of environmental changes on biodiversity (Fig. 17.1). This endeavor consists of three approaches: (1) paleo-ecogenomics for identification of long-term historical changes, (2) chronological genomics for documentation of short-term historical changes, and (3) ecological genome niche modeling for the prediction of future changes.

Paleo-Ecogenomics for Identifying Long-Term Eco-Evolutionary Changes

Genetic information from ancient DNA (aDNA) contributes powerfully to the assessment of long-term ecological and evolutionary responses to environmental change and can help predict future responses. Although the rarity of suitable, well-preserved plant fossil samples has often prevented a broader utility for aDNA studies in plant species,

DNA has been extracted successfully from ancient plant materials (Gugerli et al. 2005 and references therein), including pollen (Suyama et al. 1996; Parducci et al. 2005, 2012; Paffetti et al. 2007), subfossilized peat moss (Suyama et al. 2008), and bulk sediments from permafrost (Willerslev et al. 2003; Lydolph et al. 2005) or lakes (Coolen et al. 2004; Anderson-Carpenter et al. 2011). Because fossil pollen grains are among the best preserved and most abundant material from a variety of plants (Faegri and Iversen 1989), we have focused on these materials here for 'paleo-ecogenomics,' the study of long-term eco-evolutionary genomic changes.

Direct determination of DNA sequences from pollen grains is an efficient method of addressing questions in several research areas, such as evolutionary ecology, phylogeography, and genomics. The basic protocol for DNA amplification from pollen grains was developed more than a decade ago (Suyama et al. 1996). Recently, Suyama (2011) described modifications to the original method, resulting in development of a single-pollen genotyping procedure. Briefly, the essential prerequisites of this technique are (1) the ability to manipulate small objects in small amounts of liquid, (2) determination of the optimal conditions for minimizing the loss of DNA, (3) identification of the optimal conditions for polymerase chain reaction (PCR) amplification using extremely low-copy templates, and (4) elimination of possible contaminants. In addition, detailed methods for aDNA extraction from fossil pollen were described by Parducci et al. (2005) and Parducci and Suyama (2011). Parducci et al. (2012) reported isolation and amplification of aDNA from fossil pollen and lake sediments, as well as modern DNA. Using direct evidence, this study verified the glacial survival of boreal trees in northern Scandinavia. The authors showed the survival of a rare mitochondrial DNA haplotype of spruce in lake sediments and pollen in central Norway dating back approximately 10,300 years, as well as chloroplast DNA of pine and spruce in lake sediments adjacent to the ice-free refugium in northwestern Norway as early as approximately 22,000 and 17,700 years ago, respectively. Their findings challenged current views on the survival and spread of trees from the southern refugia in response to climate changes.

Genetic analysis of ancient pollen materials from different ages and sites allows direct evaluation of the history, dynamics, and structure of archaic plant communities, populations, and genes (Fig. 17.2). This method not only provides information on plant movement in space and time but also facilitates the detection of evolutionary responses to environmental change, especially with regard to climate change, through the analysis of genes under selection. However, the availability of suitable ancient pollen materials is limited, and analysis is hindered by several technical difficulties because of the extremely small amounts of DNA and the risk of contamination. Therefore, the impact of this technique on eco-evolutionary inferences has been limited. Arguably, a breakthrough in this field of research was the development of next-generation DNA sequencing, which can read an extensive number of DNA segments, and will undoubtedly contribute to aDNA analysis for eco-evolutionary genomic observation. In addition, analysis of different sources of materials, including bulk sediments, will help to make those inferences more quantitative.

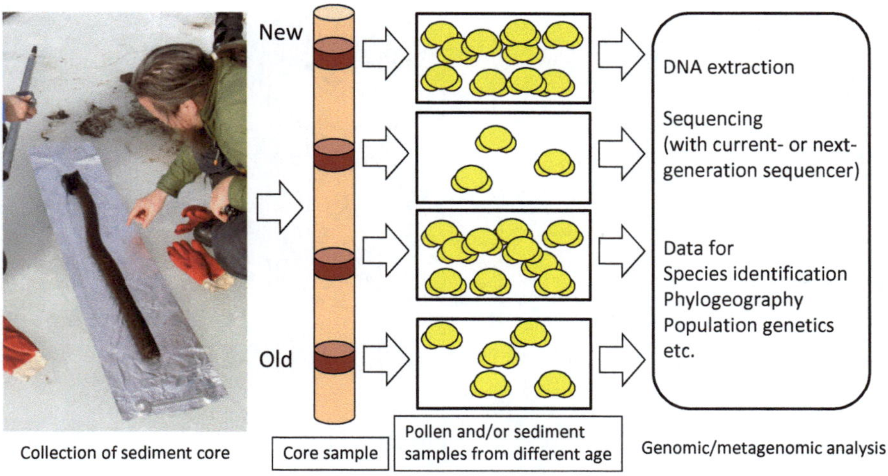

Fig. 17.2 Schematic representation of a paleo-ecogenomic study

Owing to this efficient approach, the analysis of aDNA will become the technique of choice for documenting long-term evolutionary change, and use of the paleo-ecogenomic approach will expand in the near future.

Chronological Genomics of Museum Specimens for Documenting Short-Term Eco-Evolutionary Change

Museum and herbarium specimens are meritorious resources for biology, that is, taxonomy, biogeography, ecology, morphology, and evolutionary biology. These specimens can be used to study a broad range of topics, from aspects of ecological and evolutionary theory to conservation and agriculture, by integrating their spatial data with temporal information (Graham et al. 2004). Furthermore, the specimens provide source material for genetic, biochemical, and isotopic analyses (Lister 2011).

Population genetics using museum and herbarium specimen samples have elucidated short-term evolutionary changes (Wandeler et al. 2007). Many comparisons of past and current genetic diversities in threatened, endangered, and invasive populations and species have relied on specimen samples. However, specimen sample DNA can be deteriorated and fragmented by numerous biological, physical, and chemical factors. Therefore, PCR amplification of specimen sample DNA is restricted to short fragments, usually less than 200 bp. Therefore, it is inefficient to conduct genetic analyses on specimen samples using current-generation sequencing, optimized for long-read analysis. However, next-generation DNA sequencing, which can output an extensive number of short reads, is suitable for genetic variation comparisons between past and current individuals in a species.

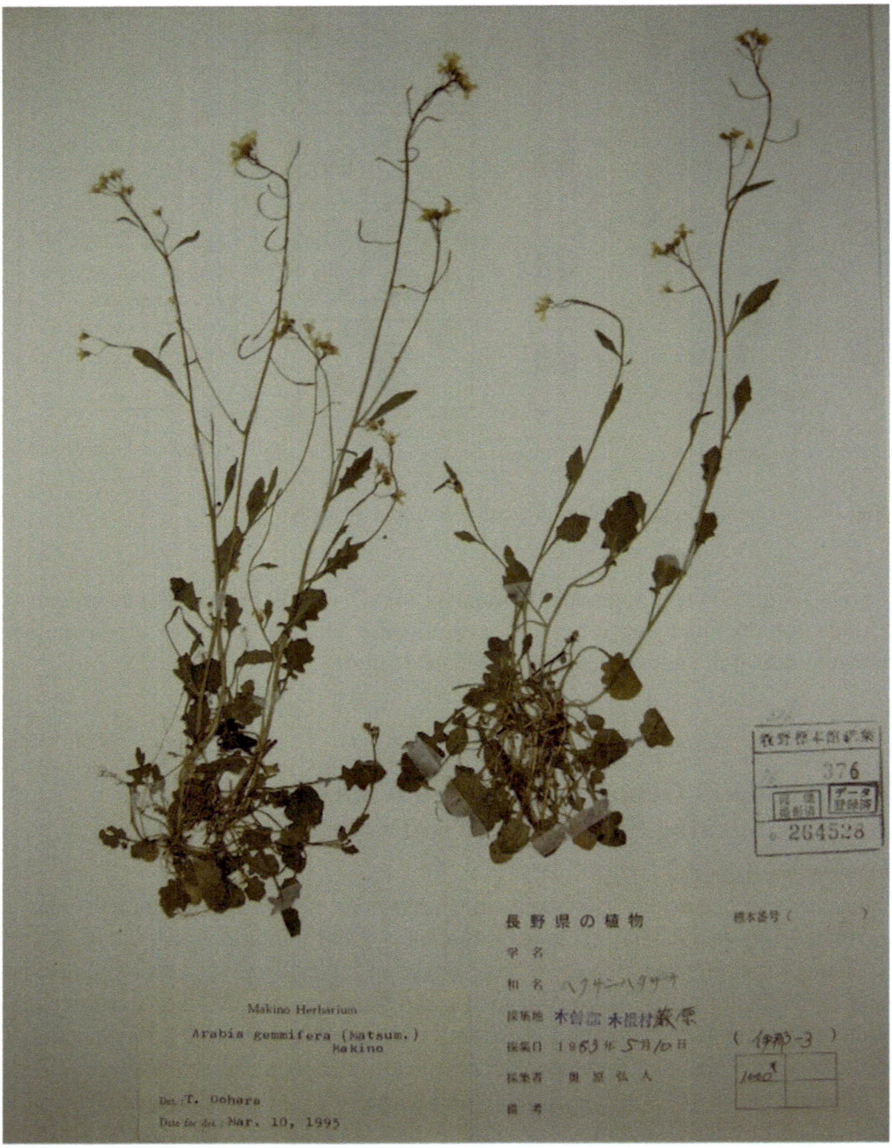

Fig. 17.3 Specimens of *Arabidopsis halleri* subsp. *gemmifera* from Makino Herbarium (MAK), Tokyo Metropolitan University

Here, we propose a population genomic approach using museum specimens to elucidate the short-term eco-evolutionary changes and detect the adaptive genes favored by natural or human-induced selection, or 'chronological genomics.' By using next-generation DNA sequencer, nucleotide variations such as single nucleotide polymorphisms (SNPs) can be detected among samples in chronological order.

Mainstream trends in genomic changes from past to current populations would indicate the demographic history, including possible bottleneck effects, founder effects, gene flow, and disturbance. In contrast, outliers of genomic changes may reveal the genes underlying adaptive evolution.

Using specimen samples of *Arabidopsis halleri* subsp. *gemmifera* (Fig. 17.3), which grows in a broad range of habitats and is currently distributed in eastern Asia, we have examined the allele frequency changes from 100 years ago to the present in 20 functional genes related to global and local environmental adaptation. Analyzing three populations, we found nonsynonymous variations in almost all genes; the allele frequencies also changed during this time (Morinaga et al., unpublished data). Great and rapid allele frequency changes during the past 20 years were detected in some important functional genes. These large genetic changes may be the consequence of natural selection, itself caused by recent rapid environmental changes. Furthermore, we sequenced the genomes of many past and current individuals by using the next-generation DNA sequencer. Such analyses can detect the dynamics of adaptive genes in chronological order, and the pattern showed ongoing adaptive and neutral evolution. By comparing the results among several populations inhabiting various environmental and historical situations, differences in the number and molecular functions of the adaptive genes will shed light on the microevolutionary mechanisms caused by rapid climate and environmental changes.

Ecological Genome Niche Modeling for Conservation of Future Adaptive Evolvability

Ecological genome niche modeling (EGNM), a new concept derived from ecological niche modeling (ENM), is based on the distribution data of adaptive variants detected by genome-wide scans and the geographic patterns of various environmental factors (Fig. 17.4). ENM, a statistical method for prediction of a potential species distribution range based on the species distribution data and the geographic patterns of environmental factors, is used widely to describe many species (Murray-Smith et al. 2009; Tinoco et al. 2009). Recent advances in genomic analyses, including next-generation DNA sequencing techniques, have facilitated genome-wide scans to identify adaptive variants or genes (Baird et al. 2008; Atwell et al. 2010; Fournier-Level et al. 2011). Based on the distribution data of such adaptive variants, EGNM allows prediction of their potential distribution ranges. Furthermore, by overlay analysis of the predicted distribution ranges, we can obtain a geographic map of the potential genetic diversity of the adaptive variants. The map indicates the pattern of local adaptation, because the diversity and distribution of such adaptive variants represent a basis for adaptive evolution. Monitoring and conservation of these variants would be one of the important goals in the next strategy for biodiversity conservation.

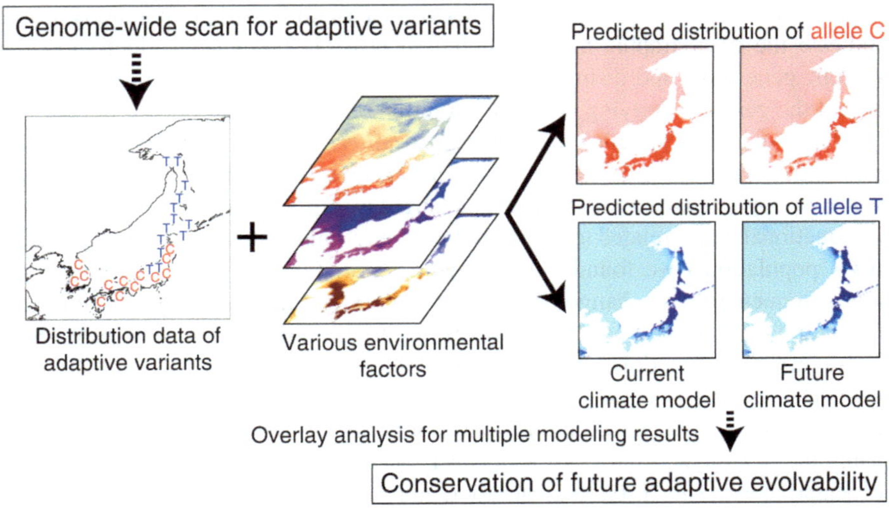

Fig. 17.4 Concept of ecological genome niche modeling

The various methods and environmental data that have been developed for ENM can also be applied to EGNM. Many types of environmental data, including current climate data, geological feature data, and land-use data, are now provided as Geographic Information System (GIS) data (summarized in Franklin and Miller 2009). Using the geographic patterns of the various environmental factors on GIS, EGNM will provide an ecological niche model and map of the potential distribution ranges for each adaptive variant. Furthermore, the model enables identification of the environmental factors that influence the distribution patterns of adaptive variants. However, EGNM cannot ignore the influence of demographic history, as measured by neutral genetic variations, and the effect of the interaction between the adaptive genes. To solve these problems, it will be necessary to develop a statistical method that takes into consideration the contributions of ecogenomics and population genetics.

We have examined the intraspecific polymorphisms in nine functional genes in a diploid herb, *Cardamine leucantha*, which grows widely on cool-temperate forest floors and margins in eastern Asia. As a result of genetic analysis of about 100 samples from more than 50 localities, we found several nonsynonymous SNPs with distribution patterns that clearly differed from those of neutral genetic differentiation (Iwasaki et al., unpublished data). Such geographic patterns might have been shaped by natural selection. However, comparison among a small number of variants (<100) is not sufficient to detect adaptive variants. Genome-wide scans by using a next-generation DNA sequencer are necessary to detect and evaluate adaptive variants, which are used for EGNM.

The most important advantage of this approach is its ability to predict the distribution range of adaptive variants not only in the present, but also in the future (e.g., the era of global warming), by projecting the current model to other environmental situations. Future climate model simulations based on the IPCC

Fourth Assessment Report (IPCC 2007) can be used as future environmental situations. Using EGNM, we can first make a geographic map that indicates the pattern of local adaptation in the present. It also means that EGNM will lead us to predict a change of the pattern from the present to the future. The change is adaptive evolution itself, as a response to the future environmental changes. Furthermore, considering the various limiting factors that influence the movement of adaptive variants (e.g., topography and dispersal speed of species and genes), we can predict which local gene pool will contribute to adaptive evolution in the future, providing information as to the relative importance of local genetic diversity. 'Conservation of future adaptive evolvability' using EGNM will be a new key concept for biodiversity conservation in this post-genome era.

Conclusions

Recent advances in genomic analysis tools, such as the next-generation DNA sequencing technique, have allowed genetic observation of ecological and evolutionary responses to environmental changes. In this review, we proposed three approaches for 'eco-evolutionary genomic observation' to assess and predict the effect of environmental changes on biodiversity. First, paleo-ecogenomics, the genomic and metagenomic analyses of ancient pollen grains or sediments, can be employed to identify long-term evolutionary changes, thereby facilitating the reconstruction of ancient biological communities and interactions. Second, chronological genomics, which uses museum or herbarium specimen samples to study population genomics, can be used to elucidate short-term evolutionary changes and detect the adaptive genes favored by natural or human-induced selection. Third, ecological genome niche modeling, based on the distribution data of adaptive variants detected by genome-wide scans and the geographic patterns of various environmental factors, will be useful for prediction of evolutionary responses to future environmental changes. These new approaches provide a basis for the observation of evolution caused by environmental changes in chronological order.

Acknowledgements The authors thank H. Tachida at Kyushu University for valuable comments on this manuscript and N. Murakami at Makino Herbarium, Tokyo Metropolitan University, for permission to use the specimen photographs. This study was partly supported by the Environment Research and Technology Development Fund (S9) of the Ministry of the Environment, Japan, KAKENHI (No. 22114504, 22658046, 22770013, 24248025, 24770014, and 25840139), Inamori Foundation, and JST, CREST.

References

Anderson-Carpenter LL, McLachlan JS, Jackson ST, Kuch M, Lumibao CY, Poinar HN (2011) Ancient DNA from lake sediments: bridging the gap between paleoecology and genetics. BMC Evol Biol 11:30

Atwell S, Huang YS, Vilhjálmsson BJ, Willems G, Horton M, Li Y, Meng D, Platt A, Tarone AM, Hu TT, Jiang R, Muliyati NW, Zhang X, Amer MA, Baxter I, Brachi B, Chory J, Dean C, Debieu M, de Meaux J, Ecker JR, Faure N, Kniskern JM, Jones JDG, Michael T, Nemri A, Roux F, Salt DE, Tang C, Todesco M, Traw MB, Weigel D, Marjoram P, Borevitz JO, Bergelson J, Nordborg M (2010) Genome-wide association study of 107 phenotypes in *Arabidopsis thaliana* inbred lines. Nature (Lond) 465:627–631

Avise JC (2000) Phylogeography: the history and formation of species. Harvard University Press, Cambridge

Baird NA, Etter PD, Atwood TS, Currey MC, Shiver AL, Lewis ZA, Selker EU, Cresko WA, Johnson EA (2008) Rapid SNP discovery and genetic mapping using sequenced RAD markers. PLoS ONE 3:e3376

Bennett KJ (1997) Evolution and ecology: the pace of life. Cambridge University Press, Cambridge

Coolen MJL, Muyzer G, Rijpstra WIC, Schouten S, Volkman JK, Sinninghe Damsté JS (2004) Combined DNA and lipid analyses of sediments reveal changes in Holocene haptophyte and diatom populations in an Antarctic lake. Earth Planet Sci Lett 223:225–239

Ekblom R, Galindo J (2011) Applications of next generation sequencing in molecular ecology of non-model organisms. Heredity 107:1–15

Faegri K, Iversen J (1989) Textbook of pollen analysis. Wiley, Chichester

Fournier-Level A, Korte A, Cooper MD, Nordborg M, Schmitt J, Wilczek AM (2011) A map of local adaptation in *Arabidopsis thaliana*. Science 334:86–89

Franklin J, Miller JA (2009) Mapping species distributions: spatial inference and prediction. Cambridge University Press, Cambridge

Gilpin ME, Soulé ME (1986) Minimum viable population: processes of extinction. In: Soulé ME (ed) Conservation biology: the science of scarcity and diversity. Sinauer Associates, Sunderland

Graham CH, Ferrier S, Huettman F, Moritz C, Peterson AT (2004) New developments in museum-based informatics and applications in biodiversity analysis. Trends Ecol Evol 19:497–503

Gugerli F, Parducci L, Petit RJ (2005) Ancient plant DNA: review and prospects. New Phytol 166:409–418

Hahs AK, McDonnell MJ, McCarthy MA, Vesk PA, Corlett RT, Norton BA, Clemants SE, Duncan RP, Thompson K, Schwartz MW, Williams NSG (2009) A global synthesis of plant extinction rates in urban areas. Ecol Lett 12:1165–1173

Hewitt GM (2000) The genetic legacy of the Quaternary ice ages. Nature (Lond) 405:907–913

Hewitt GM (2004) Genetic consequences of climatic oscillations in the Quaternary. Philos Trans R Soc Lond B Biol Sci 359:183–195

Hoffmann AA, Daborn PJ (2007) Towards genetic markers in animal populations as biomonitors for human-induced environmental change. Ecol Lett 10:63–76

Hoffmann AA, Willi Y (2008) Detecting genetic responses to environmental change. Nat Rev Genet 9:421–432

IPCC (2007) Climate change 2007: synthesis report. An assessment of the intergovernmental panel on climate change. Cambridge University Press, Cambridge

Jump AS, Penuelas J (2005) Running to stand still: adaptation and the response of plants to rapid climate change. Ecol Lett 8:1010–1020

Kinnison MT, Hendry AP (2001) The pace of modern life. II: From rates of contemporary microevolution to pattern and process. Genetica 112–113:145–164

Lister AM (2011) Natural history collections as sources of long-term datasets. Trends Ecol Evol 26:153–154

Lydolph MC, Jacobsen J, Arctander P, Gilbert MT, Gilichinsky DA, Hansen AJ, Willerslev E, Lange L (2005) Beringian paleoecology inferred from permafrost-preserved fungal DNA. Appl Environ Microbiol 71:1012–1017

Mardis ER (2008) The impact of next-generation sequencing technology on genetics. Trends Genet 24:133–141

McKinney M (2008) Effects of urbanization on species richness: a review of plants and animals. Urban Ecosyst 11:161–176

Murray-Smith C, Brummitt NA, Oliveira-Filho AT, Bachman S, Moat J, Lughadha EMN, Lucas EJ (2009) Plant diversity hotspots in the Atlantic coastal forests of Brazil. Conserv Biol 23:151–163

Paffetti D, Vettori C, Caramelli D, Vernesi C, Lari M, Paganelli A, Paule L, Giannini R (2007) Unexpected presence of *Fagus orientalis* complex in Italy as inferred from 45,000-year-old DNA pollen samples from Venice lagoon. BMC Evol Biol 16(suppl 2):S6

Parducci L, Suyama Y (2011) Single-pollen genotyping of Holocene lake sediments. In: Isagi Y, Suyama Y (eds) Single-pollen genotyping. Springer, Tokyo

Parducci L, Suyama Y, Lascoux M, Bennett KD (2005) Ancient DNA from pollen: a genetic record of population history in Scots pine. Mol Ecol 14:2873–2882

Parducci L, Jørgensen T, Tollefsrud MM, Elverland E, Alm T, Fontana SL, Bennett KD, Haile J, Matetovici I, Suyama Y, Edwards ME, Andersen K, Rasmussen M, Boessenkool S, Coissac E, Brochmann C, Taberlet P, Houmark-Nielsen M, Larsen NK, Orlando L, Gilbert MTP, Kjær KH, Alsos IG, Willerslev E (2012) Glacial survival of boreal trees in northern Scandinavia. Science 335:1083–1086

Renn SC, Siemens DH (2010) Ecological genomics: changing perspectives on Darwin's basic concerns. Mol Ecol 19:3025–3030

Simon C, Daniel R (2011) Metagenomic analyses: past and future trends. Appl Environ Microbiol 77:1153–1161

Stapley J, Reger J, Feulner PGD, Smadja C, Galindo J, Ekblom R, Bennison C, Ball AD, Beckerman AP, Slate J (2010) Adaptation genomics: the next generation. Trends Ecol Evol 25:705–712

Suyama Y (2011) Procedure for single-pollen genotyping. In: Isagi Y, Suyama Y (eds) Single-pollen genotyping. Springer, Tokyo

Suyama Y, Kawamuro K, Kinoshita I, Yoshimura K, Tsumura Y, Takahara H (1996) DNA sequence from a fossil pollen of *Abies* spp. from Pleistocene peat. Genes Genet Syst 71:145–149

Suyama Y, Gunnarsson U, Parducci L (2008) Analysis of short DNA fragments from Holocene peatmoss samples. Holocene 18:1003–1006

Tinoco BA, Astudillo PX, Latta SC, Graham CH (2009) Distribution, ecology and conservation of an endangered Andean hummingbird: the Violet-throated Metaltail (*Metallura baroni*). Bird Conserv Int 19:63–76

Wandeler P, Hoeck PEA, Keller LF (2007) Back to the future: museum specimens in population genetics. Trends Ecol Evol 22:634–642

Willerslev E, Hansen AJ, Binladen J, Brand TB, Gilbert MTP, Shapiro B, Bunce M, Wiuf C, Gilichinsky DA, Cooper A (2003) Diverse plant and animal genetic records from Holocene and Pleistocene sediments. Science 300:791–795

Williams SE, Bolitho EE, Fox S (2003) Climate change in Australian tropical rainforests: an impending environmental catastrophe. Proc R Soc Lond B Biol Sci 270:1887–1892

Young A, Boyle T, Brown T (1996) The population genetic consequences of habitat fragmentation for plants. Trends Ecol Evol 11:413–418

Part V
Ecosystem Service and Socioeconomic Aspects with Special Reference to Biodiversity

Chapter 18
Monitoring Changes in Ecosystem Services Within a Forest Ecosystem

Kimiko Okabe and Hisatomo Taki

Abstract The importance of ecosystem services has been widely accepted since the publication of both the United Nations 2005 Millennium Ecosystem Assessment and The Economics of Ecosystems and Biodiversity, which assess future scenarios with and without ecosystem services, respectively. To sustainably maintain ecosystem services, assessment of the current state of ecosystem services together with the development of sound management practices are needed at both local and national/regional scales. In the monitoring of biodiversity, coarse-filter indicators are preferred as forest biodiversity indicators, although assessments based on fine-filter indicators should be included if possible because overall forest architecture is not necessarily correlated with species and genetic diversity in the forest. A variety of computer-based tools for ecosystem service assessment are available, although none of these is specialized for the forest ecosystem and for Asian survey. To develop ecosystem service monitoring at a larger (e.g., national) scale, it is ideal to select indicators based on the analysis of monitoring results of fine-scale indicators.

Keywords Biodiversity • Decomposition • Ecosystem service • Indicator • Pest control • Pollination • Scale

Introduction

Resources and natural processes that are provided by ecosystems and benefit human society are referred to as ecosystem services (Daily 1997; Kremen 2005). Ecosystem services can be classified into the following four categories: supporting services (e.g., nutrient cycling, soil formation, and primary production), provisioning

K. Okabe (✉) • H. Taki
Forestry and Forest Products Research Institute, 1 Matsunosato, Tsukuba, Ibaraki 305-8687, Japan
e-mail: kimikook@ffpri.affrc.go.jp

S.-i. Nakano et al. (eds.), *Integrative Observations and Assessments*, Ecological Research Monographs, DOI 10.1007/978-4-431-54783-9_18,

services (e.g., food, freshwater, wood, and fuel), regulating services (e.g., climate regulation, disease regulation, and water purification), and cultural services (e.g., spiritual, educational, and recreational; MEA 2005). Biodiversity fundamentally supports ecosystem services; however, its significance differs for each service provided by any given ecosystem (Dobson et al. 2006). Although there remains a lack of quantitative data regarding ecosystem services, a meta-analysis of published work indicated that a greater degree of biodiversity improves the provision of most services (Balvanera et al. 2006).

The importance of ecosystem services has been widely accepted since the publication of both the United Nations 2005 Millennium Ecosystem Assessment (MEA) and The Economics of Ecosystems and Biodiversity (TEEB), which assess future scenarios with and without ecosystem services, respectively, and the economic values associated with them (MEA 2005; Kumar and Muradian 2009). Although the understanding of ecosystem services is increasing worldwide, services from terrestrial ecosystems tend to receive the most attention. Various ecosystem services operate over different scales. For example, although some ecosystem services, such as pollination, are provided locally, others such as water regulation can operate over distances of hundreds of kilometers from a river basin, whereas factors such as air quality regulation occur on a global scale (MEA 2005; Kremen 2005). Consequently, to assess terrestrial ecosystem services, it is essential to know the relationship between the service targeted and the scale to be assessed.

It is also important to consider trade-offs in space and time between services. A good example of this is the increase of agricultural yields (a provisional service) with fertilizer application, which may result in degradation of water quality at downstream locations (a regulatory service). Similarly, large-scale removal of natural vegetation for agricultural production can generate regional groundwater problems at a later point in time, resulting in a temporal trade-off (Rodriguez et al. 2006).

To sustainably maintain ecosystem services, assessment of the current state of ecosystem services together with the development of sound management practices are needed at both the local and the national/regional scale. This chapter reviews ecosystem services in relationship to overall biodiversity, with particular emphasis on fauna, and the monitoring of forest biodiversity.

Ecosystem Services Provided by Forest Fauna

Hunting and fishing of forest fauna are well-established provisional services in many countries. For example, a conservative estimation of hunting yields indicates that 89.224 tons of bushmeat, with a market value of USD 190.7 million, are consumed each year by the rural population in the Brazilian Amazon (Peres 2000). Furthermore, the viability of hunting and fishing of wild vertebrates depends on the presence of other species in the ecosystem. Arthropods contribute to the recreational services of hunting, fishing, and observing wild vertebrates (e.g., birdwatching) by providing a critical nutritional resource. An economic study in the United States indicated that the annual value of recreational services provided by native insects was USD 49.96 billion (Losey and Vaughan 2006). Forest fauna provide cultural, intellectual, and scientific investigation services (de Groot et al. 2002). The use of forest fauna for ornamental purposes is also considerable. Wildlife can be collected and traded as pets or used for decoration in private homes, or to supplement academic collections or scientific research, and animal products can be used and traded as souvenirs, collectable items, or academic specimens. The forest fauna also provides a sense of human connection expressed through ethical, heritage, and religious values, such as respect for species that are considered sacred by certain religions.

Forest animal species, functional groups, and guilds contribute to regulating and supporting ecosystem services (Kremen 2005). In many cases, high diversity of species, functional groups, and guilds improves the resilience of advantageous ecosystem states, and subsequently protects the assembly of essential ecosystem services, particularly regulating and supporting services (Elmqvist et al. 2003). One of the best understood regulating and supporting services is pollination as performed by bees, flies, moths, beetles, and other insects, as well as some vertebrates. The growth of more than 75 % of the world's essential crops and 35 % of global food production is dependent upon animal pollination (Klein et al. 2007). In the Japanese buckwheat agro-ecosystem, forest insects such as wild honeybees contribute significant production services through pollination (Taki et al. 2010). These services are even more important for wild plant species; indeed, it is estimated that 87.5 % of plant species are pollinated by animals, ranging from 78 % in temperate regions to 94 % in tropical regions (Ollerton et al. 2011).

Forest ecosystems also provide habitats to predators and parasitoids, thus preventing overpopulation of pests. This service includes provision of habitats for reproduction, overwintering, and feeding on a variety of plant resources such as nectar, pollen, and sap (Jonsson et al. 2008). Thies and Tscharntke (1999) demonstrated that an increasingly diverse landscape, including forested areas, is correlated with the diversity and abundance of natural enemy populations within crop plantations, as well as with the potential to enhance the ecosystem service of pest control. Seed dispersal by forest species, functional groups, and guilds of animals such as birds, mammals, ants, and other fauna is another essential regulatory or support service (Kremen 2005). An economic evaluation conducted in a Swedish park containing a large population of oak trees (*Quercus* spp.) as a hemi-boreal forest keystone species analyzed the seed dispersal service performed by the Eurasian Jay (*Garrulus glandarius*), which nests in trees or large shrubs (Hougner et al. 2006). The results demonstrated that, depending on the seeding or planting technique chosen, the replacement cost per pair of jays in the park was between USD 4,900 and USD 22,500.

Waste decomposition is another service provided by fauna that is totally or partly dependent on forest ecosystems. For example, confining large domesticated mammals such as cattle to small areas creates challenging waste management problems. In addition to forage fouling, fauna such as dung beetles provide services that promote decomposition of dung into labile forms of nitrogen that can be absorbed by plants and then function as fertilizer (Gillard 1967) as well as reduce habitats for pest species such as flies. The annual economic value of dung burial provided by primarily native insects in the United States has been estimated at USD 380 million (Losey and Vaughan 2006).

A wide range of ecosystem services is provided by animal species, functional groups, and guilds that completely or partly inhabit forest ecosystems. However, forest fauna generate both benefits and costs (ecosystem disservices) to human activities. A lack of attention to these ecosystem disservices can negatively impact decision making in environmental management (Lyytimaki and Sipila 2009). Ecosystem services and disservices confer benefits and costs, respectively (Zhang et al. 2007), which are supplied by species over a range of scales, involving intentional and unintentional human activities. A particularly relevant example is the presence of crop and timber pest species, including herbivores and vectors of fungal, bacterial, and virus pathogens, which can decrease, and occasionally lead to the complete loss of, ecosystem productivity.

Monitoring of Forest Biodiversity

The Food and Agricultural Organisation of the United Nations (FAO) initiated integrated forest assessment, referred to as global forest resource assessment (FRA). This scheme utilizes a voluntary monitoring scheme conducted by each country every 5–10 years since 1946 to promote worldwide sustainable forest management. In the FRA, forest biodiversity is assessed as one of the seven thematic elements of sustainable forest management (FAO 2010). Although the FRA 2010 indicated a global increase of protected forest area (13 % of total world forest) as well as an increase of growing stock, which could be an indication of tree species richness and relative abundance, it also found that primary forest has decreased by 0.4 % every year over the preceding 10 years (FAO 2010). Given that the FRA was not designed for the conservation of forest biodiversity, determining indicators that can be used to monitor biodiversity would be an important addition to the process (FAO 2010).

The Global Biodiversity Outlook (GBO) undertaken by the Convention on Biological Diversity (CBD) assessed forest ecosystems alongside others including inland water and marine ecosystems (CBD 2001, 2006, 2010). Continuous rapid loss of forest was reported in GBO3, although the rate of net forest loss slowed because of forest expansion in the temperate zone (CBD 2010). GBO3 also identified the main negative impacts on terrestrial biodiversity as habitat loss and degradation, climate change, pollution, overexploitation and unsustainable use, and invasive alien species based on a meta-analysis of national reports (CBD 2010). An overall global assessment was completed, although regional and national assessments were not undertaken. The national strategy for the conservation and sustainable use of biological diversity of Japan identified exploitation, abandonment of managed/used land, threats resulting from human activity such as chemical pollution, invasive alien species, and climate change as significant threats to Japanese biodiversity (The Ministry of the Environment; MOE 2007). Based on these threats, the Japanese Biodiversity Outlook (JBO) assessed the past and current states alongside drivers for each threat, in each ecosystem, using national survey data collected by the MOE, scientific publications, and responses from conservation professionals (MOE 2010). The JBO has subsequently undertaken the first national-level assessment for overall biodiversity (including species other than those classed as endangered or threatened) and identified factors driving these threats, thus enabling policy makers to determine appropriate countermeasures.

For such assessments, long-term ecological monitoring, such as the 70-year bird monitoring data held by the British Trust for Ornithology (BTO), is extremely useful. Studies including meta-data are most suitable for larger-scale analyses. Collaborative use of existing databases such as the CPB/NCEAS (Centre of Population Biology/National Centre for Ecological Analysis) Global Population Dynamics Database and the Global Biodiversity Information Facility (GBIF) is typically undertaken in large-scale studies. However, although scientific databases are generally accurate, most of the data were collected at a regional level. The available long-term statistical measurements relating to forest ecosystems include data such as the area of forest cover that can be used to estimate deforestation, as mentioned previously, but this information cannot be used to estimate species composition and biodiversity. There is a need for research to bridge the gaps between statistical information (e.g., land-use change) and ecological data (e.g., forest biodiversity), using past trends to predict future conditions. For example, in Japan it was suggested that bird distribution surveys be conducted twice in different years because of land-use changes in forests as evidenced from national statistics (Yamaura et al. 2009). For assessments such as these, it is useful to generate a relevant index, such as the Living Planet Index developed by GBO. The linkage between science and decision making is essential for assessment as well as sustainable and adaptive management (Thompson et al. 2011).

Okabe and Ogawa (2011) identified five steps to establish a new forest monitoring and assessment scheme: (1) definition of explicit objectives and goals, (2) selection of a relevant organization to conduct the monitoring, (3) selection of indicators, (4) development of monitoring methods using the indicators, and (5) analysis and practical use of monitoring results. Within these stages, the selection of indicators, including surrogate indicators, is considered the most fundamental step to developing a successful assessment. Either fine-filter or coarse-filter indicators can be selected depending on the scale of the study and its objectives and goals. Fine-filter indicators are generally used for small-scale monitoring, whereas coarse-filter indicators are used at the full ecosystem level, and in particular when the heterogeneity of the ecosystem and its connectivity to the landscape at a larger scale must be considered (Noss 1987). Coarse-filter indicators are preferred as forest biodiversity indicators because existing data are available, including national statistics

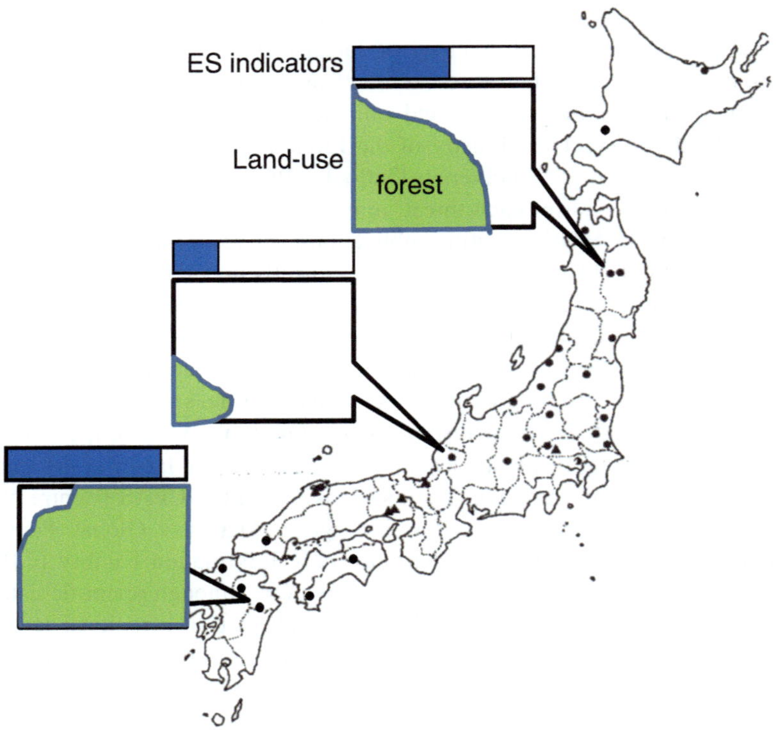

Fig. 18.1 Schematic diagram to find relevant coarse-filter indicator(s) of ecosystem services (*ES*)

and management records at the stand level, as well as remote sensing images. However, assessments based on fine-filter indicators should be included where possible because overall forest architecture is not necessarily correlated with species and genetic diversity in the forest (Redford 1992; Failing and Gregory 2003). Temporal-scale differences between a coarse-filter indicator and species and genetic diversity should also be considered because tree species might be more resistant to disturbance than organisms exploiting a microhabitat in the forest (Lindenmayer et al. 2000). For large-scale monitoring, it is ideal to select indicators based on the analysis of monitoring results of fine-scale indicators that were carefully selected within the larger scale (Fig. 18.1). For example, for the pollination and pest regulation services provided by honeybees (Fig. 18.2a) and hunting wasps (Fig. 18.2b), respectively, the area of the case study site will extend to the landscape scale to encompass the home ranges of key ecosystem service agents. By contrast, for the decomposition service provided by fungi associated with the decay of dead wood (Fig. 18.2c), coarse-filter indicators are likely difficult to determine and might require different approaches to estimate the service.

Fig. 18.2 Examples of ecosystem service agents: pollinator, *Apis cerana* (**a**); predator, *Vespa crabro* (**b**); decomposer, *Trametes hirsute* (**c**)

Monitoring of Forest Ecosystem Services

A variety of computer-based tools for ecosystem service assessment are available. InVEST (Integrated Valuation of Environmental Services and Tradeoffs) was developed to enable decision makers to assess trade-offs between human development and ecosystem conservation (Natural Capital Project 2012). The Urban Forest Effects (UFORE) model, which was developed by the USDA to improve urban forest planning, management, and design, can also assess air quality regulation and net carbon sequestration services (USDA 2009). However, improvement of these assessment tools is necessary to cover as many ecosystem services as possible and to maintain accuracy when moving from the local to larger scale. Some ecosystem services, such as air and water regulation as well as natural hazard regulation, can be monitored relatively easily because they are physically measurable, whereas ecosystem services related to organisms are relatively difficult to quantify; this is because service agents are not always well known, and the amounts of both product and outcome provided by the ecosystem services and service agents fluctuate over time. Also, the scale that a service reaches can be unclear. The mechanism for each of the ecosystem services must therefore be fully understood. The ultimate beneficiary of the ecosystem service must also be determined. A further challenge has proven to be the limited knowledge of terrestrial biodiversity in East and Southeast Asia compared with tropical ecosystems in other countries (Sodhi and Brool 2006). The spatial scale of pollination, pest regulation, and decomposition are all local and depend on a group of organisms inhabiting, and being active across, a relatively small location (Guariguata and Balvanera 2009). Local forest management assessed over a radius of a few kilometers would therefore be helpful for famers expecting these services (Taki et al. 2010, 2011).

Ecosystem services associated with forest degradation are a major concern for carbon emission (Miles and Kapos 2008). Forest degradation can be defined as the reduction of the capacity of a forest to provide goods and services (FAO 2002), although there are more than 50 alternative definitions (Lund 2009; FAO 2011). Biodiversity is considered to be safeguarded in the activities of the UN REDD (Reduced Emissions from Deforestation and forest Degradation) program, but uncertainty remains in determining an appropriate definition of forest degradation and sustainable forest management. Additionally, interactions between biodiversity and carbon management should be clearly assessed. Local indicators and measurements are required, because forest degradation is not a static state but is, rather, an ongoing process (FAO 2011).

Acknowledgments The study was partly supported by the Global Environment Research Fund S9 and D-1008.

References

Balvanera P, Pfisterer AB, Buchmann N, He J-S, Nakashizuka T, Raffaelli D, Schmid B (2006) Quantifying the evidence for biodiversity effects on ecosystem functioning and services. Ecol Lett 9:1146–1156

CBD (2001) Global biodiversity outlook. CBD, Montreal

CBD (2006) Global biodiversity outlook 2. CBD, Montreal

CBD (2010) Global biodiversity outlook 3. CBD, Montreal

Daily GC (1997) Nature's services. Island Press, Washington, DC

de Groot RS, Wilson MA, Boumans RMJ (2002) A typology for the classification, description and valuation of ecosystem functions, goods and services. Ecol Econ 41:393–408

Dobson A, Lodge D, Alder J, Cumming GS, Keymer J, McGlade J, Mooney H, Rusak JM, Sala O, Wolters V, Wall D, Winfree R, Xenopoulos MA (2006) Habitat loss, trophic collapse, and the decline of ecosystem services. Ecology 87:1915–1924

Elmqvist T, Folke C, Nystrom M, Peterson G, Bengtsson J, Walker B, Norberg J (2003) Response diversity, ecosystem change, and resilience. Front Ecol Environ 1:488–494

Failing L, Gregory R (2003) Ten common mistakes in designing biodiversity indicators for forest policy. J Environ Manag 68:121–132

FAO (2002) Proceedings: second expert meeting on harmonizing forest-related definitions for use by various stakeholders. http://www.fao.org/docrep/005/y4171e/y4171e00.HTM

FAO (2010) FAO forestry paper 163. FAO, Rome

FAO (2011) Assessing forest degradation. Forest Resources Assessment Working Paper 177. FAO, Rome.

Gillard P (1967) Coprophagous beetles in pasture ecosystems. J Aust Inst Agric Sci 33:30–34

Guariguata MR, Balvanera P (2009) Tropical forest service flows: improving our understanding of the biophysical dimension of ecosystem services. For Ecol Manag 258:1825–1829

Hougner C, Colding J, Soderqvist T (2006) Economic valuation of a seed dispersal service in the Stockholm National Urban Park, Sweden. Ecol Econ 59:364–374

Jonsson M, Wratten SD, Landis DA, Gurr GM (2008) Recent advances in conservation biological control of arthropods by arthropods. Biol Control 45:172–175

Klein AM, Vaissiere BE, Cane JH, Steffan-Dewenter I, Cunningham SA, Kremen C, Tscharntke T (2007) Importance of pollinators in changing landscapes for world crops. Proc R Soc Lond B Biol Sci 274:303–313

Kremen C (2005) Managing ecosystem services: what do we need to know? Ecol Lett 8:468–479

Kumar P, Muradian R (2009) Payment for ecosystem services. Oxford University Press, Oxford

Lindenmayer DB, Margules CR, Botkin DB (2000) Indicators of biodiversity for ecologically sustainable forest management. Conserv Biol 14:941–950

Losey JE, Vaughan M (2006) The economic value of ecological services provided by insects. Bioscience 56:311–323

Lund HG (2009) What is a degradation forest? Forest Information Service, Gainsville

Lyytimaki J, Sipila M (2009) Hopping on one leg—the challenge of ecosystem disservices for urban green management. Urban For Urban Green 8:309–315

MEA (2005) Ecosystems and human well-being. Island Press, Washington, DC

Miles L, Kapos V (2008) Reducing greenhouse gas emissions from deforestation and forest degradation: global land-use implications. Science 320:1454–1455

MOE (2007) National strategy for the conservation and sustainable use of biological diversity. http://www.biodic.go.jp/biodiversity/about/initiatives3/index.html

MOE (2010) Japan Biodiversity Outlook. www.biodic.go.jp/biodiversity/activity/policy/jbo/jbo/files/allin.pdf (in Japanese)

Natural Capital Project (2012) InVEST. http://www.naturalcapitalproject.org/InVEST.html. Accessed 21 February 2012

Noss RF (1987) From plant communities to landscapes in conservation inventories: a look at the Nature Conservancy (USA). Biol Conserv 41:11–37

Okabe K, Ogawa M (2011) History and future perspective of monitoring of forest biodiversity. Bull FFPRI 10:231–250

Ollerton J, Winfree R, Tarrant S (2011) How many flowering plants are pollinated by animals? Oikos 120:321–326

Peres CA (2000) Effects of subsistence hunting on vertebrate community structure in Amazonian forests. Conserv Biol 14:240–253

Redford KH (1992) The empty forest. Bioscience 42:412–422

Rodriguez JP, Beard TD Jr, Bennett EM, Cumming GS, Cork SJ, Agard J, Dobson AP, Peterson GD (2006) Trade-offs across space, time, and ecosystem services. Ecol Soc 11:28. http://www.ecologyandsociety.org/vol11/iss1/art28/

Sodhi NS, Brool BW (2006) Southeast Asian biodiversity in crisis. Cambridge University Press, Cambridge

Taki H, Okabe K, Yamaura Y, Matsuura T, Sueyoshi M, Makino S, Maeto K (2010) Effects of landscape metrics on *Apis* and non-*Apis* pollinators and seed set in common buckwheat. Basic Appl Ecol 11:594–602

Taki H, Yamaura Y, Okabe K, Maeto K (2011) Plantation vs. natural forest: matrix quality determines pollinator abundance in crop fields. Sci Rep. http://www.nature.com/srep/2011/111028/srep00132/full/srep00132.html

Thies C, Tscharntke T (1999) Landscape structure and biological control in agroecosystems. Science 285:893–895

Thompson ID, Okabe K, Tylianakis JM, Kumar P, Brockerhoff EG, Schellhorn N, Parrotta JA, Nasi R (2011) Forest biodiversity and the delivery of ecosystem goods and services: translating science into policy. Bioscience 61:972–981

USDA (2009) Urban Natural Resources Stewardship. http://www.nrs.fs.fed.us/urban/monitoring/tools_ecosystem_services/. Accessed 20 February 2012.

Yamaura Y, Amano T, Koizumi T, Mitsuda Y, Taki H, Okabe K (2009) Does land-use change affect biodiversity dynamics at a macroecological scale? A case study of birds over the past 20 years in Japan. Anim Conserv 12:110–119

Zhang W, Ricketts TH, Kremen C, Carney K, Swinton SM (2007) Ecosystem services and dis-services to agriculture. Ecol Econ 64:253–260

Chapter 19
Evaluating Relationships Between Biodiversity and Ecosystem Functions in Forests Using Forest Inventory and Allometry Data

Masae I. Ishihara, Hajime Utsugi, Hiroyuki Tanouchi, and Tsutom Hiura

Abstract Biodiversity loss may cause various changes in ecosystem functions of forests. This chapter presents methods and possibilities of evaluating biodiversity–ecosystem function relationships in forest ecosystems using forest inventory and allometry data. Although these data have been collected in plantations and natural forests all over the world, their use in biodiversity–ecosystem function studies is still limited. Inventory and allometry data provide various information such as productivity and carbon stock of forests, growth and demography of individual trees, tree size and mass distribution, horizontal and vertical structure of forests, and tree species composition of the local community and regional species pool. Such information is indispensable to understand the mechanisms of biodiversity effects on ecosystem functions, to predict the consequence of biodiversity loss by improving functional trait approaches toward more mechanistic and predictive paths, and to develop studies of biodiversity–ecosystem function that are relevant to conservation. Two new databases, a forest inventory database and an allometry database, and data papers will promote biodiversity–ecosystem function studies through enhanced data availability, especially in Asia where such studies are limited.

M.I. Ishihara (✉)
Field Science Center for Northern Biosphere, Hokkaido University, Kita 9, Nishi 9, Kita-ku, Sapporo 060-0809, Japan
e-mail: masaish.n@gmail.com

H. Utsugi
Hokkaido Research Center, Forestry and Forest Products Research Institute, Hitsujigaoka 7, Toyohira, Sapporo 062-8516, Japan

H. Tanouchi
Shikoku Research Center, Forestry and Forest Products Research Institute, 2-915 Asakuranishimachi, Kochi 780-8077, Japan

T. Hiura
Tomakomai Research Station, Field Science Center for Northern Biosphere, Hokkaido University, Takaoka, Tomakomai, 053-0035, Japan

S.-i. Nakano et al. (eds.), *Integrative Observations and Assessments*, Ecological Research Monographs, DOI 10.1007/978-4-431-54783-9_19, © Springer Japan 2014

Keywords Biomass • Carbon stock • Data paper • Database • Functional traits • Production • Tree census

Introduction

Loss of biodiversity caused by human activities and climate change is a global environmental issue. The ecological consequences of biodiversity loss on ecosystem functions has become one of central research agendas (Srivastava and Vellend 2005; Naeem et al. 2007). The relationship between biodiversity and ecosystem functions has been studied mainly in ecosystems that are easy to manipulate, such as grasslands and aquatic microcosms (Scherer-Lorenzen et al. 2007b). By contrast, there are comparatively few studies in forest ecosystems (Cardinale et al. 2011). It remains uncertain whether biodiversity loss causes changes in the functions of forest ecosystems (Nadrowski et al. 2010).

The purpose of this chapter is to present methods and possibilities of evaluating biodiversity–ecosystem function relationship in forest ecosystems using forest inventory and allometry data. Inventory data have been widely used to determine tree species diversity and their demography (growth, mortality, and recruitment rates). By using allometric equations relating tree size to biomass, inventory data have provided estimates of ecosystem functions, such as productivity and aboveground carbon stock. First, we review how inventory and allometric data have been used in biodiversity–ecosystem function studies. We limit our discussion to nonbiotic functions in forests although biotic functions and services, such as pollination and resistance to pathogens or invasive species, are important as well (see Okabe and Taki, Chap. 18, this volume). Second, we discuss the utility of inventory and allometry data in the development of biodiversity–ecosystem function studies that are relevant to conservation. Prerequisite for such development is data availability. In the final section, we introduce two new databases, a forest inventory database and an allometry database, and data papers as measures to archive data and increase public accessibility of data.

Biodiversity and Ecosystem Function Studies in Forests

The earlier studies of biodiversity–ecosystem function in forests compared productivity of monoculture to 2- to 4-species polyculture stands using inventory data of established plantations (Piotto 2008). These studies lacked intermediate- and high-diversity stands (Scherer-Lorenzen et al. 2007b). Effects of silvicultural management, such as thinning, made the results difficult to interpret. To overcome these limitations, forest stands differing in tree diversity (1–18 species) have been created in boreal, temperate, and tropical regions since 1999 (Scherer-Lorenzen et al. 2005). These synthetic assemblage experiments controlled environmental conditions, past land-use history, and management practices so that the effects of biodiversity on various ecosystem functions can be tested.

The relevance of these experiments for natural systems may be limited (Leuschner et al. 2009). For example, Ruiz-Jaen and Potvin (2011) could not detect the effect of species richness on aboveground carbon stock in a old-growth natural forest, although species richness was positively correlated to carbon stock in a young synthetic assemblage forest. Plot sizes in experiments are generally not large enough to obtain a significant sample size of adult trees, to reduce edge effect, and to investigate gap dynamics and resilience to disturbance (Leuschner et al. 2009). Even-aged forests in these experiments present unnatural canopy and root structures, which may change the effect of biodiversity on ecosytem functions.

Observational studies that compare existing natural forests using forest inventory data can complement the shortcomings of studies in plantations and experiments (Leuschner et al. 2009). They can test whether the relationship found in experimental studies is consistent in natural systems, especially at a regional scale. However, results of observational studies should be interpreted with caution. Correlation between biodiversity and ecosystem functions does not necessarily imply a causal relationship between the two because correlation might be a result of covarying factors, such as climatic and environmental factors, disturbance regimes, forest successional stages, and silvicultural practices (Vilà et al. 2005; Scherer-Lorenzen et al. 2007b). Statistical analysis incorporating covarying variables can partition the effect of biodiversity and other factors on ecosystem functions. For example, Vilà et al. (2007) and Paquette and Messier (2011) found a positive effect of biodiversity on tree productivity even after controlling for climatic, environmental, and silvicultural variables. They found that the biodiversity–ecosystem function relationship differs among forest types using forest inventory data collected at a regional scale. Such observational studies should be conducted in various regions to find possible factors causing various biodiversity–ecosystem function relationships in natural systems.

To estimate productivity and aboveground carbon stock from inventory data, allometric equations that relate biomass to stem diameter and tree height are necessary. Although some studies used growth rate of stem or tree height as a surrogate of productivity instead of biomass, these indices respond differently to diversity (Piotto 2008). Numerous allometric equations were created for different species at various localities (e.g., Ter-Mikaelian and Korzukhin 1997). Usually, researchers used one of these equations that was created for closely related species or at a geographically or environmentally close site because the creation of a new allometric equation for each species or each stand is labor intensive or sometimes impossible. However, allometric equations remain one of the largest sources of uncertainty in estimates of biomass and carbon stock. Difference between the estimated biomass using different equations could be as large as 1.2 fold (Chave et al. 2004) to 2 fold (Ishihara, unpublished data). Allometric equations that can reliably estimate biomass across broad regions or species are required.

New Direction in Biodiversity and Ecosystem Function Studies and the Role of Inventory and Allometry Data

Mechanisms of Relationship Between Biodiversity and Ecosystem Functions

For biodiversity–ecosystem function studies to be relevant to conservation, mechanisms of the relationship between biodiversity and ecosystem functions should be revealed. Potential mechanisms that create positive relationships have been proposed: niche complementarity, facilitation, sampling effect, and dilution effect for stock and flux of ecosystem functions; and insurance effect, portfolio effect, and compensatory dynamic effect for the stability of ecosystem functions (Srivastava and Vellend 2005). However, most studies relied on correlation, and therefore the mechanisms are still unclear (Srivastava and Vellend 2005; Cardinale et al. 2011).

Detailed analysis of the growth and demography of each tree is necessary to reveal the mechanisms of biodiversity effect on forest productivity and carbon stock. Potvin and Dutilleul (2009) suggested that the biodiversity effect can be reduced to a neighborhood effect because niche complementarity and facilitation occur between neighborhoods. Inventory data collected over the long term at many places provide a large data set of individual tree growth and demography. By using such data, the effect of species identity and diversity on tree growth and demography can be tested (Uriarte et al. 2004): this is an advantage of the tree system over a grass system where it is difficult to track each individual plant. Species richness may create higher structural diversity, which may enhance niche complementarity through vertical and horizontal environmental heterogeneity, and may result in higher forest productivity (Ishii et al. 2004; Lei et al. 2009; Morin et al. 2011). The effect of structural diversity on ecosystem function can be tested using inventory data that record the horizontal and vertical distribution of each tree.

Predictive Science from Functional Trait Approach

Studies on biodiversity–ecosystem function have been increasingly required to predict changes in ecosystem functions caused by biodiversity loss and environmental changes (Naeem et al. 2007). The functional traits approach (Naeem and Wright 2003; McGill et al. 2006; Naeem et al. 2007) has been assumed to be promising. This approach is based on the assumption that there is strong interaction between an ecosystem and species functional traits (Lavorel and Garnier 2002; Naeem et al. 2007). Species composition and abundance of a local community will be determined by environmental factors that filter species from the regional species pool according to their response traits, and then by community interaction and feedback (Lavorel and Garnier 2002; Naeem and Wright 2003; Naeem et al. 2007).

Local species composition will be translated into effect trait distributions from which ecosystem functions can be predicted. For example, Bunker et al. (2005) varied extinction probabilities of tree species according to their functional traits and simulated changes in the aboveground carbon stock of a tropical forest.

The environment–trait relationship is an essential part of this approach. The relationship has been expressed using regression models regardless of frequentist or Bayesian inferences (Webb et al. 2010). Because what regression implies is causation, at best, but not a mechanism, regression models have limitations in predicting future conditions. The functional trait approach will be more mechanistic and predictive by incorporating demography, mechanistic models, and neutral theory.

Recent study found a relationship between functional traits and demographic rates, such as growth rate and mortality (Poorter et al. 2008). Functional traits may affect species composition or ecosystem functions through changes in species demographic rates. However, demographic rates are difficult to measure compared to morphological or physiological traits. Long-term inventory data are needed to estimate demographic rates in natural systems. Demographic rates are affected by environmental factors, and thus should be expressed as a function of environmental factors. Availability of light and other resources for each tree, which is usually not measured in inventory data, can be estimated as a probability distribution conditional on measured variables in inventory data, such as crown class and neighborhood crowding. Lichstein et al. (2010) estimated light-dependent growth rate from inventory data that lack direct individual-level measurements of light availability. By using such a method, we can estimate demographic rates as a function of environmental factors. Metabolic theory may also provide the estimates of demographic rates from functional traits. Enquist et al. (2007) developed a model that predicts growth rate of a plant from functional traits, body mass, and geometry of plant architecture using physiological and metabolic scaling theories. Such a model may be used to predict the productivity of a forest, if the body mass distribution of each species composing the community is known. Inventory and allometry data can provide such data as well. However, the utility of such global- or continental-scale theory at a local scale has yet to be tested (Duffy et al. 2009).

The functional trait approach can incorporate mechanistic models or models based on principles. Webb et al. (2010) introduced a hybrid framework incorporating a dynamic population model with a conventional phenomenological Bayesian model. Beside such models at a population level, individual-based process models are also promising for the development of the predictive functional trait approach. Sterck et al. (2011) developed a biophysical model to predict net carbon gain of a sapling from measurable functional traits. Such models may be extended to simulate carbon gain of adult trees, and thus stand-level productivity, although ontogenetic physiological change and interaction between neighboring trees have to be incorporated. Individual-based models rely mostly on simulations and are useful to test the possible mechanisms of biodiversity effect on ecosystem functions. Sterck et al. (2011) tested which functional traits drove functional divergence along light and water availability. Furthermore, by comparing the results of simulation and the observed distribution along environmental gradients in the field, they found that functional divergence does not necessarily contribute to niche differentiation. Inventory data can be used to validate mechanistic models and to test various hypotheses.

It is promising to include neutral theory (Hubbell 2001) with the functional trait-based approach, which is based on niche assembly perspective. Niche differentiation is not the only process that determines species diversity. Neutral theory showed that dispersal limitation alone can create observed diversity patterns, such as species abundance distribution and species–area curves, without assuming inter-specific difference in fitness. After intense debate, it is widely accepted that niche differentiation and dispersal limitation are both important. The strength of neutral theory is that it realizes fundamental processes, such as birth, death, or immigration of an individual, that it gives a quantitative prediction, and that it is based on sampling theory, which enables us to compare model predictions to field data (Alonso et al. 2006). Furthermore, neutral theory relates the regional species pool to a local community and predicts diversity patterns at various scales. Such a theory that relates the local to the regional scale is necessary for biodiversity–ecosystem function studies to be relevant to conservation (Srivastava and Vellend 2005). Jabot et al. (2008) included environmental filtering to the neutral theory, and Jabot (2010) further developed an analytical method to estimate parameters of environmental filtering function from regional and local species abundances and a functional trait of each species. Inventory data, if pooled, can provide species abundance of a regional pool as well as that of a local community (Hubbell 2001; Jabot et al. 2008).

Stability of Ecosystem Functions and Diversity at Landscape Levels

In a conservation context, stability of ecosystem functions is more desired in general than the magnitude of flux or stock (Srivastava and Vellend 2005). However, few studies have tested the relationship between biodiversity and stability, and it is difficult to generalize these results (Griffin et al. 2009). To study the effect of biodiversity on the stability of ecosystem functions, we need long-term data over decades or centuries: the effect of perturbation on ecosystem functions may last several years, at least in forests, and time-series data are correlated with one another. Furthermore, short-term data provide few temporal replications of perturbations, particularly in a forest ecosystem where perturbations are rare. Using 60-year tree growth data obtained from stem increment cores in natural coniferous forests, DeClerck et al. (2006) found that species richness did not affect the stability of stem growth under severe droughts. As long-term inventory data along natural diversity gradients become available, biodiversity effects on stability can be tested.

The stability of ecosystem functions may depend on diversity at the landscape level. Evenness of forest types within a landscape is positively correlated to long-term stability of net primary production (Bradford 2011). Forest types can be substituted into more quantitative measures based on species abundance. The recent development of remote sensing techniques enables us to estimate species diversity (Tollefson 2011) and ecosystem functions, such as carbon stock (Saatchi et al. 2011), at a large scale. Inventory data can be used for the development and validation of such techniques.

Measures to Archive and Share Inventory and Allometry Data

To promote studies mentioned earlier, inventory and allometry data should be publicly available. Raw data can be used for new studies or meta-analyses, verification of results, resources for policy makers, education, and new uses (Whitlock et al. 2010). However, ecological data are frequently lost to science after publication because data are stored in inappropriate media, lack detailed information about methodology, or are discarded when the investigators change careers, retire, or pass away (Michener et al. 1997; Whitlock et al. 2010). Inventory and allometry data have been collected in many locations worldwide, but there are few databases that archive raw data or metadata of these data throughout different projects. Here, we introduce two new databases and data papers as measures to archive and share data.

Forest Inventory Database

We created a forest inventory database collecting metadata of 968 forest plots in Japan (Fig. 19.1). The forest plot list was generated with formal searches of journals published in Japan (journals published by the Japanese Forest Society and the Ecological Society of Japan, bulletins of university forests, bulletins published by the Forestry and Forest Products Research Institute, and those of prefectural forestry research institutes). This list was supplemented with the lists reported in Kanzaki (2006) and Ishihara et al. (2011) and with the databases of PlotNet (http://eco1.ees.hokudai.ac.jp/plotnet/home, accessed 31 March 2012), Forest Dynamic Data Base (http://fddb.ffpri-108.affrc.go.jp/, accessed 31 March 2012), and long-term observational forests of Hokkaido University (http://forest.fsc.hokudai.ac.jp/~exfor/fr/, accessed 31 March 2012), covering almost all plots that were used for publications in international journals. We limited our search to plots with area $\geq$0.04 ha and in natural forests. The forest plot list includes plots that were measured only once as well as permanent plots. Double count of the same plot

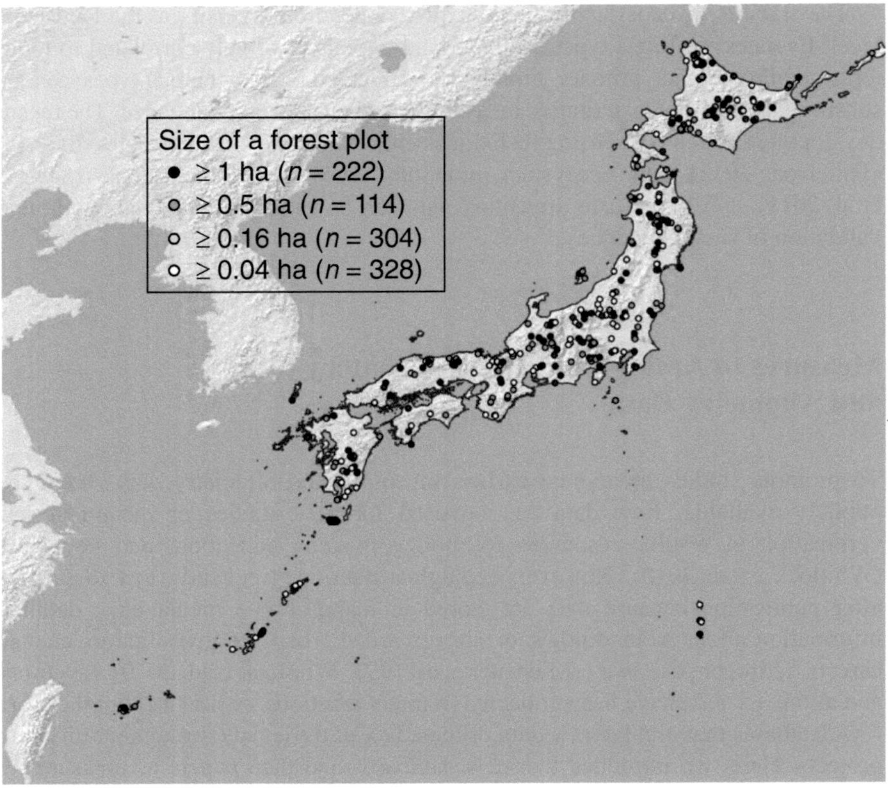

Fig. 19.1 Location of 968 forest plots in Japan recorded in the forest inventory database

reported in different publications was avoided. We collected metadata of each plot: location, plot size, data availability, census year, forest type, and minimum size of trees measured.

Although 65 % of the plots were smaller than 0.5 ha, there were 222 plots larger than or equal to 1 ha (Fig. 19.2). More than 80 % of the plots provided data in the form of either raw data or species composition (abundance, basal area, or stem wood volume). Pooling local abundance of smaller plots may provide species abundance of regional species pools while larger plots may be used for the evaluation of ecosystem functions.

Allometry Database

We created an allometry database archiving biomass data of more than 1,486 trees of 122 species belonging to 65 genera and 37 families sampled in 73 natural forests over Japan (Fig. 19.3). These data were collected from 42 studies published

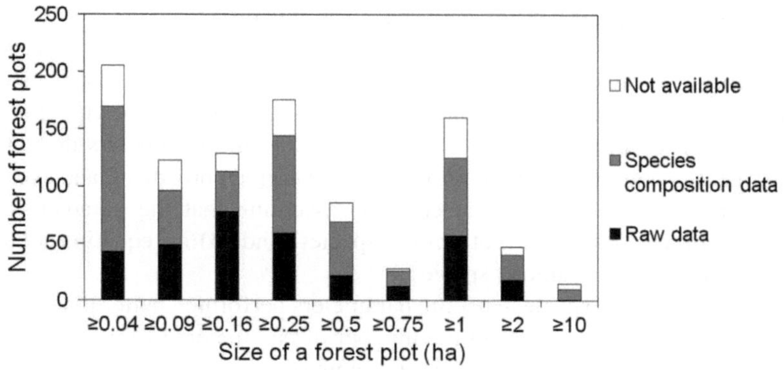

Fig. 19.2 Frequency of forest plots in different sizes in Japan ($n = 968$). Plots were classified with the data availability. Raw data: raw inventory data were available through publications or database. Species composition data: abundance, basal area, or stem volume of each species was reported in the literature. Not available: neither raw data nor species composition data were available through publications or database

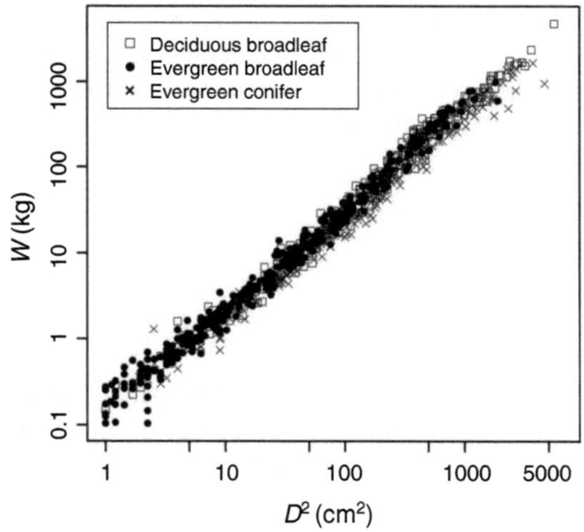

Fig. 19.3 Relationship between square of stem diameter (D^2) and aboveground biomass of a tree (W) for three different functional groups

between 1957 and 2010 using a database search (Japanese Forestry Literature Information System: http://www2.ffpri.affrc.go.jp/folis21/folis-hp-j.html, accessed 31 May, 2012) and unpublished data. Size of trees (stem diameter and tree height) and biomass of stems, branches, leaves, and roots were digitized.

Using this dataset, we tested whether the information of functional groups and species improves the estimation of aboveground biomass. We limited our analysis to species with more than nine samples. Deciduous conifer species were excluded because *Larix kaempferi* was the only species in this functional group. At the end, 932 trees were used for the analysis. Aboveground biomass (W) was regressed to stem diameter at breast height (D) and tree height (H) as $\ln(W) = \alpha + \beta \ln(D) + \gamma \ln(H)$ (Chave et al. 2005). Three versions of this allometric equation were compared: (1) an allometric equation assuming that the parameters do not vary among functional groups (deciduous broadleaf, evergreen broadleaf, and evergreen conifer) or among species; (2) an equation assuming that the parameters vary among functional groups but not among species; and (3) an equation assuming that the parameters vary among species.

The model (3) with species-specific parameters performed better than the other two models even after penalizing the number of parameters (Table 19.1). Errors between the estimated and measured total biomass of a stand reduced from 16 % to 5 % on average when the species information was included in the allometric equation. These trends were consistent when we compared three models for each functional group. Even when we compared the models in a single forest, the model with species information performed better than the other two. These results suggest that information on tree species identity improves biomass estimation. The effect of species on biomass may be reduced to the effect of functional traits such as wood density. Chave et al. (2005) and Komiyama et al. (2011) developed allometric equations with wood density as a variable that can predict biomass of broad species in tropical forests and cool temperate forests, respectively. As functional trait data are accumulated, common allometric equations that can be applied to various forest types in wide regions may be obtained from the present allometry database.

Data Paper

One mechanism to promote data sharing is data papers. Data papers are peer-reviewed articles documenting data sets of ecological significance and published by journals such as *Ecology* and *Ecological Research*. Data sets can be original data collected by authors (e.g., Clark and Clark 2000) or those obtained through a national project (e.g., Ishihara et al. 2011), or secondary data collected from publications (e.g., Cardinale et al. 2009). Only the abstract is printed in the journals, and digitized data sets and metadata are available from archives.

Data papers are promising way to archive and increase public accessibility of data while rewarding and giving credit to data collectors. Anyone can use data sets by citing data papers. For example, 51 data papers were published in *Ecology* from 2000 to 2011, and 6 of them documented forest inventory data sets. These data papers were cited as references or used as test data sets for papers arguing methodological improvements. Although the data set was not used for meta-analysis, such usage may increase as more data papers are published.

Table 19.1 Effects of functional group and species on the estimation of aboveground biomass by allometric equations in Japanese natural forests

Difference in parameters	Regression model	AIC	*df*	R^2	Error (%)
All functional groups					
(1) No difference	$\ln(W) = -2.41 + 2.07n(D) + 0.45 \ln(H)$	176.8	4	0.986	16.3 ± 14.3
(2) Among functional groups ($j = 1-3$)	$\ln(W) = \alpha_j + \beta_j \ln(D) + \gamma_j \ln(H)$[a]	24.8	10	0.996	12.3 ± 9.4
(3) Among species ($i = 1-27$)	$\ln(W) = \alpha_i + \beta_i \ln(D) + \gamma_i \ln(H)$[b]	−359.7	82	0.997	5.4 ± 4.1
Deciduous broadleaf					
(2) Among functional groups ($j = 1$)	$\ln(W) = -2.51 + 2.25 \ln(D) + 0.33 \ln(H)^{\dagger}$	−54.7	4	0.986	8.7 ± 6.0
(3) Species ($i = 1-12$)	$\ln(W) = \alpha_i + \beta_i \ln(D) + \gamma_i \ln(H)$[b]	−183.0	37	0.998	5.2 ± 4.3
Evergreen broadleaf					
(2) Among functional groups ($j = 2$)	$\ln(W) = -2.24 + 2.13 \ln(D) + 0.34 \ln(H)^{\dagger}$	37.0	4	0.986	14.8 ± 10.3
(3) Species ($i = 1-9$)	$\ln(W) = \alpha_i + \beta_i \ln(D) + \gamma_i \ln(H)$[b]	−65.6	28	0.994	8.6 ± 6.8
Evergreen conifer					
(2) Among functional groups ($j = 3$)	$\ln(W) = -2.37 + 2.15 \ln(D) + 0.27 \ln(H)^{\dagger}$	38.0	4	0.985	21.9 ± 13.2
(3) Species ($i = 1-6$)	$\ln(W) = \alpha_i + \beta_i \ln(D) + \gamma_i \ln(H)$[b]	−124.6	19	0.998	4.5 ± 3.0

[a] Parameters of each functional group are identical to the parameters of the models with †

[b] Parameters of model (3) are not shown for simplicity

Three regression models (1–3) were obtained to estimate *W* (aboveground biomass of a tree, kg) from *D* (stem diameter at breast height, cm) and *H* (tree height, m), assuming that parameters (1) do not differ among functional groups or species, (2) differ among functional groups but not among species, and (3) differ among species. The models were compared using the error between the estimated and measured total biomass of a forest for each model and those of each functional group within a forest. Shown are the models with best-fit parameters, Akaike Information Criterion (AIC), degree of freedom (*df*), and coefficient of determination (R^2) of the models, and mean ± SD of errors between the measured and estimated biomass using the model for each forest (Error)

Conclusions

Forest inventory and allometry data are indispensable for the development of biodiversity–ecosystem function studies. These data have been used to determine species diversity, productivity, and aboveground carbon stock of local communities. Furthermore, detailed analysis of individual trees recorded in inventory data can reveal the mechanisms of the biodiversity–ecosystem function relationship. Inventory and allometry data provide demographic rates of species, distribution of tree size and mass of a local community, horizontal and vertical structure of a forest, and species composition and abundance of a regional species pool. Such information is necessary for the development of biodiversity–ecosystem function studies.

Inventory data can be used to evaluate various ecosystem functions when used with other data, although our discussion mainly focused on productivity and aboveground carbon stock. For example, we can estimate the amount of emissions of species-specific biogenic volatile organic compounds from trees, which mediate not only atmospheric interactions (e.g., aerosol formation, ozone destruction) but also biological interactions (e.g., herbivory, pollination), if inventory data are used with leaf area data (Matsunaga et al. 2012). Net primary production, a function at ecosystem level that is rarely studied (Nadrowski et al. 2010; but see Scherer-Lorenzen et al. 2007a; Potvin et al. 2011), can be estimated from inventory data and litterfall data. Furthermore, we may scale up results of decomposition experiments, such as litter mixture, to nutrient cycling of a forest using inventory data. For example, the decomposition rate of litter depends on species composition of litter and standing tree species (Barantal et al. 2011), both of which can be estimated from inventory data (Ishihara and Hiura 2011). We hope that new forest inventory and allometry databases and data papers promote biodiversity–ecosystem function studies through enhanced data availability, especially in Asia where such studies are limited (Schmid et al. 2009).

Acknowledgments We thank Dr. Kaichiro Sasa and Dr. Kentarou Takagi for permission for use of data, Mr. Masahiro Nagano and the late Mr. Masahito Nishioka for the provision of biomass data, Dr. Shin-ichiro Aiba for valuable information, and Mr. Takahiro Inoue and Mr. Hayato Aoyama for data collection. The study was supported by the Environmental Research and Technology Development Fund (S-9-3) of the Ministry of the Environment, Japan.

References

Alonso D, Etienne RS, McKane AJ (2006) The merits of neutral theory. Trends Ecol Evol 21:451–457

Barantal S, Roy J, Fromin N, Schimann H, Hattenschwiler S (2011) Long-term presence of tree species but not chemical diversity affect litter mixture effects on decomposition in a neotropical rainforest. Oecologia (Berl) 167:241–252

Bradford J (2011) Divergence in forest-type response to climate and weather: evidence for regional links between forest-type evenness and net primary productivity. Ecosystems 14:975–986

Bunker DE, DeClerck F, Bradford JC, Colwell RK, Perfecto I, Phillips OL, Sankaran M, Naeem S (2005) Species loss and aboveground carbon storage in a tropical forest. Science 310:1029–1031

Cardinale BJ, Srivastava DS, Duffy JE, Wright JP, Downing AL, Sankaran M, Jouseau C, Cadotte MW, Carroll IT, Weis JJ, Hector A, Loreau M, Michener WK (2009) Effects of biodiversity on the functioning of ecosystems: a summary of 164 experimental manipulations of species richness. Ecology 90:854

Cardinale B, Matulich K, Hooper D, Byrnes J, Duffy E, Gamfeldt L, Balvanera P, O'Connor M, Gonzalez A (2011) The functional role of producer diversity in ecosystems. Am J Bot 98:572–592

Chave J, Condit R, Aguilar S, Hernandez A, Lao S, Perez R (2004) Error propagation and scaling for tropical forest biomass estimates. Philos Trans R Soc B 359:409–420

Chave J, Andalo C, Brown S, Cairns MA, Chambers JQ, Eamus D, Folster H, Fromard F, Higuchi N, Kira T, Lescure JP, Nelson BW, Ogawa H, Puig H, Riera B, Yamakura T (2005) Tree allometry and improved estimation of carbon stocks and balance in tropical forests. Oecologia (Berl) 145:87–99

Clark DB, Clark DA (2000) Tree growth, mortality, physical condition, and microsite in old-growth lowland tropical rain forest. Ecology 81:294

DeClerck FAJ, Barbour MG, Sawyer JO (2006) Species richness and stand stability in conifer forests of the Sierra Nevada. Ecology 87:2787–2799

Duffy JE, Srivastava DS, McLaren J, Sankaran M, Solan M, Griffin J, Emmerson M, Jones KE (2009) Forecasting decline in ecosystem services under realistic scenarios of extinction. In: Naeem S, Bunker DE, Hector A, Loreau M, Perrings C (eds) Biodiversity, ecosystem functioning, and human wellbeing. Oxford University Press, New York, pp 60–77

Enquist BJ, Kerkhoff AJ, Stark SC, Swenson NG, McCarthy MC, Price CA (2007) A general integrative model for scaling plant growth, carbon flux, and functional trait spectra. Nature (Lond) 449:218–222

Griffin JN, Ogorman EJ, Emmerson MC, Jenkins SR, Klein AM, Loreau M, Symstad A (2009) Biodiversity and the stability of ecosystem functioning. In: Naeem S, Bunker DE, Hector A, Loreau M, Perrings C (eds) Biodiversity, ecosystem functioning, and human wellbeing. Oxford University Press, New York, pp 78–94

Hubbell SP (2001) The unified neutral theory of biodiversity and biogeography. Princeton University Press, Princeton

Ishihara MI, Hiura T (2011) Modeling leaf area index from litter collection and tree data in a deciduous broadleaf forest. Agric For Meteorol 151:1016–1022

Ishihara MI, Suzuki S, Nakamura M, Enoki T, Fujiwara A, Hiura T, Homma K, Hoshino D, Hoshizaki K, Ida H, Ishida K, Itoh A, Kaneko T, Kubota K, Kuraji K, Kuramoto S, Makita A, Masaki T, Namikawa K, Niiyama K, Noguchi M, Nomiya H, Ohkubo T, Saito S, Sakai T, Sakimoto M, Sakio H, Shibano H, Sugita H, Suzuki M, Takashima A, Tanaka N, Tashiro N, Tokuchi N, Yoshida T, Yoshida Y (2011) Forest stand structure, composition, and dynamics in 34 sites over Japan. Ecol Res 26:1007–1008

Ishii H, Tanabe S, Hiura T (2004) Exploring the relationship between canopy structure, stand productivity and biodiversity of natural forest ecosystems: implications for conservation and management of canopy ecosystem functions. For Sci 50:342–355

Jabot F (2010) A stochastic dispersal-limited trait-based model of community dynamics. J Theor Biol 262:650–661

Jabot F, Etienne RS, Chave J (2008) Reconciling neutral community models and environmental filtering: theory and an empirical test. Oikos 117:1308–1320

Kanzaki M (2006) Long-term ecological research sites in Japanese forests (in Japanese; title translated by the authors). In: Masaki T, Tanaka H, Shibata M, The Society for the Study of

Species Biology (SSSB) (eds) Forest ecology, with long-term perspectives. Bun-ich Sogo Shuppan, Tokyo, pp 361–370

Komiyama A, Nakagawa M, Kato S (2011) Common allometric relationships for estimating tree biomass in cool temperate forests of Japan (in Japanese). J Jpn For Soc 93:220–225

Lavorel S, Garnier E (2002) Predicting changes in community composition and ecosystem functioning from plant traits: revisiting the Holy Grail. Funct Ecol 16:545–556

Lei XD, Wang WF, Peng CH (2009) Relationships between stand growth and structural diversity in spruce-dominated forests in New Brunswick, Canada. Can J For Res 39:1835–1847

Leuschner C, Jungkunst HF, Fleck S (2009) Functional role of forest diversity: pros and cons of synthetic stands and across-site comparisons in established forests. Basic Appl Ecol 10:1–9

Lichstein JW, Dushoff J, Ogle K, Chen AP, Purves DW, Caspersen JP, Pacala SW (2010) Unlocking the forest inventory data: relating individual tree performance to unmeasured environmental factors. Ecol Appl 20:684–699

Matsunaga SN, Chatani S, Nakatsuka S, Kusumoto D, Kubota K, Utsumi Y, Enoki T, Tani A, Hiura T (2012) Determination and potential importance of diterpene (kaur-16-ene) emission from dominant coniferous trees in Japan. Chemosphere 87:886–893

McGill BJ, Enquist BJ, Weiher E, Westoby M (2006) Rebuilding community ecology from functional traits. Trends Ecol Evol 21:178–185

Michener W, Brunt J, Helly J, Kirchner T, Stafford S (1997) Nongeospatial metadata for the ecological sciences. Ecol Appl 7:330–342

Morin X, Fahse L, Scherer-Lorenzen M, Bugmann H (2011) Tree species richness promotes productivity in temperate forests through strong complementarity between species. Ecol Lett 14:1211–1219

Nadrowski K, Wirth C, Scherer-Lorenzen M (2010) Is forest diversity driving ecosystem function and service? Curr Opin Environ Sustain 2:75–79

Naeem S, Wright J (2003) Disentangling biodiversity effects on ecosystem functioning: deriving solutions to a seemingly insurmountable problem. Ecol Lett 6:567–579

Naeem S, Colwell R, Díaz S, Hughes J, Jouseau C, Lavorel S, Morin P, Petchey O, Wright J (2007) Predicting the ecosystem consequences of biodiversity loss: the biomerge framework. In: Canadell JG, Pataki DE, Pitelka LF (eds) Terrestrial ecosystems in a changing world. Springer, New York, pp 113–126

Paquette A, Messier C (2011) The effect of biodiversity on tree productivity: from temperate to boreal forests. Glob Ecol Biogeogr 20:170–180

Piotto D (2008) A meta-analysis comparing tree growth in monocultures and mixed plantations. For Ecol Manag 255:781–786

Poorter L, Wright SJ, Paz H, Ackerly DD, Condit R, Ibarra-Manriques G, Harms KE, Licona JC, Martinez-Ramos M, Mazer SJ, Muller-Landau HC, Pena-Claros M, Webb CO, Wright IJ (2008) Are functional traits good predictors of demographic rates? Evidence from five Neotropical forests. Ecology 89:1908–1920

Potvin C, Dutilleul P (2009) Neighborhood effects and size-asymmetric competition in a tree plantation varying in diversity. Ecology 90:321–327

Potvin C, Mancilla L, Buchmann N, Monteza J, Moore T, Murphy M, Oelmann Y, Scherer-Lorenzen M, Turner BL, Wilcke W, Zeugin F, Wolf S (2011) An ecosystem approach to biodiversity effects: carbon pools in a tropical tree plantation. For Ecol Manag 261:1614–1624

Ruiz-Jaen MC, Potvin C (2011) Can we predit carbon stocks in tropical ecosystems from tree diversity? Comparing species and functional diversity in a plantation and a natural forest. New Phytol 189:978–987

Saatchi S, Harris N, Brown S, Lefsky M, Mitchard E, Salas W, Zutta B, Buermann W, Lewis S, Hagen S, Petrova S, White L, Silman M, Morel A (2011) Benchmark map of forest carbon stocks in tropical regions across three continents. Proc Natl Acad Sci USA 108:9899–9904

Scherer-Lorenzen M, Potvin C, Koricheva J, Schmid B, Hector A, Bornik Z, Reynolds G, Schulze E-D (2005) The design of experimental tree plantations for functional biodiversity research. In: Scherer-Lorenzen M, Körner C, Schulze E-D (eds) Forest diversity and function. Springer, New York, pp 347–376

Scherer-Lorenzen M, Bonilla JL, Potvin C (2007a) Tree species richness affects litter production and decomposition rates in a tropical biodiversity experiment. Oikos 116:2108–2124

Scherer-Lorenzen M, Schulze E-D, Don A, Schumacher J, Weller E (2007b) Exploring the functional significance of forest diversity: a new long-term experiment with temperate tree species (BIOTREE). Perspect Plant Ecol Evol Syst 9:53–70

Schmid B, Balvanera P, Cardinale J, Godbold J, Pfisterer B, Raffaelli D, Solan M, Srivastava DS (2009) Consequences of species loss for ecosystem functioning: meta-analyses of data from biodiversity experiments. In: Naeem S, Bunker DE, Hector A, Loreau M, Perrings C (eds) Biodiversity, ecosystem functioning, and human wellbeing. Oxford University Press, New York, pp 14–29

Srivastava DS, Vellend M (2005) Biodiversity–ecosystem function research: is it relevant to conservation? Annu Rev Ecol Evol Syst 36:267–294

Sterck F, Markesteijn L, Schieving F, Poorter L (2011) Functional traits determine trade-offs and niches in a tropical forest community. Proc Natl Acad Sci USA 108:20627–20632

Ter-Mikaelian MT, Korzukhin MD (1997) Biomass equations for sixty-five North American tree species. For Ecol Manag 97:1–24

Tollefson J (2011) A new eye on biodiversity. Nature (Lond) 474:13–14

Uriarte M, Condit R, Canham CD, Hubbell SP (2004) A spatially explicit model of sapling growth in a tropical forest: does the identity of neighbours matter? J Ecol 92:348–360

Vilà M, Inchausti P, Vayreda J, Barrantes O, Gracia C, Ibáñez JJ, Mata T (2005) Confounding factors in the observed productivity–diversity relationship in forests. In: Scherer-Lorenzen M, Körner C, Schulze E-D (eds) Forest diversity and function. Springer, New York, pp 65–86

Vilà M, Vayreda J, Comas L, Ibáñez JJ, Mata T, Obón B (2007) Species richness and wood production: a positive association in Mediterranean forests. Ecol Lett 10:241–250

Webb CT, Hoeting JA, Ames GM, Pyne MI, Poff NL (2010) A structured and dynamic framework to advance traits-based theory and prediction in ecology. Ecol Lett 13:267–283

Whitlock M, McPeek M, Rausher M, Rieseberg L, Moore A (2010) Data archiving. Am Nat 175:145–146

Chapter 20
Regional Comparison of the Ecosystem Services from Seagrass Beds in Asia

Masahiro Nakaoka, Kun-Seop Lee, Xiaoping Huang, Tutu Almonte, Japar Sidik Bujang, Wawan Kiswara, Rohani Ambo-Rappe, Siti Maryam Yaakub, M.P. Prabhakaran, M.K. Abu Hena, Masakazu Hori, Peidong Zhang, Anchana Prathep, and Miguel D. Fortes

Abstract Coastal ecosystems offer valuable services to human society. However, these ecosystems are facing multiple impacts of human-induced stress, including overexploitation, eutrophication, land construction, and global climate change. The prediction of long-term changes in coastal ecosystems under multiple impacts is difficult because nonlinear and cumulative effects operate simultaneously. This difficulty is especially true for Asian regions, where coastal biodiversity is the world's highest but the least studied. In this chapter, we compare ecosystem services of coastal areas of Asia based on the expert knowledge of practioners who study coastal ecosystems at each locality. We especially focused on seagrass beds, which provide important services to humans including provision of seafood

---

M. Nakaoka (✉)
Akkeshi Marine Station, Hokkaido University, Akkeshi, Japan
e-mail: nakaoka@fsc.hokudai.ac.jp

K.-S. Lee
Pusan National University, Busan, South Korea

X. Huang
South China Sea Institute of Oceanology, Guangzhou, China

T. Almonte
City Agricultural Office of Pt. Princesa, Puerto Princesa City, The Philippines

J.S. Bujang
Universiti Putra Malaysia Bintulu Sarawak Campus, Bintulu, Sarawak, Malaysia

W. Kiswara
Research Center for Oceanography, Indonesian Institute for Sciences, Jakarta, Indonesia

R. Ambo-Rappe
Hasanuddin University, Makassar, Indonesia

S.M. Yaakub
Department of Biological Sciences, National University of Singapore, Singapore, Singapore

M.P. Prabhakaran
School of Marine Sciences, CUSAT, Cochin, India

S.-i. Nakano et al. (eds.), *Integrative Observations and Assessments*, Ecological Research Monographs, DOI 10.1007/978-4-431-54783-9_20,

and regulation of water conditions. We selected the six most important ecosystem services at each of 13 seagrass beds ranging from Japan to India and evaluated the direction of changes over the past two decades. We also evaluated public awareness and data certainty for each service. Food provisioning, water purification and waste treatment, erosion regulation, recreation and ecotourism, and educational values were selected as major ecosystem services of seagrass beds. Degradation during the over past 10–20 years was reported for most provisioning and regulating services, whereas improving trend was found for cultural services in most sites. Public awareness and certainty of information were generally high for provisioning services, but low for most regulating services. Regional variation along latitude, and differences between temperate and tropical seagrass beds, were not detected for the examined variables. Regional comparisons of ecosystem services shed light on general and specific aspects of the status of seagrass beds, which will provide baseline data for planning effective conservation and management strategies under multiple human impacts.

Keywords Asia • Data certainty • Ecosystem service • Expert knowledge • Multivariate analysis • Public awareness • Seagrass bed • Temporal change

Introduction

Coastal ecosystems worldwide, such as coral reefs, mangrove forests, seagrass meadows, and algal beds, provide valuable ecosystem services and form highly productive zones in an otherwise barren ocean. For example, the net primary production of coastal ecosystems accounts for approximately 25 % of that of the entire marine ecosystem, despite occupying a mere 9 % of the world's oceans (Duarte and Cebrián 1996). The net primary production from aquatic vegetation such as seagrass and salt marsh habitats is comparable to that of tropical and temperate rainforests (Whittakar 1975; Duarte and Chiscano 1999). Coastal habitats such as coral reefs, seagrass beds, and mangrove forests are also effective habitats and nurseries because their three-dimensional structure supports numerous

M.K. Abu Hena
Universiti Putra Malaysia Bintulu Sarawak Campus, Bintulu, Sarawak, Malaysia

Institute of Marine Sciences and Fisheries, University of Chittagong, Chittagong, Bangladesh

M. Hori
National Research Institute of Fisheries and Environment of Inland Sea, Fisheries Research Agency, Hiroshima, Japan

P. Zhang
Ocean University of China, Qingdao, China

A. Prathep
Faculty of Science, Prince of Songkla University, Songkhla, Thailand

M.D. Fortes
University of the Philippines, Quezon City, The Philippines

flora and fauna, including endangered (Kikuchi and Perès 1977; Williams and Heck Jr 2001; Nakaoka 2005) and commercially important (Heck et al. 2003) species. These coastal habitats also have important roles in the cycling of nutrient and organic matter, through processes such as decomposition and accumulation, thus regulating water quality and sediment dynamics in these areas (Marbà et al. 2006).

The many functions of coastal ecosystems provide valuable services to human society, of which provision of seafood and regulation of water quality are two of the best known (Constanza et al. 1997; McArthur and Boland 2006; Unsworth and Cullen 2010). The economic valuation of regulatory services for water quality and nutrient cycling provided by coastal ecosystems such as salt marshes, mangroves, seagrass beds, and algal beds alone are estimated to be ten times that of major terrestrial ecosystems per unit area (Constanza et al. 1997). Furthermore, these values are almost certainly an underestimate in light of recent studies that highlight the importance of coastal ecosystems in their roles of climate regulation through carbon sequestration (Nellemann et al. 2009; Fourquean et al. 2012) and disaster prevention against natural disturbances such as typhoons and tsunamis (Adger et al. 2005; Chatenoux and Peduzzi 2007; Whanpetch et al. 2010). The increased recognition from the scientific and research community on the importance of coastal ecosystems and the vital services they provide makes them ideal examples for environmental education and for generating public awareness on environmental issues (Huang et al. 2006; Fortes et al. 2007; Abu Hena and Ashraful 2009).

The last century has seen, worldwide, a rapid deterioration of key coastal habitats and the valuable ecosystem services they provide, largely as a result of multiple human-induced impacts such as overexploitation, eutrophication, land construction, and global climate changes (Steneck and Carlton 2001; Harley et al. 2006). For example, since 1990, the majority of tropical reef systems are experiencing a decrease in live coral cover at rates of 1–9 % year^{-1} (Gardner et al. 2003; Bellwood et al. 2004), mangrove forests are declining at an estimated rate of 2 % year^{-1} (Valiela et al. 2001), and seagrass beds are disappearing at a rate of 7 % year^{-1} globally (Waycott et al. 2009). In most cases, the estimates were based on data collected in developed Western countries with limited input from significant coastal regions such as Asia, where a large percentage of the populations reside in coastal areas and are directly dependent on its resources. Asia, and Southeast Asia in particular, is the region with the highest concentration of marine biodiversity in the world, and it is possible that the estimated rates of decline are in fact underestimated because of a dearth of long-term quantitative scientific data from the region (Spalding et al. 2001, 2010; Green and Short 2003). This possibility brings to light the urgent need for systematic studies in Asian coastal areas to understand the status of habitats and ecosystem services, and to promote active and informed decision making and management for the conservation of marine biodiversity and sustainable use of marine resources.

A cost-effective approach in understanding the status of coastal systems where quantitative scientific data are lacking is to carry out comparative analyses based on the knowledge of experts in each locality, especially when analyzing ecosystem services for parameters such as cultural services for which quantitative data are difficult to obtain. Among different coastal habitats in Asia, seagrass beds are suitable for broad-scale comparative studies because they occur widely along the whole Asian coast, from tropical regions to the temperate and subarctic seas along the northwestern Pacific (Green and Short 2003). The network of Asian seagrass researchers has also been developed through some key international organizations and programs such as World Seagrass Association (http://wsa.seagrassonline.org), SeagrassNet (http://www.seagrassnet.org), Seagrass Watch (http://www.seagrasswatch.org), and Census of Marine Life (http://www.coml.org), which has given seagrass scientists and managers the capacity and opportunity to carry out a cross-site comparison over broad spatial scales.

The aim of this chapter is to analyze the current status of some key seagrass beds in Asia using a unified protocol focused on changes in ecosystem services of seagrass beds during the past 20 years. We analyzed inputs from local seagrass experts to identify regional variations in important ecosystem services, the direction of changes, public awareness, and data certainty. The results from this study will be used as baseline data to aid in understanding coastal marine biodiversity and in forming effective management strategies for coastal resources in Asia.

Methods

Study Sites

We targeted 13 seagrass beds from ten Asian countries spanning both temperate (Japan, Korea, north China) and tropical (south China, Philippines, Indonesia, Singapore, Malaysia, Thailand, Bangladesh, India) seagrass beds (Fig. 20.1). One to two sites were chosen from each country to avoid skews in the distribution of sites, which may cause bias in analyses. For the same reasons, each expert report (coauthors of this paper) was confined to one site, except for the Philippines where two experts cooperated to report on one site. All sites selected have been observed and/or studied by each expert and his/her colleagues over the long term (10–20 years or more), and information on temporal changes is available either quantitatively or qualitatively.

To avoid mismatches in scale of the sites selected, we defined the spatial extent of seagrass beds hierarchically according to three categories: (1) region: the area defined at extent of ~20–100 km in scale; (2) local: the area defined at extent of ~5–20 km in scale; and (3) meadow: the area defined at <5 km in scale. As a result, each site is represented either by one large seagrass bed or several moderate to relatively small seagrass beds. An exception to this rule is Singapore where six very small seagrass beds scattered over a distance of 40 km were treated as a group because the ecosystem services for these small meadows were similar to each other. Approximate seagrass bed size and seagrass species composition were described for each site.

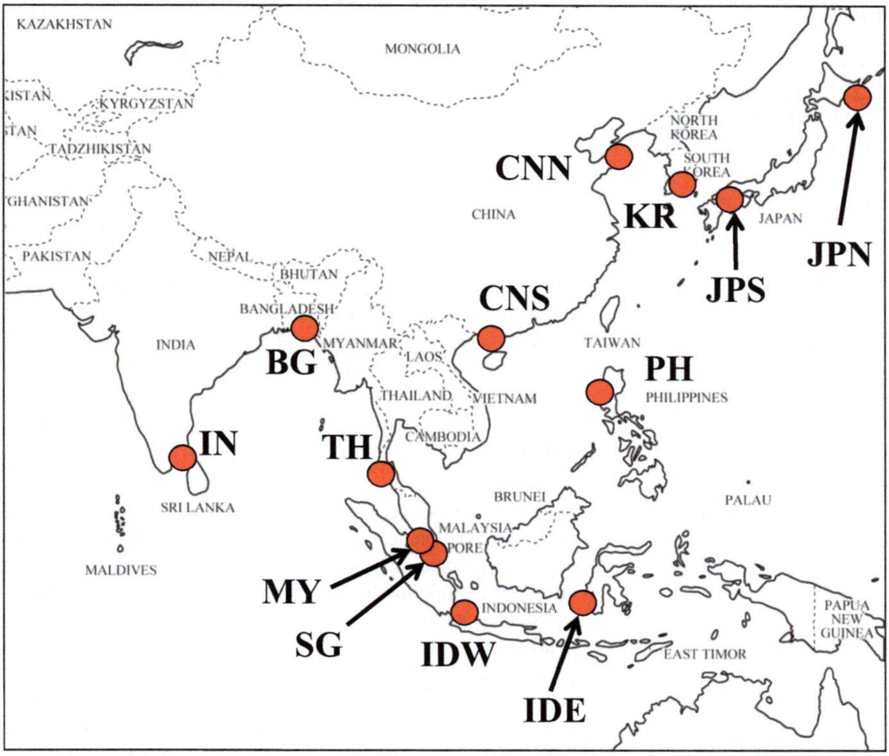

Fig. 20.1 Location of 13 study sites in Asia. See Table 20.2 for the site codes

Evaluation of Ecosystem Services

Each expert selected six ecosystem services that he/she determined was most important to that site. The selection was based on the criteria given by the Millennium Ecosystem Assessment (2003) (Table 20.1). We used categories in the provisioning, regulating, and cultural services, but did not include supporting services because they were indirect services and overlapped with three other types of services. Some categories were not relevant to marine ecosystems (e.g. freshwater supply, air quality regulation, water regulation, and pollination) and these were excluded from the list.

For each selected ecosystem service, we evaluated the direction of change based on the three categories: (1) improving; (2) degrading, and (3) no change. The temporal scale was set to track changes that occurred over the past 20 years (from the 1990s to early 2010s). We also scored public awareness of each ecosystem service by assigning a rank according to the perceived degree of awareness of the ecosystem services provided as follows: (1) high: public is keenly aware; (2) medium: informed scientists, managers, and/or leaders are aware; and (3) low: seldom considered in decision making, low awareness within local community.

Table 20.1 Classification of ecosystem services and their description based on Millennium Ecosystem Assessment (2003)

Types of ecosystem services	Description
Provisioning services	
Food	Includes the vast range of food products derived from plants, animals, and microbes
Fiber	Materials included here are wood, jute, cotton, hemp, silk, and wool
Fuel	Wood, dung, and other biological materials serve as sources of energy
Genetic resources	Includes the genes and genetic information used for animal and plant breeding and biotechnology
Biochemicals and medicines	Many medicines, biocides, food additives such as alginates, and biological materials are derived from ecosystems
Ornamental resources	Animal and plant products, such as skins, shells, and flowers, are used as ornaments, and whole plants are used for landscaping and ornaments
Regulating services	
Climate regulation	Ecosystems influence climate both locally and globally. At a local scale, for example, changes in land cover can affect both temperature and precipitation. At the global scale, ecosystems play an important role in climate by either sequestering or emitting greenhouse gases
Erosion regulation	Vegetative cover plays an important role in soil retention and the prevention of landslides
Water purification and waste treatment	Ecosystems can be a source of impurities (for instance, in freshwater) but also can help filter out and decompose organic wastes introduced into inland waters and coastal and marine ecosystems and can assimilate and detoxify compounds through soil and subsoil processes
Disease regulation	Changes in ecosystems can directly change the abundance of human pathogens, such as cholera, and can alter the abundance of disease vectors, such as mosquitoes
Pest regulation	Ecosystem changes affect the prevalence of crop and livestock pests and diseases
Natural hazard regulation	The presence of coastal ecosystems such as mangroves and coral reefs can reduce the damage caused by hurricanes and large waves
Cultural services	
Cultural diversity	The diversity of ecosystems is one factor influencing the diversity of cultures
Spiritual and religious values	Many religions attach spiritual and religious values to ecosystems or their components
Knowledge systems	Ecosystems influence the types of knowledge systems developed by different cultures
Educational values	Ecosystems and their components and processes provide the basis for both formal and informal education in many societies
Inspiration	Ecosystems provide a rich source of inspiration for art, folklore, national symbols, architecture, and advertising

(continued)

Table 20.1 (continued)

Types of ecosystem services	Description
Aesthetic values	Many people find beauty or aesthetic value in various aspects of ecosystems, as reflected in the support for parks, scenic drives, and the selection of housing locations
Social relations	Ecosystems influence the types of social relations that are established in particular cultures. Fishing societies, for example, differ in many respects in their social relations from nomadic herding or agricultural societies
Sense of place	Many people value the sense of place that is associated with recognized features of their environment, including aspects of the ecosystem
Cultural heritage values	Many societies place high value on the maintenance of either historically important landscapes (cultural landscapes) or culturally significant species
Recreation and ecotourism	People often choose where to spend their leisure time based in part on the characteristics of the natural or cultivated landscapes in a particular area

Finally, the certainty of the information provided was classified according to three ranks: (1) high: quantitative data available showing the change, and/or qualitative but reliable data available from literature; (2) moderate: no quantitative data or literature available, but each expert is quite certain about the change based on his/her own observation and/or through personal communication from colleagues and local community; and (3) low: little information available, but an expert could speculate the direction of changes based on his/her experience and previous knowledge, such as through case studies at other sites under similar conditions.

Data Analyses

In addition to qualitative comparisons of obtained data, the following univariate and multivariate statistical analyses were conducted on site information and ecosystem service evaluation.

Variation with latitude in seagrass bed area and seagrass species richness was tested by Spearman's rank correlation.

Status of ecosystem services was represented by the four multivariate parameters: (1) Top Six Ecosystem Services, (2) Direction of Change, (3) Public Awareness, and (4) Data Certainty. Dissimilarity in these parameters among the 13 sites was analyzed using the Bray–Curtis index. For ecosystem service selection, the presence/absence data on each service were used to calculate the dissimilarity. For the latter three parameters, a rank score between 1 and 3 was given, where 1 indicates degradation and 3 indicates improvement for the direction of change parameter, whereas 1 indicates low and 3 high for both public awareness and the data certainty parameters. For the ranked data, the average value for each of the three categories of ecosystem services (provisioning, regulating, and cultural) was then obtained and used to calculate the dissimilarity between all pairs of sites. Variation in dissimilarity was graphed using a nonmetric Multidimensional Scaling (nMDS) ordination method based on 20 iterations of data (Clarke and Warwick 2001).

To test if the patterns of these variables on ecosystem services vary between temperate (four sites in Japan, Korea, and Northern China) and tropical (nine sites) seagrass beds, one-way PERMANOVA (Anderson 2001) was carried out using the statistical software R with "vegan" packages (Oksanen et al. 2012; R Development Core Team 2012). Finally, the correlation among four dissimilarity matrices on ecosystem service properties was tested using the Mantel test based on 999 permutations. The correlation between each dissimilarity matrix and three accountable variables was also analyzed to test if the similarity in ecosystem service status is related to these factors. The three variables aforementioned are (1) the geographic distance between the sites; (2) the difference in latitude between sites; and (3) the dissimilarity in seagrass species composition (taken from the Bray–Curtis index on the presence/absence data).

Results

The 13 seagrass beds selected for the comparative analyses varied greatly in geography and latitude (Fig. 20.1, Table 20.2). The area of seagrass differed by more than 400 fold between the largest one at the Gulf of Manner in India and the smallest at Bakkhali Estuary in Bangladesh (Table 20.2), but there was no significant correlation between bed size and latitude (Spearman's rank correlation: $\rho = 0.242, p = 0.425$). Species richness of seagrass beds varied from a minimum of 1 species in Bakkhali Estuary (Bangladesh) to a maximum of 12 species in Singapore and India (Table 20.3). Seagrass species richness was negatively correlated with latitude ($\rho = -0.738, p = 0.004$).

Seagrass beds at the four northern sites were dominated by species belonging to Zosteraceae, whereas those in the tropics were dominated by species belonging to Hydrocharitaceae and Cymodoceaceae (Table 20.3). Some species such as *Zostera japonica*, *Halophila ovalis*, and *Ruppia maritima* were found in both regions.

Selection of Important Ecosystem Services

Many experts overlapped in their selection of the most important ecosystem services for their respective sites (Table 20.4). Food provisioning was selected at all sites, whereas other categories in provisioning services were not selected with the exception of genetic resources, biochemical, natural medicine, and pharmaceuticals in some sites.

For regulating services, many experts selected water purification and waste treatment, which are related to the role that seagrasses play in nutrient cycling.

Table 20.2 Summary of information of studied seagrass beds

Country	Site code	Region Extent of ~20–100 km	Local Extent of ~5–20 km	Meadow Extent of <5 km	Latitude	Longitude
Japan North	JPN	Eastern Hokkaido	Akkeshi	Akkeshi-ko Estuary	43.06°N	144.90°E
Japan West	JPW	Western Seto Inland Sea	Aki-wan	Ikunoshima Estuary	34.30°N	132.91°E
Southern Korea	KR	Southern coast of Korea	Koje Island	Koje Bay	34.82°N	128.58°E
China North	CNN	West Coast of Yellow Sea	Shandong Province	Yuehu (Swan Lake)	37.35°N	122.57°E
China South	CNS	Southern coast of China	Guangxi	Hepu	21.52°N	109.59°E
Philippines	PH	Northern Luzon	Province of Pangasinan	Bolinao Seagrass Demonstration Site	16.42°N	119.95°E
Indonesia East	IDE	South Sulawesi	Spermonde Archipelago	Barrang Lompo Island	5.04°S	119.33°E
Indonesia West	IDW	West Java	Serang	Banten Bay	6.00°S	106.14°E
Singapore	SG	Singapore	Singapore	Singapore	1.28°N	103.79°E
Malaysia	MY	Southwest Johore	Merambong-Tg. Adang	Sg. Pulai Estuary	1.34°N	103.57°E
Thailand	TH	SW Andaman Sea Coast	Trang	Laem Yong Lam	7.38°N	99.33°E
Bangladesh	BG	Eastern Bangladesh	Cox's Bazar	Bakkhali Estuary	21.47°N	91.97°E
India	IN	Southern India	Palk Strait	Gulf of Mannar	9.48°N	78.92°E

Other regulating services that were considered important at several sites include erosion regulation, climate regulation, and natural hazard regulation.

Among cultural services, two categories—educational values, and recreation and ecotourism—were selected at almost all sites, whereas other components such as cultural diversity, spiritual and religious values, inspiration, and aesthetic value were selected in one to three sites.

Variation in the patterns of selection of important ecosystem services among the 13 sites (represented by the dissimilarity index) did not show any tendencies among regions. The selection of ecosystem services was identical for Singapore and Koje Bay (Korea); for Barrang Lompo Island (Eastern Indonesia) and Bakkhali Estuary (Bangladesh); and for Akkeshi-ko Estuary (Northern Japan) and Banten Bay (Western Indonesia) (Fig. 20.2a). The difference in dissimilarity of ecosystem services was not significant between temperate and tropical seagrass beds [permutational multivariate analysis of variance (PERMANOVA): df 1, 11; $F = 1.176$, $p = 0.364$]. The variation in dissimilarity of ecosystem services was also not significantly correlated with either geographical distance, latitude, or seagrass species composition (Table 20.5).

Table 20.3 Species richness and species composition of seagrasses in the study sites

Site code	Site name	Area of seagrass beds (km^2)	Species richness	Species composition: *Zm*	*Zcae*	*Zcau*	*Zj*	*Hu*	*Hp*	*Cr*	*Cs*	*Si*	*Tc*	*Ea*	*Th*	*Ho*	*Hm*	*Hg*	*Hd*	*Hst*	*Hb*	*Hsp*	*Rm*	References
JPN	Akkeshi-ko estuary	12.0	3	+			+																+	Nakaoka et al., unpublished data
JPW	Ikunoshima Estuary	0.5	3	+			+									+								Hori et al., unpublished data
KR	Koje Bay	3.0	4	+	+	+	+																	Lee et al. (2005), Park et al. (2011)
CNN	Yuehu (Swan Lake)	2.5	2	+			+																	Zhang et al., unpublished data
CNS	Hepu	5.4	4				+	+								+					+			Huang et al. (2006)
PH	Bolinao Seagrass Demonstration Site	34.0	9					+	+	+	+	+		+	+	+		+					+	Fortes et al., unpublished data
IDE	Barrang Lompo Island	0.5	6					+		+		+		+	+	+								Sterrenburg et al. (1995), Ambo-Rappe (2010), Amran (2010)
IDW	Banten Bay	3.6	7					+		+	+	+		+	+	+								Douven et al. (2003)
SG	Singapore	0.3	12					+	+	+	+	+		+	+	+	+		+		+	+		Yaakub, unpublished data
MY	Sg. Pulai Estuary	1.1	10					+	+	+	+	+		+	+	+	+					+		Japar Sidik et al. (1996, 2000), Japar Sidik and Muta Harah (2003)

TH	Laem Yong Lam	10.7	10					+	+	+	+	+		+	+	+	+		+					Prathep et al., unpublished data
BG	Bakkhali Estuary	0.2	1																		+			Abu Hena et al. (2007), Rezaul (2008)
IN	Gulf of Mannar	85.5	12					+	+	+	+	+	+	+	+	+	+		+	+				Manikandan et al. (2011)

Zm, *Zostera marina*; *Zcae*, *Zostera caespitosa*; *Zcau*, *Zostera caulescens*; *Zj*, *Zostera japonica*; *Hu*, *Halodule univervis*; *Hp*, *Halodule pinifolia*; *Cr*, *Cymodocea rotundata*; *Cs*, *Cymodocea serrulata*; *Si*, *Syringodium isoetifolium*; *Tc*, *Thalassodendron ciliatum*; *Ea*, *Enhalus acoroides*; *Th*, *Thalassia hemprichii*; *Ho*, *Halophila ovalis*; *Hm*, *Halophila minor*; *Hg*, *Halophila gaudichaudii*; *Hd*, *Halophila decipiens*; *Hst*, *Halophila stipulacea*;, *Hb*, *Halophila beccarii*; *Hsp*, *Halophila spinulosa*; *Rm*, *Ruppia maritina*

Table 20.4 Summary of the selection of important ecosystem services and the direction of their changes

Types of ecosystem services	JPN	JPW	KR	CNN	CNS	PH	IDE	IDW	SG	MY	TH	BG	IN
Provisioning services													
Food	Improving	Degrading	Degrading	Improving	Degrading	Degrading	Degrading	Degrading	Degrading	Degrading	Degrading	Improving	Improving
Genetic resources	–	–	–	–	Degrading	–	–	–	–	Degrading	–	–	Improving
Biochemicals and medicines	–	Degrading	–	–	–	–	–	–	–	Degrading	–	–	–
Regulating services													
Climate regulation	Degrading	–	–	–	–	Degrading	–	Degrading	–	–	–	–	Degrading
Erosion regulation	Degrading	–	Degrading	–	–	Degrading	Improving	Degrading	Degrading	Degrading	No Change	Degrading	–
Water purification and waste treatment	Degrading	Improving	Degrading	Improving	Degrading	–	Degrading	Degrading	Degrading	Degrading	Degrading	Degrading	Degrading
Natural hazard regulation	–	–	–	Improving	Degrading	Degrading	Degrading	–	–	–	–	Degrading	Improving
Cultural services													
Cultural diversity	–	Degrading	–	–	–	–	–	–	–	–	–	–	–
Spiritual and religious values	–	–	–	–	–	–	–	–	–	–	No Change	–	–
Educational values	Improving	Improving	Improving	Improving	Improving	Improving	Improving	Improving	Improving	Degrading	Improving	Improving	–
Inspiration	–	–	–	–	–	–	–	–	–	–	Improving	–	–
Aesthetic values	–	–	Improving	Improving	–	–	–	–	Improving	–	–	–	–
Recreation and ecotourism	Improving	Improving	Improving	Improving	Improving	Improving	Degrading	Degrading	Improving	–	–	Degrading	Improving
References	(1–4)	(5–9)	(10–12)	(13–14)	(15)	(16–18)	(19–20)	(21–26)	(27–29)	(30–39)	(40–41)	(42–44)	(45–47)

(1) Hokkaido Prefecture (1991–2009), (2) Iizumi et al. (1995), (3) Hokkaido Enironmental Science Center (2005), (4) Mukai (2005), (5) Hori (2006), (6) Innami (2006), (7) Hori et al. (2007), (8) Yoshida et al. (2010), (9) Hori (2011), (10) Kang et al. (2000), (11) Yoon et al. (2009), (12) Park et al. (2011), (13) Jia et al. (2003), (14) Liu et al. (2011), (15) Huang et al. (2006), (16) Fortes et al. (2007), (17) Holmer et al. (2002), (18) San Diego-McGlone et al. (2008), (19) Erftemeijer and Allen (1993), (20) Ambo-Rappe (2010), (21) Douven (1999), (22) Douven et al. (2003), (23) Glimmerveen (2001), (24) Kiswara (2009), (25) Lindeboom et al. (2000), (26) Tiwi (1999), (27) Loo et al. (1996), (28) Low and Chou (1994), (29) Davison et al. (2008), (30) Japar Sidik and Muta Harah (2003), (31) Sasekumar et al. (1989), (32) Arshad et al. (2001), (33) Arshad et al. (2008), (34) Choo and Liew (2005), (35) Muta Harah and Japar Sidik (2011), (36) Mohd Hanafi et al. (2008), (37) Mohd Hanafi et al. (2009), (38) Cob et al. (2009), (39) Roushon et al. (2010), (40) Adulyanukosol (2002), (41) Rattanachot and Prathep (2011), (42) Abu Hena et al. (2007), (43) Abu Hena and Ashraful (2009), (44) Chowdhury et al. (2011), (45) Wilson et al. (2005), (46) Kumaraguru et al. (2005), (47) Gangal et al. (2012)

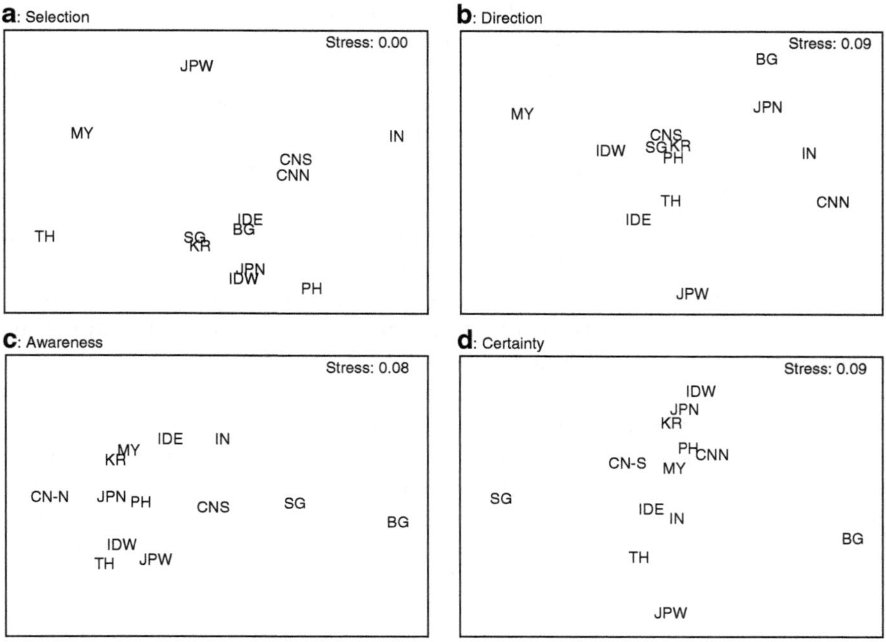

Fig. 20.2 Results of nonmetric multidimensional scaling (nMDS) showing dissimilarity in the characteristics of four ecosystem services (**a–d**) among the 13 seagrass beds in Asia. See Table 20.2 for the site codes

Table 20.5 Results of Mantel test on the correlation of the dissimilarity matrices on seagrass ecosystem services with distance, latitude, and seagrass species composition

	Independent variables		
Dependent variables	Distance	Differences in latitude	Dissimilarity in seagrass species composition
Selection of ecosystem services	$r = -0.092$	$r = -0.203$	$r = -0.214$
	$p = 0.729$	$p = 0.959$	$p = 0.938$
Direction of changes	$r = 0.170$	$r = 0.138$	$r = 0.169$
	$p = 0.089$	$p = 0.102$	$p = 0.098$
Public awareness	$r = 0.000$	$r = -0.139$	$r = 0.246$
	$p = 0.464$	$p = 0.878$	$p = 0.072$
Data certainty	$r = -0.065$	$r = -0.093$	$r = 0.189$
	$p = 0.655$	$p = 0.754$	$p = 0.086$

Direction of Changes and Their Driver

The direction of temporal changes in ecosystem services also varied greatly among sites and selected types (Table 20.4). Four sites indicated an improvement in food provisioning, indicating that the amounts harvested from these beds have increased, whereas the other nine sites indicated a degradation in food provisioning, showing decreasing levels of harvest from the seagrass beds.

For regulation services, overall decline was reported for almost all services and sites. The exceptions to this trend were two sites at Ikunoshima Estuary (western Japan) and Yuehu (northern China) where there was improvement in the regulating services from water purification and waste treatment at both sites and additionally in natural hazard regulation in the latter.

Among cultural services, educational value showed improvement at all the selected sites except for Southwest Johor (Malaysia). Recreation and ecotourism also showed improvement across all sites, with the exception of Barrang Lompo Island (eastern Indonesia), Banten Bay (western Indonesia), and Bakkhali Estuary (Bangladesh), where there was a decline.

Patterns in the direction of ecosystem service changes are similar for Koje Bay (Korea), Hepu (southern China), Bolinao (the Philippines), and Singapore. No obvious regional tendency was found among the 13 sites (Fig. 20.2b), with no significant differences detected between temperate and tropical seagrass beds (PERMANOVA: $F = 2.324$, $p = 0.176$). Again, the variation in dissimilarity was not significantly correlated with distance, latitude, or seagrass species composition (Table 20.5).

Public Awareness and Data Certainty

Public awareness of the selected ecosystem services varied among different types of services (Table 20.6). It was generally high for food provisioning in most sites except for Singapore, Bakkhali Estuary (Bangladesh), and Gulf of Manner (India). Awareness of local communities on regulation services was generally low, whereas that of cultural services varied greatly, with most sites indicating moderate awareness. Awareness of all the services was low in Bakkhali Estuary whereas it was high to medium in Yuehu (northern China) and Sg. Pulai Estuary (Malaysia).

Scores for data certainty followed the similar patterns to those for public awareness, with high to medium certainty for food provisioning (except for Singapore), and medium to low certainty for most regulating services (Table 20.7). The score was highly variable for cultural services. MDS plots for the awareness data and the certainty data showed similar patterns of variation with Bakkhali Estuary (Bangladesh) and Singapore separated from other sites. The similarity matrices of these two parameters were highly correlated with each other (Mantel test; $r = 0.637, p < 0.001$).

There was no latitudinal or regional tendency in the patterns found for awareness and data certainty as shown by nonsignificant variation between temperate and tropical regions (PERMANOVA; $F = 2.114$, $p = 0.154$ for the awareness score; $F = 1.173$, $p = 0.372$ for the certainty score) and by nonsignificant correlation with distance, latitude, and species composition (Table 20.5).

Table 20.6 Summary of the public awareness of ecosystem services in seagrass beds

Types of ecosystem services	JPN	JPW	KR	CNN	CNS	PH	IDE	IDW	SG	MY	TH	BG	IN
Provisioning services													
Food	High	High	High	High	Medium	High	Medium	High	Low	High	High	Low	Low
Genetic resources	–	–	–	–	Medium	–	–	–	–	Medium	–	–	Medium
Biochemicals and medicines	–	Medium	–	–	–	–	–	–	–	High	–	–	–
Regulating services													
Climate regulation	Low	–	–	–	–	Medium	–	Low	–	–	–	–	Low
Erosion regulation	Low	–	Low	–	–	Low	High	Low	Low	Medium	Low	Low	–
Water purification and waste treatment	Medium	Low	High	High	Low	–	Medium	Low	Low	Medium	Low	Low	Medium
Natural hazard regulation	–	–	–	Medium	Low	Low	Low	–	–	–	–	Low	Medium
Cultural services													
Cultural diversity	–	Medium	–	–	–	–	–	–	–	–	–	–	–
Spiritual and religious values	–	–	–	–	–	–	–	–	–	–	High	–	–
Educational values	Medium	High	Medium	High	Medium	Medium	High	High	Medium	Medium	High	Low	–
Inspiration	–	–	–	–	–	–	–	–	–	–	High	–	–
Aesthetic values	–	–	Medium	High	–	–	–	–	Medium	–	–	–	–
Recreation and ecotourism	Low	High	Medium	High	Medium	Medium	Low	Medium	Medium	–	–	Low	Medium

Table 20.7 Summary on data certainty of ecosystem services in seagrass beds

Types of ecosystem services	JPN	JPW	KR	CNN	CNS	PH	IDE	IDW	SG	MY	TH	BG	IN
Provisioning services													
Food	High	High	High	High	High	High	Medium	High	Low	High	Medium	Medium	High
Genetic resources	–	–	–	–	Medium	–	–	–	–	Medium	–	–	Low
Biochemicals and medicines	–	Low	–	–	–	–	–	–	–	High	–	–	–
Regulating services													
Climate regulation	Low	–	–	–	–	Medium	–	Low	–	–	–	–	Medium
Erosion regulation	Low	–	Low	–	–	Medium	Medium	Low	Medium	Medium	Medium	Low	–
Water purification and waste treatment	Medium	High	Medium	High	Medium	–	Medium	Low	Low	Medium	Medium	Low	Medium
Natural hazard regulation	–	–	–	Medium	Low	Medium	Low	–	–	–	–	Low	Medium
Cultural services													
Cultural diversity	–	Medium	–	–	–	–	–	–	–	–	–	–	–
Spiritual and religious values	–	–	–	–	–	–	–	–	–	–	Medium	–	–
Educational values	Medium	High	Medium	Medium	Medium	Medium	High	High	Medium	Medium	High	Low	–
Inspiration	–	–	–	–	–	–	–	–	–	–	High	–	–
Aesthetic values	–	–	Medium	Medium	–	–	–	–	Medium	–	–	–	–
Recreation and ecotourism	Medium	High	Medium	Medium	High	Medium	Low	Low	High	–	–	Low	Medium

Discussion

Coastal ecosystems offer various types of ecosystem services that are important both economically and culturally (Constanza et al. 1997; Hein et al. 2006). However, long-term quantitative data are mostly lacking in Asian countries, despite the high proportion of marine diversity that is found in this region. The present study is a first attempt to analyze the status of ecosystem services provided by seagrass beds in Asia, encompassing both temperate and tropical regions. It gives an overview of the status of seagrass beds in Asia based on expert assessment where quantitative information is lacking and provides a baseline and directions for future studies.

The seagrass beds chosen in this study varied greatly in latitude, meadow size, and species composition. Furthermore, there were no significant geographic trends in the selection of important ecosystem services and in the pattern of temporal changes. Food provisioning was the overriding ecosystem function across all sites that was identified by local experts, followed by services such as water purification and waste treatment, erosion regulation, educational value, and recreation and ecotourism, identified based on the general list by Millennium Ecosystem Assessment (2003). This result indicates that the recognition of important ecosystem services of seagrass beds is shared internationally among experts who are scientists and ecosystem managers.

The direction of changes in these ecosystem services showed some common trends across different sites, with provisioning and regulating services showing a downward or degrading trend whereas cultural services of seagrass beds showed the opposite trend. For food provisioning, the declines in most sites were related to both overfishing and degradation of the seagrass bed by pollution and land development, which is a theme that is identified repeatedly in coastal habitats throughout Asia (Fortes 1988; Duarte 2002; Silvestre et al. 2003; Unsworth and Cullen 2010). Even at sites where food provisioning is reported to be improving, it is important to understand the context in which this change is occurring. For example, an increase in food provisioning services in terms of net harvest is not necessarily positive as the improvement may be driven by an increased demand for seafood, hence a corresponding increase in production through increased or intensified fisheries and aquaculture activity. There is no indication of whether the increased production is sustainable, and hence these sites could see deterioration in the future should production continue above ecologically sustainable levels.

The decreasing trend observed in regulating services at many sites reflects the fact that human-induced impact is a serious problem for seagrass beds throughout Asia, and this hold true when we explore the factors affecting specific sites in this study. For example, high rates of sedimentation and eutrophication of coastal waters caused by environmentally unfriendly practices in aquaculture have been observed in Akkeshi Estuary in northern Japan (Hokkaido Environmental Science Center 2005), Koje Bay in southern Korea (Yoon et al. 2009), and Bolinao in the Philippines (Holmer et al. 2002; San Diego-McGlone et al. 2008), which may lead to seagrass loss from decreased light availability. Another very serious threat is the

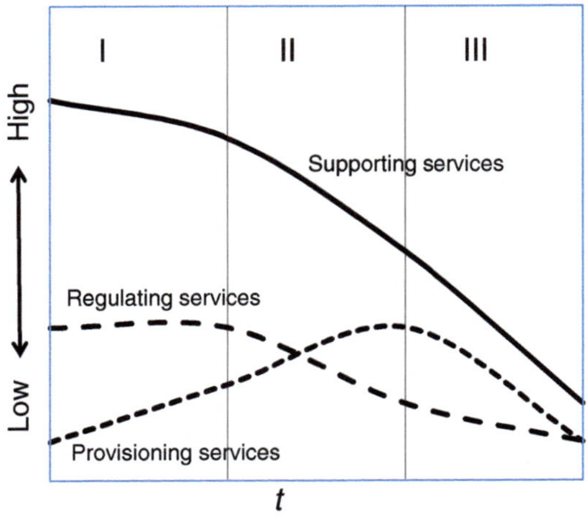

Fig. 20.3 A schematic model describing temporal changes in the relationship between provisioning and regulating services with the degradation of supporting services. See text for the explanation of the three phases (*I*, *II*, *III*)

direct loss of seagrass beds from land development and reclamation, which occurred in Banten Bay, western Indonesia (Douven et al. 2003), and Sg. Pulai Estuary, Malaysia (Japar Sidik and Muta Harah 2003; Muta Harah and Japar Sidik 2011). The exceptions are found in Seto Inland (western Japan) and in Yuehu (northern China), where regulating services have been improving over the past two decades from better water quality and some successful restoration projects, resulting in an increase in seagrass area (Jia et al. 2003; Yamamoto 2003).

Studies on terrestrial and freshwater ecosystems have shown that trade-offs often arise when it comes to management decisions on the utilization of ecosystem services and conservation (Rodríguez et al. 2006). In most cases, unsustainable use of ecosystems for purposes of maximizing direct services (e.g., provisioning services such as food production) inevitably results in the degradation of not only said service, but also in indirect services such as regulating (e.g., nutrient cycling) and cultural (e.g., educational values) services (Bennett et al. 2009; Carpenter et al. 2009). In our analyses, negative relationships between food provisioning and regulation services were found only in a few sites. In the case of Akkeshi-ko Estuary (northern Japan), it is likely that intensive use of the estuary for aquaculture caused a decrease in water quality (Nakaoka and Matsuda 2011). However, for other sites such as Ikunoshima Estuary (western Japan) and Bakkhali Estuary (Bangladesh), the relationship was less clear as the cause may have been one direct impact or several compounding factors, although no studies have investigated their links. In the other sites, both provisioning and regulating services were experiencing decline and thus positively correlated except for Yuehu (northern China), where both services were improving after the recovery of the eelgrass bed (Jia et al. 2003).

The variable relationship between provisioning, regulating, and supporting services observed at many sites may be viewed as phases, which describe changes in the balance of each service along a spectrum (Fig. 20.3). Phase I shows high levels of regulating services such as primary productivity and nutrient cycling, over that of provisioning services, which in turn can increase while maintaining regulating services in good condition (Fig. 20.3). Phase II shows intensive exploitation of provisioning services over sustainable levels that would lead to a decrease in both supporting and regulating services, leading to trade-offs between the two. Continued and sustained exploitation of provisioning services over the threshold level of regulating services leads to a decrease in all three ecosystem services (Phase III).

It is likely that the current situation in each study site reflects the different phases of these gradients: Yuehu (northern China) in Phase I, Akkeshi-ko Estuary (northern Hokkaido) in Phase II, and most other sites in Phase III. In terrestrial and freshwater ecosystems, the shift in the status of ecosystem services can occur nonlinearly with a threshold (Kinzig et al. 2005). It is unknown how these temporal changes occur in seagrass beds during the process of ecosystem degradation and what the associated thresholds for the ecosystem services in the seagrass beds might be.

Changes in cultural services were generally positive at most sites, especially for the educational value of seagrass beds. In each Asian country, there is increased recognition within the population of the value of natural ecosystems, as shown by the middle to high scores of public awareness. The improvement is partly from translation of research results into outreach and advocacy at both educational and institutional levels (Fortes et al. 2007), and through community-based monitoring programs such as Seagrass Watch (McKenzie et al. 2000). Improvement in the recreation and ecotourism industries in most regions may also be related to increased awareness of the value of coastal ecosystems, although it should be noted that excess and unregulated recreational use can also degrade habitat quality (Milazzo et al. 2002; Davenport and Davenport 2006). In fact, the recreational use of seagrass beds is in decline for sites where there was habitat loss and deterioration, such as in Banten Bay (western Indonesia) and Bakkhali Estuary (Bangladesh).

As expected, public awareness of seagrass ecosystem services was high for food provisioning in most countries, which is tightly related to the economic status of local communities. In contrast, the awareness of regulating services was low, as these services are intangible and thus difficult to comprehend without specific scientific outreach from specialists. It has been pointed out that there is a large gap between scientific knowledge and public awareness on the role of seagrass ecosystems compared to other types of coastal habitats such as coral reefs and mangroves (Orth et al. 2006; Duarte et al. 2008). The variation in data certainty is likely to be related to the needs of local communities and the direction of scientific endeavor, which translates to their level of awareness. For example, the amount of reliable data for food provisioning services was relatively high as these services are directly related to human needs and activity; this translates to more scientific resources channeled toward this service, which then contributes to both awareness of the public (through outreach) and the certainty of data from local experts.

One of the unexpected findings is that the level of awareness and data certainty did not differ significantly between temperate seagrass beds in economically developed countries (such as Japan and Korea), and tropical areas in South and

Southeast Asia, which have a higher proportion of developing countries. It should be noted, however, that the results of the present analyses were based on data from one or two seagrass beds from each country, and that the obtained regional variation may not be necessarily indicative of variation in economic situation or decision-making processes among countries. Nevertheless, the lack of scientific data on functions and services of ecosystems, especially on those of regulating services, and insufficient outreach to local communities and stakeholders remains a major concern in all Asian countries. Promotion of more systematic and effective scientific monitoring and research on biodiversity and the ecosystem functioning of coastal areas are both highly needed to address the paucity of data and to reverse the direction of changes in ecosystem services, which are still being degraded in most Asian regions.

In conclusion, the comparative analyses on ecosystem services based on expert knowledge clarified some general and specific aspects of the status of seagrass beds in Asian regions for which quantitative scientific data are lacking. The approach is a positive step in the attempt to deepen our understanding of coastal ecosystems, and the results obtained will be used as baseline data for planning effective conservation and management strategies under multiple human impacts. The current dataset can be further improved by covering more sites in Asia and by replicating the analysis in other regions of the world and on other coastal habitats such as mangroves and coral reefs. This approach allows for within-group, cross-regional, and cross-habitat comparisons, which may delineate interesting trends that are not yet apparent.

Acknowledgments This paper was written as a product of the workshop in the 9th International Seagrass Biology Workshop (at Trang, Thailand, in November 2010) entitled "Towards an integrated coastal ecosystem conservation and adaptive management of coastal areas of Southeast Asia." We thank participants of the workshop who gave us valuable comments. This work is supported by Environment Research and Technology Development Fund, Ministry of the Environment Government of Japan (B-0910 and S9), Flexible International Exchanges program (JSPS), the grants-in-aid from JSPS (Nos. 19405009, 21254002), and JST-JICA SATREPS (CECAM).

References

Abu Hena MK, Ashraful MAK (2009) Coastal and estuarine resources of Bangladesh: management and conservation issues. Maejo Int J Sci Technol 3:313–342

Abu Hena MK, Short FT, Sharifuzzaman SM, Hasan M, Rezowan M, Ali M (2007) Salt marsh and seagrass communities of Bakkhali Estuary, Cox's Bazar, Bangladesh. Estuar Coast Shelf Sci 75:72–78

Adger WN, Hughes TP, Folke C, Carpenter SR, Rockström J (2005) Social-ecological resilience to coastal disasters. Science 309:1036–1039

Adulyanukosol K (2002) Dugong in Phuket Province. Thai Fish Gaz 55:47–50

Ambo-Rappe R (2010) Fish community structure in different seagrass beds of Barrang Lompo Island. J Trop Mar Sci Technol 2:62–73

Amran MA (2010) Estimation of seagrass coverage by depth invariant indices on quickbird imagery. Biotropia 17:42–50

Anderson MJ (2001) A new method for non-parametric multivariate analysis of variance. Aust Ecol 26:32–46

Arshad A, Japar Sidik B, Muta Harah Z (2001) Fishes associated with seagrass habitat. In: Japar Sidik B, Arshad A, Tan SG, Daud SK, Jambari HA, Sugiyama S (eds) Aquatic resource and environmental studies of the Straits of Malacca: current research and reviews. Malacca Straits Research and Development Centre (MASDEC), Universiti Putra Malaysia, Serdang, pp 151–162

Arshad A, Jimmy A, Nurul Amin SM, Japar Sidik B, Muta Harah Z (2008) Length-weight and length-length relationships of five fish species collected from seagrass beds of the Sungai Pulai estuary, Peninsular Malaysia. J Appl Ichthyol 24:328–329

Bellwood DR, Hughes TP, Folke C, Nystrom M (2004) Confronting the coral reef crisis. Nature (Lond) 429:827–833

Bennett EM, Peterson GD, Gordon LJ (2009) Understanding relationships among multiple ecosystem services. Ecol Lett 12:1394–1404

Carpenter SR, Mooney HA, Agard J, Capistrano D, DeFries RS, Díaz S, Dietz T, Duraiappah AK, Oteng-Yeboah O, Pereira HM, Perrings C, Reid WV, Sarukhan J, Scholes RJ, Whyte A (2009) Science for managing ecosystem services: beyond the millennium ecosystem assessment. Proc Natl Acad Sci USA 106:1305–1312

Chatenoux B, Peduzzi P (2007) Impacts from the 2004 Indian Ocean Tsunami: analyzing the potential protecting role of environmental features. Nat Hazards 40:289–304

Choo CK, Liew HC (2005) Exploitation and trade in seahorses in Peninsular Malaysia. Malayan Nat J 57:57–66

Chowdhury MSN, Hossain MS, Das NG, Barua P (2011) Environmental variables and fisheries diversity of the Naaf River Estuary, Bangladesh. J Coast Conserv 15:163–180

Clarke KR, Warwick RM (2001) Change in marine communities: an approach to statistical analysis and interpretation, 2nd edn. PRIMER-E, Plymouth

Cob ZC, Arshad A, Japar Sidik B, Mazlan AG (2009) Species description and distribution of *Strombus* (Molusca, Strombidae) in Johor Straits and its surrounding areas. Sains Malays 38:39–46

Constanza R, d'Arge R, de Groot R, Farber S, Grasso M, Hannon B, Limburg K, Naeem S, O'Neill RV, Paruelo J, Raskin RG, Sutton P, van den Belt M (1997) The value of the world's ecosystem services and natural capital. Nature (Lond) 387:253–260

Davenport J, Davenport JL (2006) The impact of tourism and personal leisure transport on coastal environments: a review. Estuar Coast Shelf Sci 67:280–292

Davison GWH, Ng PKL, Ho HC (2008) The Singapore red data book: threatened plants and animals of Singapore, 2nd edn. Nature Society Singapore, Singapore

Douven WJAM (1999) Human pressure on marine ecosystems in the Teluk Banten coastal zone: recent situation and future prospects. Teluk Banten Research Program. Report Series No. 3. IHE-Delft, WOTRO

Douven WJAM, Buurman JJG, Kiswara W (2003) Spatial information for coastal zone management: the example of the Banten Bay seagrass ecosystem, Indonesia. Ocean Coast Manag 46:615–634

Duarte CM (2002) The future of seagrass meadows. Environ Conserv 29:192–206

Duarte CM, Cebrián J (1996) The fate of marine autotrophic production. Limnol Oceanogr 41:1758–1766

Duarte CM, Chiscano CL (1999) Seagrass biomass and production: a reassessment. Aquat Bot 65:159–174

Duarte CM, Dennison WC, Orth RJW, Carruthers TJB (2008) The charisma of coastal ecosystems: addressing the imbalance. Estuar Coast 31:233–238

Erftemeijer PLA, Allen GR (1993) Fish fauna of seagrass beds in South Sulawesi, Indonesia. Rec West Aust Mus 16:269–277

Fortes MD (1988) Mangrove and seagrass beds of East Asia: habitats under stress. Ambio 17:207–213

Fortes MD, Sour K, Huang X, Kuriandewa TE, Van Tien N, Vibol O, Montano NM, Almonte T, Sstumantpan S (2007) The UNEP/GEF seagrass demonstration sites in South China Sea:

Milestones in seagrass research and biodiversity management in East Asia. Publ Seto Mar Biol Lab Spec Publ Ser 8:125–137
Fourquean JW, Duarte CM, Kennedy H, Marbà N, Holmer M, Mateo MA, Apostolaki ET, Kendrick GA, Krause-Jensen D, Glathery KJ, Serrano O (2012) Seagrass ecosystems as a globally significant carbon stock. Nat Geosci 5:505–509
Gangal M, Arthur R, Alcoverro T (2012) Structure and dynamics of South East Indian seagrass meadows across a sediment gradient. Aquat Bot 98:34–39
Gardner TA, Cote IM, Gill JA, Grant A, Watkinson AR (2003) Long-term region-wide declines in Caribbean corals. Science 301:958–960
Glimmerveen M (2001) Modelling interactions between natural and socio-economic systems: the catch and trade of live fish food in Teluk Banten, West Java, Indonesia. Master of Science thesis, School of Systems Engineering, Policy Analysis and Management, Delft University of Technology. Teluk Banten Research Program Report Series No. 5
Green EP, Short FT (2003) World atlas of seagrasses. University of California Press, Berkeley
Harley CDG, Hughes AR, Hultgren KM, Miner BG, Sorte CJB, Thornber CS, Rodriguez LF, Tomanek L, Williams SL (2006) The impacts of climate change in coastal marine systems. Ecol Lett 9:228–241
Heck KL Jr, Hays G, Orth RJ (2003) Critical evaluation of the nursery role hypothesis for seagrass meadows. Mar Ecol Prog Ser 253:123–136
Hein L, van Koppen K, de Groot RS, van Ierland EC (2006) Spatial scales, stakeholders and the valuation of ecosystem services. Ecol Econ 57:209–228
Hokkaido Prefecture (1991–2009) Current situation in fisheries in Hokkaido. Hokkaido Prefecture, Sapporo
Hokkaido Environmental Science Center (2005) Report on the results of survey of classified water systems (total nitrogen and phosphate) on the environmental criteria for conservation of life environment. Hokkaido Prefecture, Sapporo
Holmer M, Marbà N, Terrados J, Duarte CM, Fortes MD (2002) Impacts of milkfish (*Chanos chanos*) aquaculture on carbon and nutrient fluxes in the Bolinao area, Philippines. Mar Pollut Bull 44:685–696
Hori M (2006) Some problems toward evaluation and conservation of ecosystem functions of shallow macrophyte beds. Seto Inland Sea 47:21–24
Hori M (2011) Ecosystem services in shallow water: the roles of biological production and biodiversity. In: Shoji J, Hori M, Yamashita Y (eds) Ecosystem services of coastal waters: benefits of the sea and its sustainable use. Kouseisha Kouseikaku, Tokyo, pp 11–25
Hori M, Kamimura S, Nakaoka M (2007) The importance of ecosystem-functioning research for conservation and sustainable utilization of coastal seas. Jpn J Benthol 62:46–51
Huang X, Huang L, Li Y, Xu Z, Fong CW, Huang D, Han Q, Huang H, Tan Y, Liu S (2006) Main seagrass beds and threats to their habitats in the coastal sea of South China. Chin Sci Bull 51:136–142
Iizumi H, Taguchi S, Minami T, Mukai H, Maekawa S (1995) Distribution and variability nutrients, chlorophyll a, particulate organic matters, and their carbon and nitrogen contents, in Akkeshi-ko, an estuary in northern Japan. Bull Hokkaido Nat Fish Res Inst 59:43–67
Innami T (2006) Ethnic culture in Seto Inland Sea. 2: Stone sauna and eelgrass. Seto Inland Sea 45:40–44
Japar Sidik B, Muta Harah Z (2003) The seagrasses of Malaysia. In: Green EP, Short FT (eds) World atlas of seagrasses. University of California Press, Berkeley, pp 152–160
Japar Sidik B, Arshad A, Hishamuddin O, Muta Harah Z, Misni S (1996) Seagrass and macroalgal communities of Sungai Pulai estuary, south-west Johore, Peninsular Malaysia. In: Kuo J, Walker DI, Kirkman H (eds) Seagrass biology: scientific discussion from an international workshop. Faculty of Science, The University of Western Australia, Rottnest Island, Nedlands, Western Australia, pp 3–12
Japar Sidik B, Muta Harah Z, Fazrullah Rizally AR, Kamin B (2000) New observations on *Halophila spinulosa* (R. Br.) Aschers. in Neumayer, Malaysia. Biol Mar Medit 7:75–78

Jia J-J, Gao S, Xue Y-C (2003) Sediment dynamic processes of the Yuehu inlet system, Shandong Peninsula. Chin Estuar Coast Shelf Sci 57:783–801

Kang C-K, Park MS, Lee P-Y, Choi W-J, Lee W-C (2000) Seasonal variations in condition, reproductive activity, and biochemical composition of Pacific oyster, *Crassostrea gigas* (Thunberg) in suspended culture in two coastal bays of Korea. J Shellfish Res 19:771–778

Kikuchi T, Perès JM (1977) Consumer ecology of seagrass beds. In: McRoy CP, Helfferich C (eds) Seagrass ecosystem. A scientific perspective. Dekker, New York, pp 147–193

Kinzig AP, Ryan P, Etienne M, Allison H, Elmqvist T, Walker BH (2005) Resilience and regime shifts: assessing cascading effects. Ecol Soc 11:20

Kiswara W (2009) Preliminary study: economic valuation of the snail (*Pyrene versicolor*) in seagrass bed of Kepuh, Banten Bay-Indonesia. In: Coral Reef Management Symposium in Coral Triangle Area, Jakarta, October 2009, pp 12–13

Kumaraguru AK, Jayakumar K, Wilson JJ, Ramakritinan CM (2005) Impact of the tsunami of 26 December 2004 on the coral reef environment of Gulf of Mannar and Palk Bay in the southeast coast of India. Curr Sci 89:1729–1741

Lee K-S, Park SR, Kim J-B (2005) Production dynamics of the eelgrass, *Zostera marina*, in two bay systems on the south coast of the Korean peninsula. Mar Biol 147:1091–1108

Lindeboom HJ, de Boois H, van Kessel-Hagesteijn RR, Hoekstra P, Heun JC, Douven WJAM, Kiswara W (2000) Coastal research in Teluk Banten: Integrated Science for Management-Research Program Indonesia, 1996–2001. In: Proceedings of the international conference interactive north–south research for development with special attention for natural resources management. KNAW, Amsterdam, pp 29–45

Liu J, Zhang P, Guo D, Niu S, Zhang X (2011) Annual change in photosynthetic pigment contents of *Zostera marina* L. in Swan Lake. Afr J Biotechnol 10:18194–18199

Loo MGK, Tun KPP, Chou LM (1996) Environmental status of seagrass communities in Singapore. J Singapore Nat Acad Sci 22–24:97–102

Low JKY, Chou LM (1994) Sedimentation rates in Singapore waters. In: Sudara S, Wilkinson CR, Chou LM (eds) Proceedings of the 3rd ASEAN-Australia symposium on living coastal resources, vol 2: research papers. Chulalongkorn University, Bangkok, pp 697–701

Manikandan S, Ganesapandian S, Parthiban K (2011) Distribution and zonation of seagrasses in the Palk Bay, southeastern India. J Fish Aquat 6:178–185

Marbà N, Holmer M, Gacia E, Barrón C (2006) Seagrass beds and coastal biogeochemistry. In: Larkum AWD, Orth RJ, Duarte CM (eds) Seagrasses: biology, ecology and conservation. Springer, Dordrecht, pp 135–157

McArthur LC, Boland JW (2006) The economic contribution of seagrass to secondary production in South Australia. Ecol Model 196:163–172

McKenzie LJ, Lee Long WJ, Coles RG, Roder CA (2000) Seagrass-watch: community based monitoring of seagrass resources. Biol Mar Medit 7:393–396

Milazzo M, Chemello R, Badalamenti F, Camarda R, Riggio S (2002) The impact of human recreational activities in marine protected areas: what lessons should be learnt in the Mediterranean Sea? P S Z N Mar Ecol 23(S1):280–290

Millennium Ecosystem Assessment (2003) Ecosystems and human well-being: a framework for assessment. Island Press, Washington, DC

Mohd Hanafi I, Arshad A, Japar Sidik B, Mazlan AG, Siti Khalijah D (2008) Biodiversity and distribution of pen shells (Bivalvia: Pinnidae) from the seagrass beds of Sungai Pulai, Peninsular Malaysia. Res J Fish Hydrobiol 3:54–62

Mohd Hanafi I, Arshad A, Japar Sidik B, Mazlan AG, Siti Khalijah D (2009) Morphological characteristics of *Pinna bicolor* Gmelin and *Pinna deltodes* Menke in seagrass area of Sg. Pulai, Johor, Peninsular Malaysia. Sains Malays 38:333–339

Mukai H (2005) Effects of bioelements from agricultural ecosystems on coastal ecosystems. Research Report for Grant-in-Aid by JSPS (no. 14209002), Akkeshi Marine Station, Hokkaido University, Akkeshi

Muta Harah Z, Japar Sidik B (2011) Disturbances in seagrasses ecosystem in Malaysia. In: Ogawa H, Japar Sidik B, Muta Harah Z (eds) Seagrasses: resource status and trends in Indonesia, Japan, Malaysia, Thailand and Vietnam. Seizando-Shoten, Tokyo, pp 67–78

Nakaoka M (2005) Plant–animal interactions in seagrass beds: ongoing and future challenges for understanding population and community dynamics. Popul Ecol 47:167–177

Nakaoka M, Matsuda H (2011) Elucidating the link among global climate change, ecosystem service and human activity. In: Shoji J, Hori M, Yamashita Y (eds) Ecosystem services of coastal waters: benefits of the sea and its sustainable use. Kouseisha Kouseikaku, Tokyo, pp 129–143

Nellemann C, Corcoran E, Duarte CM, Valdes L, DeYoung C, Fonseca L, Grimsditch, G (2009) Blue carbon. A rapid response assessment. United Nations Environment Programme GRID-Arendal, Birkeland Trykkeri

Oksanen J, Blanchet FG, Kindt R, Legendre P, Minchin PR, O'Hara RB, Simpson GL, Solymos P, Stevens MHH, Wagner H (2012) vegan: community ecology package. R package version 2.0-3. http://CRAN.R-project.org/package=vegan

Orth RB, Carruthers TJB, Dennison WC, Duarte CM, Fourqurean JW, Heck KL Jr, Hughes AR, Kendrick GA, Kenworthy WJ, Olyarnik S, Short FT, Waycott M, Williams SL (2006) A global crisis for seagrass ecosystems. Bioscience 56:987–996

Park SR, Kim YK, Kim J-H, Kang C-K, Lee K-S (2011) Rapid recovery of the intertidal seagrass *Zostera japonica* following intense Manila clam (*Ruditapes philippinarum*) harvesting activity in Korea. J Exp Mar Biol Ecol 407:275–283

R Development Core Team (2012) R: a language and environment for statistical computing. R Foundation for Statistical Computing, Vienna

Rattanachot E, Prathep A (2011) Temporal variation in growth and reproduction of *Enhalus acoroides* (L.f.) Royle in a monospecific meadow in Haad Chao Mai National Park, Trang Province, Thailand. Bot Mar 54:201–207

Rezaul KD (2008) Sea grass *Halophila beccarii* from Bakkhali River Estuary, Cox's Bazar. BSc (Hons) Term Paper, Institute of Marine Sciences and Fisheries, University of Chittagong, Bangladesh

Rodríguez JP, Beard TD Jr, Bennett EM, Cumming GS, Cork S, Agard J, Dobson AP, Peterson GD (2006) Trade-offs across space, time, and ecosystem services. Ecol Soc 11:28

Roushon A, Aziz A, Nurul Amin SM, Siti Khalijah D, Japar Sidik B, Mazlan AG (2010) Feeding habits of larval fishes of the family Gobiidae (Actinopterygii: Perciformes) in seagrass beds of Sungai Pulai estuary, Johor Strait, Malaysia. Coast Mar Sci 34:123–128

San Diego-McGlone ML, Azanza RV, Villanoy CL, Jacinto GS (2008) Eutrophic waters, algal bloom and fish kill in fish farming areas in Bolinao, Pangasinan, Philippines. Mar Pollut Bull 57:295–301

Sasekumar A, Leh CMU, Chong VC, Rebecca D, Audery ML (1989) The Sungai Pulai (Johore): a unique mangrove estuary. In: Phang SM, Sasekumar A, Vickineswary S (eds) Proceedings of the 12th annual seminar of the Malaysian Society of Marine Sciences. Research priorities for marine sciences in the 90's. Malaysian Society of Marine Science, Kuala Lumpur, pp 191–211

Silvestre GT, Garces LR, Stobutzki I, Ahmed M, Valmonte-Santos RA, Luna CZ, Zhou W (2003) South and south-east Asian coastal fisheries: their status and directions for improved management: conference synopsis and recommendations. In: Silvestre G, Garces L, Stobutzki I, Ahmed M, Valmonte-Santos RA, Luna CZ, Lachica-Aliño L, Munro P, Christensen V, Pauly D (eds) Assessment, management and future directions for coastal fisheries in Asian countries. WorldFish Center Conference Proceedings, vol 67, pp 1–40

Spalding MD, Ravilious C, Green EP (2001) World atlas of coral reefs. University of California Press, Berkeley

Spalding M, Kainuma M, Collins L (2010) World atlas of mangroves. EarthScan, London

Steneck RS, Carlton JT (2001) Human alterations of marine communities: students beware! In: Bertness MD, Gaines SD, Hay ME (eds) Marine community ecology. Sinauer, Sunderland, pp 445–468

Sterrenburg FAS, Erftemeijer PLA, Nienhuis PH (1995) Diatoms as epiphytes on seagrasses in South Sulawesi (Indonesia): comparison with growth on inert substrata. Bot Mar 38:1–7
Tiwi DA (1999) Description and analysis of institutional problems in the Banten Bay coastal zone. Teluk Banten Research Program, International Institute for Infrastructural, Hydraulic and Environmental Engineering (IHE), Delft
Unsworth RKF, Cullen LC (2010) Recognising the necessity for Indo-Pacific seagrass conservation. Conserv Lett 3:63–73
Valiela I, Bowen JL, York JK (2001) Mangrove forests: one of the world's threatened major tropical environments. Bioscience 51:807–815
Waycott MW, Duarte CM, Carruthers TJB, Orth RJ, Dennison WC, Olyarnik S, Calladine A, Fourqurean JW, Heck KL Jr, Hughes AR, Kendrick GA, Kenworthy WJ, Short FT, Williams SL (2009) Accelerating loss of seagrasses across the globe threatens coastal ecosystems. Proc Natl Acad Sci USA 106:12377–12381
Whanpetch N, Nakaoka M, Mukai H, Suzuki T, Nojima S, Kawai T, Aryuthaka C (2010) Temporal changes in benthic communities of seagrass beds impacted by a tsunami in the Andaman Sea, Thailand. Estuar Coast Shelf Sci 87:246–252
Whittakar RH (1975) Communities and ecosystems, 2nd edn. Macmillan, London
Williams SL, Heck KL Jr (2001) Seagrass community ecology. In: Bertness MD, Gaines SD, Hay ME (eds) Marine community ecology. Sinauer, Sunderland, pp 317–337
Wilson J, Marirnuthu N, Kumaraguru AK (2005) Sedimentation of silt in the coral reef environment of Palk Bay, India. J Mar Biol Assoc India 47:83–87
Yamamoto T (2003) The Seto Inland Sea–eutrophic or oligotrophic? Mar Pollut Bull 47:37–42
Yoon S-P, Jung R-H, Kim Y-J, Hong S-J, Oh H-T, Lee W-C (2009) Spatio-temporal changes in macrobenthic community structure and benthic environment at an intensive oyster culturing ground in Koje-Hansan Bay, Korea. J Kor Soc Oceanogr 14:213–228
Yoshida G, Hori M, Sakiyama K, Hamaguchi M, Kajita J, Nishimura K, Shoji J (2010) Distribution of seaweed bed and tidal flat, and their correlations with fisheries catches in the nine sea areas of the Seto Inland Sea. Fish Eng 47:19–29

Chapter 21
Environmental Disclosure in Japanese Wood-Related Companies for 2005 and 2010

Ryo Kohsaka, Satoshi Tachibana, and Mari Iwata

Abstract Environmental and sustainability reports by private sectors are primarily aimed at information disclosure for individual entities. Comparison of such reports in two different years provides us with a unique opportunity to explore the underlying changes. This chapter compares Japanese wood-related sectors in 2005 and in 2010. During this half-decade, the economy was sluggish, with a number of dramatic meltdowns symbolized by the "Lehman shock" in 2008. The period also experience upheaval in environmental movements with a major UN COP on biodiversity taking place in Nagoya in 2010, making comparison of these 2 years more interesting. Environmental disclosure from 51 companies, mainly from seven wood-related sectors, was reviewed. The trends in disclosed information were reviewed based on the following criteria: publishing of environmental reports, reference to illegal logging, definition of sustainable forest management, and membership in the Global Compact. The results indicated an increase in many of the indicators for large-sized companies, particularly for trading companies. The majority of small- and mid-sized companies did not conduct basic activities such as publishing reports.

Keywords Environmental report • Forest certification scheme • Global Compact • Illegal logging • SFM (Sustainable Forestry Management) • Wood-related sector

---

R. Kohsaka (✉)
Kanazawa University, Graduate School of Human and Socio-Environmental Studies, Kakumacho, Kanzawa, Ishikawa 920-1192, Japan
e-mail: kohsaka@hotmail.com

S. Tachibana
Faculty of Life and Environmental Sciences, University of Tsukuba, Tennodai 1-1-1, Tsukuba, Ibaraki 305-8572, Japan

M. Iwata
Mitsubishi Research Institute, Nagata-cho 2-10-3, Chiyoda-ku, Tokyo 100-8141, Japan

S.-i. Nakano et al. (eds.), *Integrative Observations and Assessments*, Ecological Research Monographs, DOI 10.1007/978-4-431-54783-9_21,

Introduction

Since the 1990s, Japan has been experiencing a transitional period of corporate accountability stemming from expanding public concern and action toward serious global environmental issues. One of the issues is loss of biodiversity, which has captured significant public attention since the Kyushu-Okinawa G8 Summit in 2000 (Tsuru 2005) and more recently the tenth Conference of the Parties to the Convention on Biological Diversity held in Nagoya, Japan, in 2010.

Forests, rainforests in particular, are critical for the health of biodiversity and increasingly important as carbon sinks. Yet, measures to conserve and fight their losses have been difficult. Illegal logging and related underground activities are identified as one of the key drivers for the loss of biodiversity. Under poor governance, illegal logging takes place in highly valuable forest areas (Navarrete et al. 2011). Government and institutional failures, such as corruption and weak enforcement capacity, are one of the major causes of illegal logging. Political actions and policy measures such as rules or treaties are needed but have been only belatedly and insufficiently introduced (Smith et al. 2003; Palmer 2001).

On the other hand, the private sector is arguably the most influential actor in activities related to biodiversity conservation. The behavior of private companies changes in response to pressure from civil societies including protests, media campaigns, and boycotts in extreme cases. The change in the behavior of private companies can be traced by the degree of their information disclosures, because information disclosure is a prerequisite to build trust with society. The publication of environmental or Corporate Social Responsibility (CSR) reports by companies is a method through which corporations demonstrate their accountability. In addition, environmental labeling on their products can serve a similar function.

These two aspects of reporting and certified wood products are examined here to clarify the extent and development of disclosure trends in Japanese wood-related sectors. In this study, we reviewed and analyzed CSR or environmental reports published in 2005 and 2010 by major companies in our focus area. There were notable changes in the international policy environment from 2005 to 2010, such as revision of the Lacey Act in the United States in 2008 and implementation of Voluntary Partnership Agreements (VPA) between the EU and some developing countries. The tenth Conference of the Parties to the Convention on Biological Diversity, held in Japan in 2010, provided a major driver for the Japanese private sector to initiate their information disclosure and reflect upon their influence related to biodiversity, mainly through their supply chains. The global economic environment also changed remarkably by means of deregulation, leading to the collapse of Lehman Brothers in 2008. These changes provide the rationale for comparing the environmental disclosure trends of Japanese companies in 2005 and 2010.

In this survey, we assessed environmental disclosure from 51 companies from seven wood-related sectors. Wood-related sectors are those directly related to forestry, including forestry, pulp and paper, trading of wood products, supplying office equipment using wood, stationery, retail of wood products, and the housing sector. The companies from wood-related sectors are mostly small- and mid-sized companies (SMEs). In addition, seven companies from two other sectors, automobile and utility, were assessed as "non-wood-related sectors." All the companies reviewed are based in Japan. Furthermore, we included telephone and e-mail surveys for some companies and, additionally, website information as available in the review and analysis.

Review

Debate over Corporate Social Responsibility (CSR) and Sustainable Development

Scientists are struggling with the question of CSR. To date, there has not been any internally agreed definition of the term CSR (Taneja et al. 2011). There are regional differences as well. CSR has been discussed mainly by scholars examining North American issues whereas "European interest in CSR is a relatively recent phenomenon" (Falck and Heblich 2007:248). Loew et al. (2004) made a similar observation and noted that the CSR discussion mainly focused on the United States and that similar discussions in the European context were largely associated with "sustainable development." From an application perspective, the two concepts have been approaching each other since the 1990s, with CSR increasingly being recognized as a part of sustainable development (Loew et al. 2004). In other words, there is a progressive transition toward the private sector as being part of the problems as well as part of the solutions, with CSR activities contributing toward the realization of sustainable development.

What exactly is the social responsibility of a corporation in concrete terms? There are various codes and standards for CSR-related corporate activities. Illustrative codes and standards are shown in Fig. 21.1, including various codes, standards, and legal instruments in socio-ethical, environment, and financial realms, showing at which phases the activities are checked. The phases state whether the companies are covering their activities at levels of policy, accounting, assurance, or reporting. For example, the ISO14001 series focuses mostly on environmental phases but does not cover any of the social or financial realms. The Global Compact is more wide ranging and covers social and ethical matters but focuses on the policy realm. This chapter focuses on corporate activities at the reporting phase.

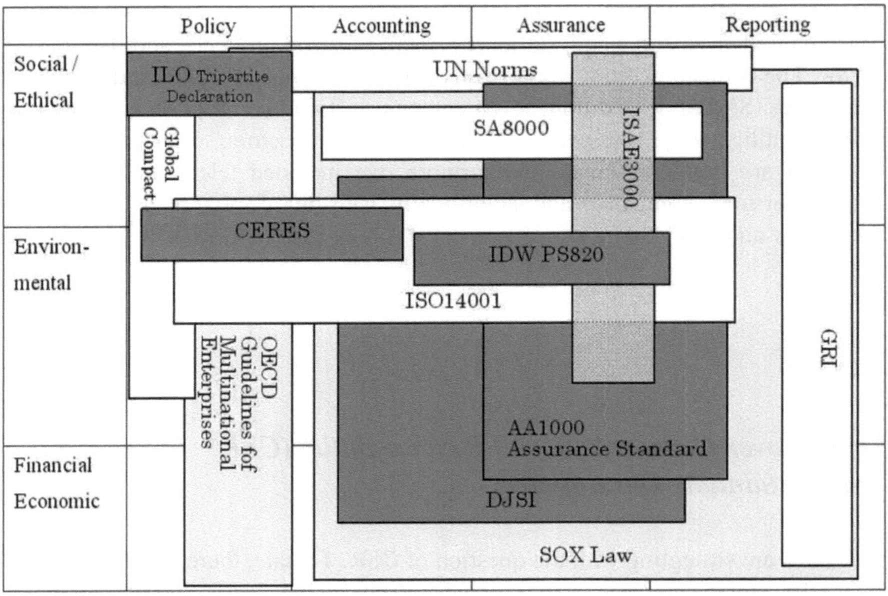

Fig. 21.1 Range of Corporate Social Responsibility (CSR)-related standards and criteria. Original: Humboldt-Universität zu Berlin/Institut für Management "Standards, Codes of Conduct und Berichterstattung." Revised by the authors

Studies in Wood-Related Sectors

Literature specifically for wood-related sectors is reviewed here. In general, wood-related sectors are categorized as "environmentally sensitive industries," and it was suggested in the context of the UK and USA that these companies are more likely to increase disclosure of environmental information than companies in other sectors (Aerts et al. 2008:645). Sectors in the same categories are oil and gas, chemicals, and utilities, and they are often subject to stricter governmental regulations and increased public pressure than are wood-related sectors.

The question of whether wood-related sectors are more willing to disclose environmental information than other sectors is addressed in two parts. First, findings from the existing studies are examined according to the study area, variables (or measurable points), and any characteristics examined (Table 21.1). Second, the Japanese wood- and forestry-related companies and other related sectors are examined from perspectives of environmental reporting.

Studies specific to the paper and pulp industry have been published (Mikkilä et al. 2005; Mikkilä and Toppinen 2008). Mikkilä et al. (2005) analyzed four sawmills in China, Finland, Germany, and Portugal for specified trends in their social performance. Reports by the world's largest Japanese, North American, and Scandinavian pulp and paper companies "shows little regional variation" (Mikkilä and Toppinen 2008:505). These results were found true despite an orientation

Table 21.1 Review of major existing literature in forestry science

Author	Study area	Variables/measure	Characteristics
Kärnä et al. (2001)	Finnish forest industry		
Mikkilä et al. (2005)	Pulp and paper: China, Finland, Germany, and Portugal	Economic, legal, ethical, and discretionary responsibility	Social performance
Mikkilä and Toppinen (2008)	World Largest 10 Pulp and Paper (US, Finland, Sweden, Japan)	Merits for sustainability performance (Economic, Environmental, Social)	Focus on pulp and paper
Vidal and Kozak (2008)	World Largest 20 Pulp and Paper companies	Content analysis methodology, environmental reports: (1) no. pages, (2) type of reports, (3) topics	Focus on pulp and paper period 2000–2005

Compiled by the author, modifying Table 21.1 of Mikkilä and Toppinen 2008

toward profit making (Japan), philanthropy (North America), and compliance (Scandinavian), as pointed out in the existing literature.

Kärnä et al. (2003) suggest that the value chain closest to end-users emphasizes both "redirecting towards sustainability" and "profitability orientation" in their values more than companies at the beginning of the chain. This statement suggests that the environmental activities in the sector are reactions to consumer and other stakeholder responses.

Vidal and Kozak (2008) reviewed the top 20 paper and pulp companies using a content analysis methodology, comparing the trends and content of environmental reports in 2000 and 2005. For example, they demonstrated that "sustainable forestry" was one of the best covered topics in both 2000 and 2005. Economics, human resources, and certification-related elements are among the topics that increased in coverage from 2000 to 2005.

Kärnä et al. (2001) investigated the substance behind the green advertisement of the Finnish forestry sector by checking environmental activities at various phases of the supply chain (decision making, recruitment, planning, etc.) to see whether activities in the advertisements are implemented on the ground. They found that "there is clear environmental substance" (Kärnä et al. 2001) and that it is not mere green advertisement; practices are actually in place.

It should, however, be noted that the Finnish forestry sector is not SMEs but mostly large companies. In Japan, SMEs are an important factor as the majority of companies in the Japanese forestry sector belong to this category; this provides us with a unique opportunity to explore the disclosure activities and notion of CSR. Existing studies indicate that external pressures from stakeholders, including those of civil societies and foreign capital, are the drivers for CSR activities (Lynch-Wood et al. 2009; Oh et al. 2011). The pressures from neighborhoods, environmental groups, community members, or foreign capital can actually positively affect the behavior of the corporation to act in a more socially responsible manner.

On the other hand, these pressures are unlikely to motivate the SMEs as the potential influences to their reputations and their stakeholders would be less significant (Lynch-Wood et al. 2009).

Thus, there is a need to capture the factors and possible motivations for CSR initiatives with specific conditions for SMEs.

Analysis and Results

Overview

A review of CSR or environmental reports was conducted on 55 companies from nine sectors. The first aspect of the analysis focused on whether the selected companies disclose information in the form of CSR or environmental reports. Second, by comparing 2005 and 2010 reports with other related sources of information, references to illegal logging and their changes over time were examined. Illegal logging, that is, harvesting timber in contravention of a country's laws, and the associated trade of wood and wood products are worldwide problems that have widespread economic and environmental impacts (FFPRI 2005). Finally, the trend toward the use of certified wood products was assessed. Forest certification schemes have an important role in realizing sustainable forest management (SFM). The results are summarized in Table 21.2. In addition, basic information on the reviewed companies is also provided (Table 21.3). Because three companies had not published a 2010 report when we conducted our survey, their 2009 reports were used instead.

We found that the majority of companies in the forestry sector (mostly wood handling and processing companies) had not published any CSR or environmental reports in 2005 and in 2010. The size of these companies and the scale of their budgets probably had the most influence on whether a company was disclosing environment-related information. The amount of capital in the forestry sector is low compared to that in other sectors.

Because few companies in the forestry sectors published CSR or environmental reports, surveys were conducted by telephone and e-mail. There are several trends in the wood-related sectors between 2005 and 2010, with the major trends summarized as follows.

- A continued lack of or limited disclosure of environmental information from companies in the Japanese wood-related sectors (2 companies in the 2005 reports and 1 in 2010, of 18 companies).
- A shift toward providing certified products (from 1 company in 2005 to 9 in 2010, of 18 companies).
- A major increase in references to illegal logging, especially for wood, paper, and pulp, and trading companies (from 1 in 2005 to all 18 companies in the sectors, if we include references to supply chain and certified products to combat illegal logging).

Table 21.2 Trend of each sector's performance on Corporate Social Responsibility (CSR) activities and sustainable forestry (comparison between 2005 and 2010)

Sector	CSR/environmental reporting	Reference to illegal logging	Sustainable forest management	Global compact
Forestry	↓ (2/18 → 1/18)	↑↑ (1/18 → 18/18)	↑ (1/18 → 9/18)	→ (0/18 → 0/18)
Pulp and paper	→ (6/6 → 6/6)	↑↑ (3/6 → 6/6)	→ (6/6 → 6/6)	→ (2/6 → 2/6)
Trading	→ (5/5 → 5/5)	↑↑ (1/5 → 4/5)	→ (5/5 → 5/5)	↑↑ (1/5 → 5/5)
Supplying office equipment	↓ (6/6 → 5/6)	↑ (1/6 → 4/6)	↑↑ (1/6 → 5/6)	↑ (2/6 → 4/6)
Stationery	→ (3/7 → 3/7)	↑ (1/7 → 2/7)	→ (2/7 → 2/7)	→ (0/7 → 0/7)
Retail	↓ (4/4 → 3/4)	↑ (0/4 → 1/4)	↑ (0/4 → 2/4)	→ (0/4 → 1/4)
Housing	→ (5/5 → 5/5)	↑ (0/5 → 2/5)	↑ (1/5 → 3/5)	↑ (0/5 → 3/5)
Automobile	→ (2/2 → 2/2)	→ (0/2 → 0/2)	↑ (0/2 → 1/2)	↑ (0/2 → 1/2)
Utility	↓ (2/2 → 1/2)	→ (0/2 → 0/2)	→ (0/2 → 0/2)	→ (0/2 → 0/2)

Note: Numbers beneath the *arrows* indicate "the number of companies who performed listed environmental activities in 2005/(of) the total number of companies subject to research, and the number of companies that performed in 2010/(of) the total number of companies subject to research." ↑↑ indicates a sector that improved their performance significantly with more than 80 % of companies performing listed activities. ↑ indicates a sector that improved their performance to some extent with less than 80 % of companies performing listed activities. → indicates a sector whose performance remained the same in 2005 and 2010

Of 18 companies, the reference to and provision of certified products were limited to only one company in 2005, but by 2010 half the companies were providing certified products. There were more options for third-party certification in the wood-related sectors in 2010 than in 2005. The certificate scheme includes forest management certification and chain of custody certification.

References to illegal logging issues are now common, and the frequency of references increased during the 5-year period. In Japan, the Ministry of Environment revised the "Law Concerning the Promotion of Procurement of Eco-Friendly Goods and Services by the State and Other Entities" (known as "the Law on Promoting Green Purchasing") in 2006, introducing guidelines for procuring legal wood and wood products. Under the guidelines, many industrial associations in the wood-related sectors have been registered as certifiers since 2006, certifying companies supplying eco-friendly goods. Few companies made references to illegal logging issues in 2005, but the issue became more prevalent in 2010, and all companies in the wood-related sectors made some references or activities to prevent illegal logging. The key components of CSR or environmental reports in reference to illegal logging issues and certification actions for SFM are reviewed individually next.

Table 21.3 Review results on a company basis

Company			CSR/ environmental reporting		Reference to illegal logging		Sustainable forest management (SFM) certificate		Global compact	
No.	Capital (million JPY)	Employees	2005	2010	2005	2010	2005	2010	2005	2010
Forestry										
1	NA	18	Yes	No	Yes	Yes	Yes	Yes	No	No
2	100	1,149	No	No	No	Yes	No	Yes	No	No
3	42	NA	No	No	No	Yes	No	Yes	No	No
4	7,324	1,579	Yes	Yes	No	Yes	No	Yes	No	No
5	100	NA	No	No	No	Yes	No	Yes	No	No
6	9,000	1,016	No	No	No	Yes	No	Yes	No	No
7	50	NA	No	No	No	Yes	No	Yes	No	No
8	100	1,646	No	No	No	Yes	No	Yes	No	No
9	NA	110	No	No	No	Yes	No	Yes	No	No
10	NA	NA	No	No	No	Yes	No	No	No	No
11	100	102	No	No	No	Yes	No	No	No	No
12	403	427	No	No	No	Yes	No	No	No	No
13	41	73	No	No	No	Yes	No	No	No	No
14	24	25	No	No	No	Yes	No	No	No	No
15	160	60	No	No	No	Yes	No	No	No	No
16	52	93	No	No	No	Yes	No	No	No	No
17	190	67	No	No	No	Yes	No	No	No	No
18	50	NA	No	No	No	Yes	No	No	No	No
Pulp and paper										
1	103,380	4,021	Yes	Yes	Yes	Yes	Yes	Yes	Yes	Yes
2	55,730	13,834	Yes	Yes	Yes	Yes	Yes	Yes	Yes	Yes
3	32,756	4,304	Yes	Yes	Yes	Yes	Yes	Yes	No	No
4	5,140	89	Yes	Yes	No	Yes	Yes	Yes	No	No
5	42,021	4,023	Yes	Yes	No	Yes	Yes	Yes	No	No
6	30,400	3,238	Yes	Yes	No	Yes	Yes	Yes	No	No
Trading										
1	202,241	4,358	Yes	Yes	Yes	Yes	Yes	Yes	No	Yes
2	341,482	6,118	Yes	Yes	No	Yes	Yes	Yes	Yes	Yes
3	203,598	5,665	Yes	Yes	No	No	Yes	Yes	No	Yes
4	219,300	5,218	Yes	Yes	No	Yes	Yes	Yes	No	Yes
5	262,686	4,020	Yes	Yes	No	Yes	Yes	Yes	No	Yes
Supplying office equipment										
1	20,000	9,911	Yes	Yes	Yes	Yes	Yes	Yes	Yes	Yes
2	135,300	109,014	Yes	Yes	No	Yes	No	Yes	No	Yes
3	53,240	13,311	Yes	Yes	No	Yes	No	Yes	Yes	Yes
4	174,762	26,019	Yes	Yes	No	Yes	No	Yes	No	Yes
5	12,000	13,109	Yes	Yes	No	No	No	No	No	No
6	10,375	6,760	Yes	No	No	No	No	Yes	No	No

(continued)

Table 21.3 (continued)

Company			CSR/ environmental reporting		Reference to illegal logging		Sustainable forest management (SFM) certificate		Global compact	
No.	Capital (million JPY)	Employees	2005	2010	2005	2010	2005	2010	2005	2010
Stationery										
1	15,800	4,991	Yes	Yes	Yes	Yes	Yes	Yes	No	No
2	3,535	469	Yes	No	No	Yes	Yes	Yes	No	No
3	4,500	826	Yes	Yes	No	No	No	No	No	No
4	90	972	Yes	Yes	No	No	No	No	No	No
5	200	394	No	No	No	No	No	No	No	No
6	2,341	1,181	No	No	No	No	No	No	No	No
7	4,497	551	No	No	No	No	No	No	No	No
Retail										
1	199,054	NA	Yes	Yes	No	No	No	Yes	No	Yes
2	100	17,026	Yes	No	No	No	No	No	No	No
3	58,566	5,706	Yes	Yes	No	No	No	No	No	No
4	17,200	5,729	Yes	Yes	No	Yes	No	Yes	No	No
Housing										
1	1,101,205	13,482	Yes	Yes	No	No	No	Yes	No	No
2	103	25,085	Yes	Yes	No	No	No	No	No	Yes
3	186,554	15,032	Yes	Yes	No	Yes	No	Yes	No	No
4	100,000	19,770	Yes	Yes	No	No	No	No	No	Yes
5	27,672	13,601	Yes	Yes	No	Yes	Yes	Yes	No	Yes
Automobile										
1	397,050	320,590	Yes	Yes	No	No	No	Yes	No	No
2	605,813	29,878	Yes	Yes	No	No	No	No	No	Yes
Utility										
1	676,400	38,671	Yes	No	No	No	No	No	No	No
2	489,300	22,143	Yes	Yes	No	No	No	No	No	No

NA, not available

CSR or Environmental Reports

Table 21.4 summarizes reporting in 2005 and 2010.

Trends are clearly specific to each sector. In the forestry and stationery sectors, it is not common to publish CSR or environmental reports. In the sectors with large companies, the majority of companies published reports in both 2005 and 2010. Because most of the companies in the wood-related sector are small to medium size, they did not publish a report in 2005. The situation remained the same in 2010; one company published a report in 2005 but could not be confirmed for 2010.

Table 21.4 Number of companies publishing CSR or environmental reports in 2005/2010

Sector	Total no. of companies reviewed	No. of companies publishing a report in 2005	No. of companies publishing a report in 2010	Note
Forestry	18	2	1	–
Pulp and paper	6	6	6[a]	–
Trading	5	5	5	–
Supplying office equipment	6	6	5	One company started web-based reporting instead of printing a report.
Stationery	7	3	3[a]	One company started web-based reporting instead of printing a report, another started to report.
Retail	4	4	3	One company ceased to publish its report because of change of business circumstances (M&A).
Housing	5	5	5	
Automobile	2	2	2	
Utility	2	2	1	TEPCO ceased to publish its CSR report consequent to nuclear accident (at the time of writing).

[a]Two companies in the pulp and paper industry and one company in stationery had not published their 2010 reports yet at the time of writing, so we used 2009 reports

In general, companies operating in the wood-related sectors were less forthcoming to disclose information compared to other industries. When interviewed by telephone for illegal logging and SFM, SMEs were especially cautious. The so-called Sick House Syndrome (sickness caused by air pollutants in newly built houses) was one of the major issues in the media around 2005, and the companies were very sensitive about the issue when interviewed (this issue was not a focus of our study).

Some staff members of the forest- or housing-related companies were nervous and so did not communicate clearly. They responded that their products did not contain any chemical elements related to the sickness even though the question was on other topics such as SFM.

Reference to Illegal Logging

Although the publication of CSR or environmental reports has made little progress in some areas, references to illegal logging have significantly increased during the past 5 years (Table 21.5). Because not all the companies publish CSR or environmental reports, we confirmed the references in the following way:

Table 21.5 Number of companies referring to illegal logging issues in 2005/2010

Sector	Total no. of companies reviewed	No. of companies referring to illegal logging in 2005	No. of companies referring to illegal logging in 2010	Actions taken by companies that refer to illegal logging
Forestry	18	1	18	Obtaining certification for legal wood supplier and for sustainable forest management
Pulp and paper	6	3	6	Using certified wood products, making procurement rules
Trading	5	1	4	Using certified wood products, making procurement rules
Supplying office equipment	6	1	4	Using certified wood products, making procurement rules, funding conservation projects
Stationery	7	1	2	Making procurement rules
Retail	4	0	1	Funding conservation projects
Housing	5	0	2	Using certified wood products
Automobile	2	0	0	–
Utility	2	0	0	–

1. If the company published a report, we examined whether some references to illegal logging issues were made in the reports.
2. If the company does not publish a report, we looked at their website and reviewed the related materials published online.
3. In addition to the website, we examined if companies have obtained a certificate as a "legal woods supplier" issued by a wood industry association. We assumed that taking such a certificate indicates awareness of the impact of illegal logging and that they are willing to take action to prevent it (Table 21.5).

In the wood-related sectors, only one company referred to illegal logging issues in 2005. That company obtained a Forest Stewardship Council (FSC) Forest Management (FM) certificate in 2005. In 2010, all the companies surveyed had been certified as suppliers of legal wood products by a wood industry association. Normally such certificates were only available after 2006 under the revised "Law on Promoting Green Purchasing." Also, an increasing number of companies had obtained FSC FM and other sustainability certificates in 2010. In the pulp and paper sector, we observed changes between 2005 and 2010: these companies were making more frequent references to illegal logging in 2010 than in 2005. In 2010, companies took some actions such as using certified wood products as raw materials and embedding "legality" into their procurement policies. Similar trends were

Table 21.6 Number of companies obtaining a Sustainable Forest Management certificate or retailing certified products in 2005/2010

Sector	Total no. of companies reviewed	No. of companies in 2005	No. of companies in 2010	Note
Forestry	18	1	9	–
Pulp and paper	6	6	6	–
Trading	5	5	5	–
Supplying office equipment	6	1	5	Supplying Forest Stewardship Council (FSC) paper
Stationery	7	2	2	–
Retail	4	0	2	Supplying FSC paper, building their shop with FSC wood products
Housing	5	1	3	Supplying FSC wood products, using FSC paper in their operations (such as catalogs, business cards)
Automobile	2	0	1	FSC certificate in company-owned forest
Utility	2	0	0	–

identified for trading companies. Although referring to "illegal logging" issues in the reports has become a new norm for the paper and pulp and trading sectors, it is less common among the stationery, retail, and housing sectors during the same survey period.

Certification for Sustainable Forest Management

The trend of certified wood products is assessed in Table 21.6.

The number of certified wood products increased remarkably in the product lineups of wood-related companies. Larger-scale companies are especially obtaining FSC or PEFC (Programme for the Endorsement of Forest Certification) certificates or both. Most of them obtained the certificates in the year 2010, but some began earlier, in 2006 or 2009. The companies indicated that the driving force for obtaining certification is demand from the customers for more environmentally friendly products or questions regarding the risk of their products containing illegally logged wood. These customer questions were channeled through the branches to company headquarters, leading to changes in their products in many cases. In the case of one company, the PEFC-certified products were introduced in response to the concern of customers against illegally logged wood, but they also mentioned that the customers are not willing to pay a higher price.

One SME indicated why they have not yet introduced forest-certified products. First, the cost of obtaining such a certificate is yet to be reflected in product prices. In other words, "there are needs for certified products but the products are not yet pricing in the costs." The other reason for not obtaining certification is that the area of forest that is owned is not large enough to warrant paying the costs for certification.

Summary

A significant increase in references to illegal logging was found, especially for wood, pulp and paper, and trading companies. Those sectors also started to provide certified products. What was the driving force behind these changes? Government policies were behind these changes in 2010. The Ministry of Land, Infrastructure, Transport and Tourism (MLIT) started a program to support "Long-Life Quality Housing" by providing a 200,000 yen/house subsidy if legal timber, locally harvested, is used for construction. Furthermore, as the Japanese "Law on Promotion of Use of Wood in Public Buildings" is enforced, public buildings will shift toward using more sustainable wood products. In addition to these policy changes, several companies responded that their customers are requesting certified products (even though they might not be ready to pay more).

These two elements are estimated to be the underlying forces for the changes between 2005 and 2010. In general, the companies in wood-related sectors are making more frequent reference to illegal logging issues in 2010 than in 2005 and are providing more certified products such as FSC, PEFC, SGEC (Sustainable Green Ecosystem Council) and legal wood products. Yet, not all trends were positive. The number of companies publishing environmental or sustainability reports actually decreased.

Conclusions

How can we change the behavior of the private sector? Scientific projects to evaluate ecosystem services in quantitative terms are underway. In this volume with other authors, the visualization of ecological footprints is discussed. Such quantification and mapping measures are important tools for the future, but probably not rigorous enough for monitoring purposes. In this chapter we have examined the changes in information disclosure of the private sector in 2005 and 2010. This practical review is an important source to actually monitor the changes of the private sector.

First, we reviewed the existing literature in forestry science and found some evidence indicating that the wood-related sectors are disclosing substantial information and that it is not a mere advertisement, such as in the case of the Finnish forest industry. The need to explore SMEs in a quantitative manner, instead of case studies, was identified as an area to be explored.

Second, we reviewed reporting by companies in the Japanese wood-related sectors. We found that the majority of the companies in the Japanese wood-related sectors were publishing environmental reports less frequently than in other sectors. The size of the company and the scale of their budget probably had more influence on whether a company was disclosing more environment-related information. Our work confirms the findings from Lynch-Wood et al. (2009) that SMEs have different motivations than do large-sized companies.

We found an important factor of more exposure to overseas consumers, and partners were critical in companies disclosing information and joining CSR initiatives. The latter result was especially apparent for sectors such as paper and pulp or trading with involvement with certified products or memberships with the Global Compact.

Further research is necessary to see what could potentially motivate Japanese wood-related sectors to disclose more information. The sectors do not have much exposure to foreign partners and will need different types of drivers and pressure to disclose more information.

Acknowledgment This study was conducts as a part of the research project, "Trade off Analysis and Local Governance Model of Satoyama Ecosystem Services" (ES-Tradeoff, FY2013–FY2015; 1-1303), which is supported by the Ministry of the Environment (MoE), Japan through its Environment Research and Technology Development Fund. Part of the project was funded by the Japan Society for the Promotion of Science (Project Leader R. Kohsaka). The authors' thanks are extended to the valuable discussions at the advisory meeting of Environment Research and Technology Development fund (S9-1: Project Leader T. Yahara) of the MOE, Japan.

References

Aerts W, Cormier D, Magnan M (2008) Corporate environmental disclosure, financial markets and the media: an international perspective. Ecol Econ 64:643–659

Falck O, Heblich S (2007) Corporate social responsibility: doing well by doing good. Bus Horiz 50:247–254

Forestry and Forest Products Research Institute (FFPRI) (2005) Current activities to combat illegal logging and associated trade in illegally sourced wood products in Japan. Prepared for DFID, p 54. http://www.illegal-logging.info/uploads/Japan_illegal_logging_action_survey_2005[1].pdf. Accessed 25 Sep 2011

Kärnä J, Juslin H, Ahonen V, Hansen E (2001) Green advertising: greenwash or a true reflection of marketing strategy? Greener Manag Int 33:59–70

Kärnä J, Hansen E, Juslin H (2003) Social responsibility in environmental marketing planning. Eur J Mark 37:848–871

Loew T, Ankele K, Braun S, Clausen J (2004) Bedeutung der CSR-Diskussion für Nachhaltigkeit und die Anforderungen an Unternehmen (Significance of the CSR Debate for Sustainability and the Requirements for Companies). IÖW, p 126

Lynch-Wood G, Williamson D, Jenkins W (2009) The over-reliance on self-regulation in CSR policy. Bus Ethics 18:52–65

Mikkilä M, Toppinen A (2008) Corporate responsibility reporting by large pulp and paper company. For Policy Econ 10:500–506

Mikkilä M, Kolehmainen O, Pukkala T (2005) Multi-attribute assessment of acceptability of operations in the pulp and paper industries. For Polocy Econ 7:227–243
Navarrete J, Ramirez M, Perez-Salicrup D (2011) Logging within protected areas: special evaluation of the monarch butterfly biosphere reserve, Mexico. For Ecol Manag 262:646–654
Oh W, Chang Y, Martynov A (2011) The effect of ownership structure on corporate social responsibility: empirical evidence from Korea. J Bus Ethics 104(2):283–297. doi:10.1007/s10551-011-0912-z
Palmer C (2001) The extent and causes of illegal logging: an analysis of a major cause of tropical deforestation in Indonesia. Center of Social and Economic Research on the Global Environment Working Paper, London
Smith J, Obidzinski K, Subarudi, Suramenggala I (2003) Illegal logging, collusive corruption and fragmented governments in Kalimantan, Indonesia. Int For Rev 5(3):293–302
Taneja S, Taneja P, Gutpa R (2011) Researches in corporate social responsibility: a review of shifting focus, paradigms, and methodologies. J Bus Ethics 101:343–364
Tsuru S (2005) Current activities to combat illegal logging and associated trade in illegally sourced wood products in Japan. Prepared for DFID, UK, Forestry & Forest Products Research Institute
Vidal NG, Kozak RA (2008) The recent evolution of corporate responsibility practices in the forestry sector. Int For Rev 10:1–13

Chapter 22
Greening the Cities with Biodiversity Indicators: Experience and Challenges from Japanese Cities with CBI

Ryo Kohsaka and Shigefumi Okumura

Abstract For scientists, and other stakeholders in biodiversity monitoring systems (including AP-BON), capturing and understanding the status and trends of biodiversity and ecosystem services are a main focus. In the policy–science interface, communicating the complex results in comprehensible ways has been one of the key challenges. Development of indicators, maps, and other visualization tools is instrumental for identification, understanding, and support of the relevant policy decisions and processes.

In recent years, different cities have explored the development of such indicators in the urban context through negotiation. The development of indicators for urban ecosystems and biodiversity is shown here. The potential challenge of the application and use of such indicators in Japanese urban contexts is reviewed based on interviews and existing data. This chapter discusses and reviews the advantages and limitations of urban biodiversity indicators. The review focuses on applying the newly developed City Biodiversity Index (CBI), modifying the Singapore city biodiversity index adjusted so Japanese local municipalities can easily use it practically. The data are based on a research project implemented by the Ministry of Land, Infrastructure, Transport and Tourism, Japan (MLIT).

The existing literature points out that policy makers tend to emphasize ecosystem services for justification of their policies whereas scientists tend to focus on biodiversity. Kohsaka R (2010) Developing biodiversity indicators for cities: applying the DPSIR model to Nagoya and integrating social and ecological aspects. Ecol Res 25:925–936. Such twists are not a major problem if the status of biodiversity correlates with ecosystem services: this is true at a global or at a regional

R. Kohsaka (✉)
Kanazawa University Graduate School of Human and Socio-Environmental Studies,
Kakuma, Kanazawa, Ishikawa 920-1192, Japan
e-mail: kohsaka@hotmail.com

S. Okumura
Mitsubishi Research Institute, Nagata-cho 2-10-3, Chiyoda-ku, Tokyo 100-8141, Japan

S.-i. Nakano et al. (eds.), *Integrative Observations and Assessments*, Ecological Research Monographs, DOI 10.1007/978-4-431-54783-9_22, © Springer Japan 2014

scale, but may be different at the local level. For example, the results of studies by the city of Nagoya indicate that ecosystem services correlate with the size of green or open spaces and not with the status of biodiversity. As such, applying biodiversity indicators at different scales can be a contentious issue. In addition, the integration of biodiversity relevant elements to ecological footprint maps is often discussed from the perspectives of local governments.

Keywords Concepts and terminologies • Definition • Ecosystem service • Indicators • Urban biodiversity

Introduction to Urban Biodiversity Indicators

The year 2007 marked the era of having more than half of the world's population living in cities for the first time in history (UN-HABITAT 2008). It is also the year when the three major metropolitan areas (Tokyo, Kansai, and greater Nagoya) contained half of the Japanese population.

The use and conservation of urban ecosystems is one of the key challenges facing the globe. The development of indicators is explored in urban contexts to monitor and identify trends. Such indicators are instrumental for policy makers and the civilian society as well for decision making. In addition, indicators are useful for communication for scientists, policy makers, and citizens alike. If the trends can be visualized, trends and scenarios can be shown, in figures, tables, and maps, as also explored in this project.

The OECD (Organisation for Economic Co-operation and Development) led the initial development of indicators in the 1970s (OECD 1978). It was pointed out that the indicators were frequently separated from one another depending on their administrative units or data source. Such examples are the domain of economics (employment, production, energy), the social domain (housing, education), and the environmental domain (green areas, water quality). In the context of biodiversity and ecosystems, the Convention on Biological Diversity (CBD) started the discussion of urban indicators in collaboration with the United Nations Human Settlements Programme (UN-HABITAT).

At the Ninth Conference of the Parties of the CBD in 2008, the role of local authorities (such as cities and mayors) was discussed. Their roles in implementing the goals of the Convention were adopted (Decision IX/28; see SCBD 2008).

During such discussions, the government of Singapore initiated a round of expert meetings to develop a biodiversity indicator specific to the urban context, titled the Singapore Index on Cities' Biodiversity (CBI). The government of Singapore used its unique position of being city and a state. It was an open process for external experts to comment (Chan and Djoghlaf 2009). The indicator became an integral part of the Plan of Action at the tenth Conference of the Parties held in Nagoya, Japan (Decision X/22). International organizations such as ICLEI

(Local Governments for Sustainability) and the IUCN (International Union for Conservation of Nature) collaborated in the development in addition to the municipalities.

Japan initiated a domestic process in the legal framework as well. Under Article 13 of the Basic Act on Biodiversity (*Seibutsu Tayousei Kihon-ho*), local municipalities are encouraged to develop their version of a local biodiversity action plan (LBCP). However, these processes remain voluntary. The municipalities are not necessarily aware of the LBCP, particularly among the smaller-scale cities (Chiba et al. 2012).

Integration and mainstreaming of biodiversity elements are underway with more fundamental plans such as the Greening Plan (*Midori no Kihon Keikaku*) at the individual Japanese municipality level. The Ministry of Land, Infrastructure, Transport and Tourism (MLIT) has provided technical guidelines relative to biodiversity within the Greening Plan since October 2011 (MLIT 2011). In the guidelines, the elements of ecological networks, such as corridors, buffer zones, establishment of core, central zones, and monitoring in the urban context, are illustrated.

The problem of urban biodiversity is interlinked with land-use changes and population in the rural areas. In the ninth Delphi Survey conducted in 2009, a survey for future technologies, the rural–urban dichotomy and technology was regarded as an urgent issue for Japanese society (MEXT 2010). The urgency was shared by the majority of scientists. The research is periodically conducted every 5 years by the Ministry of Education, Culture, Sports, Science and Technology. The technology related to global warming and risk assessment ranked high as important technologies both globally and nationally. The rural–urban dichotomy and technology ranked as uniquely important for the domestic context in Japan, which included technologies such as "commuting agriculture system" (farming system by people commuting daily who are living in urban areas).

In this chapter, experiences from municipalities in Japan were reviewed with a specific focus on applying the newly developed CBI. The data are based on a research project implemented by the MLIT. The Ministry has taken over the trial of Nagoya City for a city biodiversity index and is modifying the Singapore city biodiversity index, adjusting it so that Japanese local municipalities can use it easily and apply it in a practical way. The MLIT conducted interviews with officials in local municipalities to investigate the applicability of the tentative MLIT version CBI to Japanese local municipalities.

As described here, challenges lie at different levels of terminology (fragmentation, differentiation of human/nature influences), methodology (application at different scales, quantification of ecosystem services), and institutional levels (data split with administrative unit or competing goals).

Last, it is to be noted that the purpose of CBI is not to make comparisons or ranking of different municipalities but rather to deepen our understanding of the science underlying the indicators and to improve the CBI in different contexts.

Review

Global Review

There is much existing literature on values of ecosystems at global, regional, or national levels. A few illustrative examples are listed in Table 22.1. The widely quoted work by Costanza et al. (1997) was one of the first attempts to capture global ecosystem services at a global scale.

Reflecting these evaluations of ecosystems, the need arises for specific evaluations of urban ecosystem services.

Under the framework of the CBD, the first comprehensive report for such urban ecosystems, titled "The Cities and Biodiversity Outlook (CBO-1)," has been published in 2013. The essence of experiences and the lessons learned are shared from international cities. A chapter is devoted for the development of CBI applications in range of international cities (Kohsaka et al. 2013). Several key messages are presented in the report, summarized as follows:

1. Unsustainable urbanization is a critical driver behind global biodiversity loss and ecosystem change.
2. Rich biodiversity can exist in cities.
3. Biodiversity and ecosystem services represent critical natural capital.
4. Urban ecosystems contribute significantly to improved human health.
5. Incorporating biodiversity and ecosystems in urban planning and design helps reduce carbon emissions and enhances adaptation to climate change.
6. Food and nutrition security depend on local and biodiversity-based food systems.
7. Ecosystem functions must be integrated in urban policy and planning.
8. Successful management of biodiversity and ecosystem services includes all levels and all sectors.
9. Cities offer unique opportunities for learning and education about a resilient and sustainable future.
10. Cities have a large potential to generate innovations and governance tools and therefore can—and must—take the lead in sustainable development.

Source: Key message from CBO-1

Table 22.1 Existing literature on values of ecosystems

Objects	Value of ecosystem services	Authors
Global ecosystem services	Average of USD 33 trillion per year	Costanza et al. (1997)
Ecosystem services of wetlands	USD 3.4 billion per year	Schuyt and Brander (2004)
Forests in Japan	JPY 70 trillion per year	Japan Science Council (2001)
Paddy in Japan	JPY 8,200 billion	Japan Science Council (2001)

Table 22.2 Empirical studies on economic valuation of ecosystem service

Habitat	Economic valuation of ecosystem service	Authors
Value of park and biotope in school in Japan	Values of park and biotope in school are estimated as JPY 1.29 billion (23 sites, total 2.4 ha)	Yokota et al. (2004)
Park in city area	Choice-based conjoint analysis was conducted targeting 2,000 households in Setagaya-ku, Tokyo	Fujiwara et al. (2004)
	Marginal willingness to pay for a urban park in Setagaya is estimated to be JPY 7,865/household/year	
Development of green zone	Willingness to pay for developing green zone in Rokko Cordillera in Hyogo Prefecture is estimated to be JPY 8,872/household/year. Multiplied by number of households (849,364), the overall willingness to pay is JPY 75 billion/year	Tsuge (2001)

The messages are mixed, with a tone of optimism for the potential and a sense of urgency for the status quo. In the context of Japanese cities, urbanization is frequently pointed out as a cause of destruction. Urbanization is a key element for the future of Satoyama (cf. UNU-IAS & JSSA 2012: 35 Fig. 11.13).

Examples from Individual Cities

There are discussions with individual cities as well. Brack (2002), quoted in the TEEB report, analyzed the benefits and costs of plantations in Canberra, Australia. The benefits were estimated to be 20–68 million AUS$ for the urban plantation.

The City of Nagoya provides a discounted interest rate for private landowners with the condition that the owners leave certain portion of their lands as a green area or as a reserve.

As illustrated in Table 22.2, there are empirical studies with the contingent valuation method (CVM) and conjoint analysis.

Profiles of Cities and Methodology

Structured interviews were conducted to identify the challenges of city biodiversity indicators. Eight different types of cities were selected for interviewees, in consideration of variety of size, availability of data, population density, etc. The eight cities are Yokohama, Nagoya, Fukuoka, Kitakyushu, Hamamatsu, Kashiwa, Minami-Alps, and Itami.

Yokohama has a population of 3.69 million, and its population density is 8,433.8/km^2. Yokohama is making an effort to implement the policies of low carbon, greening, and biodiversity. The coastal areas are urbanized, although there are large-scale forestlands and farmlands remaining in the suburbs; the city and the wild areas are woven together in a mosaic-like pattern.

Nagoya has a population of 2.26 million, and its population density is 6,935.3/km^2. The city is made up of the metropolitan area in the center and green outskirts as follows: the western area faces the sea and remains in tidal mudflats and paddy fields, and the eastern area has thickets, irrigation ponds, and wetlands, although development of land for housing has been progressing. As Nagoya hosted Aichi Expo and COP10, citizens are conscious of the environment and biodiversity.

Fukuoka has a population of 1.46 million, and its population density is 4,288.5/km^2. Although metropolitan, it remains rich in nature. The urban area is set against the backdrop of the Sefuri and Sangun Mountains, and the mountains and forest mingle with the urban area. Fukuoka also boasts a coastline and the islands of Hakata Bay, the rivers and streams that join the mountains to the Hakata coast, and the irrigation ponds and farms that dot the suburban areas. More than half the municipal area is a natural area, 80 % of which is a protected natural area.

Kitakyushu has a population of 980,000, and its population density is 2,002.2/km^2. Although a manufacturing and industrial city, the city is blessed with rich nature. It is surrounded by the sea on three sides, bordered by the Sea of Hibiki, Suou, and the Kanmon Straits; approximately 40 % of the municipal area is forest.

Hamamatsu has a population of 800,000, and its population density is 514/km^2. Being the second largest municipal area in Japan, from the Pacific Ocean to the Southern Alps, the city covers a wide variety of environments such as reservoirs, forests, villages, rivers, and the sea. It enables us to understand the ties of life in one city. Thanks to the large waterfront areas such as Lake Hamana and the Tenryu River, waterfront vegetation has developed so that abundant fish and birds inhabit the area, which makes Lake Hamana a potential candidate for the Ramsar Convention. Although the city has urban areas, 68 % of the municipal area is woodlands. Moreover, the city has much farmland, many rice paddies, and abundant Satoyama.

Kashiwa has a population of 400,000, and its population density is 3,516.2/km^2. Various natural environments remain there, such as waterside areas along the Tone River and other rivers, paddy fields, valleys, spring water areas, forests around shrines and temples, woods around residences, hillside woods, and the site of an old castle. On the other hand, urbanization continues, and the loss of nature such as water, greenery, and soil caused by the development of land for housing and road building has led to fears that once-familiar wild creatures may disappear.

Minami-Alps has a population of 70,000, and its population density is 275.1/km^2. The city consists of two regions that are unified geographically and topographically. One region is the basin of the Kamanashi River and the alluvial fan of the Midai River on the western region of the Kofu basin, and the other is the Southern Alps mountain range in the upstream part of those rivers. Minami-Alps is famous as a city of mountains and gardens, robed in the verdure of the majestic Southern Alps.

Itami has a population of 200,000, and its population density is 7,877.6/km^2. Located in the southeastern part of Hyogo Prefecture, the municipal area spreads around the elevated terrace of an alluvial plain. In former times, the region was abundantly blessed with nature, with spreading rural landscapes of Satoyama, and a wide variety of creatures living in its ponds and woods. However, hand in hand with the increase of the population during the era of high economic growth, the landscape has been rapidly replaced by buildings such as business establishments, factories, housing, and public facilities.

Results

The results of interviews revealed the challenges and opportunities for developing the Japanese version of CBI, with the following points. The results are illustrated under Needs by Cities (Expressed Preferences, Comments for Specific Index, Economic Valuation of Ecosystem Services, Quantitative Analysis) and Challenge and Solutions. It is to be noted that the cities are at the early stage of their applications and trials. The results are of tentative nature and some of the results and needs are changing rapidly, especially those related to ecosystem services and economic valuations.

Needs

- Local government expects that CBI can be used for assessing the progress of local biodiversity strategy, and for giving background for the biodiversity-related projects by understanding the strength or weakness of their own cities, capturing the changes of biodiversity.
- On the other hand, some cities are negative about introducing CBI. They are concerned about difficulties in improving the results in some indices.
- Local governments expect the guideline to be published by the national government. They also request the collaboration among different ministries such as the MLIT and the Ministry of the Environment.

Expressed Preferences

- Local governments prefer the easily understandable indices that do not require data collection and are easily explainable to other departments than the department in charge of biodiversity conservation.
- It is pointed out that the indices using common data (e.g., data owned by national government) are understandable and easily used. Local governments want the national government to provide data for CBI.

- The index should not uniformly valuate and compare the status of biodiversity among different cities, but should reflect the situation of each city.
- The index should be used for self-evaluation and should not be used for comparing different cities.

Comments for Specific Index

In addition, the interviews collected the opinions on each specific index as shown in Table 22.3. First, regarding "Native Biodiversity," local governments mentioned that they do not have the skill and capacity for defining the index, such as difficulty in setting the appropriate distance of patches. They also indicated that it is difficult to collect information on number of species, ratio of native species, etc. In some cases, improvement in these indices are not the result of biodiversity conservation, but just the result of improved monitoring skills or statistical measures. Local government advocated that they cannot use the index in a policy-making process because they cannot understand the relationship between conservation activities and the number of species.

Needs for Economic Valuation of Ecosystem Services

Many local governments showed interest in quantitative valuation or economic valuation of ecosystem services. In particular, regarding economic valuation of ecosystem service in green areas, they expect to use it to capture the effect of the projects and also to reserve the budget. Recently, transparency is more and more often requested. Accordingly, the importance of showing the effect of the projects in a quantitative manner is increasing. The importance of biodiversity is not well known in local governments. If they could show the economic valuation of biodiversity, the departments in charge would acquire the budget easily. Local governments also have a need to show economic loss related to degradation of natural resources.

On the other hand, some cities are anxious that they will be criticized for affecting land price because of showing the data they have provided. In such cases, other actors such as scientists are expected to conduct an economic valuation.

There are developments with local governments and private sector collaboration. The city of Nagoya, for example, has provided economic incentive for greening its privately owned lands, in collaboration with banking sectors (Kohsaka 2009).

Quantitative Analysis

Nagoya city conducted a trial of quantitative assessment of ecosystem services in their city area. Nagoya city conducted it for two model areas. For the first model area, the "Hachiryu" green area, among regulatory services, water storage and climate change mitigation are valuated quantitatively. As a result of the estimation, 5,352–7,765 m^3 of water is stored as an ecosystem service of the "Hachiryu" green

Table 22.3 Comments on specific category of index

Native biodiversity	Proportion of natural areas in city	The unit 100 m is too long for the distance between different patches. However, local governments do not have the skill and capacity for defining the index.
	Connectivity measures or ecological networks to counterfragmentation	
	Native biodiversity in built-up areas (bird species)	It is difficult to collect information on number of species, ratio of native species, etc. In some cases, improvement in these indices are not the result of biodiversity conservation, but just the result of improved monitoring skills or statistical measures.
	Change in number of native species	
	Proportion of protected natural areas	Local governments do not monitor the status of protected natural areas. In another case, there are no natural areas under control, or with control policies, and appropriately managed.
	Proportion of invasive alien species (as opposed to native species)	Many local governments do not have data related to alien species. In another case, cities only have data on specific alien species under specific legal regulations.
Ecosystem service	Regulation on quantity of water	It is difficult to calculate the cooling effects. Many local governments want to use data on provisional services as an index. They expect to grasp the cultural services to users living outside the cities.
	Climate regulation: carbon storage and cooling effects of vegetation	
	Recreational and educational services	
Governance and Management	Budget allocated to biodiversity	It is difficult for local governments to judge to what extent they must include this index into their budget. If the scope of the index is clearly defined, the local government can work on it.
	Number of biodiversity projects implemented by the city annually	If the index has clear definitions, local government can work with this index.
	Rules, regulations, and policy: existence of local biodiversity strategy and action plan	–
	Institutional capacity	The scope should be defined clearly. The index cannot be used as a policy-making tool.
	Participation and partnership	If the scope is clearly defined, they can work on this index, but it requires a lot of steps to collect information. The number of NPOs promoting biodiversity conservation can be summarized but the number of companies promoting it is difficult to capture.
	Education and awareness	It is hard to collect information because they have to contact all schools. The number of participants of events related to biodiversity can be used as an index on increasing awareness.

area; 48,110 kg CO_2 (which is equal to the CO_2 emission from twenty-first century cars) is absorbed annually. Regarding mitigation of the heat islands effect, therapy effect, and provision of habitat to living organism, these could be valuated quantitatively; however, these ecosystem services were analyzed qualitatively.

For the second model area, a paddy in "Nanyo-cho," groundwater recharge, prevention of floods, and water purification were valuated quantitatively. It was revealed that paddy in "Nanyo-cho" has the ability of groundwater recharge of 2.57–3.48 million m^3, which corresponds to the amount of water used by 8,540–11,560 households (four people per household). It also holds a water volume of 400 m^3 to prevent flooding, which corresponds to the amount of water used by 1,317 households. A biotope paddy in "Nanyo-cho" (721 m^2) also absorbs 88.4 kg nitrogen for one season. Similar to the first model area, mitigation of the effect of heat islands and provision of habitat to living organisms could not be valuated quantitatively. However, these ecosystem services were analyzed qualitatively.

Challenges and Solutions

The study of Nagoya city demonstrates the difficulty of quantitative evaluation of ecosystem services as a practical index of city biodiversity management. If they want to utilize the results as a policy-making tool, very locally specific information or assumptions are indispensable. Upon quantification of ecosystem services to a particular region, actual measurement of all the indices is very difficult financially for local governments. Therefore, estimates will be carried out for some indices alone. However, the problem of estimation accuracy occurs because of differences in factors such as growth rate of green vegetation and trees, the nature of the soil, and the temperature in the region, which are originally different from region to region.

This point is also applicable to the economic evaluation of ecosystem services. The MLIT summarized measures for the economic evaluation of ecosystem services for local governments. The result indicated that when using the alternative method, the accuracy of the evaluation of ecosystem services, which is a prerequisite for economic evaluation, can be problematic. When using conjoint analysis and CVM, there arises the problem that evaluation results in one region cannot be directly used in a different region.

The MLIT, in consideration of the previous trial of Nagoya-city and the results of interviews, is now trying to develop the Japanese version of CBI. A tentative version published in May 2013, is shown in Table 22.4. The table is shown only with the essential points. Detailed information such as purposes and evaluation criteria of the indexes, etc are provided in the MLIT website (cf; MLIT, (2013) in Japanese).

The current version is just a tentative one and the MLIT has a plan to improve it by pilot implementation with collaboration with local governments. It is expected for the MLIT to revise it to be more useful for practical city biodiversity management.

Table 22.4 The Japanese version of CBI

No.	Index	Definition	Measure for calculation (example)
Category: profile of the cities			
–	Size	Administrative district area, city planning area, urban areas	
	Population	Total population, urban planning area population, population density	
	terrain	Annual mean temperature, annual rainfall, elevation	
Category: diversity of ecosystem and habitat			
1	Percentage of green space contributing to conservation of biodiversity in urban areas	The percentage of green space capable of offering habitat for animals and plants, in urban areas	The proportion of "plant-covered and water area in green space" capable of contributing to conservation of biodiversity in the size of city planning area (based on survey conducted by cities) I_1 = (size of plant-covered and water area in green space capable of contributing to conservation of biodiversity) ÷ (size of city planning area) × 100 *Interim method (in the case a city cannot apply the above-mentioned measure)* The proportion of "green space" capable of contributing to conservation of biodiversity in the size of city planning area Calculate according to current condition of land use based on existing Land Use Status Survey in Urban Planning Basic Survey
2	*Condition of legally secured green spaces, etc.* (Percentage of green space contributing to conservation of biodiversity in urban areas)	The percentage of *legally secured* green space capable of offering habitat for animals and plants in urban areas	The proportion of *legally secured* "plant-covered and water area in green spaces" capable of offering habitat for animals and plants in the size of city planning area I_2 = (size of *legally secured* plant-covered and water area in green spaces capable of contributing to conservation of biodiversity)in urban areas *Interim method (in the case a city cannot apply the above-mentioned measure)* The proportion of *legally secured* "green spaces" capable of contributing to conservation of biodiversity in the size of city planning area

(continued)

Table 22.4 (continued)

No.	Index	Definition	Measure for calculation (example)
3	Situation of ecological network in urban areas	The situation of development in ecological network consisting of green space capable of offering habitat for animals and plants	Calculate the situation of development in ecological network consisting of green spaces capable of contributing to biodiversity conservation according to following formula $I_3 = \frac{1}{A_{total}}\left(A_1^2 + A_2^2 + A_3^2 + \ldots + A_n^2\right)$ n: total number of green space capable of contributing to biodiversity conservation in urban area A_{total}: total size of green space A_n: size of each green space *Two green spaces with less than 100 m distance, not separated by roads, artificially modified rivers or houses or other artificial obstacles, are treated as one space *Interim method* Define plant and animal species as monitoring species in the situation of ecological network development, and calculate the situation of development consisting of continuous green spaces which offer habitat for such species, using the formula above
4	Situation of animal and plant species (Situation of the number of animal and plant species living in urban areas)	Secular change in the number of animal and plant species living in urban areas	Establish monitoring sites according to major ecosystem in that city and set habitant species as reference species, and calculate the change in numbers between reference times *Interim method* Calculate the change between reference times in numbers of endangered species confirmed to have lived in administrative territory of local government or corresponding city, or in ecosystem or habitats which are vital to conserving biodiversity in the city
Category: ecological services			
5	Situation of ecological services	The situation of ecological services accrued by conserving urban biodiversity, and by conservation, reclamation, creation of animal and plant habitat	Estimate change between reference times in the quantity of ecological services listed below and additional unique indexes defined by local governments by considering local characteristic. (i) Effect on global warming (absorption of greenhouse gas from urban greening)

		(ii) Cooling effect of green spaces (canopy covered area by green spaces, etc.) (iii) Water regulation (filtration effect of green spaces) (iv) Purification of water and impact on ecology (water quality of rivers) (v) Provision of cultural services (per person size of index 2) (vi) Utilization of green area (annual trend of citizen's visit to green spaces) (vii) Educational use of green spaces (annual numbers of visit to green spaces by children under 16 hosted by schools) *Interim method* On more than or equal to 2 items from above mentioned ecosystem services and calculate the change in results between two given points in times
Category: cities' activities		
6 Cities activities for conserving biodiversity (Situation of consideration in urban administrative planning for conserving biodiversity)	Situation of consideration on conserving urban biodiversity in plans formulated by local governments on green conservation, greening promotion, conservation of biodiversity and natural environment conservation, situation of actual implementation, check and evaluation of measures for conserving urban biodiversity	Count the numbers of following activities implemented in the cities Three activities on current situation study, analysis and evaluation, e.g. (1) The local government do own survey concerning distribution and habitat situation of green covered areas, animals and plants (2) Conduct surveys on situation of legal regulations for protecting ecosystem, animals and plants, and on plans and projects related to developing ecological network Six activities on goals, policy of green spaces settings and formulation of policies, e.g. (5) Set policies on ecological network development for conserving biodiversity and on green spaces distribution (6) Developed more than or equal to one policy for conservation, reclamation and creation of green covered area and water which make up ecological networks Five activities on implementation, check and evaluation of plans, e.g. (14) Reflect the results of check and evaluation of policies for revising plans and policies

(continued)

Table 22.4 (continued)

No.	Index	Definition	Measure for calculation (example)
7	(Situation of participation by residents, etc. in urban administrative planning) (Situation of participation by residents and corporations in conservation of biodiversity in urban administrative planning)	Situation of participation by residents and corporations in each step of formulation, publication, implementation, check and evaluation on plans positioning considerations on conservation of urban biodiversity	Count the numbers of following activities implemented in the cities Two activities on formulation and publication of plans, e.g. (1) Implement efforts to reflect opinions from residents Six activities on implementation, check, evaluation of the plans, e.g. (3) Conserve, reclaim, create and manage green spaces offering habitats for animals and plants according to plans in cooperation with various stakeholders such as NPOs, residents, corporations, education and research institutions and experts, etc. (7) Grasp continuously implementation status of policies as well as performing check and evaluation on effects in cooperate with various stakeholders such as NPOs, residents, corporations, education and research institutions and experts, etc.

* (as of December 2011)

Conclusion

Biodiversity conservation in cities is one of the key elements to achieve the Aichi Biodiversity Targets. As mentioned in Decision X/28, an indicator is indispensable to promote biodiversity management in cities. However, the trials of Nagoya-city and the MLIT revealed that development and implementation of a practical index is very difficult. One of the obstacles is difficulty of quantitative valuation of biodiversity. City management required dependable and robust indices as evidence of valuation for explanation to stakeholders. As this study shows, there are many existing studies on the quantitative valuation of biodiversity, but the results of such valuation are diversified because of the different assumptions. The assumptions depend on local conditions. Many indices, in particular indices for ecosystem services in the Japanese version of CBI developed by the MLIT, required estimates based on local specific conditions. Accordingly, various studies for different local conditions definitely contribute to the development of index.

However, the study shows that the inconsistency of the interests between policy makers and scientists also prevents improvement of indices. The existing literature points out that the policy makers tend to emphasize ecosystem services for justification of their policies whereas scientists tend to focus on biodiversity per se.

The CBI should be improved for practical implementation. The MLIT also declared that they would modify the Japanese version of CBI to be applicable to Japanese conditions. Scientists are requested to conduct more studies on valuation of ecosystem services in local specific conditions and contribute more for the conservation of city biodiversity.

Acknowledgment We would like to thank Parks, Green Spaces and Landscape Division, City Bureau, Ministry of Land, Infrastructure, Transport and Toursim, Mr. Suzuki Takehiko and Ms. Tsukamoto Aya, for in particular for their valuable input.

This study was conducted as a part of the research project, the Japan Society for the Promotion of Science (Grants-in-Aid for Scientific Research 26360062 Project Leader R. Kohsaka, FY2014–FY2016). This study was conducts as a part of the research project, the Japan Society for the Promotion of Science (Grants-in-Aid for Scientific Research 26360062 Project Leader R. Kohsaka, FY2014–FY2016). The authors' thanks are extended to the valuable discussions at the advisory meeting of Environment Research and Technology Development fund (S9-1: Project Leader T. Yahara) of the MOE, Japan.

References

Brack CL (2002) Pollution mitigation and carbon sequestration by an urban forest. Environ Pollut 116(suppl 1):S195–S200

Chan L, Djoghlaf A (2009) Correspondence: invitation to help compile an index of biodiversity in cities. Nature (Lond) 460:33. doi:10.1038/460033a

Chiba T, Nishida T, Kiyotani K, Abe T, Nagai K (2012) Current status and issues of regional biodiversity strategy: the results of surveys that target local government. Jpn J Conserv Ecol 1:37–47

Costanza R, d'Arge R, de Groot R, Farber S, Grasso M, Hannon B, Limburg K, Naeem S, O'Neill S, Paruelo J (1997) The value of the world's ecosystem services and natural capital. Nature (Lond) 387:253–260

Fujiwara N, Takeda Y, Yonezawa N (2004) Research on economical evaluation of city parks using conjoint analysis. J Jpn Inst Landsc Arch 67(5):709–712

Kohsaka R (2009) Economics and the convention on biodiversity: financial incentives for encouraging biodiversity in Nagoya. In: Mueller N, Werner P, Kelcey GJ (eds) Urban biodiversity and design. Blackwell, London, pp 593–607

Kohsaka R, Pereira MH, Elmqvist T, Chan L, Moreno-Peñaranda R, Morimoto Y, Inoue T, Iwata M, Nishi M, Mathias M, Cruz SC, Cabral M, Brunfeldt M, Parkkinen M, Niemelä J, Kulkarni-Kawli Y, Pearsall G (2013) Indicators for management of urban biodiversity and ecosystem services: City Biodiversity Index. In: Elmqvist TH, Fragkias M, Goodness J, Güneralp B, Marcotullio PJ, McDonald RI, Parnell S, Haase D, Sendstad M, Seto KC, Wilkinson C (eds) Urbanization, biodiversity and ecosystem services: challenges and opportunities. Springer, New York, pp 699–718. doi:10.1007/978-94-007-7088-1_32

MEXT (2010) The 9th Science and Technology Foresight Contribution of Science and Technology to Future Society: The 9th Delphi Survey. Science and Technology Foresight Center National Institute of Science and Technology Policy NISTEP REPORT No. 140, Tokyo

MLIT (2011) Technical guideline related to biodiversity in the Greening Plan. http://www.mlit.go.jp/crd/park/joho/houritsu/pdf/H231001hairyojikou.pdf. Accessed 1 July 2012

MLIT (2013) City Biodiversity Index (Draft). http://www.mlit.go.jp/toshi/park/toshi_parkgreen_tk_000022.html. Accessed 31 March 2013

OECD (1978) Urban environmental indicator. OECD, Paris

SCBD (2008) Decision IX/28. Promoting engagement of cities and local authorities. UNEP/CBD/COP/DEC/IX/28 URL: http://www.cbd.int/decision/cop/?id=11671. Retrieved 31 August 2009

Schuyt K, Brander L (2004) The economic values of the world's wetlands. WWF, Gland/Amsterdam

Science Council of Japan (2001) Chikyuu-kankyou, Ningen-seikatsu-ni-kakawaru Nougyu Oyobi Shinrin No Tamententeki-na-kinou No Hyouka-ni-tsuite (Evaluation of multifunctions of agriculture and forestry relevant to human life). Science Council of Japan, Tokyo

Tsuge (2001) Possibility of assessment and its use as policy on its social functions of forests based on the preferences of citizens, empirical study by selected type experiment

United Nations University-Institute of Advanced Studies (UNU-IAS) and Japan Satoyama Satoumi Assessment (JSSA) (2012) Satoyama–Satoumi Ecosystems and Human Well-Being: Socio-Ecological Production Landscapes of Japan United Nations University Press, Tokyo

UN-HABITAT (2008) State of the World's Cities 2008/2009 UN-HABITAT, Nairobi

Yokota et al (2004) Studies on the environment and economic value of biotope (no. 2). Estimates of the social benefits of biotope development projects. Summaries of technical papers of Annual Meeting of the Architectural Institute of Japan, August 2004

Index

S.-i. Nakano et al. (eds.), *Integrative Observations and Assessments*, Ecological Research Monographs, DOI 10.1007/978-4-431-54783-9,

MIX
Papier aus verantwortungsvollen Quellen
Paper from responsible sources
FSC® C105338

If you have any concerns about our products,
you can contact us on
ProductSafety@springernature.com

In case Publisher is established outside the EU,
the EU authorized representative is:
Springer Nature Customer Service Center GmbH
Europaplatz 3, 69115 Heidelberg, Germany

Printed by Libri Plureos GmbH
in Hamburg, Germany